T0319180

Principles of Translational Science in Medicine

Principles of Translational Science in Medicine

From Bench to Bedside

Second Edition

Edited by

Martin Wehling
Department of Clinical Pharmacology Mannheim
University of Heidelberg, Mannheim, Germany

AMSTERDAM • BOSTON • HEIDELBERG • LONDON • NEW YORK • OXFORD • PARIS
SAN DIEGO • SAN FRANCISCO • SINGAPORE • SYDNEY • TOKYO
Academic Press is an imprint of Elsevier

Academic Press is an imprint of Elsevier
125 London Wall, London EC2Y 5AS, UK
525 B Street, Suite 1800, San Diego, CA 92101-4495, USA
225 Wyman Street, Waltham, MA 02451, USA
The Boulevard, Langford Lane, Kidlington, Oxford OX5 1GB, UK

Notices
Knowledge and best practice in this field are constantly changing. As new research and experience broaden our understanding, changes in research methods, professional practices, or medical treatment may become necessary.

Practitioners and researchers must always rely on their own experience and knowledge in evaluating and using any information, methods, compounds, or experiments described herein. In using such information or methods they should be mindful of their own safety and the safety of others, including parties for whom they have a professional responsibility.

To the fullest extent of the law, neither the Publisher nor the authors, contributors, or editors, assume any liability for any injury and/or damage to persons or property as a matter of products liability, negligence or otherwise, or from any use or operation of any methods, products, instructions, or ideas contained in the material herein.

Library of Congress Cataloging-in-Publication Data
A catalog record for this book is available from the Library of Congress

British Library Cataloguing-in-Publication Data
A catalogue record for this book is available from the British Library

ISBN: 978-0-12-800687-0

For Information on all Academic Press publications
visit our website at http://store.elsevier.com/

Publisher: Christine Minihane
Acquisition Editor: Christine Minihane
Editorial Project Manager: Shannon Stanton
Production Project Manager: Caroline Johnson
Designer: Greg Harris

Typeset by TNQ Books and Journals
www.tnq.co.in

Printed and bound in the United States of America

Contents

List of Contributors xi
Preface xiii

1. Introduction and Definitions

Martin Wehling

What is Translational Medicine? 1
Primary Translation versus Secondary Translation 2
The History of Translational Medicine, Obstacles, and Remits 3
What Translational Medicine Can and Cannot Do 6
The Present Status of Translational Medicine (Initiatives and Deficiencies) 8
New Pathways to Discovery 8
Research Teams of the Future 8
Re-Engineering the Clinical Research Enterprise 9
Translational Science in Medicine: The Current Challenge 11
References 12

2. Target Identification and Validation

2.1.1 "Omics" Translation: A Challenge for Laboratory Medicine

Mario Plebani, Martina Zaninotto and Giuseppe Lippi

Introduction 15
"Omics": What does it mean? 15
Proteomics as a Paradigm of Problems in Translational Medicine 15
Development of Biomarkers: From Discovery to Clinical Application 18
Discovery 18
Identification/Characterization 18
Validation 19
Standardization/Harmonization 19
Clinical Association and Clinical Benefit 20
Translating Omics into Clinical Practice 21
Continuum of Translation Research and Omics 21
Conclusions 23
References 23

2.1.2 "Omics" Technologies: Promises and Benefits for Molecular Medicine

David M. Pereira, João C. Fernandes, Patrícia Valentão and Paula B. Andrade

Introduction 25
Genomics 25
Genomic Tools 25
Applications of Genomics in Molecular Medicine 31
Metabolomics 33
Metabolomics or Metabonomics? 33
Analytical Techniques in Metabolomics/Metabonomics 34
"Omics" and Biomarkers 35
Metabolomics/Metabonomics in Clinical Use 36
Conclusion 37
References 37

2.1.3 Potency Analysis of Cellular Therapies: The Role of Molecular Assays

David F. Stroncek, Ping Jin, Ena Wang, Jiaqiang Ren, Luciano Castellio, Marianna Sabatino and Francesco M. Marincola

Potency Testing 42
Complexities Associated with Potency Testing of Cellular Therapies 43
Factors Affecting the Potency of Cellular Therapies 44
Measuring Potency of Cellular Therapies 44
Gene Expression Arrays for Potency Testing 45
Potential Applications of Gene Expression Profiling for Potency Testing 46
Predicting the Confluence of Human Embryonic Kidney 293 Cells 46
Cell Differentiation Status Analysis of Embryonic Stem Cells 46
Potency Testing of Hematopoietic Stem Cells 47
Potency Testing of Dendritic Cells 48
Cultured CD4+ Cells 52
Bone Marrow Stromal Cell 52

MicroRNAs as Potency Assays 54
Conclusions 54
References 55

2.1.4 Translational Pharmacogenetics to Support Pharmacogenetically Driven Clinical Decision Making

Julia Stingl

Introduction 59
Pharmacogenetics as a Tool for Improving Individual Drug Therapy 59
Types of Drug Therapies that Might Profit from Pharmacogenetic Diagnostics 60
The Status of Translational Pharmacogenetics in Various Drug Therapy Fields 60
Depression 60
Cardiovascular Disease 61
Statins and Proton Pump Inhibitors 63
Pain Treatment 65
Malignant Diseases 65
Translational Pharmacogenetics and the Need for Clinical Studies to Support Pharmacogenetically Driven Prescribing 67
References 70

2.1.5 Tissue Biobanks

W. Peeters, F.L. Moll, R.J. Guzman, D.P.V. de Kleijn and G. Pasterkamp

Introduction 75
Principles and Types of Tissue Biobanks: Pros and Cons 75
Developments in Vascular Biobanking Research and Clinical Relevance 77
Atherosclerosis 77
Clinical Value of Vascular Biobanks and Diagnostic Imaging 78
Vascular Biobank Example: The Athero-Express Biobank 78
Challenges for Future Biobanks 79
Summary 79
References 80

2.1.6 Animal Models: Value and Translational Potency

Philipp Mergenthaler and Andreas Meisel

What is the Value of Animal Models? Pathophysiological Concepts 83
What is a Good Animal Model for Translational Research? 83
Modeling Comorbidities 84
Modeling Care of Patients 84
What is the Translational Value of Animal Models? 85
Remedies for Failed Translation: Improving Preclinical Research 87
Improving Models 87
Improving Rigor of Preclinical Studies 87
Summary 89
References 89

2.1.7 Localization Technologies and Immunoassays: Promises and Benefits for Molecular Medicine

Estelle Marrer, Frank Dieterle and Jacky Vonderscher

Introduction 91
Localization Technologies 91
Genetics and Genomics 91
Protein Localization 93
Immunoassays 93
Enzyme-Linked Immunosorbent Assays 93
Diagnostic Immunoassay Devices: High-Throughput Immunoassay Platforms 94
Diagnostic Immunoassay Devices: Bedside Devices 95
Case Study: Screening of a Biomarker for Kidney Injury Using Localization and Immunoassays 96
References 99

2.1.8 Biomarkers in the Context of Health Authorities and Consortia

Frank Dieterle, Estelle Marrer and Jacky Vonderscher

The Critical Path Initiative 101
New Technologies, Health Authorities, and Regulatory Decision Making 101
Voluntary Genomics Data Submission and Voluntary Exploratory Data Submission 102
Current Guidances for Biomarker Qualification and Validation 102
Private–Public Partnerships (Cooperative R&D Agreements) 102
Consortia 103
The Critical Path Institute's Predictive Safety Testing Consortium 103
The Innovative Medicines Initiative in Europe 104
The PhRMA Biomarkers Consortium 104
The International Serious Adverse Event Consortium 105
References 105

2.1.9 **Human Studies as a Source of Target Information**

Martin Wehling

Using Old Drugs for New Purposes: Baclofen 107
Serendipity: Sildenafil 109
Reverse Pharmacology 110
References 113

2.2 **Target Profiling in Terms of Translatability and Early Translation Planning**

Martin Wehling

Essential Dimensions of Early Translational Assessment 115
A Novel Translatability Scoring Instrument: Risk Balancing of Portfolios and Project Improvement 117
Case Studies: Applying the Novel Translatability Scoring Instrument to Real-Life Experiences 120
References 123

3. Biomarkers

3.1 **Defining Biomarkers as Very Important Contributors to Translational Science**

Martin Wehling

References 130

3.2 **Classes of Biomarkers**

Martin Wehling

References 144

3.3 **Development of Biomarkers**

Martin Wehling

References 149

3.4 **Predictivity Classification of Biomarkers and Scores**

Martin Wehling

References 154

3.5 **Case Studies**

Martin Wehling

References 157

3.6 **Biomarker Panels and Multiple Readouts**

Andrea Padoan, Daniela Bernardi and Mario Plebani

Introduction 159
Source of Errors in Proteomics Studies 159
Statistical and Computational Methods in Clinical Proteomics 160
Supervised and Unsupervised Computational Methods 160
Receiver Operating Characteristic Analysis and Net Reclassification Improvement 161
Overfitting Issues 161
Multiparameter Approach 161
Ovarian Cancer 161
Breast Cancer 162
Prostate Cancer 163
Pancreatic Cancer 164
Liver Fibrosis 164
Conclusions 165
References 165

3.7.1 **Cardiovascular Biomarkers: Translational Aspects of Hypertension, Atherosclerosis, and Heart Failure in Drug Development**

A. Rogier van der Velde, Wouter C. Meijers and Rudolf A. de Boer

Hypertension 167
Introduction 167
Animal Models of Hypertension 167
Biomarkers of Hypertension: Clinical Markers of Hypertension 168
Biomarkers of Hypertension: (Blood-Borne) Biomarkers 168
Conclusion 170
Atherosclerosis 170
Introduction 170
Animal Models 171
Biomarkers for Atherosclerosis 171
Imaging 172
Conclusion 173
Heart Failure 173
Introduction 173
Animal Models in Heart Failure 174
Clinical Markers for Heart Failure 176
(Blood-Borne) Biomarkers of Heart Failure 176
Conclusion 178
References 178

3.7.2 **Biomarkers in Oncology**

Jon Cleland, Faisal Azam, Rachel Midgley and David J. Kerr

Other Pathways That May Be Suitable for Biomarker-Assisted Development 186
Tissue Biopsies 187
References 188

3.7.3 **Translational Imaging Research**

Lars Johansson

Why Imaging? 189
Differences between Translational Imaging and Conventional Imaging 189
Validation of Imaging and Back-Translation 189
Imaging Modalities 190
X-Ray 190
Computed Tomography 190
Ultrasound 190
Magnetic Resonance Imaging 190
Magnetic Resonance Spectroscopy 191
Gamma Camera and SPECT 191
Positron Emission Tomography 191
Characteristics of Various Imaging Modalities 191
Examples of Translational Imaging in Various Disease Areas 192
Cardiovascular Medicine 192
Metabolic Medicine 192
Oncology 192
Neuroscience 193
Conclusion 193
References 193

3.7.4 **Translational Medicine in Psychiatry: Challenges and Imaging Biomarkers**

Andreas Meyer-Lindenberg, Heike Tost and Emanuel Schwarz

Biological Treatment of Psychiatric Disorders 195
Specific Challenges of Translation in Psychiatry 196
Unknown Pathophysiology 196
Stigma and the Second Translation 197
New Biomarkers for Translation in Psychiatry 197
Imaging Biomarkers in Schizophrenia 198
Structural Brain Biomarkers 198
Functional Imaging Markers in Schizophrenia 200
Auditory and Language Processing 200
Motor Functioning 201
Working Memory 201
Selective Attention 202
Imaging of Genetic Susceptibility Factors 203
Characterization of Antipsychotic Drug Effects 205
Conclusions and Future Directions 206
References 206

4. Early Clinical Trial Design

4.1 **Methodological Studies**

Faisal Azam, Rhiana Newport, Rebecca Johnson, Rachel Midgley and David J. Kerr

Conventional Phase I Trial Methodology 217
Aims 217
Design 217
Patient Entry Criteria 218
Special Drug Administration or Procedures 219
Patient Consent 219
Calculation of the Starting Dose 219
Dose Escalation 219
Number of Patients Required for Dose Administration 220
Stopping Rules 220
Measuring Endpoints 220
Toxicity 220
Pharmacodynamic Endpoints 221
Mechanism-Oriented Trial Design 221
Proof-of-Mechanism 221
Proof-of-Principle 222
Proof-of-Concept 222
Can We Make Go-or-No-Go Decisions at the End of Phase I? 222
Phase II Trials 223
Personalized Medicine 223
Open Access Clinical Trials 224
References 224

4.2 **The Pharmaceutical R&D Productivity Crisis: Can Exploratory Clinical Studies Be of Any Help?**

Cecilia Karlsson

Traditional Drug Development 227
IND Application 228
Opportunities for Earlier Decision Making 228
References 228

4.3 **Exploratory Clinical Studies ("Phase 0 Trials")**

Cecilia Karlsson

Types of Studies 229
Microdosing 230
Accelerator Mass Spectrometry Microdosing 230
Positron Emission Tomography Microdosing 231
Repeated Dosing 232
Should Exploratory Clinical Studies Be Performed in All Projects? 232

Practical Applications 232
References 233

4.4 Adaptive Trial Design
Martin Wehling
References 238

4.5 Combining Regulatory and Exploratory Trials
Martin Wehling
References 242

4.6 Accelerating Proof-of-Concept by Smart Early Clinical Trials
Martin Wehling
References 246

5. Pharmaceutical Toxicology
Steffen Ernst
Introduction 249
Basic Principles of Toxicology 250
Elements of Toxicity 250
Viewpoints of Toxicity 254
Risk Assessment 255
Regulatory Toxicology 255
Introduction 255
Regulatory Toxicity Studies 256
The Importance of Good Laboratory Practice 258
Animal Models 258
New Approaches in Regulatory Toxicology: The Exploratory Investigational New Drug Approach 259
Biomarkers 260
Links 260
Regulatory Agencies and Testing Guidelines: Pharmaceuticals 260
Societies 260
Related Sites 260
The Practice of Discovery Safety Assessment 261
Target-Based Safety Assessment 261
Safety-Directed Drug Design 262
Summary 264
Preclinical Safety from a Translational Perspective 264
References 265

6. Translational Science Biostatistics
Georg Ferber and Ekkehard Glimm
Statistical Problems in Translational Science 267
Statistical Models and Statistical Inference 268
Design and Interpretation of an Experiment 270
Multiplicity 273
Biomarkers 274
Biological Modeling 275
Example 1: Pharmacodynamics 276
Example 2: Pharmacokinetics 276
Statistical Models 278
References 279

7. Intellectual Property and Innovation in Translational Medicine
Palmira Granados Moreno and Yann Joly
Introduction 281
Context 281
General Description of Translational Medicine 281
Intellectual Property and TM 282
Open Science 285
Public–Private Partnership Models 287
Trends in Translational Intellectual Property 288
Patents and Research Tools 288
Patents on Genetic Tests and Personalized Therapies 289
Patents on Risk Prediction Models 290
Patents on New and Repositioned Drugs 290
Secrecy 291
Discussion 291
A Perspective on the Future of Genetic Patents 291
Trade Secrecy as an Option (Pros and Cons) 293
Toward Balanced Innovation Environment 294
Conclusion 294
References 295
Legal Citation 295
Scientific Citations 295

8. Translational Research in the Fastest-Growing Population: Older Adults
Kevin P. High and Stephen Kritchevsky
Introduction 299
Why Study Aging? 299
Lifespan versus Healthspan 299
Gerontology Versus Geriatrics 300
Animal Models of Aging 301
Human Approaches to Translational Aging Research 302
Identifying Determinants of Longevity and Healthspan Using Epidemiologic Approaches 303
Testing Treatments to Extend Health- and Lifespan 304
Limitations for Both Animal and Human Models 305
Example of Translational Research in Aging: Calorie Restriction 306

Translational Aging Resources 309
Animals and Animal Tissues 309
Cohorts/Populations 309
Tools and Toolboxes 309
Conclusion 309
References 310

9. Translational Medicine: The Changing Role of Big Pharma

C. Simone Fishburn

Introduction 313
History: How Did We Get Here? 313
Causes of Change 314
Business-Focused Solutions: M&A and Licensing 315
The Productivity Crunch 316
Translational Solutions 316
Role of the National Institutes of Health (NIH) 317
Integrated Discovery Nexuses 317
Public–Private Partnerships: The New Mantra 317
Pharma and Academia 317
Biotech Companies and Academia 319
Precompetitive Consortia: The Road Ahead 323
Summary 324
References 324

10. Translational Science in Medicine: Putting the Pieces Together

Martin Wehling

Reference 329

11. Learning by Experience

Martin Wehling

Example of a Smart, Successful Translational Process 331
Example of a Failed Translational Process 332
References 336

Index 339

List of Contributors

Paula B. Andrade REQUIMTE/Laboratório de Farmacognosia, Departamento de Química Faculdade de Farmácia Universidade do Porto, Porto, Portugal

Faisal Azam Radcliffe Department of Medicine, University of Oxford, Oxford, UK

Daniela Bernardi Department of Laboratory Medicine, University-Hospital of Padova, Padova, Italy

Rudolf de Boer Department of Cardiology, University of Groningen, University Medical Center Groningen, Groningen, Netherlands

Luciano Castellio Department of Transfusion Medicine, Clinical Center, National Institutes of Health, Bethesda, USA

Jon Cleland Nuffield Division of Clinical and Laboratory Sciences, John Radcliffe Infirmary, Headington, Oxford, UK

Frank Dieterle Novartis Pharma AG, Basel, Switzerland

Georg Ferber Riehen, Switzerland

João C. Fernandes Laboratory of Pharmacology and Experimental Therapeutics, IBILI Faculty of Medicine, University of Coimbra, Coimbra, Portugal

C. Simone Fishburn BioCentury Publications Inc., Redwood City, CA, USA

Ekkehard Glimm Novartis Pharma AG, Basel, Switzerland and Medical Faculty, Otto-von-Guericke-University, Magdeburg, Germany

Palmira Granados Moreno Centre of Genomics and Policy, McGill University, Montreal, QC, Canada

R.J. Guzman Department of Vascular Surgery, Beth Israel Deaconess Medical Center, Harvard Medical School, Boston MA, USA

Kevin P. High Department of Internal Medicine, Section on Infectious Disease, Wake Forest School of Medicine, Winston-Salem, NC, USA

Ping Jin Department of Transfusion Medicine, Clinical Center, National Institutes of Health, Bethesda, USA

Lars Johansson Institutionen för onkologi, radiologi och klinisk immunologi, Akademiska sjukhuset, Uppsala, Sweden

Rebecca Johnson Radcliffe Department of Medicine, University of Oxford Headley Way, Oxford, UK

Yann Joly Department of Human Genetics, McGill University, Montreal, Quebec, Canada; Centre of Genomics and Policy, McGill University, Montreal, Quebec, Canada

Cecilia Karlsson Translational Medicine Unit, Early Clinical Development, Innovative Medicines, AstraZeneca R&D Mölndal, Mölndal, Sweden

David J. Kerr Radcliffe Department of Medicine, University of Oxford, Oxford, UK

D.P.V. de Kleijn Experimental Cardiology Laboratory, University Medical Center Utrecht, Utrecht, The Netherlands; Surgery & Cardiovascular Research Institute, National University (NUS) & National University Hospital (HUS) Singapore; and School of Biological Sciences, Nanyang Technological University, Singapore

Stephen Kritchevsky Sticht Center on Aging, Wake Forest School of Medicine, Winston-Salem, NC, USA

Giuseppe Lippi Laboratory of Clinical Chemistry and Hematology, Academic Hospital of Parma, Parma, Italy

Francesco M. Marincola Sidra Medical and Research Centre, Al Nasr Towe, Qatar Foundation, Doha, Qatar

Estelle Marrer Novartis Pharma AG, Basel, Switzerland

Wouter C. Meijers Department of Cardiology, University of Groningen, University Medical Center Groningen, Groningen, Netherlands

Andreas Meisel NeuroCure Clinical Research Center NCRC, Center for Stroke Research Berlin CSB, Department of Neurology, Department of Experimental Neurology, Charité Universitätsmedizin Berlin, Berlin, Germany

Philipp Mergenthaler Department of Experimental Neurology, Department of Neurology, NeuroCure Clinical Research Center, Center for Stroke Research Berlin, Charité Universitätsmedizin Berlin, Berlin, Germany

Andreas Meyer-Lindenberg Central Institute of Mental Health, Medical Faculty Manneheim, Heidelberg University, Mannheim, Germany

Rachel Midgley Radcliffe Department of Medicine, University of Oxford, Oxford, UK

F.L. Moll Department of Vascular Surgery, University Medical Center Utrecht, Utrecht, The Netherlands

Rhiana Newport Radcliffe Department of Medicine, University of Oxford, Oxford, UK

Andrea Padoan Department of Medicine, DiMED University of Padova, Padova, Italy

G. Pasterkamp Experimental Cardiology Laboratory, University Medical Centre Utrecht, Utrecht, The Netherlands

W. Peeters Interuniversity Cardiology Institute of the Netherlands and Experimental Cardiology Laboratory, University Medical Center Utrecht, Utrecht, The Netherlands

David M. Pereira REQUIMTE/Laboratório de Farmacognosia, Departamento de Química Faculdade de Farmácia Universidade do Porto, Porto, Portugal

Mario Plebani Department of Laboratory Medicine, University-Hospital of Padova, Via Guistiniani, Padova, Italy

Jiaqiang Ren Department of Transfusion Medicine, Clinical Center, National Institutes of Health, Bethesda, MD, USA

Marianna Sabatino Department of Transfusion Medicine, Clinical Center, National Institutes of Health, Bethesda, MD, USA

Emanuel Schwarz Central Institute of Mental Health, Medical Faculty Manneheim, Heidelberg University, Mannheim, Germany

Julia Stingl Translational Pharmacology, Medical Faculty University Bonn, Bonn, Germany; Research Division, Federal Institute for Drugs and Medical Devices, Bonn, Germany

David F. Stroncek Cell Processing Section, Department of Transfusion Medicine, NIH Clinical Center, Bethesda, USA

Heike Tost Central Institute of Mental Health, Medical Faculty Mannheim, Heidelberg University, Mannheim, Germany

Patrícia Valentão REQUIMTE/Laboratório de Farmacognosia, Departamento de Química Faculdade de Farmácia, Universidade do Porto, Porto, Portugal

A. Rogier van der Velde Department of Cardiology, University of Groningen, University Medical Center Groningen, Groningen, Netherlands

Jacky Vonderscher Molecular Medicine Labs, Roche, Basel, Switzerland

Ena Wang Sidra Medical and Research Centre, Qatar Foundation, Doha, Qatar

Martin Wehling Clinical Pharmacology Mannheim, University of Heidelberg, Mannheim, Germany

Martina Zaninotto Department of Laboratory Medicine, University-Hospital of Padova, Padova, Italy

Preface

Despite tremendous efforts and despite the cloning of the entire human genome, innovations at the patient level are becoming rare events, and insufficiencies in predicting the human efficacy or safety of new drugs from early discovery and development work are blamed for many failures. "Translational medicine" has become a fashionable phrase, but this is just not enough. It is a complex science that is still in its infancy and needs careful development both as a generic science and in concrete projects. As a generic science, the principles of translational activities need to be further explored, standardized, and developed. Important milestones in the development of translational science are the classification of biomarkers in regard to their predictive value for cross-species efficacy and safety extrapolations, and the overall grading of translatability of a given biomedical project by scoring major translational assets, thereby predicting risk and potential. The institutionalization of translational science and its integration into large networking structures at sites of prime research and clinical facilities seem to be a timely investment. In the United States, an increasing number of universities have undertaken those efforts and built up institutes of translational medicine, and many are now supported by the National Institutes of Health (NIH) by Clinical and Translational Science Awards (CTSA). The NIH has founded a dedicated institute, the National Center for Advancing Translational Sciences (NCATS), running on an annual budget approaching a billion U.S. dollars. Governmental and regulatory bodies (e.g., the Food and Drug Administration [FDA]) have called for increased activities under those auspices (e.g., the Critical Path Initiative), and the era of using phrases and fashionable titles without true translational content should end as soon as possible. This text aims at addressing the scientific aspects of translational medicine and teaching the essential components of a complex network of activities that should lead to successful and reliable translation of preclinical results into clinical results.

This text deals with major preclinical and clinical issues relevant to the translational success of pharmaceutical or medical device or diagnostic innovations. This includes target risk assessment, biomarker evaluation and predictivity grading for both efficacy and toxicity, early human trial design adequate to guide go-or-no-go decisions on grounds of biomarker panels, translatability assessment, and biostatistical methods to analyze multiple readout situations and quantify risk projections. The text provides guidance to design smart profiling strategies for new approaches and aims at showing its readers how to cut timelines and concentrate on quality issues early on in the developmental processes. Translational efforts are benchmarked against patients' needs and integrative strategies to optimize yield and cost ratios. Recent progress in the field, new institutions as mentioned previously, and new technology and assessment approaches mandated a new edition after 5 years; the former publisher, Cambridge University Press, handed the project over to Elsevier as its new publisher.

The text comprises state-of-the-art knowledge in translational medicine, with emphasis on its scientific backbone and its strengths, but also on its weaknesses as a young discipline. Under didactic auspices, it is hoped that this text will promote the substantiation of this emerging science, create awareness about its potential to promote urgently needed innovations in clinical practice, but also inform about the threat implied by the empty phraseology inherent to the present hype in this area.

Martin Wehling
Mannheim
December 2014

Motto
Though this be madness, yet there is method in't.
William Shakespeare, *Hamlet, Act 2, scene 2s, 193–206*

Chapter 1

Introduction and Definitions

Martin Wehling

WHAT IS TRANSLATIONAL MEDICINE?

The definition of *translational medicine* found on the main website of a leading scientific journal, *Science*, is as follows:

> *Often described as an effort to carry scientific knowledge "from bench to bedside," translational medicine builds on basic research advances—studies of biological processes using cell cultures, for example, or animal models—and uses them to develop new therapies or medical procedures.* (*Science Translational Science, 2013*)

In Wikipedia, an online encyclopedia, the definition is as follows:

> ***Translational medicine*** *(also referred to as translational science) is a discipline within biomedical and public health research that aims to improve the health of individuals and the community by "translating" findings into diagnostic tools, medicines, procedures, policies and education.* (*Wikipedia, 2013*)

In an earlier version on Wikipedia, some explanatory sentences were given:

> *In the case of drug discovery and development, translational medicine typically refers to the "translation" of basic research into real therapies for real patients. The emphasis is on the linkage between the laboratory and the patient's bedside, without a real disconnect. This is often called the "bench to bedside" definition.* (*Wikipedia, 2007*)
>
> *Translational medicine can also have a much broader definition, referring to the development and application of new technologies in a patient driven environment—where the emphasis is on early patient testing and evaluation. In modern healthcare, we are seeing a move to a more open, patient driven research process, and the embrace of a more research driven clinical practice of medicine.* (*Wikipedia, 2007*)

Although these attempts at a definition are probably the most accurate and concise ones at present, a simpler definition may serve the purpose even better: Translational medicine describes the transition of in vitro and experimental animal research to human applications (see Figure 1.1).

Other names for the same entity are *experimental medicine*, *discovery medicine*, and *clinical discovery*. Translational medicine shares major aspects of clinical pharmacology when it relates to drugs, as early clinical trials are major components of translational processes.

The need to develop this discipline reflects the cleft that has been brought about by the separation of medical teaching and pharmaceutical research into preclinical and clinical categories. Bridging this gap is crucial to success in curing diseases in humans. It is obvious that the term is born out of a situation in which the transition—the prediction or extrapolation, respectively—from basic findings to human findings has been disappointing. This difficulty is simply a reflection of the differences among in vitro conditions (e.g., cell cultures or test tube experiments), the wide variety of animal species, and, finally, humans. For example, cell cultures of vascular smooth muscle cells are artificial, as they grow only in the presence of serum (e.g., fetal calf serum). In contrast, while in vivo, nature does everything to ensure that vascular smooth muscle cells do not encounter serum; endothelium protects them against it. If the cells become damaged and are exposed to serum, all types of vascular pathology commence: hypertrophy, hyperplasia, dedifferentiation, inflammation, and finally atherosclerosis. It is very conceivable that results from vascular smooth muscle cells in culture may not reflect even basal physiological in vivo conditions, and projections from such experiments into human pathology may be fruitless or misleading, especially as cells change their phenotypes with increasing culture time or passage numbers (Chamley-Campbell et al., 1979).

Principles of Translational Science in Medicine. http://dx.doi.org/10.1016/B978-0-12-800687-0.00001-3

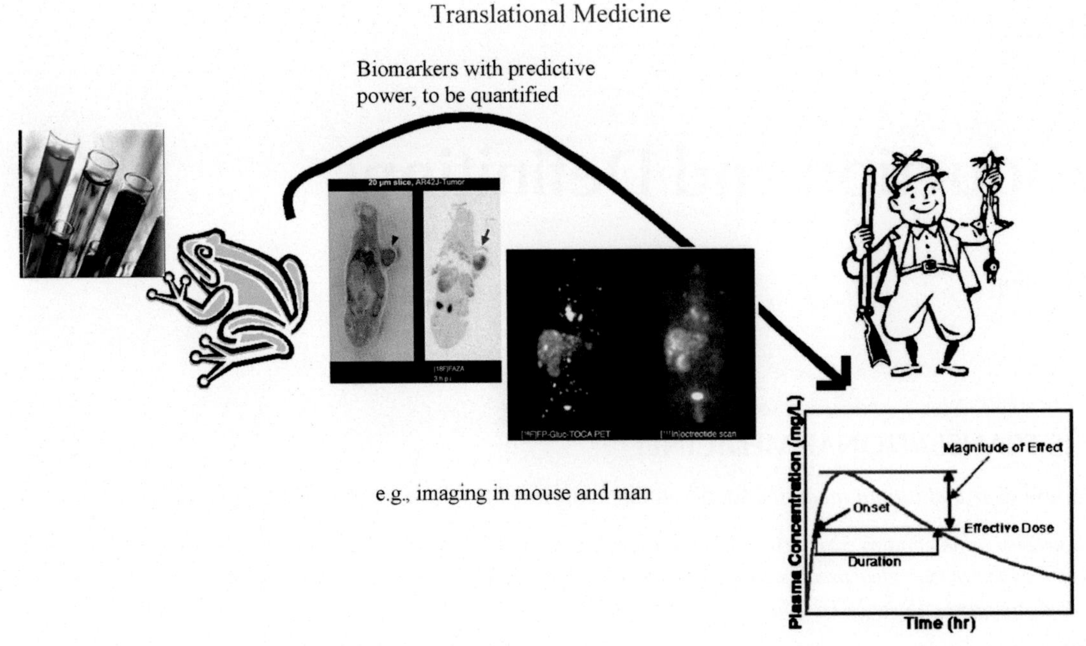

FIGURE 1.1 The main aspects of translational medicine: biomarkers as major tools for the transition from test tube/animal experiments to human trials, with imaging as a major biomarker subset. *(From Wehling, 2006, with kind permission from Springer Science and Business Media.)*

Such artifacts can only raise hypotheses that may or may not be corroborated in animal or, finally, human experiments. The artifact character of test tube systems is obvious, and differences among species are profound at both the genotype and phenotype levels, so no one is surprised if an intervention works in one species but not another. Although morphine is a strong emetic in dogs, it does not have this effect in rats. It is apparent that this variability applies even more when dealing with human diseases, which may or may not have any correlates in animal models. This especially concerns neuropathologic diseases for which animal models are either lacking or misleading (e.g., psychiatric diseases such as schizophrenia).

Thus, the difficulty of predicting the beneficial or toxic effects of drugs or medicinal devices, or the accuracy and value of diagnostic tests, is a major problem that prevents innovations from being useful for treating human diseases. From this end, the following is an operational definition of translational medicine:

> By optimization of predictive processes from preclinical to clinical stages, translational medicine aims at improving the innovative yield of biomedical research in terms of patient treatment amelioration.

PRIMARY TRANSLATION VERSUS SECONDARY TRANSLATION

In the definitions mentioned previously, the focus is clearly concentrated on translation in development courses from preclinical to clinical stages, in particular as applied to the development of new drugs. These developments would bring innovation to the patients who receive the new drug, test, or device. It seems odd to underscore that some patients may receive the innovation and, thus, benefit from it, whereas others may not. However, there is yet another gap that prevents innovations from flourishing to their full potential. Even if innovative drugs have changed clinical guidelines and rules and thus been undoubtedly proven to represent beneficial options to suitable patients, they may not be applied in what is commonly termed "real life."

Undertreatment may result from ignorance, budget restrictions, or patient or doctor noncompliance and often has severe socioeconomic implications: Though potentially correctable in all patients, arterial hypertension is treated to guideline targets only in 20%–50% of patients (Boersma et al., 2003); LDL-cholesterol in cardiovascular high-risk patients is at target levels in 12%–60% of patients (Böhler et al., 2007). This means that innovations that have successfully passed all translational hurdles in the developmental process from bench to bedside still may not reach the patients at large, as there is a second barrier between guideline recommendations and real-life medicine (see Figure 1.2).

This translational aspect of innovation is sometimes called *secondary translation* (as opposed to the developmental primary translation). Because problems in secondary translation mainly reflect insufficiency at the level of patient care, socioeconomic structures, education and society, and habits, the scientific challenge is secondary to social and political tasks and obligations. Therefore, this textbook is entirely devoted to the scientific aspects of primary translation and does not deal with secondary translation, although its impact on patient care may also be crucial.

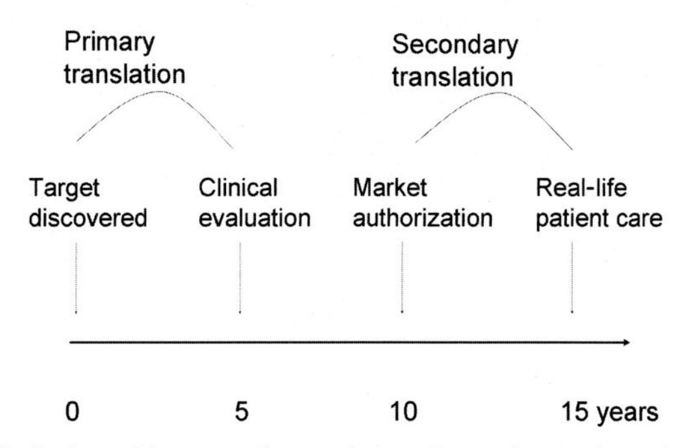

FIGURE 1.2 Scheme depicting the two principal transition zones for translation of, e.g., drug projects: from preclinical ("target discovery") to clinical development = primary translation, and from market approval to real-life patient care = secondary translation.

Some authors propose an even more detailed labeling of translational stages. T1 describes translation from basic genome-based discovery into a candidate health application (e.g., genetic test/intervention); T2 from application for health practice to the development of evidence-based guidelines; in T3 evidence-based guidelines are moved into health practice; and T4, finally, seeks to evaluate the "real-world" health outcomes of an application in practice (Khoury et al., 2007).

THE HISTORY OF TRANSLATIONAL MEDICINE, OBSTACLES, AND REMITS

As described previously, the main feature of translational medicine is the bridging function between preclinical and clinical research. It aims at answering the simple but tremendously important question, if a drug *X* works in rats, rabbits, and even monkeys, how likely is it that it will be beneficial to humans? Historically, how did this simple and straightforward question, which is naturally inherent to all drug development processes, become of prime relevance in biomedical research?

If all drug, device, or test development components were closely connected within a common structure, the necessity to develop this discipline would probably not have become apparent. As it stands now, however, the new emphasis on translational medicine reflects the wide and strict separation of biomedical research into preclinical and clinical issues, a situation best illustrated by the acronym *R&D*, which is used in pharmaceutical companies to describe their active investments into science as opposed to marketing. *R* stands for research, which largely means preclinical drug discovery, and *D* stands for development, which is largely identical to clinical drug development. It is obvious that even the words behind R&D arbitrarily divide things that share a lot of similarities: Clinical development and clinical research are very congruent terms, and compounds are *developed* within the preclinical environment, for example, from the lead identification stage to the lead optimization stage.

In the drug industry, the drug discovery and development process follows a linear stage progression; a major organizational transition occurs when a candidate drug is delivered from discovery (R) to clinical development (D), which is synonymous with trials in humans. When this happens, it is often said that the discovery department has "thrown a compound over the fence." This ironic or cynical expression exposes the main concern in this context: clinical issues—that is, the human dimension of a drug project—are not properly and prospectively addressed in the early stages of preclinical discovery or even at the level of target identification or validation. Clinical researchers are then surprised or even upset by what has been sent to be developed in humans. A chemical that had been shaped years earlier with too little or no clinical input or projections may turn out to be impractical for swallowing (e.g., the compound dose may be too large or measured in grams instead of milligrams) or may quickly prove to be too short-lived, requiring multiple dosing schemes that are far out of scope in many therapeutic areas.

Why is this interface problem relevant? Bridging this divide or improving the interface performance is a major prerequisite for success if laboratory or animal data are to finally lead to treatment of diseases in humans. There is an old dispute over free and basic sciences versus applied sciences, and universities in particular take pride in being independent and free in their choice of research areas and scientific strategies. This *l'art-pour-l'art* approach is thought to still yield "useful" discoveries—namely by serendipity or simply by chance findings. Even worse, it is thought that big, applicable discoveries can only flourish in unrestricted, free scientific settings.

Unfortunately, drug discovery and development have to assume that a restricted, structured, and therapy-driven process is the only way to cope with modern standards of drug-approval requirements. Chance findings may trigger the initial steps of drug discovery, but those are rare in clinical stages. (One famous exception is sildenafil, which had been clinically developed as an antianginal agent when its effects on erectile function were incidentally discovered.) The typical R&D process has to rely on projections across this interface, and, thus, it has to focus its early discovery stages on later applications, that is, the treatment of human disease.

This implies that "throwing a drug over the fence" is not optimal if the final output is to be measured in terms of the number of approved new drugs being sold on the market. Unfortunately, output *is* in fact the major concern: Complaints

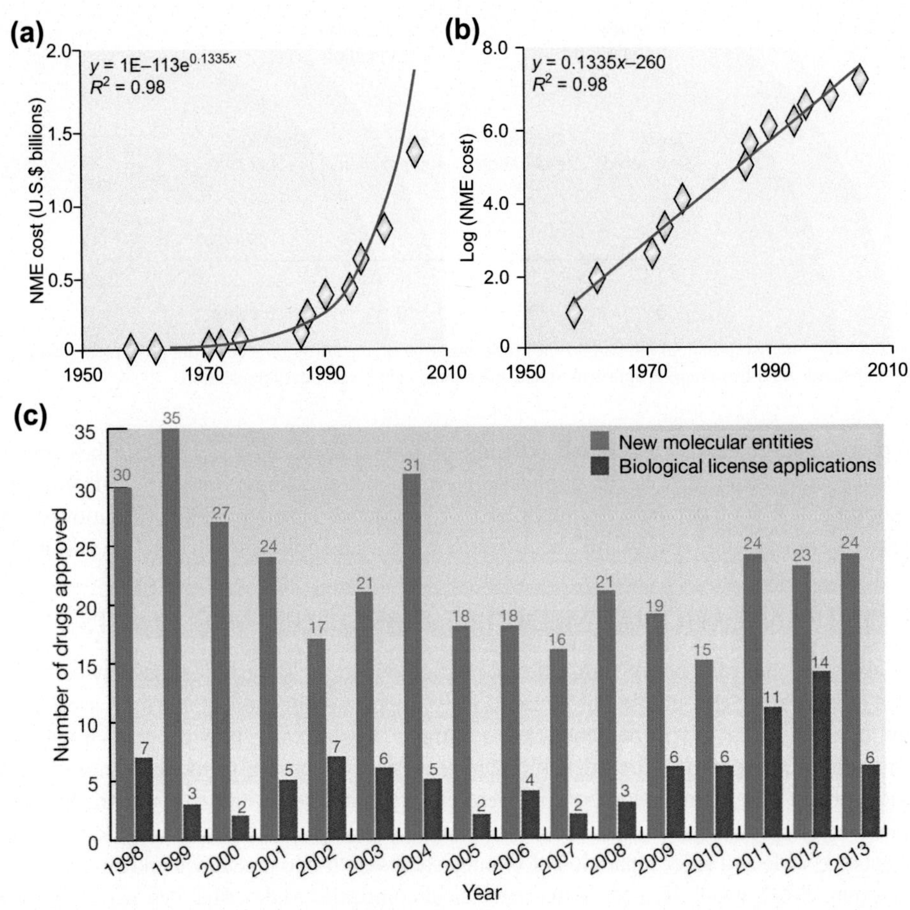

FIGURE 1.3 Increasing R&D costs (a, b) versus decline in numbers of new drug approvals (c). (*From Munos, 2009 and Kling, 2014 by kind permission.*)

about this interface problem have largely been driven by the widening gap between surging R&D costs and the steadily and dramatically decreasing output of drugs from shrinking pipelines (Figure 1.3).

Shrinkage correlates with high late-stage attrition rates, meaning that many drug projects die after billions of dollars and 5–10 years of investment. This attrition problem particularly applies to expensive clinical phase IIb trials and especially phase III trials. Attrition can be largely attributed to the inability to predict the efficacy and/or safety of a new candidate drug from in vitro, animal, or early human data. From 1991 to 2000, only 11% of all drugs delivered to humans for the first time were successfully registered (Figure 1.4).

It is obvious that there are huge differences among therapeutic areas; for example, success rates in central nervous system (CNS) or oncology drugs are particularly low (7% or 5% versus a 20% success rate in cardiovascular drugs). This means that in CNS only 1 out of about every 14 compounds that have passed all hurdles to be applied to humans for the first time will ever reach the market and, thus, the patient. In more than 30% of cases, attrition was related to either clinical safety or toxicology, just fewer than 30% were efficacy-related, and the remainder were caused by portfolio considerations and other reasons (Figure 1.5) (Kola and Landis, 2004). Attrition caused by portfolio considerations means that the company producing the project has lost interest in it because, for example, a competitor has reached related goals before the project was finished and thus the project no longer has a unique selling position.

Late-stage attrition is a problem for all large companies, and lack of innovation is a major reason for the recent stagnation in progress in the treatment of major diseases. If the tremendous costs of drug development continue to rise, companies may resort to concentrating on the relatively safe "me-too" approach. This approach aims at minimally altered compounds that are patentable but resemble their congeners as much as possible in terms of efficacy and safety. These compounds are (sometimes erroneously) thought to be without pharmaceutical risk; their main disadvantage is the fact that they are not innovative.

Thus, tackling the translational challenges in the R&D process may become essential to the struggle for the survival of the pharmaceutical industry in an increasingly adverse environment. This adverse environment includes reduced remunerations for smaller innovative gains (such as those made by the aforementioned me-too compounds) and ethical issues that continuously undermine the reputation of the drug industry, which is now seen as similar to the reputations of the oil and

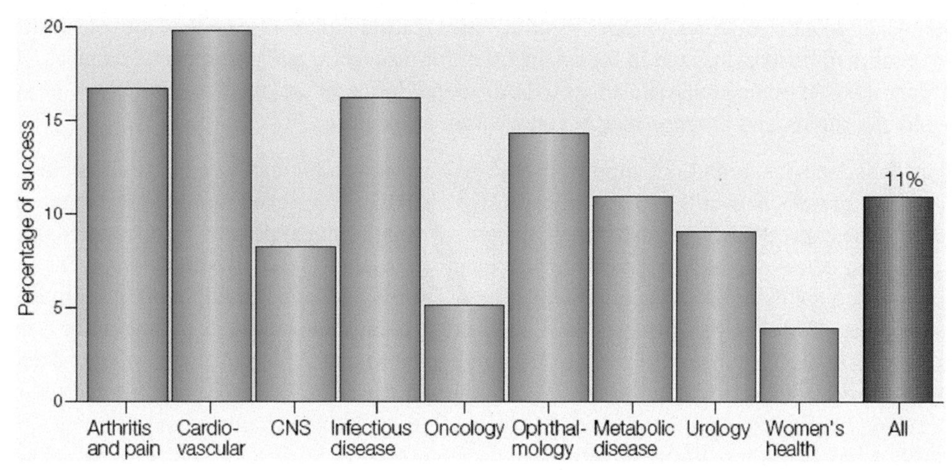

FIGURE 1.4 Success rates from first-in-man to registration. (*From Kola and Landis, 2004, reprinted by permission from Macmillan Publishers Ltd: [*Nature Rev. Drug Discov.*] (3: 711–715) ©2004.*)

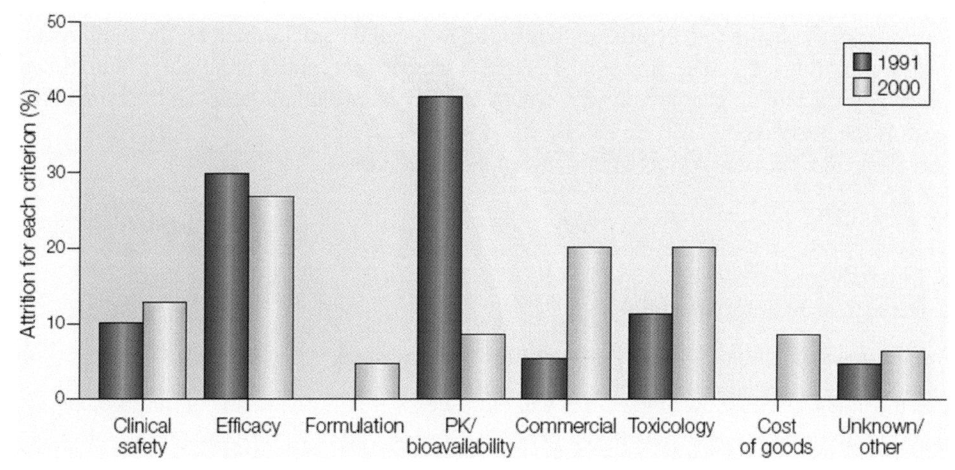

FIGURE 1.5 Main reasons for termination of drug development—for "wasted investment" 1991–2000. (*From Kola and Landis, 2004, reprinted by permission from Macmillan Publishers Ltd: [*Nature Rev. Drug Discov.*] (3: 711–715) ©2004.*)

tobacco industries (Harris Interactive, 2006). Thus, translational medicine, if successfully applied, appears to be an important remedy for improving the ethical (i.e., patient-oriented) and financial success of the R&D process. It could also help the battered reputation of the drug industry by improving the treatment of major diseases.

It is important to note that translational medicine problems do not pertain only to the drug industry; they are inherent to all developmental biomedical processes and include device and diagnostic tool development as well. They also exist in academia, in which translation is not the primary goal of research; at least it is not perceived as such. However, in academia there is also growing awareness of the fact that public funding of expensive biomedical research will not continue forever if this funding is not seen to lead to patient-oriented results. Thus, academic research utilizes this phraseology increasingly as well.

It is obvious that the persistence of the low-output syndrome in terms of true medical innovations is a threat to the existence of

- Big pharmaceutical companies (known collectively as "big pharma"): Big pharmaceutical companies are laying off tens of thousands of people. For example, Pfizer laid off 10,000 in 2007. Further cuts in the workforce are being announced almost every week. For example, AstraZeneca has shut down its Charnwood and Lund facilities. It is feared that 30%–50% of all jobs in big pharma R&D will be axed within the next 5–10 years. In a 2010 Reuters review (Reuters Special Report, 2010), 200,000 jobs in big pharma are feared to be lost by 2015.
- Academia: Taxpayers will not tolerate expenditures of billions of dollars or euros without measurable treatment improvement; the U.S. parliament has asked researchers what happened to the $100 billion invested into cancer research from the mid-1980s to mid-1990s in terms of measurable outcome.

- Society: If biomedical research does not improve its utility and create an impressive track record of substantial innovations, biomedical research will be marginalized in the competition for resources, as environmental changes, such as climate or energy catastrophes, create tremendous challenges to humankind. In the future, medicine may become static, executed by robots fed by old algorithms, and progress may become a term of the past.

All these negative statements should not suppress some positive developments that became obvious only recently. If one looks at the number of new medical/biological entities (NMEs/NBEs) approved by the FDA (Figure 1.3c), some hope seems to loom around the corner. Regarding NMEs/NBEs, only 21 marketing approvals were counted in 2010. This figure would be back to or even below average observed for the past 5 years. In 2011, 35 NMEs/NBEs were approved; in 2012, 37; and in 2013, 30, which are essential signs of hope. Did translational medicine finally work after more than 10 years of "hype"? It is obvious that numbers by themselves do not tell much about innovation, as redundant or minimally divergent drugs may precipitate excessive optimism. Conversely, the surge in biologicals (humanized monoclonal antibodies) as demonstrated in Figure 1.3 reflects a sound principle with convincing features of translatability, limited toxicity, and targeted efficacy; biologicals may, thus, represent one successful approach to tackle the challenge of shrinking (or even vanishing) pipelines. The question remains where this relatively narrow avenue ends or results get saturated; antibodies have no access to intracellular structures, which caps their potential considerably.

WHAT TRANSLATIONAL MEDICINE CAN AND CANNOT DO

Proponents of translational medicine feel that the high attrition rate can be ameliorated by the main remits of translational medicine, as illustrated in Table 1.1. The first goal is target identification and validation in humans. Identification has already been achieved by the human genome project, which literally identified all genes in the human body. Thus, validation of known genes is the next task.

Genetics is one of the most powerful tools in this regard, because it tests

- Disease association genes
- Normal alleles
- Mutant genes, especially in oncology
- Susceptible genes such as BCRabl (imatinib)

To this end, we must ask and attempt to answer the following questions:

- In general, does the target at least exist in the target cell or tissue, or is expression low or undetectable?
- Is it dysregulated in diseased tissues? Functional genomics, for example, Her2neu expression (trastuzumab) or K-Ras (Parsons and Myers, 2013)

Another approach utilizes test or probe molecules:

- Can we test the hypothesis with a probe molecule?
    - Using a substandard candidate drug or the side effects of a drug used for something else
    - Monoclonal antibodies
    - Antisense technology
    - Fluorescent probes
- Has someone else tested the hypothesis?
    - Antegrin for VLA4 antagonists in multiple sclerosis

This is just a small fraction of the possible target validation or identification approaches. The basic principle is the early testing of human evidence at a preclinical stage of the drug development process. The reverse could be true as well: Knowledge of the side effects of drugs can be utilized to discover new drugs by exposing this side effect as a major

TABLE 1.1 Main remits of translational medicine
Target investigation and target validation in man.
Early evaluation of efficacy and safety using biomarkers in man.
Use the intact living human as our ultimate screening test system.

effect. Minoxidil was developed as an anti–hair loss agent until its ability to lower blood pressure was clinically detected. Although this *reverse pharmacology* approach has been utilized to find pure blood pressure drugs and pure anti–hair loss drugs, most attempts have failed so far. The principle however—human target identification and validation with subsequent feedback into preclinical stages (see Chapter 2.1.9)—has been proven to be a successful strategy in general.

Another important focus in translational activities is on predicting as early as possible the safety and efficacy of a new compound in humans, mainly by the identification, development, and smart utilization of biomarkers. Several chapters of this book are devoted to biomarkers, which describe physiological, pathophysiological, and biological systems and the impact of interventions in those systems, including those of drugs. This is the most important translational work, and 80% of translational efforts are devoted to finding or developing the right biomarker to predict subsequent success across species, including humans. Biomarker work includes the smart design of the early clinical trials in which those experimental biomarkers are most suitably exploited. This work may also include the validation work necessary to establish the predictive value of novel biomarkers; thus, it may include a developmental program (for the biomarker) that is embedded in the drug development program.

The remit of biomarker work goes far beyond early efficacy and safety prediction, but is increasingly seen as a necessary tool for profiling compounds to better fit the needs of individual patients. The fashionable term in this context is *personalized medicine*, which is a term as old as drugs are. Renal drugs (excreted by kidneys) have always necessitated tests to assess kidney function and thus require personalized medicine; otherwise, poisoning in renal impairment is inevitable. The novelty in this regard is the use of profiling to achieve better matches between success rates (responder concentration) and thus increase cost-effectiveness. It is thought that this approach will save billions of U.S. dollars in revenue (Figure 1.6) when the blockbuster is new.

Another remit of translational medicine is its facilitation of early testing of principles in humans without directly aiming at the market development of the compound tested. These human trials are called exploratory trials, and they may involve experimental investigational new drugs, which are compounds that are known to have shortcomings (e.g., a compound with a half-life that is too short for the compound to become a useful drug) but could be ideal test compounds to prove the basic hypothesis of efficacy in the ultimate test system, the human being. Such tests could validate the importance of, for example, a particular receptor in the human pathophysiology; could substantiate investment decisions; and could speed up developmental processes at early stages. Examples are given in Chapter 4.

This short list of remits is incomplete, but it should demonstrate that the major tool of translational medicine is the early, intensive, and smart involvement of humans as the ultimate test system in discovery and development processes. Its scope reaches from straightforward translation power through reverse pharmacology to personalized medicine.

In an ideal world, translational medicine creates forward-signaling loops and reverse-signaling loops along the artificially linear development line of drugs (Figure 1.7). It can speed up the process, allow for parallel processing, and generate knowledge for other projects as well (e.g., generic biomarker tools and side effects as target starting points).

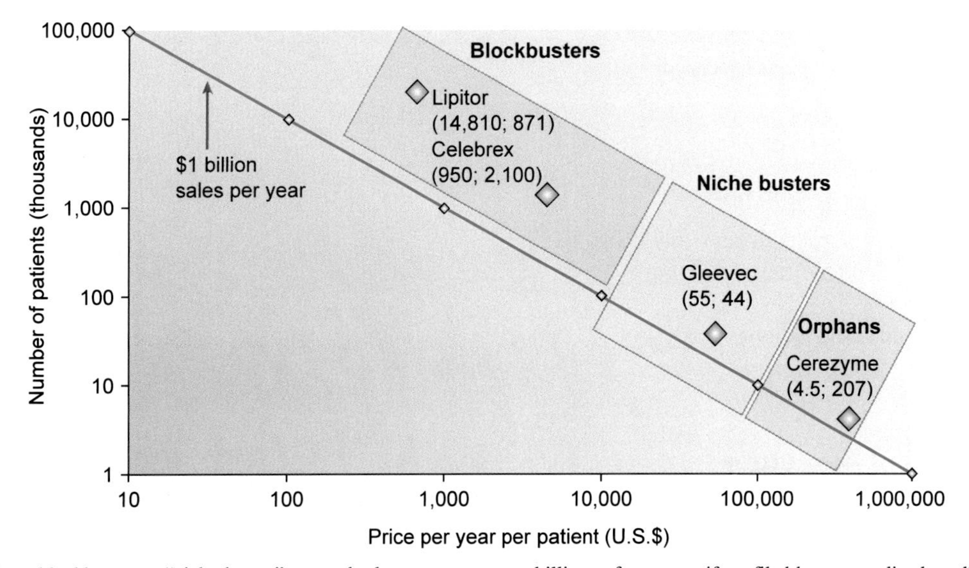

FIGURE 1.6 From blockbuster to "niche buster": even the latter can generate billions of revenues if profiled by personalized medicine approaches; highly effective and high-prized. (*From Trusheim et al., 2007, reprinted by permission from Macmillan Publishers Ltd: [*Nature Rev. Drug Discov.*] (6(4): 287–293) © 2007.*)

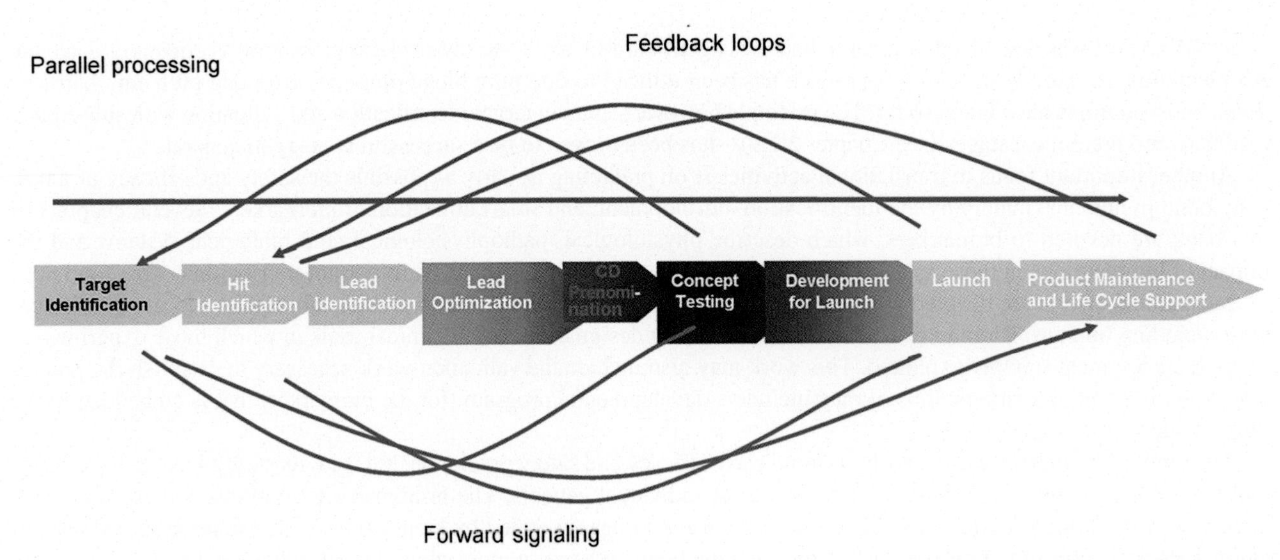

FIGURE 1.7 The "pseudo"-linear model of drug development, translational medicine creates forward- and reverse-signaling loops and speeds up processes and allows for parallel processing.

Translational medicine cannot replace the most expensive study—the pivotal phase III (safety) trial. However, it can increase the likelihood of success in phase III trials. It cannot invent new targets (all potential targets are gene-related and all genes have, meanwhile, been "invented" and described), but it can significantly help to assess the validity of targets and reduce lapses due to "unimportant" targets at the human level. For these reasons, translational medicine might be the key to preventing biomedical research and medicine from falling into oblivion because of transfer of funding to more successful areas of innovation such as energy and climate survival technologies.

THE PRESENT STATUS OF TRANSLATIONAL MEDICINE (INITIATIVES AND DEFICIENCIES)

The term *translational medicine* was rarely used in the 1990s. Its inflationary use was caused by increasing efforts to focus on translational issues in all areas of biomedical research; conversely, its inflationary use also caused awareness and attention, including some truly innovative initiatives.

One of the first major initiatives was the NIH Roadmap, announced in September 2003 (National Institutes of Health, 2007b). The Roadmap is a series of initiatives intended to "speed the movement of research discoveries from the bench to the bedside" and introduced by the new head of the NIH, Dr. Elias Zerhouni, who took over in 2002. The Roadmap outlined a series of goals that were put into action in 2004 or 2005. Figure 1.8 illustrates these goals.

The following areas have been specified.

New Pathways to Discovery

The implementation groups in this area are

- Building blocks, biological pathways, and networks
- Molecular libraries and molecular imaging
- Structural biology
- Bioinformatics and computational biology
- Nanomedicine

Research Teams of the Future

The implementation groups in this area are

- High-risk research
- Interdisciplinary research
- Public–private partnerships

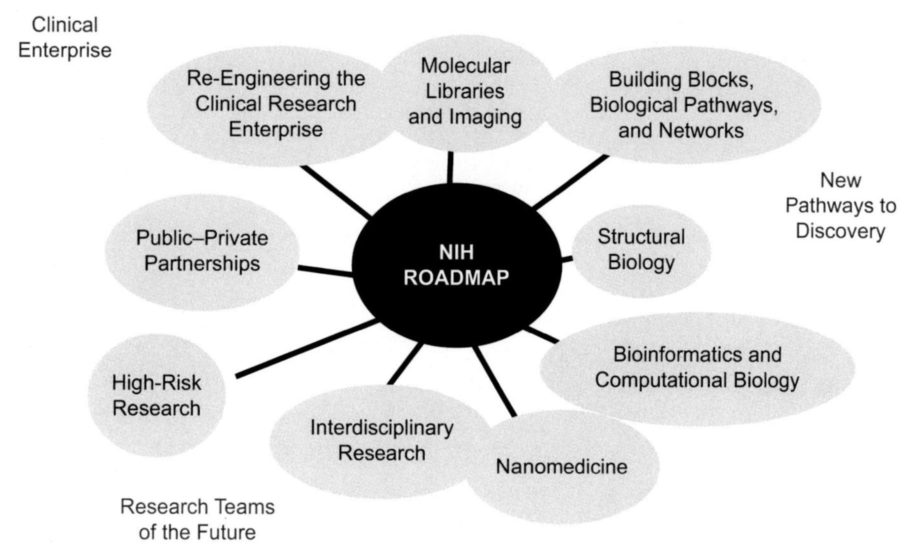

FIGURE 1.8 The NIH Roadmap initiatives. (*From National Institutes of Health Roadmap, 2007a.*)

Re-Engineering the Clinical Research Enterprise

The implementation groups in this area are

- Clinical research networks/NECTAR
- Clinical research policy analysis and coordination
- Clinical research workforce training
- Dynamic assessment of patient-reported chronic disease outcomes
- Translational research

Part of the strategy of the NIH Roadmap involves funding of about 60 centers for clinical translation (clinical and translational science awards, or CTSA) in the United States (Clinical and Translational Science Awards, 2014a), an initiative started in 2006. The consortium formed from this initiative (about 62 members) has five strategic goals:

- National Clinical and Translational Research Capability
- The Training and Career Development of Clinical and Translational Scientists
- Consortium-Wide Collaborations
- The Health of Our Communities and the Nation
- T1 Translational Research (Clinical and Translational Science Awards, 2014b)

Critical minds are not convinced that their funding will truly be spent translationally, as most of it could be used to finance isolated clinical trials.

The pressure of increasing R&D costs and low output in terms of critically novel drugs forced the Food and Drug Administration (FDA) to reconsider its own actions and those of major players in biomedical research in terms of timelines, costs, design, and, ultimately, success. The Critical Path Initiative (the official title of which is "Challenge and Opportunity on the Critical Path to New Medical Products") published in March 2004 represented a milestone in this context. It reflects a concerted action initiated by a regulatory authority criticized for its "retarding" activities, which were claimed to have caused the low-output syndrome described previously. Although certainly not the first public initiative to address translational medicine issues as a major concern, it was one of the most influential and respected ones.

There is major overlap between the Critical Path Initiative and translational medicine (Figure 1.9), although the Critical Path Initiative reaches further into industrialization. However, its first two major goals—safety and medical utility (efficacy)—are entirely dependent on translation (Figure 1.10).

American universities have also addressed the challenge of translational medicine; many have established centers for translational medicine such as those at Duke University and Pennsylvania University. The single most important development regarding translational medicine in the United States was the foundation of the National Center for Advancing Translational Sciences (NCATS) in December 2011; its budget request for the fiscal year 2014 is U.S.$665 million (National Center for Advancing Translational Sciences, 2014). Chapter 2 explicitly deals with this exemplary institution.

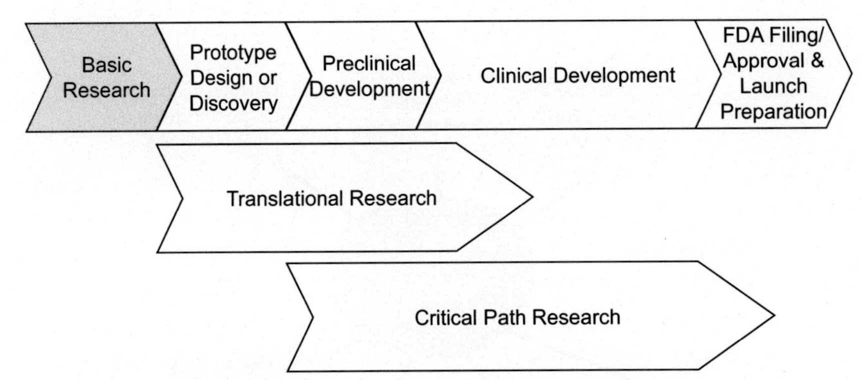

FIGURE 1.9 Essential overlap between the "Critical Path" by FDA and translational medicine. (*Modified from U.S. Food and Drug Administration, 2004.*)

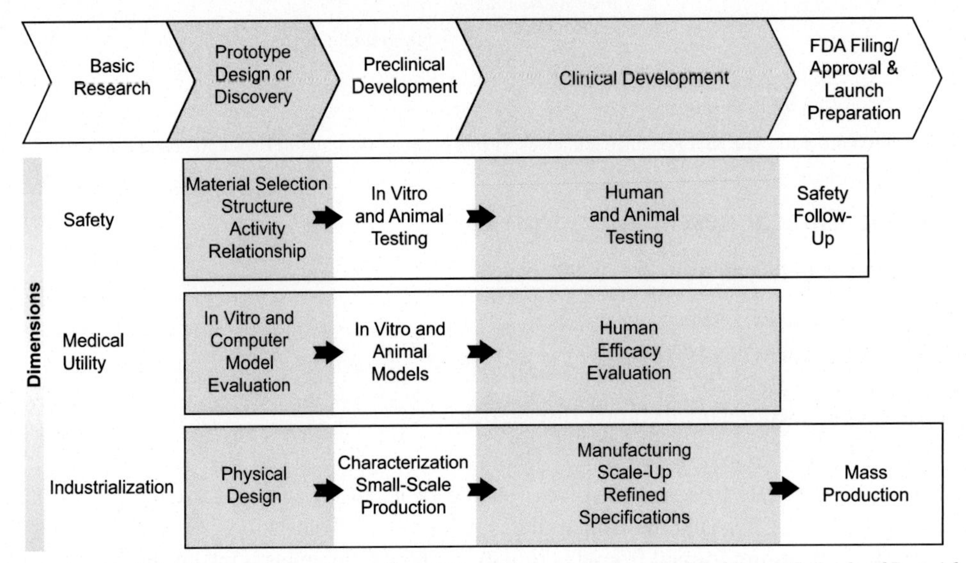

FIGURE 1.10 Three major tasks in the "Critical Path": safety, medical utility, and industrialization. (*Modified from U.S. Food and Drug Administration, 2004.*)

In addition, there are initiatives in Europe that are worth mentioning, including the European Organization for Research and Treatment of Cancer (EORTC), which is committed to making translational research a part of all cancer clinical trials, and the National Translational Cancer Research Network, which was announced by the British government to facilitate and enhance translational research in the United Kingdom.

In general, research funding programs in Europe such as the Horizon 2020 Program of the European Union (EU) use the phrase "translational activities" in most of their topics, but institutionalization is rare on this continent. In the largest EU member state, Germany, there is almost no structural activity (e.g., university departments or independent institutes) covering this most important subject when it comes to prognosis of medical research survival although courses for translational medicine exist in some medical faculties. If one screens the scientific programs of disease-specific institutions such as cancer institutes, the term *translational medicine* will always appear in a prominent position; as mentioned previously, in the absence of dedicated structures, however, this appears as phraseology rather than a sound approach. The U.K. and maybe the Netherlands seem most advanced in Europe, although distinct structures are still rare.

A major EU investment is the Innovative Medicines Initiative (IMI) fostering private–public partnerships for the development of drugs. Its budget of 2 billion EUR is the largest in the field worldwide, with equal contributions of 1 billion from both the EU and industry (Innovative Medicines Initiative, 2014).

Almost all major drug companies have addressed the issue of translational medicine in one way or another. The institutional structures range from independent departments of translational or discovery medicine to entirely embedded dependent structures that are part of the drug discovery or development teams without central facilities. Special interest groups are the least specific modification of the R&D process, although they are not necessarily ineffective. Specific examples cannot be given here for secrecy reasons, but it is obvious that the challenge has been collectively identified by industry and met in widely variable ways.

In general, U.S. biomedical players have invested the most (NIH has announced that it plans to spend a total of up to $10 billion) and are at the forefront of translational institutionalization, with the individual countries within Europe lagging behind at very different distances. Commercialization of academic inventions has traditionally been much more efficient in the United States than in the EU, and success in translational medicine has a lot to do with expertise accrued during commercialization processes. Later processes can be lucrative only if translational steps have been added to a biomedical invention, and facilitating such early developmental investments has been on the to-do list of successful universities (e.g., University of California, San Francisco, and Harvard University) and also smaller companies for many years. The rapid pace of business development of currently medium-sized or large companies such as Amgen or Genentech may have to do with their early and effective understanding and instrumentalization of translational issues.

Translational medicine in BRIC countries (Brazil, Russia, India, and China) will be developed as their pharmaceutical and biomedical markets develop, although particular structures are not yet easily identifiable. It would come as a surprise if innovation in those countries were to ignore the potential of translational medicine. In fact, the opposite could well become reality; for example, China might even develop leadership in segments of this area. But this is hypothetical right now. So far, there is comparably strong Chinese participation in, for example, the editorial board of the *Journal of Translational Medicine* (editor-in-chief: Francesco Marincola), and this may indicate that, at least, awareness has already been seeded in this huge country.

The aforementioned journal, founded in 2003, so far represents one of the only two journals explicitly devoted to translational medicine. On its 2007 home page, it stated that it

> *aims to improve the communication between basic and clinical science so that more therapeutic insights may be derived from new scientific ideas—and vice versa. Translation research goes from bench to bedside, where theories emerging from preclinical experimentation are tested on disease-affected human subjects, and from bedside to bench, where information obtained from preliminary human experimentation can be used to refine our understanding of the biological principles underpinning the heterogeneity of human disease and polymorphism(s)* (*Journal of Translational Medicine*, 2007).

In 2009, a new journal was established, *Science Translational Medicine*, by the American Association for the Advancement of Science, belonging to the *Science* family of high-impact journals. In its fourth year, 2012, the journal already had an impact factor of 10.757.

In conclusion, it appears that translational medicine has been basically understood and substantially funded mainly in the United States, whereas "old" Europe is lagging behind. BRIC countries, especially China, seem to be aware of this challenge and may undertake serious efforts to catch up with the United States, thus leaving Europe behind if its investments remain minor or unstructured. The IMI of Europe, although it is the most expensive program in the world, lacks clear and institutional translational structures.

TRANSLATIONAL SCIENCE IN MEDICINE: THE CURRENT CHALLENGE

The title of this book (*Principles of Translational Science in Medicine*) contains the main challenge in this context as seen by the authors. It does not use the common phrase "translational medicine: from bench to bedside" but rather introduces the term *science* as the second, and thus very important, word. Why?

The simpler term *translational medicine* seems to reflect the wishful thinking of people who use it to denote the appropriate direction that should be taken. As with all biomedical science, there is no nor has there ever been substantial doubt about the direction to be taken: all efforts should, in a more or less direct sequence, ultimately lead to improvement of patient care. If one accepts this proposition, translational medicine, in a more philosophical sense, is not new: it simply describes the final direction and destination of all biomedical activities (with the exception of forensic or insurance activities in medicine, which certainly serve purposes other than patient care, but this is only a very minor segment of the total pie). All modern drugs whose discovery and development followed the classical path from the test tube through animal experiments to human application must have gone through a more or less efficient translational process.

It is obvious that, for many—largely ethical—reasons, systematic development of drugs, devices, and medical tests cannot be performed in humans from day one. Thus, translation processes have been inherent to all biomedical research ever done under ethical auspices. Therefore, it is obvious that translational medicine was "invented" (although not identified as such) by all the biomedical disciplines that have attempted to introduce new treatments for human diseases for the past thousands of years. There are famous examples of translational processes in many historic documents (e.g., Alexander Fleming's translation of the in vitro observation of fungi that inhibit bacterial growth into clinically used penicillin). So, what is new then?

As said earlier, the fame of the term *translational medicine* has been brought about by the increasing demand for a successful, reliable, reproducible, and efficient way of translating results from animals or test tubes to humans, as the translational paradigms that have successfully worked in the past seem to be failing. Thus, neither the claim nor the procedure as such is new, but the quality of the process is not sufficient and needs to be improved if major innovations in medicine are to come back at an appealing pace. The pivotal question is thus how this can be achieved, not how we can convince others of the need for translation. The methods of how translation should take place—rather than the fact that it should take place—should be the true claim of the present translational movement.

This claim, however, has generally been neglected in terms of structured approaches and at the level of concept development. The clear formulation of this contemporary challenge is to claim that this aim can be achieved only by the establishment and development of a novel science. The *Encyclopedia Britannica* (2007) defines science as "any system of knowledge that is concerned with the physical world and its phenomena and that entails unbiased observations and systematic experimentation. In general, a science involves a pursuit of knowledge covering general truths or the operations of fundamental laws." The science of translational medicine is truly innovative and genuine in that it has not yet been explicitly named, supported, promoted, established, or even recognized. It is thus termed "translational science in medicine." The specification "in medicine" reflects the fact that translational sciences may be established in other scientific areas as well, for example, physics and chemistry.

If it is declared a science, what claims and remits should this science deal with? The experimental processes and tools or methods used for translational processes in medicine should be clearly defined and used in reproducible, objective, and measurable translational algorithms. Thus, toolboxes (as named in the Critical Path Initiative, discussed previously) need to be developed; the strategies described in translational development plans need to be standardized; and decision trees need to be developed, tested, validated, and exercised. The early clinical trial program should be structured according to translational needs (early efficacy and safety testing). Thus, in an ideal world, the methodology of translational science would comprise a canon of widely applicable, generic procedures that reliably generate quantifiable prediction quality. This includes toolboxes with appropriate biomarkers, their validation, their grading for predictivity, smart early human trial designs, biostatistics methods to cope with multiple readout problems, decision tree (go-or-no-go decision) algorithms, and many other activities, which together could give translational science the right to be called a true science.

The basic goals of this book are to help readers start thinking in these terms, to trigger concept development, and to collect available pieces—such as research in the biomarker arena—that can be incorporated into the nascent plot of translational science in medicine. It is hoped that this book will become a seminal effort to launch this science and, thus, contribute to its projected establishment and related benefits.

REFERENCES

Boersma, E., Keil, U., De Bacquer, D., De Backer, G., Pyörälä, K., Poldermans, D., et al., 2003. EUROASPIRE I and II Study Groups. Blood pressure is insufficiently controlled in European patients with established coronary heart disease. J. Hypertens. 21, 1831–1840.

Böhler, S., Scharnagl, H., Freisinger, F., Stojakovic, T., Glaesmer, H., Klotsche, J., et al., 2007. Unmet needs in the diagnosis and treatment of dyslipidemia in the primary care setting in Germany. Atherosclerosis 190, 397–407.

Chamley-Campbell, J., Campbell, G.R., Ross, R., 1979. The smooth muscle cell in culture. Physiol. Rev. 59, 1–61.

Clinical and Translational Science Awards, 2014a. Home page. https://www.ctsacentral.org (accessed 07.02.14.).

Clinical and Translational Science Awards, 2014b. About the CTSA Consortium. https://www.ctsacentral.org/about-us/ctsa (accessed 07.02.14.).

Encyclopedia Britannica [online], 2007. http://www.britannica.com/eb/article-9066286 (accessed 07.02.14.).

Harris Interactive [online], 2006. Reputation of pharmaceutical companies, while still poor, improves sharply for second year in a row. http://www.harrisinteractive.com/news/allnewsbydate.asp?.NewsID=1051 (accessed 07.02.14.).

Innovative Medicines Initiative, 2014. Mission. http://www.imi.europa.eu/content/mission (accessed 07.02.14.).

Journal of Translational Medicine [online], 2007. What is Journal of Translational Medicine? http://www.translational-medicine.com/info/about (accessed 18.12.07.).

Khoury, M.J., Gwinn, M., Yoon, P.W., Dowling, N., Moore, C.A., Bradley, L., 2007. The continuum of translation research in genomic medicine: how can we accelerate the appropriate integration of human genome discoveries into health care and disease prevention? Genet. Med. 9, 665–674.

Kling, J., 2014. Fresh from the biotech pipeline 2013. Nature Biotechnol. 32, 121–124.

Kola, I., Landis, J., 2004. Can the pharmaceutical industry reduce attrition rates? Nature Rev. Drug Discov. 3, 711–715.

Munos, B., 2009. Lessons from 60 years of pharmaceutical innovation. Nat. Rev. Drug Discov. 8, 959–968.

National Center for Advancing Translational Sciences, 2014. Budget. http://www.ncats.nih.gov/about/budget/budget.html (accessed 07.02.14.).

National Institutes of Health, 2007a. NIH roadmap for medical research—NIH roadmap fact sheet. Division of Program Coordination, Planning, and Strategic Initiatives. http://www.nihroadmap.nih.gov/pdf/NIHRoadmap-FactSheet-Aug06.pdf (accessed 17.12.07.).

National Institutes of Health, 2007b. NIH roadmap for medical research—Overview of the NIH roadmap. Division of Program Coordination, Planning, and Strategic Initiatives. http://www.nihroadmap.nih.gov/overview.asp (accessed 17.12.07.).

Parsons, B.L., Myers, M.B., 2013. Personalized cancer treatment and the myth of KRAS wild-type colon tumors. Discov. Med. 15, 259–267.

Reuters Special Report, 2010. Big pharma. small R&D. http://static.reuters.com/resources/media/editorial/20100621/Big_Pharma.pdf (accessed 20.02.14.).

Science Translational Science, 2013. What is translational medicine? http://www.sciencemag.org/site/marketing/stm/definition.xhtml (accessed 07.02.14.).

Trusheim, M.R., Berndt, E.R., Douglas, F.L., 2007. Stratified medicine: strategic and economic implications of combining drugs and clinical biomarkers. Nat. Rev. Drug Discov. 6, 287–293.

U.S. Food and Drug Administration, March 2004. Challenge and opportunity on the critical path to new medical products. http://www.fda.gov/downloads/ScienceResearch/SpecialTopics/CriticalPathInitiative/CriticalPathOpportunitiesReports/ucm113411.pdf (accessed 07.02.14.).

Wehling, M., 2006. Translational medicine: can it really facilitate the transition of research "from bench to bedside"? Eur. J. Clin. Pharmacol. 62, 91–95.

Wikipedia, 2007. Definition of "Translational Research." http://en.wikipedia.org/wiki/Translational_research (accessed 07.12.07.).

Wikipedia, 2013. Definition of "Translational Medicine." http://en.wikipedia.org/wiki/Translational_medicine (accessed 07.02.14.).

Chapter 2

Target Identification and Validation

Chapter 2.1.1

"Omics" Translation: A Challenge for Laboratory Medicine

Mario Plebani, Martina Zaninotto and Giuseppe Lippi

INTRODUCTION

The rapid advances in medical research that have occurred over the past few years have allowed us to dissect molecular signatures and functional pathways that underlie disease initiation and progression, as well as to identify molecular profiles related to disease subtypes in order to determine their natural course, prognosis, and responsiveness to therapies (Dammann and Weber, 2012). The "omics" revolution of the past 15 years has represented the most compelling stimulus in personalized medicine that, in turn, should be simply defined as "getting the right treatment to the right patient at the right dose and schedule at the right time" (Schilsky, 2009). As a matter of fact, among the 20 most-cited papers in molecular biology and genetics that have been published in the past decade, 13 entail omics methods or applications (Ioannidis, 2010).

"OMICS": WHAT DOES IT MEAN?

Omics is an English-language neologism that refers to a field of study in biology focusing on large-scale and holistic data, as derived from its root of Greek origin which refers to wholeness or to completion. Initially, the suffix *omics* had been used in the word *genome*, a popular word for the complete genetic makeup of an organism, and later, in the term *proteome*. Genomics and proteomics succinctly describe a new way of holistic analysis of complete genomes and proteomes, and the success of these terms led to more emphasis in the trend of using *omics* as a convenient term to describe holistic ways of looking at complex systems, particularly in biology.

Fields with names like *genomics* (genetic complement), *transcriptomics* (gene expression), *proteomics* (protein synthesis and signaling), *metabolomics* (concentration and fluxes of cellular metabolites), *metabonomics* (systemic profiling through the analysis of biological fluids), and *cytomics* (the study of cell systems—cytomes—at a single cell level) have been introduced in medicine with increasing emphasis (Plebani, 2005). However, beyond these terms, multiple "omics" fields, with names like epigenomics, ribonomics, epigenomics, oncopeptidomics, lipidomics, glycomics, spliceomics, and interactomics, have been similarly explored regarding molecular biomarkers for the diagnosis and prognosis of human diseases.

Each of these emerging disciplines grouped under the umbrella of the term *omics* shares the simultaneous characterization of dozens, hundreds, or thousands of genes (genomics), gene transcripts (transcriptomics), or proteins (proteomics) and other molecules, that in aggregate and in parallel should be coupled with sophisticated bioinformatics to reveal aspects of biological function that cannot be culled from traditional linear methods of discovery (Finn, 2007). While an increasing body of literature has been produced to prove that "omics" will irrevocably modify the practice of medicine, that change has yet to occur and its precise details are still unclear. The reasonable assumption that the application of "omics" research will be riddled with difficulties has led to a much better appreciation of concepts of knowledge translation, translational research, and translational medicine.

PROTEOMICS AS A PARADIGM OF PROBLEMS IN TRANSLATIONAL MEDICINE

The paradigm of obstacles in translating new "omics" insights into clinical practice is a study reporting that a blood test, based on pattern-recognition proteomics analysis of serum, was nearly 100% sensitive and specific for detecting

Principles of Translational Science in Medicine. http://dx.doi.org/10.1016/B978-0-12-800687-0.00002-5

ovarian cancer and was possibly useful for screening (Petricoin et al., 2002a). The approach involved the analysis of a drop of blood using mass spectrometry, resulting in a large number of mass-to-charge ratio peaks (15,000 to 300,000 peaks, depending on technology), that were then subjected to pattern-recognition analysis to derive an algorithm that discriminates patients with cancer from those without. Which substances cause the peak (e.g., proteins, peptides, or something else) was yet unknown, as it was unclear whether these substances were released by tumor cells or by their microenvironment.

After this and further supporting work (Adam et al., 2002; Drake et al., 2003; Petricoin et al., 2002b; Vlahou et al., 2001; Zhu et al., 2003), commercial laboratories planned to market a test in late 2003 or early 2004, but plans were delayed by the U.S. Food and Drug Administration (FDA). Questions were raised about whether technological results were reproducible and reliable enough for widespread application in clinical practice.

During the past few years, a large number of scientists have been able to identify other candidate protein disease biomarker profiles using patient research study sets and to achieve high diagnostic sensitivity and specificity in blinded test sets. Nevertheless, translating these research findings to useful and reliable clinical tests has been the most challenging accomplishment. Clinical translation of promising ion fingerprints has been hampered by "sample collection bias, interfering substances, biomarker perishability, laboratory-to-laboratory variability, surface-enhanced laser desorption ionization chip discontinuance and surface lot changes, and the stringent dependence of the ion signature on the subtleties of the reagent composition and incubation protocols" (Liotta and Petricoin, 2008, p.3). Systematic biases arising from preanalytical variables seem to represent a relevant issue. Examples of non-disease-associated factors include (1) within-class biological variability, which may comprise unknown subphenotypes among study populations; (2) preanalytical variables, such as systematic differences in study populations and/or sample collection, handling, and preprocessing procedures; (3) analytical variables, such as inconsistency in instrument conditions, resulting in poor reproducibility; and (4) measurement imprecision (Hortin et al., 2006). Biological variability, in particular, may entail potential diurnal variation in protein expression, thus making standardization of sample collection time virtually mandatory. An evaluation of the effects of gender, age, ethnicity, pathophysiological conditions, and benign disorders is also crucial for understanding other possible effects on protein profiling expression. Regarding preanalytical conditions (e.g., collection practices, sample handling, and storage), these may differ from institution to institution, thus influencing the detection of proteins present in biological fluids. Standardization and use of specimens from multiple institutions are hence necessary to reliably demonstrate efficiency and reproducibility of protein profiling (Lippi et al., 2006; Banks, 2008). Although these preanalytical influences have been recognized for a long time, their impact is likely to be greater in proteomics studies, given the simultaneous analysis of several proteins, resolution of multiple forms of proteins, and detection of peptide fragments arising from active cleavage processes. Moreover, relatively few studies have been performed in such a way that quality control, an essential and quality-related feature, should be incorporated in proteomic experimental protocols (Hortin, 2005). Reproducibility studies performed with adequate control materials are prerequisites for safe introduction of proteomic techniques in clinical laboratory practice. Table 2.1 summarizes the major problems in translating proteomics insights into clinical practice.

It is now clearly accepted that the lack of standardization in how specimens are collected, handled, and stored represents one of the major hurdles to progress in the hunt for new and effective biomarkers (Poste, 2011). Nevertheless, the significance of assay technical quality has recently been underpinned. Diamandis, for example, has elegantly demonstrated that the assay for a new promising marker for prostate carcinoma (Diamandis, 2007) was strongly affected by severe methodological drawbacks, including its dependence on the total protein content, namely the albumin concentration in serum. The major limitations of this assay are even more important when considering the apparently spectacular clinical results that have been highly publicized to the media, whereas potential following failures were not. A review of the literature on translational research in oncology has revealed that most of the 939 publications on prognostic factors for patients with breast cancer that have appeared over a 20-year period were based on research assays with poor evidence of robustness or analytical validity (Simon, 2008). This fact should lead journal editors to ask for more robustness of the analytical techniques used for quantification of novel, putative biomarkers (Anderson et al., 2013), since problems such as data manipulation, poor experimental design, reviewer's bias, and overinterpretation of results are reported with increasing frequency (Diamandis, 2006).

Current limitations and open questions regarding clinical proteomics reflect a lack of appreciation of the many steps involved, thus including evaluation of pre-, intra-, and postanalytical issues; inter-laboratory performance; standardization; harmonization; and quality control, which are all needed to progress from method discovery to clinical practice (Plebani and Laposata, 2006; Plebani and Marincola, 2006).

TABLE 2.1 Obstacles in translation of proteomics research

Preanalytical factors
a) Patient selection bias b) Lack of standardization of sample collection, handling, and storage
Analytical factors
a) Technological variability within and across proteomic platforms b) Different performance of prototype and routine assays c) Lack of appropriate quality control and external quality assurance data
Statistical/Bioinformatics
a) Inappropriate statistical analysis b) Data overfitting c) Multiple hypothesis testing d) Overlapping of training and validation patient cohorts
Clinical validation
a) Poor study design, particularly as regards the verification phase b) Shortcomings in identifying the true clinical use (e.g., diagnosis vs prognosis or monitoring) c) Poor evaluation of the effects of the test on clinical outcomes and patient management

It now seems unquestionable that the problem is more complex, not only involving practical considerations but—more interestingly—a philosophical reasoning. It has been emphasized that the linear process of hypothesis-driven discovery that characterized the past decades of science has recently been replaced by the hypothesis-generating power of high-throughput, array-based technologies that provide vast and complex datasets that may be mined in various ways using emerging tools of computational biology. David F. Ransohoff has dedicated a series of papers for deepening problems related to this new approach, which has been called "discovery-based research," and in which there is no need to identify targets ahead. Instead, large portions of the genome or proteome may be examined for markers that can be used in diagnosis or prognosis (Ransohoff, 2004, 2005, 2007).

The evidence that many promising initial results have then appeared unreliable or scarcely reproducible has "resuscitated" the interest for a more consistent evaluation of result reliability. *Reliability* refers to whether study results are reproducible and are not explained by chance or bias, so that a conclusion can be used as a solid building block or foundation to ask other questions. Reliability is determined by proper attention to study design and interpretation. No single study can be considered almost perfect, but each one must be fairly interpreted. Particularly, results should not be overinterpreted. The primary responsibility for interpretation and specifying study limitations belongs to the investigator(s), although reviewers and editors have some role in the final diffusion of data. According to this concept, compliance with guidelines developed for studies of diagnostic tests (STARD, Standards for Reporting of Diagnostic Accuracy) persuasively emerged (Bossuyt et al., 2003). In fact, when compliance with STARD standards is low for traditional tests, it is reportedly even lower in "omics"-based tests (Lumbreras et al., 2006). Although Ransohoff puts forward some solutions to improve "omics-research," including guidelines, recommendations, and use of "phases" in the development of a diagnostic marker (Ransohoff, 2007), a fundamental "remedy" is to facilitate collaboration and communication among diverse "audiences," including basic and laboratory researchers, clinical epidemiologists, and clinicians involved in interdisciplinary research on biomarkers. When developing and evaluating new technologies, biologists, technologists, epidemiologists, and physicians need to understand the existence of distinct steps and the limits of each of their fields. Biological and technological reasoning are necessary to develop a promising approach such as pattern-recognition mass spectroscopy for analyzing serum proteins. Generally, the first step should give a reliable answer to the question "Might it work?" After it is shown that a given technology, in fact, does work, biological and technological reasoning are again necessary to provide insights about the further reasonable question "How does it work?" However, the step in between—which is reflected by the question "Does it work?"—requires a multifaceted reasoning that is not genuinely biological or technical. That step also requires an epidemiological and clinical reasoning, entailing stringent criteria in study design, analysis of data, and—finally—straightforward evaluation of clinical results (Ransohoff, 2005).

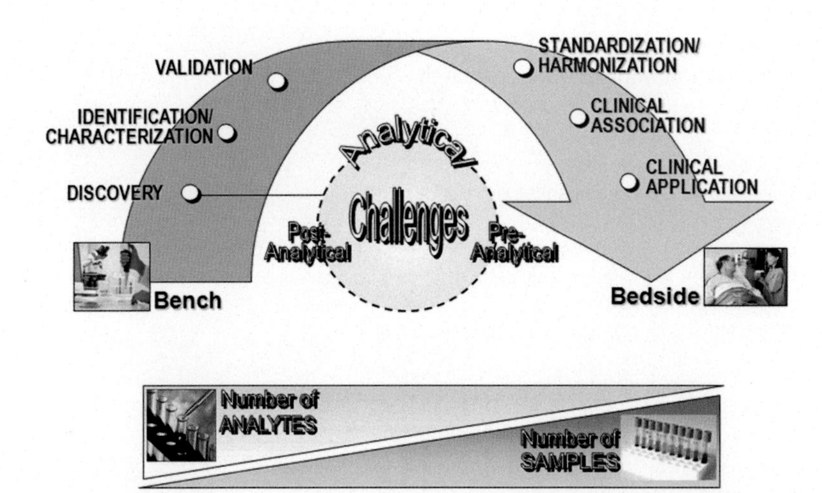

FIGURE 2.1 The six phases of the roadmap for translational research of biomarkers.

DEVELOPMENT OF BIOMARKERS: FROM DISCOVERY TO CLINICAL APPLICATION

The maturation of high-throughput-omics technologies has set the pace for biomarker discovery to a point whereby it has evolved from being an observational by-product of clinical practice, toward a large-scale systematic process, often referred to as a "pipeline" (Pavlou et al., 2013). Despite growing interest and investments, the number of biomarkers receiving U.S. Food and Drug Administration (FDA) clearance has however declined substantially over the past 10 years to less than one protein biomarker per year (Anderson and Anderson, 2002). Despite a large literature of more than 150,000 papers documenting thousands of putative biomarkers, fewer than 100 have been validated for clinical practice, like a drop in the ocean (Poste, 2011). The journey of a new biomarker from the bench to bedside is hence long and challenging, and every step must be meticulously planned and not overlooked in this pipeline.

The Early Detection Research Network (EDRN), established by the Division of Cancer Prevention, National Cancer Institute, has identified a widely accepted model, consisting of five separate phases, for development and testing of disease biomarkers (Pepe et al., 2001). This model, however, should be integrated by some particular steps and considerations that are specifically useful for utilization of biomarkers in every-day clinical practice. Figure 2.1 shows a roadmap for the translational process of biomarkers from discovery to clinical application consisting of six phases (discovery, identification, validation, standardization, clinical association, and clinical application) that better describe both the overall process and the sites involved in the steps from discovery to effective clinical application of putative biomarkers (Plebani et al., 2008).

DISCOVERY

The discovery of new biomarkers should be based on a deep knowledge of a given disease, pathophysiological mechanisms, and clinical needs, both in terms of diagnostic and therapeutic goals. The search for biomarkers should be based on possible targets for screening, diagnosis, prognostication, and therapeutic monitoring. Clinical genomics and proteomics research should follow two possible paths: the former (i.e., traditional approach) is based on the "one at a time, lead candidate biomarker approach" (Stanley et al., 2004) through the identification of new, candidate markers using "omic" techniques. The latter (i.e., innovative approach) is based on the so-called proteomic/genomic pattern diagnostics or proteomic/genomic profiling (Anderson, 2005). The rationale for this latter approach is that a pattern of multiple biomarkers may contain a higher level of discriminatory information than a single biomarker alone across large heterogeneous patient populations and for complex multistage diseases. Although the search of a unique biomarker has been compared to "finding a needle in a haystack," the complex proteomic signature of the disease-host microenvironment may represent a "biomarker amplification cascade" (Semmes, 2005). These two approaches are complementary, rather than mutually exclusive.

IDENTIFICATION/CHARACTERIZATION

The molecular/biochemical characterization of a new single biomarker rather than a biomarker pattern is needed to both provide a better understanding of its nature and to develop reliable test systems. In the post-genome era, the importance of post-translational modifications (PTMs) including phosphorylation, glycosylation, acetylation, and oxidation has been increasingly recognized since they are linked to disease pathology and are useful targets for therapeutics (Lopez et al., 2012). The recognition of different molecular forms of some biomarkers (e.g., human chorionic gonadotropin [HCG] and prostate-specific antigen [PSA]) has allowed us to better elucidate their nature as well as to develop different and more specific (immuno)assays (Morris, 2009). Therefore, the accurate characterization of the nature of each novel biomarker is an essential prerequisite for further steps in the development pipeline.

VALIDATION

Validation of biomarkers is a challenging process that requires multisite clinical studies. The initial validation process is intended to perform a preliminary evaluation of clinical performances (in particular, sensitivity and specificity) in selected groups of patients and a reference population (i.e., "controls") using a research method (technique). The term *test research* refers to this type of study that follows a single-test or univariable approach, which means studies focusing on a particular test to quantify its sensitivity, specificity, likelihood ratio, or area under the receiver operating characteristic (ROC) curve (Moons et al., 2004). The most widely acknowledged limitation of test research is that some studies are often based on an improper patient recruitment and study design (Ransohoff and Feinstein, 1978). A typical example here is the use of cardiospecific troponins for diagnosing myocardial infarction in the emergency department (ED). In this setting, the definition of the upper limit of the reference range of a given troponin immunoassay should not be based on a population of healthy individuals, wherein a differential diagnosis of myocardial infarction in the ED is made on patients who are inherently unhealthy (Lippi et al., 2013). In the case of clinical proteomics, an additional and severe bias is represented by the lack of studies aiming to demonstrate the transferability of data obtained in one site to the others.

STANDARDIZATION/HARMONIZATION

The pipeline for translation of new biomarkers into routine laboratory tests seems to have a bottleneck at the early stage of translation of research markers into clinical tests (Hortin, 2006). Research groups performing discovery and initial validation rarely have enough resources and skills to develop prototype analyzers or reagent sets, to manufacture them, or to proceed with other steps in commercialization. This is also due to the increasingly complex system of evaluation and regulation required for any diagnostic test by regulatory agencies such as the FDA and similar European authorities. These steps usually rely on the in vitro diagnostic (IVD) industry, which, however, has reduced funding for development of new biomarkers as a result of increased competition on cost and decreased revenue of laboratory products/services.

With specific emphasis on clinical proteomic profiling (although this has provided breakthroughs in the discovery of new disease markers), initial discovery methods have been typically poorly suited for clinical application. Glen Hortin, in an interesting editorial, has emphasized the "lack of appreciation of the many steps, such as evaluation of inter-laboratory performances, from method discovery to clinical practice" (Hortin, 2005, p.4).

The need to achieve standards of practice is not limited to the analytical step, but also to pre- and postanalytical issues. In particular, optimal materials and methods for sample collection and processing should be determined, including the selection of appropriate sample matrix, protease inhibitors, and collection tubes (Ayache et al., 2006).

Studies should address patient preparation, evaluating the effects of diurnal variation, fasting versus nonfasting, along with additional aspects. For any individual, many analytes present remarkable fluctuations according to the time of day, fasting state, or age. Although these changes may not be clinically relevant, they do add an additional level of complexity in elucidating disease-induced protein changes from modification due to biorhythm.

Similarly, analytical and postanalytical issues such as quality control, external quality assurance, reference intervals, and decisional levels should be carefully considered (Plebani, 2013) The definition of standards of practice and quality indicators for pre-, intra- and postanalytical phases is a prerequisite for evaluating a new biomarker in the clinical setting (Plebani et al., 2013).

A reliable evaluation of clinical utility of newly developed assays, particularly concerning the "omics," must take into account the careful examination of analytical performances. Ideally for established analytes, the performance characteristics limits are determined by several factors, including outcome studies, clinical requirements, published professional recommendations, and goals set by regulatory agencies. For a novel analyte, no such performance specification may be available, but analytical performance comparable to an established, similar assay is expected. Briefly, fundamental analytical performances are grouped as indicators of accuracy, indicators of precision, and indicators of analytical measurement range.

- *Indicators of accuracy.* Trueness and accuracy are similar, but not identical concepts to describe this aspect of assay performance. *Trueness* is the closeness of agreement between the average analyte value of different samples and the true concentration value, whereas *accuracy* is the closeness of the agreement between the value of a measurement and the true concentration of the analyte in that sample (Dybkaer, 1997; International Standards Organization, 1994). Trueness reflects bias, a measure of systematic error, whereas accuracy reflects uncertainty, which comprises both random and systematic errors. For a novel assay, the evaluation of bias is challenging because reference methods and materials do not usually exist, so that an alternative approach should be applied.

- *Indicators of precision.* Repeatability and reproducibility are measures of precision, and are used to quantitatively express the closeness of agreement among results of measurements performed under stipulated conditions (Dybkaer, 1997; NCCLS, 1999). Repeatability refers to measurements that are performed under the same conditions, whereas reproducibility is used to describe measurements performed under different test conditions. Imprecision strongly affects interpretation and reliability of a single result. For example, it is now accepted that the analytical imprecision goal for cardiac troponins is to have a coefficient of variation (CV%) < 10% at the decisional level (Thygesen et al., 2012). This represents a challenge both for manufacturers and clinical laboratories. The former is required to commercialize a diagnostic system reliably, and to be able to satisfy this goal, the latter have to select the most appropriate method/diagnostic system with regular basis monitoring in order to guarantee both the achievement of satisfactory performance and the appropriate utilization of the marker in clinical practice. Precision and accuracy are intimately related, and they strongly affect the clinical reliability of a test result, as expressed by the concept of total error that is the sum of bias and imprecision.
- *Indicators of analytical measurement range.* Linearity and limits of detection and quantification must be assessed to determine the range of analyte values over which measurements can be performed with acceptable precision and accuracy. In particular, the limit of detection is defined as the lowest value that significantly exceeds the measured value obtained with a sample that does not contain the protein or analyte of interest (International Standards Organization, 2000).

These simple concepts, and the more complex rules for assuring a good analytical performance, are typically neglected and poorly recognized by experienced and trained physicians, and sometimes by basic scientists. However, all laboratory results should assure consistency and comparability. The magic word *standardization* well underlines the importance of assuring reliability, reproducibility, and accuracy of data released in a laboratory report, particularly for tests that strongly affect clinical reasoning and patient management.

CLINICAL ASSOCIATION AND CLINICAL BENEFIT

When a reliable method/technique is available, in a format that should allow inter-laboratory agreement and results comparability, retrospective and prospective studies are needed in order to evaluate the effective clinical performances in the "real world" and not in selected groups of patients. The predictive values of a given test vary across patient populations, and a particular test may have different sensitivities and specificities within different patient subgroups, underscoring the need to move from "test research" to "diagnostic research." By "diagnostic research," we refer to studies that aim to quantify a test's added contribution beyond test results already available to the physician for determining presence or absence of a particular disease (Reid et al., 1995; Rutjes et al., 2005). The focus is on the value of the test in combination with other, previous tests, including patient history and physical examination. As any test result in real life is always considered in view of other patient characteristics and data, diagnostic accuracy studies that address only a particular test have limited relevance to practice. Therefore, well-designed studies are needed that should take into account possible spectrum effects and that, therefore, should be based on well-defined and evidence-based inclusion and exclusion criteria, reference and gold standards, as well as reliable statistical analyses. The traditional statistical methods that are used by epidemiologists to assess etiologic associations are inappropriate to assess the potential performance of a marker for classifying or predicting risk (Kattan, 1989). While this aspect is not widely appreciated, some authors have recently emphasized the limitations of some traditional methods, including the odds ratio, in gauging the performance of a marker (Baker et al., 2002; Pepe et al., 2004). Techniques that directly address classification accuracy (e.g., ROC curves) rather than traditional logistic regression techniques should be used. The need to evaluate the test performance in the setting of the specific clinical application has been highlighted, particularly when the gold standard is not appropriately defined (Pepe and Longton, 2005; Pfeiffer and Castle, 2005). Therefore, as we proceed to develop technologically sophisticated tools for individual-level prediction and classification, we must be careful to use appropriate statistical techniques when evaluating research studies aimed at assessing their performance. Rigorous criteria should also be identified and used in assessing the economical and clinical outcomes when these new biomarkers are translated into clinical practice and used alone or in combination in public health practice. The evidence on the importance of translating a new biomarker into clinical practice can hence be considered in a hierarchy, all elements of which are important to making the decision.

The technical performance is the foundation of any evidence, also because this has an important bearing on diagnostic performance. In laboratory medicine, there are well-defined and accepted guidelines for evaluating technical performances of a diagnostic system/method, including precision, accuracy, analytical range, and interferences (Linnet and Boyd, 2006). Other recommendations exist for evaluating preanalytical factors including the sample matrix of choice (plasma or serum), potential type and concentration of anticoagulants, storage conditions, etc. (Young et al., 2006). Diagnostic performance represents the second level of the hierarchy, and provides an assessment of the test in terms of objective use, namely sensitivity and

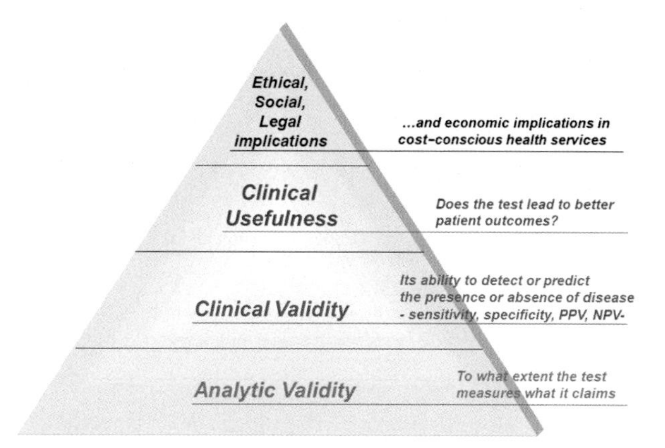

FIGURE 2.2 The ACCE framework for laboratory tests evaluation.

specificity. The evaluation of the clinical impact or benefit of the test and its contribution to the clinical decision making, which represents the third level of the hierarchy, is rarely accomplished, and this represents the main reason for the poor appropriateness of test request/interpretation in clinical practice. The evaluation of organizational impact and cost-effectiveness, including both clinical and economical outcomes (Figure 2.2), lay at the top of the hierarchy of the framework called ACCE (analytical validity, clinical validity, clinical usefulness, and ethical implications) (Lumbreras, 2006; Plebani, 2010).

TRANSLATING OMICS INTO CLINICAL PRACTICE

Omics techniques generate multiparametric datasets in which the number of dimensions/variables usually greatly exceeds the number of samples. Consequently, differences between the datasets (potential biomarkers) that enable the discrimination between any arbitrary combinations of datasets can easily be defined. This sample heterogeneity makes it virtually impossible to thoroughly validate any biomarker or combination of biomarkers only based on the training set (initial group of selected samples used to identify biomarkers), even with cross-validation. Accordingly, each clinical "omic" study must include, as a mandatory step, the validation of findings using an appropriate set of blinded samples analyzed independently. Table 2.2 summarizes some suggested steps in clinical proteomics that might be used, in general, for clinical application of "omics."

Table 2.2 does not include ethical considerations such as the need to obtain informed consent from participants and/or approval by ethical committees, as well as other issues like the time to deposit datasets in a public database or patenting and licensing.

CONTINUUM OF TRANSLATION RESEARCH AND OMICS

In a study of the "natural history" of promising diagnostic or therapeutic interventions over a 15-year period, Contopolous-Ioannidis et al. showed that only 5% of "highly promising" basic science findings were licensed for clinical use and only 1% were actually used for the licensed indication (Contopolous-Ioannidis et al., 2003). In a cited editorial, Lenfant lamented that basic sciences and clinical research findings are usually "lost in translation," and questioned about the evidence that if old insights related to traditional laboratory tests and/or well-known drugs are not used in clinical practice, the gap could be even larger with new insights related to more complex issues such as "omics" (Lenfant, 2003). In addition, "it takes an average of 17 years for only 14% of new scientific discoveries to enter the day-to-day clinical practice" (Westfall et al., 2007, p.403). A review of the literature on translational research in oncology has revealed that most of the 939 publications over a 20-year period on prognostic factors for patients with breast cancer were based on research assays without sufficient demonstration of robustness or analytical validity (Simon, 2008).

The magnitude and nature of the work required to translate findings from basic research into valid and effective clinical practice have been underestimated for a long time, and the meaning of the term *translational research* seems to be controversial despite the increasing interest in the issue, established translational research programs, and now fewer than 40 journals listed in Medline with the term *Translational* somewhere in their title. This is partly due to the fact that different stakeholders look at distinct aspects of this issue (Woolf, 2008). For academia, translational research represents a general desire to test novel ideas generated from basic investigation, with the hope of turning them into useful clinical applications. For academic purposes, translational research also responds to the need of identifying novel scientific hypotheses

TABLE 2.2 Recommended steps for clinical proteomics

1)	Define a clear clinical question and how the eventual biomarker/panel of biomarkers would improve the diagnosis and/or treatment of the disease
2)	Define patient and control populations, clinical data to be collected, as well as protocols for sampling and sample preparation/storage
3)	Define types of samples needed for discovery and validation phases
4)	Define and validate analytical platform(s) for discovery
5)	Perform a pilot study on validated discovery platform(s)
6)	Evaluate data from the pilot study with appropriate statistical treatments and calculate the number of cases and controls for training set
7)	Perform study of training set and evaluate findings with appropriate statistical treatments
8)	Transfer the assay to the diagnostic platform and evaluate technical and clinical performances
9)	Standardize pre- and post analytical procedures, and define reference ranges and interpretation criteria
10)	Transfer into clinical use to demonstrate whether the findings may improve clinical outcomes

(Modified from Mischak et al., 2007)

that are relevant to human pathology through direct observation of humans and their diseases. For the commercial sector, translational research refers more to a process aimed at expediting the development of known entities particularly in early phases and/or identifying ways to make early "go" or "no-go" decisions when the cost of product development is still relatively contained. For people more directly involved in clinical practice (physicians, clinical laboratory professionals, and patients), translational research responds to the need of accelerating the capture of benefits of research, closing the gap between "what we know and what we practice" (Littman et al., 2007).

For many, the term refers to the "bench-to-bedside" enterprise of harnessing knowledge from basic sciences to produce new drugs, devices, and treatment options for patients. For others, especially health-care scientists, laboratory professionals, and physicians, translational research refers to translating research into practice and ensuring that new diagnostic tools, new treatments, and research knowledge actually reach the patients or populations for whom they are intended and are correctly implemented. The distinction between these two definitions was articulated by the Institute of Medicine's Clinical Research Roundtable, which described two "translational blocks" in the clinical research enterprise that are labeled as T1 and T2. The first roadblock (T1) was described as "the transfer of new understandings of disease mechanisms gained in the laboratory into the development of new methods for diagnosis, therapy, and prevention and their first testing in humans," whereas the second block (T2) has been described as "the translation of results from clinical studies into the everyday clinical practice and health decision making" (Zerhouni, 2003, p.64). They include translation of basic science laboratory work in animals into an understanding of basic human medical biochemistry and physiology, and translation of basic human biochemistry and physiology into improved diagnostic tests, which, in turn, should improve patient management. The final crucial step is the delivery of recommended care to the right patient at the right time, resulting in improvement in individual patient and community health. The research roadmap is a continuum, with an overlap between sites of research and translational steps. Regarding the "omics," in the T2 roadblock, some relevant aspects include assay validation, development of appropriate diagnostic platforms, standardization, quality control, and quality assurance protocols.

In particular, diagnostic platforms to be used in clinical practice are seldom the same used in discovery and validation steps. In addition to diagnostic accuracy, clinical laboratories should assure standardization, reproducibility, robustness, appropriate turnaround time, efficiency, and consistency in everyday practice. The construction of a coherent biomarker pipeline with a higher likelihood of success than past approaches calls for a stronger involvement of laboratory professionals and for a collaborative rather than competitive relationship between clinicians and scientists (Anderson et al., 2013; Lippi et al., 2007; Pavlou et al., 2013; Rifai et al., 2006).

If the pipeline and roadmap seem to be straightforward, the expansion of practice-based research, which is grounded in, informed by, and intended to improve practice, is mandatory. Practice-based research should help to (a) identify the problems that arise in daily practice and that may create the gap between recommended and actual care; (b) provide an interdisciplinary context in which the development and validation of new biomarkers advance in a linear way; (c) demonstrate if and how the new tests improve clinical and economical outcomes; and (d) evaluate the need to change diagnostic and therapeutic clinical pathways.

CONCLUSIONS

The reality about "omics" is still uncertain, but expectations are not commensurate with the current data. After many thousands of studies using microarrays, only a handful of tests for breast cancer prediction have moved into clinical practice, and even these tests lack definitive evidence for clinical effectiveness (Marchionni et al., 2008). An even larger literature on pharmacogenetics/pharmacogenomics has left very few applications with proven improvement in clinical outcomes (Roden et al., 2006). Careful roadmaps should be implemented to include all steps that must be followed before successful translation, to demonstrate that new diagnostic tests are able to change clinical practice and reduce costs, by either improving people's health or eliminating ineffective and expensive treatments. This holds particularly true in the increasing cost-conscious health care environment (Meckley and Neumann, 2010). The complexity of translating basic discoveries into clinical trials and, finally, into clinical practice requires, in addition to well-designed pipelines and roadmaps, some changes in training programs. In fact, evidence has been gathered to demonstrate the existence of gaps in the knowledge on how to use and interpret "omics" technologies in everyday clinical practice (Ioannidis, 2006).

REFERENCES

Adam, B.L., Qu, Y., Davis, J.W., Ward, M.D., Clements, M.A., Cazares, L.H., et al., 2002. Serum protein fingerprinting coupled with pattern-matching algorithm distinguishes prostate cancer from benign prostate hyperplasia and healthy men. Cancer Res. 62 (13), 3609–3614.

Anderson, L., 2005. Candidate-based proteomics in the search for biomarkers of cardiovascular disease. J. Physiol. 563, 23–60.

Anderson, N.L., Anderson, N.G., 2002. The human plasma proteome: history, character, and diagnostic prospects. Mol. Cell Proteomics 1, 845–867.

Anderson, N.L., Ptolemy, A.S., Rifai, N., 2013. The riddle of protein diagnostics: future bleak or bright. Clin. Chem. 59, 194–197.

Ayache, S., Panelli, M., Marincola, F.M., Stroncek, D.F., 2006. Effects of storage time and exogenous protease inhibitors on plasma protein levels. Am. J. Clin. Pathol. 126, 74–84.

Baker, S.G., Kramer, B.S., Srivastava, S., 2002. Markers for early detection of cancer: statistical guidelines for nested case-control studies. BMC Med. Res. Methodol. 2, 4.

Banks, R.E., 2008. Preanalytical influences in clinical proteomic studies: raising awareness of fundamental issues in sample banking. Clin. Chem. 54, 6–7.

Bossuyt, P.M., Reistma, J.B., Bruns, D.E., Gatsonis, C.A., Glasziou, P.P., Inwig, L.M., et al., 2003. The STARD statement for reporting studies of diagnostic accuracy: explanation and elaboration. Clin. Chem. 49, 7–18.

Contopolous-Ioannidis, J.P., Ntzani, E., Ioannidis, J.P., 2003. Translation of highly promising basic science research into clinical application. Am. J. Med. 114, 477–484.

Dammann, M., Weber, F., 2012. Personalized medicine: caught between hope, hype and the real world. Clinics 67, 91–97.

Diamandis, E.P., 2006. Quality of the scientific literature: all that glitters is not gold. Clin. BioChem. 39, 1109–1111.

Diamandis, E.P., 2007. EPCA-2: A promising new serum biomarker for prostatic carcinoma? Clin. BioChem. 40, 1437–1439.

Drake, R.R., Manne, U., Bao-Ling, A., Ahn, C., Cazares, L., Semmes, O.J., et al., 2003. SELDI-TOF-MS profiling of serum for early detection of colorectal cancer. Gastroenterology 124 (11), A650.

Dybkaer, R., 1997. Vocabulary for use in measurement procedures and description of reference materials in laboratory medicine. Eur. J. Clin. Chem. Clin. BioChem. 35, 141–173.

Finn, W.G., 2007. Diagnostic pathology and laboratory medicine in the age of "omics." J. Mol. Diagn. 9, 431–436.

Hortin, G.L., 2005. Can mass spectrometry protein profiling meet desired standards of clinical laboratory practice? Clin. Chem. 51, 3–5.

Hortin, G.L., Jortani, S.A., Ritchie, J.C., Valdes, R., Chan, D.W., 2006. Proteomics: a new diagnostic frontier. Clin. Chem. 52, 1218–1222.

International Standards Organization, 1994. I.O.F.S. Accuracy (Trueness and Precision) of measurement methods and results (ISO 5725)—Part 1: General principles and definitions. ISO, Geneva.

International Standards Organization, 2000. I.O.F.S. Capability of detection—Part 2: Methodology in the linear calibration case (11843–2). ISO, Geneva.

Ioannidis, J.P., 2006. Evolution and translation of research findings: from bench to where? PLoS. Clin. Trials. 1, e36.

Ioannidis, J.P.A., 2010. Genetics, personalized medicine, and clinical epidemiology. J. Clin. Epidemiol. 63, 945–949.

Kattan, M.W., 1989. Judging new makers by their ability to improve predictive accuracy. J. Natl. Cancer Inst. 81, 1879–1886.

Lenfant, C., 2003. Shattuck lecture: clinical research to clinical practice—lost in translation? N. Engl. J. Med. 349, 868–874.

Linnet, K., Boyd, J.C., 2006. Selection and analytical evaluation of methods with statistical techniques. In: Burtis, C.A., Ashwood, E.R., Bruns, D.E. (Eds.), Tietz Textbook of Clinical Chemistry and Molecular Diagnostics. Elsevier Saunders, St. Louis, MO, pp. 353–409.

Liotta, L.A., Petricoin, E.F., 2008. Putting the "Bio" back into biomarkers: orienting proteomic discovery toward biology and away from the measurement platform. Clin. Chem. 54, 3–5.

Lippi, G., Guidi, G.C., Mattiuzzi, C., Plebani, M., 2006. Preanalytical variability: the dark side of the moon in laboratory testing. Clin. Chem. Lab. Med. 44, 358–365.

Lippi, G., Margapoti, R., Aloe, R., Cervellin, G., 2013. Highly-sensitive troponin I in patients admitted to the emergency room with acute infections. Eur. J. Intern. Med. 24. e57–e58.

Lippi, G., Plebani, M., Guidi, G.C., 2007. The paradox in translational medicine. Clin. Chem. 53, 1553.

Littman, B.H., Di Mario, L., Plebani, M., Marincola, F.M., 2007. What's next in translational medicine? Clin. Sci. 112, 217–227.

Lopez, E., Madero, L., Lopez-Pacual, J., Latterich, M., 2012. Clinical proteomics and OMICS clues useful in translational medicine research. Proteome Sci. 10, 35–41.

Lumbreras, B., Jarrin, I., Hérnandez-Aguado, I., 2006. Evaluation of the research methodology in genetic, molecular and proteomic tests. Gac. Sanit. 20, 368–373.

Marchionni, L., Wilson, R.F., Wolff, A.C., Marinopoulos, S., Parmigiani, G., Bass, E.B., et al., 2008. Systematic review: gene expression profiling assays in early-stage breast cancer. Ann. Intern. Med. 148, 358–369.

Meckley, L.M., Neumann, P.J., 2010. Personalized medicine: factors influencing reimbursement. Health Policy 94, 91–100.

Mischak, H., Apweiler, R., Banks, R.E., Conaway, M., Coon, J., Dominiczak, A., et al., 2007. Clinical proteomics: a need to define the field and to begin to set adequate standards. Proteomics Clin. Appl. 1, 148–156.

Moons, K.G.M., Biesheuvel, C.J., Grobbee, D.E., 2004. Test research versus diagnostic research. Clin. Chem. 50, 473–476.

Morris, H.A., 2009. Traceability and standardization of immunoassays: a major challenge. Clin. BioChem. 42, 241–245.

NCCLS, 1999. 1–42. Evaluation of precision performance in clinical chemistry devices: Approved guideline. NCCLS Document EP6–A (NCCLS, Wayne, PA).

Pavlou, M.P., Diamandis, E.P., Blasuting, I.M., 2013. The long journey of cancer biomarkers from the bench to the clinic. Clin. Chem. 59, 147–157.

Pepe, M.S., Etzioni, R., Feng, Z., Potter, J.D., Thompson, M.L., Thornquist, M., et al., 2001. Phases of biomarker development for early detection of cancer. J. Natl. Cancer Inst. 93, 1054–1061.

Pepe, M.S., James, H., Longton, G., Leisenring, W., Newcomb, P., 2004. Limitations of the odds ratio in gauging the performance of diagnostic, prognostic, or screening marker. Am. J. Epidemiol. 159, 882–890.

Pepe, M.S., Longton, G., 2005. Standardizing diagnostic markers to evaluate and compare their performance. Epidemiology 16, 598–603.

Petricoin, E.F., Ardekani, A.M., Hitt, B.A., Levine, P.J., Fusaro, V.A., Steinberg, S.M., et al., 2002a. Use of proteomic patterns in serum to identify ovarian cancer. Lancet 359, 572–577.

Petricoin 3rd, E.F., Ornstein, D.K., Paweletz, C.P., Ardekani, A., Hackett, P.S., Hitt, B.A., et al., 2002b. Serum proteomic patterns for detection of prostate cancer. J. Natl. Cancer Inst. 94 (20), 1576–1578.

Pfeiffer, R.M., Castle, P.E., 2005. With or without a gold standard. Epidemiology 16, 595–597.

Plebani, M., 2005. Proteomics: the next revolution in laboratory medicine? Clin. Chim. Acta. 357, 113–122.

Plebani, M., 2010. Evaluating laboratory diagnostic tests and translational research. Clin. Chem. Lab. Med. 48, 983–988.

Plebani, M., 2013. Harmonization in laboratory medicine: the complete picture. Clin. Chem. Lab. Med. 51, 741–751.

Plebani, M., Chiozza, M.L., Sciacovelli, L., 2013. Towards harmonization of quality indicators in laboratory medicine. Clin. Chem. Lab. Med. 51, 187–195.

Plebani, M., Laposata, M., 2006. Translational research involving new biomarkers of disease. A leading role for pathologists. AJCP 126, 169–171.

Plebani, M., Marincola, F.M., 2006. Research translation: a new frontier for clinical laboratories. Clin. Chem. Lab. Med. 44, 1303–1312.

Plebani, M., Zaninotto, M., Mion, M.M., 2008. Requirements of a good biomarker: translation into the clinical laboratory. In: Van Eyk, J.E., Dunn, M.J. (Eds.), Clinical Proteomics. Viley-VCH, Weinheim, pp. 615–632.

Poste, G., 2011. Bring on the biomarkers. Nature 469, 156–157.

Ransohoff, D.F., 2004. Evaluating discovery-based research: when biologic reasoning cannot work. Gastroenterology 127, 1028.

Ransohoff, D.F., 2005. Bias as a threat to the validity of cancer-molecular-marker research. Nature Rev. Cancer 5, 142–149.

Ransohoff, D.F., 2007. How to improve reliability and efficiency of research about molecular markers: roles of phases, guidelines, and study design. J. Clin. Epidemiol. 60, 1205–1219.

Ransohoff, D.E., Feinstein, A.R., 1978. Problems of spectrum and bias in evaluating the efficacy of diagnostic tests. N. Engl. J. Med. 299, 926–930.

Reid, M.C., Lachs, M.S., Feinstein, A.R., 1995. Use of methodological standards in diagnostic test research. Getting better but still not good. JAMA 274, 645–651.

Rifai, N., Gillette, M.A., Carr, S.A., 2006. Protein biomarker discovery and validation: the long and uncertain path to clinical utility. Nat. Biotechnol. 24, 971–983.

Roden, D.M., Altman, R.B., Benowitz, N.L., Flockhart, D.A., Giacomini, K.M., Johnson, J.A., et al., 2006. Pharmacogenetics research network. Pharmacogenomics: challenges and opportunities. Ann. Intern. Med. 145, 749–757.

Rutjes, A.W.S., Reitsma, J.B., Vandenbroucke, J.P., Glas, A.S., Bossuyt, P.M.M., 2005. Case-control and two-gate designs in diagnostic accuracy studies. Clin. Chem. 51, 1335–1341.

Schilsky, R.L., 2009. Personalizing cancer care: American Society of Clinical Oncology presidential address 2009. J. Clin. Oncol. 27, 3725–3730.

Semmes, O.J., 2005. The "omics" haystack: defining sources of sample bias in expression profiling. Clin. Chem. 51, 1571–1572.

Simon, R., 2008. Lost in translation: problems and pitfalls in translating laboratory observations to clinical utility. Eur. J. Cancer 44, 2707–2713.

Stanley, B.A., Gundry, R.L., Cotter, R.J., Van Eyk, J.E., 2004. Heart disease, clinical proteomics and mass spectrometry. Dis. Markers 20, 167–178.

Thygesen, K., Mair, J., Giannitsis, E., Mueller, C., Lindahl, B., Blankenberg, S., et al., 2012. How to use high-sensitivity cardiac troponins in acute cardiac care. Eur. Heart J. 33 (18), 2252–2257.

Vlahou, A., Schellhammer, P.F., Mendrinos, S., Patel, K., Kondylis, F.I., Gong, L., et al., 2001. Development of a novel proteomic approach for the detection of transitional cell carcinoma of the bladder in urine. Am. J. Pathol. 158, 1491–1502.

Westfall, J.M., Mold, J., Fagnan, L., 2007. Practice-based research—blue highways on the NIH road map. JAMA 297, 403–406.

Woolf, S.H., 2008. The meaning of translational research and why it matters. JAMA 299, 211–213.

Young, D.S., Bermes, E.W., Haverstick, D.M., 2006. Specimen collection and processing. In: Burtis, C.A., Ashwood, E.R., Bruns, D.E. (Eds.), Tietz Textbook of Clinical Chemistry and Molecular Diagnostics. Elsevier Saunders, St. Louis, MO, pp. 41–56.

Zerhouni, E., 2003. The NIH Roadmap. Science 302, 63–72.

Zhu, W., Wang, X., Ma, Y., Rao, M., Glimm, J., Kovach, J.S., 2003. Detection of cancer-specific markers amid massive mass spectral data. Proc. Natl. Acad. Sci. USA 359, 14666–14671.

Chapter 2.1.2

"Omics" Technologies: Promises and Benefits for Molecular Medicine

David M. Pereira, João C. Fernandes, Patrícia Valentão and Paula B. Andrade

INTRODUCTION

"Omics" technologies play a key role in several dimensions of human health nowadays. The application of "omics" ranges from drug discovery and development to diagnosis of diseases, namely by following their progression, improving efficacy and safety of treatments, optimizing patient selection, and adapting dose regimens of drugs, as well as helping to decide which therapy is most appropriate.

In systems biology it is widely accepted that no single "omics" approach suffices when studying complex organisms and biological processes. As so, the study can focus on several levels of information, namely the genome (genomics), RNA (transcriptomics), proteins (proteomics), and metabolites (metabolomics) (Chung et al., 2007; Collins et al., 2003; Dettmer and Hammock, 2004; Nordström and Lewensohn, 2010; Weckwerth, 2003; Zhang et al., 2013).

For obvious reasons, an integrated approach should be followed as a way to avoid misinterpretations of data. For example, changes found at the gene level may not necessarily translate into altered levels of proteins. Many authors defend that, when considering the "omics" cascade, the metabolome is most predictive of the phenotype, as it is located downstream in this cascade (Dettmer and Hammock, 2004; Weckwerth, 2003); however, it is modulated by events located upstream.

In the next pages we address the basis of genomics and metabolomics, highlighting the most used techniques in each case and its current and future applications in human health and disease.

GENOMICS

Cells are the fundamental building blocks of every living system, and all the instructions required to control cellular activities are encoded within the DNA molecules, the total amount of DNA in a single cell being called the genome. The human genome includes approximately 3 billion nucleotide pairs of DNA, while bacteria possess the smallest known genome for a free-living organism (Pray, 2008; see also Table 2.3). The study and use of genomic information and technologies, associated with biological methodologies and computational analyses in order to assess genes function, may be considered the basis of modern genomics. The main purpose is to understand how the instructions coded in DNA lead to a functioning human being, deriving meaningful knowledge from the DNA sequence.

Specialists in genomics strive to determine complete DNA sequences and perform genetic mapping to evaluate the purpose of each gene and the associated regulatory elements. Modern research also aims to find disparities in the DNA sequence between people and assess their relevance at a biological level. The findings from these investigations are currently being applied in the development of genome-based strategies for the early detection, diagnosis, and treatment of disease. Furthermore, it will enable researchers to build up novel technologies to analyze genes and DNA on a large scale and to store genomic data efficiently (Collins et al., 2003). To achieve such goals, genomic research has a number of solutions available that integrate sequencing, arrays, and polymerase chain reaction (PCR)-based systems.

Genomic Tools

Sequencing

Sequencing is the process of determining the order of nucleotide bases within a stretch of DNA or RNA. Sequencing the entire genome of many animals, plants, and microbial species is indispensable for basic biological, forensics, and medical

Principles of Translational Science in Medicine. http://dx.doi.org/10.1016/B978-0-12-800687-0.00003-7

TABLE 2.3 Comparative genome sizes of humans and other model organisms

Organism	Genome Size (Base Pairs)	Estimated Number of Genes
Human (*Homo sapiens*)	3.2 billion	25,000
Mouse (*M. musculus*)	2.6 billion	25,000
Fruit fly (*D. melanogaster*)	137 million	13,000
Roundworm (*C. elegans*)	97 million	19,000
Yeast (*S. cerevisiae*)	12.1 million	6,000
Bacterium (*E. coli*)	4.6 million	3,200
Human immunodeficiency virus (HIV)	9,700	9

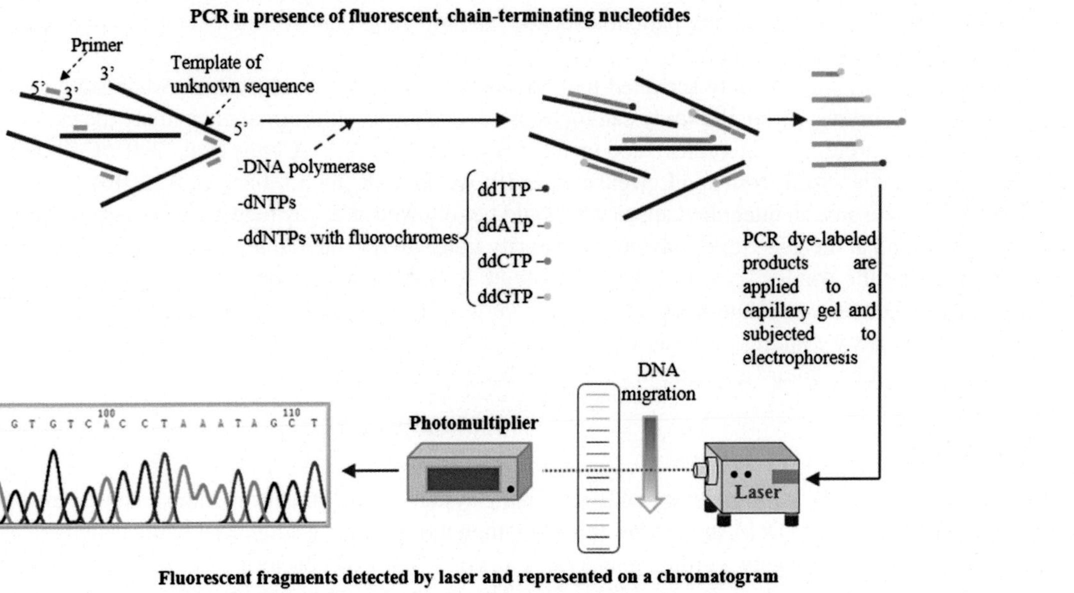

FIGURE 2.3 Illustration of the strategy for automating Sanger DNA sequencing system. *[Adapted from Lehninger Principles of Biochemistry (Nelson et al. 2008).]*

research. Since the early 1990s, DNA sequencing has almost exclusively been carried out with capillary-based, semi-automated implementations of the Sanger biochemistry (Chen, 2014). This method, developed by Sanger and colleagues and also known as dideoxy sequencing or chain termination methods, relies on denaturating DNA into single strands using high temperatures and further annealing to an oligonucleotide primer (Sanger et al., 1977). Afterward, the oligonucleotide is extended by means of a DNA polymerase, using a mixture of normal deoxynucleotide triphosphates (dNTPs) and modified dideoxynucleotide triphosphates (ddNTPs), which terminate the DNA strand elongation process (Collins et al., 2003). These modified ddNTPs lack the 3′ OH group to which the next dNTP of the emergent DNA sequence would be added. Therefore, without the 3′ OH, the process is terminated. The resultant newly synthesized DNA chains are a blend of lengths, depending on how long the chain was when a ddNTP was randomly added (Figure 2.3). This technology reached its peak in the development of single tube chemistry with fluorescently marked termination bases, heat-stable polymerases, and automated capillary electrophoresis, after which a plateau in technical development was reached (Nelson et al., 2008). The main problems of the Sanger sequencing dealing with larger sequence output were the use of gels or polymers as separation media, the limited number of samples that could be handled in parallel, and the difficulties with complete automation of the sample preparation.

Since completion of the first human genome sequence in April 2003, the demand for cheaper and faster sequencing methods has increased greatly. This demand has driven the development of second-generation sequencing methods, or next-generation sequencing (NGS). NGS platforms perform massively parallel sequencing, during which millions or even billions of DNA fragments (50–400 bases each) from a single sample are sequenced in just one run (Rehm et al., 2013). This technology enables high-throughput sequencing, which in turn allows a whole small genome to be sequenced in just a few hours (Grada and Weinbrecht, 2013). In the past decade, several NGS platforms have been developed that provide low-cost, high-throughput sequencing. Furthermore, NGS allows the identification of disease-causing mutations with application in diagnosis by targeted sequencing and offers a robust alternative to microarrays in gene expression studies (Pareek et al., 2011).

Among the most successful DNA sequencing methodologies used by different NGSs, two should be highlighted: sequencing by ligation and sequencing by synthesis.

Sequencing by Ligation

Sequencing by ligation is a DNA sequencing method that harnesses the mismatch sensitivity of DNA ligase to determine the underlying sequence of nucleotides in a given DNA sequence (Ho et al., 2011). Platforms based on this method use a pool of oligonucleotide probes of varying lengths, which are labeled with fluorescent tags, depending on the nucleotide to be determined. The fragmented DNA templates are primed with a short, known anchor sequence, which allows the probes to hybridize. DNA ligase is added to the flow cell and joins the fluorescently tagged probe to the primer and template, and thus, the preferential ligation by DNA ligase for matching sequences results in a signal informative of the nucleotide at that position. After a single position is sequenced, the query primer and anchor primer are stripped from the DNA template, effectively resetting the sequencing (Ho et al., 2011). The process begins again, sequencing a different position by using a different query primer, and is repeated until the entire sequence of the tag has been determined. The support oligonucleotide ligation detection (SOLiD™) platform from Life Tech is the main representative of sequencing by ligation. In such a platform, fragmented or mate-paired, primed libraries are enriched by means of emulsion PCR on microbeads, which are afterward adhered onto a glass slide. A set of four 1,2-probes (each tagged with a different fluorophore) composed by eight bases is added to the flow cell, competing for ligation to the sequencing primer (Figure 2.4) (Egan et al., 2012). The first two positions of the probe encompass a known di-base pair specific to the fluorophore; these two bases query every first and second base in each ligation reaction. Bases 3 to 5 are degenerate bases separated from bases 6 to 8 by a phosphorothiolate linkage. A matching 1,2-probe is linked to the primer by DNA ligase. After fluorescence imaging to assess which 1,2-probes were connected, silver ions break the phosphorothiolate link, thus regenerating the 5′ phosphate group for subsequent ligation. This procedure—primer hybridization, selective ligation of the probes, four-color imaging, and probe cleavage—is repeated continuously, the number of cycles determining the eventual read length (Metzker, 2009). After a satisfactory length is reached, the extended product is separated, the procedure begun anew, and the template reset with a primer complementary to the $n-1$ position of the previous round of primers. The template is elongated through successive ligations, and then reset four more times (five rounds of primer reset are completed for each sequence tag). This primer reset procedure results in each base being queried in two independent ligation reactions by two different primers, a check-and-balance system that is determined through the creation and alignment of a series of four-color images analyzed through space and time to assess the DNA sequence.

Sequencing by Synthesis

Sequencing by synthesis includes a group of methodologies that make use of a DNA polymerase enzyme to incorporate a single nucleotide or short oligonucleotides (provided either one at a time or fluorescently labeled), containing a reversible terminator. This allows identification of the base type of the incorporated nucleotide without interrupting the extension process (i.e., the sequencing process). Sequencing by synthesis may follow a single molecule- or ensemble-based approach, the former involving the sequencing of various identical copies of a DNA molecule. In addition, it may also follow real-time sequencing or synchronous-controlled strategies. The synchronous-controlled strategy involves *a priori* temporal knowledge to assist in the identification procedure in a *stop-and-go* iterative fashion (Fuller et al., 2009). One of the most employed techniques is denominated pyrosequencing, which is a simple, synchronous-controlled robust method that quantitatively monitors the real-time nucleotide incorporation through the enzymatic conversion of released pyrophosphate into a proportional light signal (Fuller et al., 2009). The light signal detected is then used to establish the sequence of the template strand. The wide application of this methodology stems from its accuracy, flexibility, parallel processing capacity, easy automation, as well as from avoiding the use of labeled primers, labeled nucleotides, and gel electrophoresis (Novais and Thorstenson, 2011). The Illumina HiSeq systems represent another widely used sequencing by synthesis platform, based on a reversible terminator approach

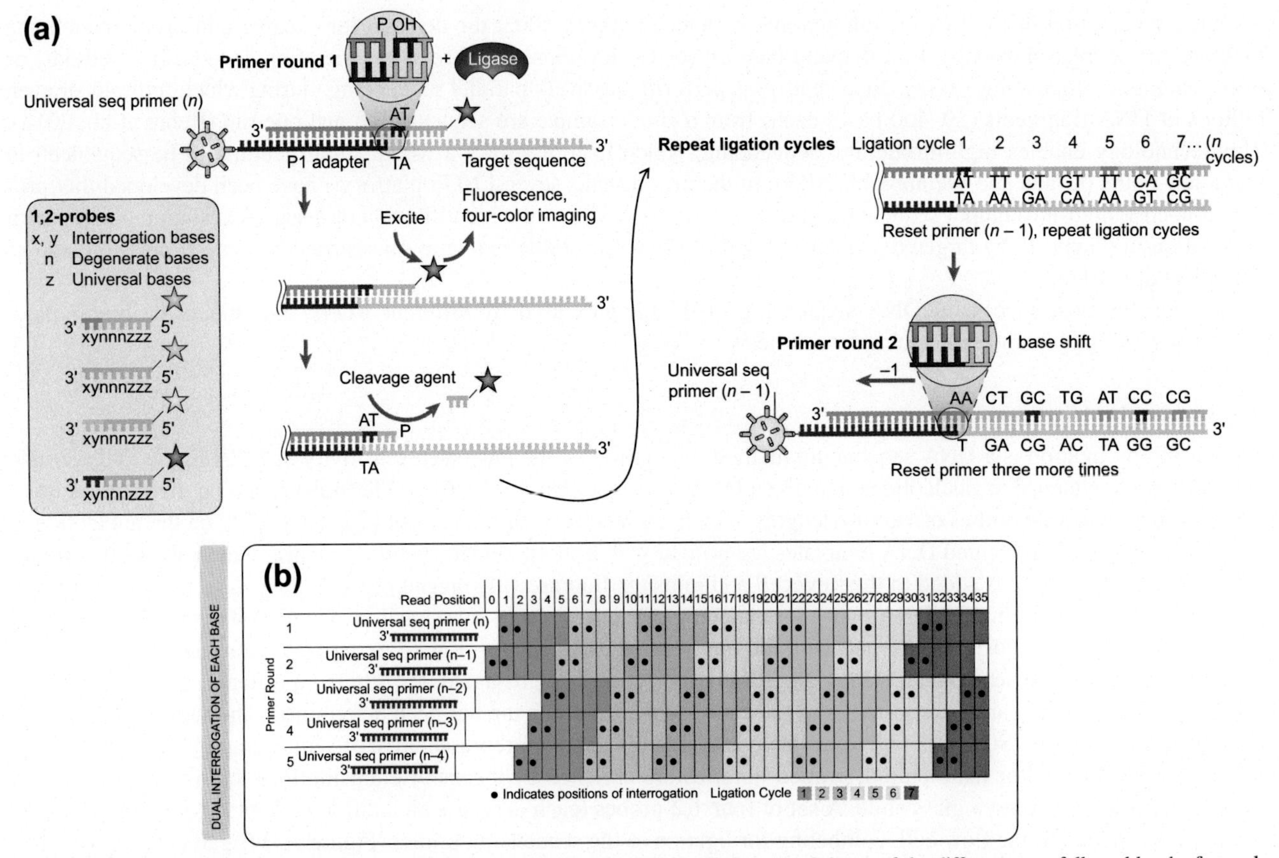

FIGURE 2.4 Illustration of the sequencing by ligation method using the SOLiD platform. (a) Scheme of the different steps followed by the four-color ligation SOLiD method—primer hybridization, selective ligation of the probes, four-color imaging, and probe cleavage. The SOLiD cycle is repeated nine more times. The extension product is removed and the template is reset with a primer complementary to the $n-1$ position for a second round of ligation cycles. (b) Five rounds of primer reset are accomplished for each sequence tag. Through the primer reset procedure, practically every base is queried in two independent ligation reactions by two different primers. *(Metzker, 2009, adapted, with permission from Macmillan Publishers Ltd. and Applied Biosystems website.)*

(Figure 2.5). This technology relies on the attachment of randomly fragmented genomic DNA to a flat, transparent surface on a flow cell. These fragments are sequenced by a four-color DNA sequencing-by-synthesis technology that makes use of reversible terminators with removable fluorescence (Minoche et al., 2011). Tagged nucleotides are incorporated at each cycle, and high-sensitivity fluorescence detection is achieved using laser excitation and total internal reflection optics. Images are compiled and processed to produce base sequences for each DNA template (Raffan and Semple, 2011).

The most commonly used platforms in research and clinical labs besides those already mentioned include the GS FLX Titanium XL+ (Roche Applied Science) and PacBio RS (Pacific Biosciences). A comparison of the features, chemistry, and performance of each mentioned platform is presented in Table 2.4. The current limitations of NGS platforms include the high cost of the platforms, inaccurate sequencing of homopolymer regions on various platforms, time required, and special know-how needed to analyze the data.

Arrays

The fundamental principle of all microarray procedures is that tagged nucleic acid molecules in solution hybridize, with high sensitivity and specificity, to complementary sequences immobilized on a solid substrate, thus easing parallel quantitative measurement of numerous sequences in a complex mixture (Cummings and Relman, 2000). DNA microarrays are a well-established technology for measuring gene expression levels (potential to measure the expression level of thousands of genes within a particular mRNA sample) or to genotype multiple regions of a genome. Microarrays designed for this purpose use relatively few probes for each gene and are biased toward known and predicted gene structures. Although several methods for building microarrays have been developed, two high-density microarrays have prevailed: oligonucleotide microarrays and cDNA microarrays (Figure 2.6).

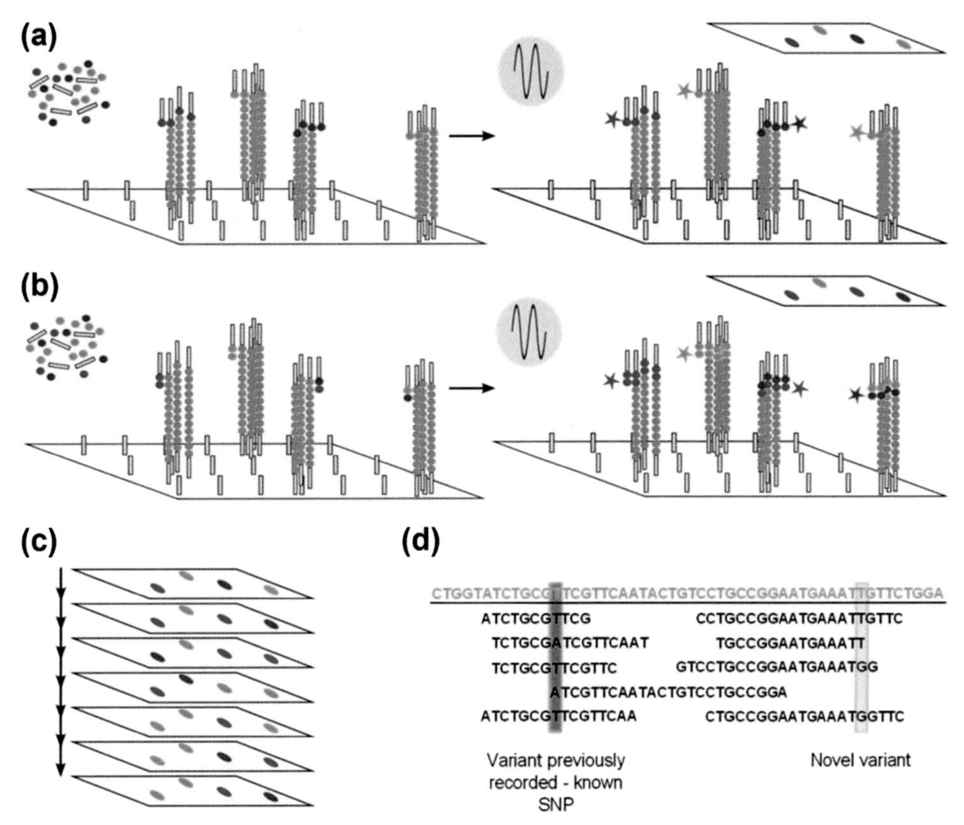

FIGURE 2.5 Illustration of the Illumina sequencing process. (a) The first sequencing cycle starts by adding a solution with four labeled reversible terminators, primers, and DNA polymerase. Temperature cycling allows just one base extension *per* cycle. Bases are integrated into the sequence, complementary to the amplified template. After laser excitation, the emitted fluorescence from each cluster is detected and the first base is identified. (b) During the next cycles, new labeled reversible terminators, primers, and DNA polymerase are added, to generate a fluorescence pattern that will become a fingerprint of the sequences of bases in the fragment (c). (d) Data are aligned and compared to a reference sequence, and sequencing differences are identified. When a difference is detected, its frequency is used to assess whether the variant is heterozygous (as represented in this illustration) or homozygous. Reference databases are searched to assess whether variants are novel or previously recognized as SNPs. *(Raffan and Semple 2011, reprinted with permission from Oxford University Press.)*

TABLE 2.4 Commercial next-generation sequencing platforms for human whole genome sequencing

	Platforms			
Main Features	**Titanium XL+**	**HiSeq 2500/2000**	**5500xl W**	**PacBio RS**
Method	Polymerase (Pyrosequencing)	Polymerase (Reversible terminator)	Ligase (Octomer)	Polymerase Single molecule
Read length (bp)	700	50 to 250	75 + 35	3000
Reads *per* run	1 billion	up to 3 billion	1.2–1.4 billion	35–75 thousand
Gb/Run	0.7	600	320	3
Template prep	Emulsion PCR	Bridge PCR	Emulsion PCR	None
Advantages	Read length	High throughput	Low error rate	No artifacts Read length
Disadvantages	High error rate	Short reads Time-consuming	Short reads Time-consuming	High error rate

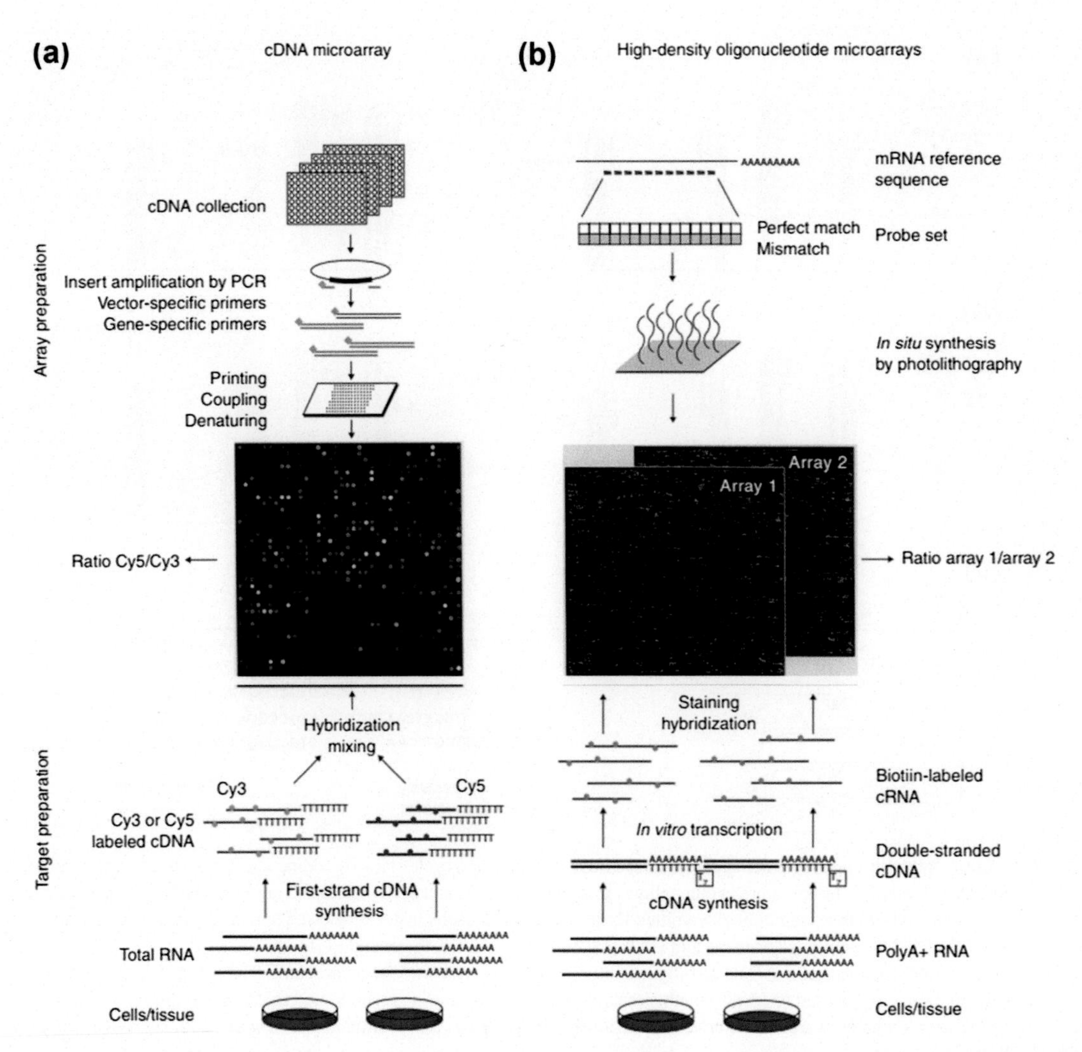

FIGURE 2.6 Illustration of arrays and target preparation. (a) cDNA microarrays: amplification of cDNA inserts by vector-specific or gene-specific primers. PCR products are mechanically attached onto a glass slide. Target preparation is performed by extracting RNA from two different tissues (sample and control), which will then be converted into cDNA in the presence of nucleotides labeled with different tracking molecules (e.g., Cy3 and Cy5). After competitive hybridization of the mixed labeled cDNA against the PCR products spotted on the array surface, a ratio of RNA abundance for the genes represented on the array is determined by high-resolution confocal fluorescence scanning. Array software often uses a green color to symbolize upregulated genes in the sample, red for downregulated genes, and yellow to represent those genes of equal abundance in both experimental and control samples. (b) Oligonucleotide arrays: sequences of 16–20 short oligonucleotides are chosen from the mRNA reference sequence of each gene and are *in situ* synthesized, by photolithography, to create high-density arrays. Target preparation is performed by extracting polyadenylated RNA from two different tissues (control and sample), which will be used to generate the corresponding cDNA. During *in vitro* transcription, biotinylated nucleotides are incorporated into the synthesized cRNA molecules. After hybridization, biotinylated cRNA bounded to the array is stained with a fluorophore conjugated to streptavidin and detected by laser scanning. Fluorescence intensities of probe array element sets on different arrays are used to determine relative mRNA abundance for the genes represented on the array. *(Schulze and Downward, 2001, reprinted with permission from Macmillan Publishers Ltd.)*

Oligonucleotide Microarrays

Oligonucleotide microarrays have emerged as a preferred platform for genomic analysis beyond simple gene expression profiling. These microarrays have several relatively small probes (typically 25 mers) for evaluating the transcript abundance of each gene. Probes are directly synthesized onto the surface of the array using *in situ* chemical synthesis technology (photolithography and combinatorial chemistry). Each chip may contain over 6 million features, each feature comprising millions of copies of a distinct probe sequence (Gregory et al., 2008; Mockler and Ecker, 2005).

cDNA Microarrays

cDNA microarrays are made by mechanically printing/attaching probes, generally amplified PCR products, oligonucleotides, or cloned DNA fragments onto a solid substrate, typically glass with desired physico-chemical characteristics (e.g., excellent mechanical stability and chemical resistance). This type of array generally possesses a lower feature density than the *in situ* synthesized oligonucleotide arrays, typically of about 10,000–40,000 elements *per* microscope slide (Gregory et al., 2008; Mockler and Ecker, 2005; Zhou and Thompson, 2002).

Although the latter method is relatively affordable and offers higher flexibility, without requiring primary sequence information to print a DNA element, it is laborious in what concerns synthesizing, purifying, and storing DNA solutions. In addition, cross-hybridization phenomena may be regarded as a major disadvantage. Nowadays, oligonucleotide arrays are the preferred platform for whole-genome analysis. This technology offers higher specificity, speed, and reproducibility. It requires only the DNA sequence under study to be known, thus eliminating the need to collect and handle cloned DNA or PCR products. Furthermore, the relatively small probe length, together with the flexibility of using several overlapping probes representing the same genomic area, makes oligonucleotide arrays the best option to detect the wide range of genomic features, such as small polymorphisms or splice variants. The disadvantages related with oligonucleotide arrays include the associated costs and low sensitivity due to short sequences used in this technology (Schulze and Downward, 2001).

Quantitative Reverse Transcription PCR and Low-Density Arrays

The real-time reverse transcription PCR is a cost-effective, robust, highly sensitive, easy-to-use, and reproducible technology, suited for validation of microarray-generated data, especially for low-expressed genes (Hui and Feng, 2013; Wong and Medrano, 2005). This technique uses fluorescent indicator molecules to screen the amount of amplification products during each cycle of the PCR reaction, thus eliminating the requirement for post-PCR processing. This enables nucleic acid amplification and detection steps to be combined into one homogeneous assay, which allows the quantitative evaluation of the expression of single genes in multiple samples, while eliminating the necessity of performing gel electrophoresis to detect the amplification products. The high sensitivity of this technique regarding mRNA detection and quantification makes it a viable option for samples with a limited number of RNA copies. Some of the newest systems based on this technology, low-density arrays, allow the simultaneous quantification of large numbers of target genes in single samples (multiplexing of PCR assays) or to measure single markers in parallel on microfluidics cards, while retaining the sensitivity of qPCR (Chung et al., 2007).

Applications of Genomics in Molecular Medicine

Through the use of conventional genetic methodologies (e.g., Southern blot or loss of heterozygosity techniques), almost 2,000 genes for single gene disorders had been identified by the year 2000. These discoveries enabled the accurate diagnosis of single gene disorders, such as Huntington's disease, Marfan syndrome, hemophilia, cystic fibrosis, hereditary spherocytosis, and sickle cell anemia (Rubin and Reisner, 2009). Many of these discoveries have been used in the development of essential newborn blood spot screening programs, which allows testing of all newborns for early signs of a number of treatable congenital disorders (e.g., cystic fibrosis and sickle cell anemia) (Levy, 2010). However, the most recent advances in genomic science, as well as the falling costs, have provided scientists and clinicians the tools to assess the role of genetic and environmental factors in disorders like diabetes, coronary heart disease, asthma, obesity, and several types of cancer. This information has been quickly integrated into clinical practice, enabling earlier and more precise diagnostics, even prevention in some cases, improved therapeutic strategies, evidence-based approaches for demonstrating clinical efficacy, side effect prediction, as well as better decision-making tools for patients and providers. The current accessibility of over 1,000 genetic assessments, targeted therapies, and pharmacogenomic data for drug and dosage selection shows that genomics is already integrated into health care and that it will be a game changer.

Molecular Diagnostics

Methodologies such as NGS platforms or microarrays are currently used to assess genomic expression patterns. The study of such patterns allows researchers to distinguish between normal and abnormal genetic processes, thus enabling the discovery of mutations and targeting of genes with known heterogeneous distribution of mutations. Furthermore, it is possible to target larger segments of the genome to identify known or novel variations, thus making these new methodologies invaluable additions to laboratory testing and clinical evaluation, yielding diagnostic, therapeutic, and prognostic information. These methodologies are presently applied in prenatal/postnatal, cancer, or neurological diagnoses, among other disorders. The evolution of prenatal diagnosis on disorders, such as trisomies, represents one of the examples of the use of genomics as a tool for diagnosis. Until a few years ago, such a diagnosis would be time-consuming and require a chorionic villus sampling or second-trimester amniocentesis, with its associated risks, to assess those chromosomal abnormalities (Newberger, 2000). Nowadays, there are several noninvasive clinical diagnostic tests in the market, which use maternal DNA and require just a blood sample to run in an NGS platform (e.g., Verify® by Illumina). These tests allow detection of trisomy 21, 18, and

13 chromosome abnormalities (along with other genetic disorders), without any risk for the pregnant woman or her unborn child, while providing results with a sensitivity similar to that of the invasive tests (Jiang, 2013).

The evolution of cancer diagnosis represents another example of the use of genomics as a diagnostic tool. The extensive diversity of cancer types sequenced and studied using genomic tools has revealed many novel genetic mechanisms governing cancer initiation, progression, and maintenance. For example, about 90% of breast cancers are due to genetic abnormalities in high-penetrance genes, such as *p53, BRCA1, BRCA2, NBS1, PTEN, ATM, LKB1*, etc. Microarray technology has been used to characterize a number of gene variations associated with breast cancer, leading to a new molecular taxonomy, which is currently employed to identify and classify breast cancer (Kumar et al., 2012). Another example of how genomics may contribute to breast cancer diagnosis was provided by Glass et al. (2006); it established a 70-gene tumor expression profile as a powerful predictor of disease outcome in young breast cancer patients. This microarray is currently available in the market as the MammaPrint© (by Agendia) diagnostic kit; it claims to determine the probability of early-stage breast cancer distant recurrence in the first 5 years following diagnosis (Buyse et al., 2006). The resulting profiles are scored to determine the risk of recurrence and, with it, the need for adjuvant therapy.

Pharmacogenomics

The occurrence of large population differences at the genetic level is responsible for different responses to pharmacotherapy. In fact, it is projected that genetics may be responsible for 20%–95% of variability in drug disposition and response (Wood et al., 2003).

Pharmacogenomics addresses the question of how an individual's genetic variation for specific drug-metabolizing factors, such as enzymes, may affect the body's response to drugs. This is attained by correlating gene expression or single-nucleotide polymorphisms with drug absorption, distribution, metabolism, and elimination, as well as drug receptor target effects (Crews et al., 2012; Evans and Relling, 2004). Knowledge of whether a patient carries any genetic variation that may influence his or her response to drugs can help prescribers to individualize drug therapy, thus decreasing the chance for adverse drug events and increasing the effectiveness of drugs. Pharmacogenomics has been applied in disorders like HIV, diabetes, cancer, depression, and cardiovascular complications (Calcagno et al., 2014; Sabunciyan et al., 2012; Siest et al., 2007; Siest et al., 2005) by successfully guiding drug selection, which results in therapeutic benefits and minimizes the potential for toxicity.

The positive impact of pharmacogenomics in the evolution of cardiovascular disorders treatment may be regarded as an accurate illustration of this technology's impact on current medicine.

A high number of enzymes are directly linked with the pharmacokinetics and pharmacodynamics of cardiovascular drugs. For instance, in the case of antihypertensive drugs, the difference in blood pressure response is markedly different between patients. The key proteins involved in patients' response to a drug are cytochrome *P*450 (CYP) 2D6, CYP2C19, CYP3A4, and the ABCB1 transporter (Siest et al., 2007). These enzymes/transporters are responsible, to a great extent, for the variability in cardiovascular drug response. Nowadays, genotyping of the most important CYP is easily performed by microarrays or NGS platforms. The use of these technologies to analyze, for example, the highly polymorphic CYP2D6 gene provides information regarding whether the patient is an extensive metabolizer (possessing a fully functional enzyme), a poor metabolizer (low or no CYP2D6 enzyme activity), or an ultrarapid metabolizer (excessive enzyme activity). This information is particularly important if we consider that poor metabolizers display an exaggerated drug response and may be at greater risk for toxicity if a drug is mainly metabolized by CYP2D6 or presents a narrow therapeutic index such as propafenone, carvedilol, or metoprolol (Siest et al., 2005).

Recent research on novel cancer therapies has focused on developing agents with a capacity to interfere with critical molecular reactions responsible for tumor phenotypes. This novel approach has been transforming the treatment of various cancers, improving the translational impact of genetic information on clinical practice (Lee et al., 2005). Molecular targeted agents hold great potential to higher treatment efficacy compared with conventional cyto- and genotoxic therapies. The success of this approach is centered on distinguishing responders from nonresponders in clinical practice, so that the drug is targeted to patients possessing a particular molecular abnormality. In the past two decades, several studies have demonstrated that dihydropyrimidine dehydrogenase (DPD, also known as dihydrouracil dehydrogenase and uracil reductase) is an important regulatory enzyme in the metabolism of one of the most prescribed chemotherapy agents, 5-fluorouracil (5-FU). This enzyme is responsible by the pyrimidine catabolic pathway, as is the rate-limiting step in 5-FU catabolism (Diasio and Johnson, 1999). Variability in this enzyme activity widely influences systemic exposure to fluorodeoxyuridine monophosphate and, consequently, the incidence of adverse effects to 5-FU, such as death. DPD activity is entirely or partly abnormal in 0.1% and 3%–5% of individuals in the general population, respectively. This deficiency has been correlated with multiple polymorphisms in the DPYD gene, which results in a decreased activity by this enzyme (Stoehlmacher and

Lenz, 2003). Quantification of DPD mRNA by reverse transcription PCR as a means of determining intratumor DPD levels and analysis by microarrays of the variation in the DPYD gene are currently being proposed as promising approaches to identify high-risk patients for severe 5-FU toxicity.

In targeted anticancer treatment directed to non-small-cell lung cancer, a great portion of the oncologic patients do not respond to gefitinib, a tyrosine kinase inhibitor that targets oncogenic epidermal growth factor receptor and is used as monotherapy when patients do not respond to standard chemotherapy (Ma and Lu, 2011). Nonetheless, about 10% of the patients have a rapid and often dramatic response to such therapy, owing to somatic mutations clustered around the ATP-binding pocket of the tyrosine kinase domain of epidermal growth factor receptor (Lynch et al., 2004). These are responsible by causing enhanced tyrosine kinase activity in response to epidermal growth factor and increased sensitivity to inhibition by gefitinib. One solution provided by genomics may be to assess patients who will have a response to gefitinib by using a screening for such mutations among lung cancer patients.

METABOLOMICS

Metabolomics or Metabonomics?

Following the earlier introduction of the terms *genome, transcriptome,* and *proteome* the word *metabolome* was introduced by Oliver in 1998 (Oliver et al., 1998). In the following year, Nicholson introduced *metabonomics* (Nicholson et al., 1999) and, in 2000, Fiehn coined the term *metabolomics* (Fiehn et al., 2000). A case can be made regarding what is considered a metabolite. Following a literal definition, all molecules that result from metabolism are metabolites. However, this definition would include several types of compounds, from peptides to other macromolecules. This is not the case, as it is generally accepted that metabolomics addresses small molecules, thus leaving out peptides, proteins, and other biological entities.

In the field of "omics," there is a fair amount of discussion regarding the nomenclature used, with different authors adopting different definitions. In this chapter, the authors chose to use the following interpretation: *Metabolomics* addresses the metabolome, that is, the complete set of molecules that constitute an organism. As so, metabolomics can be seen as a "snapshot" of the chemical composition of a tissue, organ, or organism in a given moment. Differently, *metabonomics* addresses the temporal and spatial changes in the chemical composition of the sample. In the field of biomedical science and molecular medicine, metabonomics is frequently used to profile endogenous metabolites in biological fluids (urine, blood, tissue homogenates, cerebrospinal fluid, broncho-alveolar lavage, among others) in order to characterize the metabolic phenotype and its response to stimulation or disease (Orešič, 2009; Robertson et al., 2007; Weckwerth, 2003; Zhang et al., 2013).

For obvious reasons, nowadays it still is not possible to study the complete set of all metabolites within an organism (Sumner et al., 2003). This arises as a consequence of the high number of different small molecules present in organisms and the low concentration at which they may occur. In addition, this objective is further hindered by the fact that no single analytical technique or combination of several techniques allows the identification of all metabolites, a consequence of their diversified physical and chemical properties and the wide concentration range in which they occur.

Nevertheless, from a biological point of view, the *metabolome* is the biological endpoint that links genotype to function. In addition to genotypes, environmental factors, such as nutrition, circadian rhythm, aging, stress, and hormonal changes, are known to have an impact on phenotype through the metabolome, thus being key players in the assessment of health and disease and the impact of pharmacotherapy.

Several experimental approaches can be used in the field of metabolomics. In the following sections, we highlight a few.

Metabolite Profiling

In metabolite profiling, the emphasis is placed on compounds that share chemical similarity (e.g., fatty acids, sugars, eicosanoids) or that are related through a common metabolic pathway. In this design, selective extraction procedures are commonly used to remove other metabolites that are regarded as interferences, thus improving the analysis of the target molecules. From an analytical point of view, liquid chromatography-mass spectrometry (LC-MS), gas chromatography-mass spectrometry (GC-MS) (Pereira et al., 2012b), and nuclear magnetic resonance (NMR) are the most used techniques (Hollywood et al., 2006).

Metabolite Fingerprinting

Fingerprinting analysis involves collecting spectra of unpurified and chemically complex matrices and initially ignoring the problem of making individual assignments of peaks, which frequently overlap. Multivariate statistical methods are then used to identify clusters of similarity or difference (Pereira et al., 2010). In subsequent studies, only peaks that are different between groups are analyzed, thus avoiding the time-consuming identification of all peaks. In this metabolomic strategy, sample preparation is reduced because the objective is a global, rapid, and high-throughput analysis of crude samples (Dunn et al., 2005).

The most frequently used technique for metabolic fingerprinting is NMR and several types of applications can be found in the literature; multivariate analysis of unassigned ^{1}H NMR spectra is among the most used approaches.

In addition to NMR, MS-based investigations are also possible. Here, the metabolite fingerprints are represented by *m/z* values and corresponding intensities of the detected ions. When a separation step takes place prior to the MS analysis, retention times may also be used to index metabolites.

Metabolic Footprinting

The widespread and increasingly robust use of cell cultures in several research areas has created a new approach, metabolic footprinting, which addresses the exometabolome. In this case, the analysis is performed on the culture media or other extracellular environments, thus allowing the study of compounds excreted during cellular metabolism without the need for cell disruption or lysis.

This approach may also present some advantages in what regards data complexity and analysis time. For instance, the intracellular metabolism is more dynamic and thus, the turnover of most metabolites is extremely fast, requiring an efficient quenching of cell metabolism, followed by an effective separation of intra- and extracellular metabolites and subsequent extraction of intracellular compounds (Suhre et al., 2010; Villas-Bôas et al., 2006).

Analytical Techniques in Metabolomics/Metabonomics

Nuclear Magnetic Resonance

Nuclear magnetic resonance spectroscopy explores the magnetic properties of the nuclei of certain atoms. From an instrumental point of view, it relies on the phenomenon of nuclear magnetic resonance, which can provide a wide range of information, including structure, reaction state, and chemical environment. Molecules containing at least one atom with a nonzero magnetic moment are potentially detectable by NMR, such isotopes including ^{1}H, ^{13}C, ^{14}N, ^{15}N, and ^{31}P. These signals are characterized by their frequency (chemical shift), intensity, fine structure, and magnetic relaxation properties, all of which reflect the environment of the detected nucleus. NMR is the analytical method that provides the most comprehensive structural information, including stereochemical details.

Most applications of NMR are simple one-dimensional NMR experiments of the ^{1}H nuclei; other techniques are able to provide more information, such as ^{13}C or 3–4 dimensional techniques for the study of complex molecules. Studies of 1D NMR are very useful in classifying similar groups of samples; however, when there are several overlapping peaks, identification of metabolites can be hindered. In this situation, 2D NMR studies can prove to be very useful because they provide much more information (Pereira et al., 2012a).

The high amount of structural information provided by NMR is countered by its relatively low sensitivity, with samples in the range of 1–50 mg usually being required, in opposition to MS, which can use samples in the picogram/nanogram range. For this reason, extensive research has been conducted to improve NMR sensitivity, one of the most promising findings being the use of cryoprobes. In this approach, the detection system is cooled while the sample remains at room temperature (Pereira et al., 2012a), which limits the noise voltage associated with signal detection and, when compared with regular probes, the signal-to-noise ratio is improved by a factor of 3–4 (Krishnan et al., 2005).

Despite the issues related with sensitivity, NMR presents a number of advantages, including the fast and simple sample preparation required, quick measurements, and quantification of analytes without the need for standards. NMR coupled with LC is an even more powerful analytical tool, but its prohibitive price has prevented its widespread use.

One important advantage separates NMR from MS: it is not a destructive technique, thus allowing further analysis of the same sample.

Mass Spectrometry

MS relies on the fragmentation of molecules following exposure to a high energy input. Several degrees of information can be drawn from this action, from molecular weight to the presence of a certain functional group, sometimes even differentiation of certain types of isomers. The fragmentation products are impelled and focused through a magnetic mass analyzer, to be further collected and each selected ion measured in a detector (Ferreres et al., 2010). The choice of the ionization method to be employed is dependent on the physical–chemical properties of the analyte(s) of interest, namely volatility, molecular weight, and thermolability, and on the complexity of the matrix in which the analyte is contained, among others.

MS can also detect molecules that comprise "NMR-invisible" moieties, such as sulfates (Dettmer et al., 2007). For some chemical classes, the fragmentation pattern (both peak mass and abundance) can allow the characterization of the unknown compound's structure. However, with a few exceptions (Ferreres et al., 2010), MS does not allow differentiation of isomers, which are particularly important in a biological context, in which the activity of molecules is highly influenced by the isomer and its conformation.

In a general way, MS is used in combination with a prior separation step of analytes, either LC or GC. In LC, several variables can be controlled, for example, column length, particle size, flow, and solvent system, among others. In GC, ideal candidates are molecules with high to mid volatility, although nonvolatiles can sometimes be studied if a derivatization step is introduced. Still, GC-MS offers excellent reproducibility and low detection limits. Given the fact that several databases of compound identification exist, a tentative identification in the absence of standards is possible.

In what regards ion sources, widely used options include electron ionization (EI), chemical ionization (CI), electrospray ionization (ESI), atmospheric pressure chemical ionization (APCI), and matrix-assisted laser desorption/ionization (MALDI). The bases of each ion source have been reviewed elsewhere (Douglas et al., 2005; Huang et al., 2010; Kebarle and Verkerk, 2009; March and Todd, 2005; Watson and Sparkman, 2007) and are beyond the scope of this chapter.

Among the several mass analyzers available we can refer the widespread use of single-quadrupoles, triple quadrupoles, time-of-flight (TOF), ion traps (IT), and Fourier transform ion cyclotron resonance (FT-ICR).

When MS is used, two levels of compound identification can be achieved: provisional and positive. Positive identification can be achieved when reference compounds are available, thus allowing comparison of both retention time and mass spectra. However, in the case of new or rare compounds, no standards are available, in which case provisional identification takes place.

In recent years, ultra performance liquid chromatography (UPLC), especially when in tandem with MS, has been increasingly used in pharmaceutical analysis and biomarkers discovery. This technique retains many of the principles of HPLC while requiring lower analysis time and solvent consumptions. Specially designed columns, which incorporate particles of lower sizes, typically <2 μm, are used, and the system operates at higher pressures than HPLC, results showing marked improvements in resolution and sensitivity (Nováková et al., 2006; Swartz, 2005).

MS-based metabonomics offers a few advantages when compared to NMR, namely higher sensitivity (pM-μM), rapid profiling and, in some cases, databases for structure assignments. There are also disadvantages, such as extensive sample preparation in some cases, challenging quantifications with ion-suppression when internal standards are not available and lack of standardized data structures.

Several approaches have been developed in order to combine the information provided by MS and NMR analysis. Such is the case of statistical heterospectroscopy, which allows the combination of the information provided by MS and NMR, thus increasing the number of detectable significantly changed metabolites. This approach allows the identification of relevant molecules that could not have been identified by either method separately (Crockford et al., 2006).

"Omics" and Biomarkers

A biomarker is a measurable indicator that provides status of biological state of a subject. The use of biomarkers in health and disease relies on the changes in their concentrations or flux that corresponds to a particular phenotype when compared to the control phenotype. It is important to highlight that the use of biomarkers goes beyond disease, as they can be used across several fields, such as toxicology (Lee et al., 2007), nutrition (Orešič, 2009; Rezzi et al., 2007), pharmacokinetics and pharmacodynamics (Chen et al., 2007; Orešič, 2009; Robertson, Reily, and Baker, 2007), and biotechnology (Buchholz et al., 2001), among others.

After discovery, a potential new biomarker has to undergo validation to assure that the molecule can be associated with the biological state under study in a specific way (Koulman et al., 2009; Vuckovic, 2012). Contrary to what happens in the discovery step, in which 100 samples are frequently enough, for validation it is advisable that at least 1,000 samples are used. The purpose of this approach is to assure that the sensitivity and specificity of the biomarker are robust enough to withstand the marked variation found in biological samples of a given population. Between discovery and validation, it is common to address the verification phase, in which a few hundred samples are used, the difference residing in the few (tens of) compounds under study, unlikely discovery phase, which requires more molecules.

In the cohorts used, it is important to use not only healthy individuals but also patients suffering from pathologies similar to the one under study. In this way, it may be possible to discharge the lack of specificity that could arise, for example, from a shared pathway, such as the cascades of the inflammatory pathway, which are common to several different pathological conditions.

The typical samples used are serum or plasma, owing to the relative ease with which they can be obtained and also the general rule that they represent the current state, either physiological or pathological, of the human body as a whole. Due to the high number of proteins in these matrices, immune depletion of highly abundant proteins is frequent, the reproducibility and efficiency of this technique having already been evaluated and demonstrated. Further fractionation techniques, such as cation exchange chromatography, are also common (Dardé et al., 2007; Whiteaker et al., 2007).

Taking into account the several steps involved in the acceptance of new biomarkers, nowadays the verification phase is regarded as one of the major bottlenecks.

Metabolomics/Metabonomics in Clinical Use

Several works are available in the literature showing the application of metabolomics as a tool for diagnosis in several pathologies. Cancer is one of the most widely studied diseases, namely due to the pivotal role of an early diagnosis. As so, MS and NMR-based techniques have been successfully used in several types of cancer. Here we briefly present some examples of the application of metabolomics in diagnosis; we also present our apologies in advance for the authors whose work we are unable to quote.

Nishiumi et al. (2010) used GC-MS to analyze 60 metabolites in 20 pancreatic cancer patients, 18 molecules being changed when compared with control group. In addition, the authors were able to discriminate between stage III, stage IVa, and stage IVb groups by using Multiple Classification Analysis.

UPLC-MS has been successfully used to characterize serum profiles from hepatocellular carcinoma (HCC), liver cirrhosis (LvC), and healthy subjects. The analysis of the UPLC-MS profiles of the subjects yielded 13 potential biomarkers involved in metabolic pathways involving organic acids, phospholipids, fatty acids, and bile acids (Wang et al., 2012). The use of the metabolomic profile was compared with the classical marker alpha-fetoprotein (AFP), the former being able to discriminate patients from controls and also HCC from LvC with a sensitivity and specificity of 100%. Canavaninosuccinate was decreased in LvC and increased in HCC, while glycochenodeoxycholic acid was pointed as an indicator for HCC diagnosis and disease prognosis (Wang et al., 2012).

In another study, using a nontargeted method, three metabolites from rat HCC cancer models yielded taurocholic acid, lysophosphoethanolamine 16:0, and lysophosphatidylcholine 22:5 as potential metabolites with application in grading the different stages of hepatocarcinogenesis; they also represent the abnormal metabolism during the progress of HCC in patients (Tan et al., 2012). For a review of this topic, see Wang, Zhang, and Sun (2013).

In the case of kidney cancer, urine is the sample of choice because it displays metabolic signatures of many biochemical pathways. Kind et al. (2007) described a pilot study using the combined information of LC-MS, UPLC-MS, and GC-TOF-MS. Overall, statistic analysis led to several significant components that were able to discriminate between renal cell carcinoma patients.

In a study by Wen et al. (2010) aiming to use NMR-based metabolomics in the diagnosis of biliary tract cancer, bile was collected from patients with cancer and benign biliary tract diseases. The metabolomic 2-D score plot revealed good separation between cancer and benign groups, the signals contributing to these differences being further studied using a statistical TOCSY approach. The diagnostic performance assessed by leave-one-out analysis exhibited 88% sensitivity and 81% specificity, better than the conventional markers carcinoembryonic antigen, carbohydrate antigen 19-9, and bile cytology.

Wu et al. (2009) used GC-MS for studying biomarkers useful in distinguishing between the profile of esophageal cancer and the corresponding normal mucosa, over 20 molecules scoring a statistical significance with $p < 0.05$, most of them amino acids and fatty acids.

While many of the biomarkers that result from the aforementioned studies are still under evaluation and not widely used in clinics, significant advances have already been reached. For example, in the case of prostate cancer, the most frequent tests used for diagnosis rely on prostate-specific antigen (PSA), which suffers from poor accuracy (Lokhov et al., 2009). Lokhov et al. (2010) used an MS-based metabolite fingerprinting approach to analyze blood plasma of patients. This new technique displayed sensitivity, specificity, and accuracy of 95%, 96.7%, and 95.7%, respectively, in opposition to the enzyme-linked immunosorbent assay (ELISA) PSA test, which exhibited 35%, 83%, and 52%, respectively.

The range of application of metabolomics platforms is not confined to cancer, being increasingly used in neurological disorders, such as Parkinson's (Bogdanov et al., 2008; Sinclair et al., 2010), metabolic diseases like diabetes (Salek et al., 2007; Suhre et al., 2010), or some types of obesity (Zhang et al., 2013).

Nowadays metabolomics can also be used as a tool for monitoring patients' responses to pharmacotherapy, allowing the evaluation of both efficacy and eventual toxicological phenomena. For example, Nicholson and coworkers correlated endogenous metabolite profiles before drug treatment with treatment efficacy and safety (pharmacometabonomics) (Clayton et al., 2006). A recent publication investigated the renal impairment induced by treatment with bisphosphonates (ibandronate and zoledronate), looking at urinary metabolites. In particular, the work describes *N*-acetylfelinine, a new endogenous metabolite that is directly related to the mechanism of action of these drugs. This biomarker can thus be clinically applied to monitor efficacy and safety of bisphosphonates (Dieterle et al., 2007).

For a detailed review of the application of metabolomics in biomarker discovery for neurological, virological, and cardiac diseases and cancer, see Nordström and Lewensohn (2010).

CONCLUSION

There is no doubt that "omics" technologies, in their multiple dimensions, are now regarded as a powerful tool for diagnosis, disease monitoring, and personalized medicine. While many gaps are still present, the remarkable rate at which several technological solutions are evolving will result in "omics" approaches paving their way as an unavoidable reality in the near future.

In what regards instrumental apparatus, improved MS and NMR hardware will yield faster analysis with higher sensitivity. The challenge is believed to be the integration of the information generated by the different "omics" approaches. Thus, when information from genomics, transcriptomics, proteomics, and metabolomics is combined, a true integrated view of organisms in a systems biology context can be attained. Statistical tools are likely to play an important role in this area, as only robust statistics can shed a light on the relevant information to be extracted from the enormous datasets generated by the aforementioned techniques.

REFERENCES

Bogdanov, M., Matson, W.R., Wang, L., Matson, T., Saunders-Pullman, R., Bressman, S.S., Beal, M.F., 2008. Metabolomic profiling to develop blood biomarkers for Parkinson's disease. Brain 131, 389–396.

Buchholz, A., Takors, R., Wandrey, C., 2001. Quantification of intracellular metabolites in *Escherichia coli* K12 using liquid chromatographic-electrospray ionization tandem mass spectrometric techniques. Anal. Biochem. 295, 129–137.

Buyse, M., Loi, S., Van't Veer, L., Viale, G., Delorenzi, M., Glas, A.M., et al., 2006. Validation and clinical utility of a 70-gene prognostic signature for women with node-negative breast cancer. J. Natl. Cancer Inst. 98, 1183–1192.

Calcagno, A., Lamorde, M., D'Avolio, A., Bonora, S., 2014. Personalizing HIV therapeutics in resource-limited rural communities: lessons learned from the use of new tools in Africa. Curr. Pharmacogenom. 12, 1–14.

Chen, C., 2014. DNA polymerases drive DNA sequencing-by-synthesis technologies: both past and present. Evolut. Gen. Microbiol. 5, 305.

Chen, C., Gonzalez, F.J., Idle, J.R., 2007. LC-MS-based metabolomics in drug metabolism. Drug Metab. Rev. 39, 581–597.

Chung, C.H., Levy, S., Chaurand, P., Carbone, D.P., 2007. Genomics and proteomics: emerging technologies in clinical cancer research. Crit. Rev. Oncol. Hemat. 61, 1–25.

Clayton, T.A., Lindon, J.C., Cloarec, O., Antti, H., Charuel, C., Hanton, G., et al., 2006. Pharmaco-metabonomic phenotyping and personalized drug treatment. Nature 440, 1073–1077.

Collins, F.S., Green, E.D., Guttmacher, A.E., Guyer, M.S., 2003. A vision for the future of genomics research. Nature 422, 835–847.

Crews, K.R., Hicks, J.K., Pui, C.-H., Relling, M.V., Evans, W.E., 2012. Pharmacogenomics and individualized medicine: translating science into practice. Clin. Pharmacol. Ther. 92, 467–475.

Crockford, D.J., Holmes, E., Lindon, J.C., Plumb, R.S., Zirah, S., Bruce, S.J., et al., 2006. Statistical heterospectroscopy, an approach to the integrated analysis of NMR and UPLC-MS data sets: application in metabonomic toxicology studies. Anal. Chem. 78, 363–371.

Cummings, C., Relman, D.A., 2000. Using DNA microarrays to study host-microbe interactions. Emerg. Infect. Dis. 6, 513–525.

Dardé, V.M., Barderas, M.G., Vivanco, F., 2007. Depletion of high-abundance proteins in plasma by immunoaffinity subtraction for two-dimensional difference gel electrophoresis analysis. In: Vivanco, F. (Ed.), Cardiovascular proteomics. Springer, Heidelberg, New York, pp. 351–364.

Dettmer, K., Aronov, P.A., Hammock, B.D., 2007. Mass spectrometry-based metabolomics. Mass Spectrom. Rev. 26, 51–78.

Dettmer, K., Hammock, B.D., 2004. Metabolomics—a new exciting field within the "omics" sciences. Environ. Health Persp. 112, A396.

Diasio, R.B., Johnson, M.R., 1999. Dihydropyrimidine dehydrogenase: its role in 5-fluorouracil clinical toxicity and tumor resistance. Clin. Cancer Res. 5, 2672–2673.

Dieterle, F., Schlotterbeck, G., Binder, M., Ross, A., Suter, L., Senn, H., 2007. Application of metabonomics in a comparative profiling study reveals N-acetylfelinine excretion as a biomarker for inhibition of the farnesyl pathway by bisphosphonates. Chem. Res. Toxicol. 20, 1291–1299.

Douglas, D.J., Frank, A.J., Mao, D., 2005. Linear ion traps in mass spectrometry. Mass Spectrom. Rev. 24, 1–29.

Dunn, W.B., Bailey, N.J., Johnson, H.E., 2005. Measuring the metabolome: current analytical technologies. Analyst 130, 606–625.

Egan, A.N., Schlueter, J., Spooner, D.M., 2012. Applications of next-generation sequencing in plant biology. Am. J. Bot. 99, 175–185.

Evans, W.E., Relling, M.V., 2004. Moving towards individualized medicine with pharmacogenomics. Nature 429, 464–468.

Ferreres, F., Pereira, D.M., Valentão, P., Andrade, P.B., 2010. First report of non-coloured flavonoids in *Echium plantagineum* bee pollen: differentiation of isomers by liquid chromatography/ion trap mass spectrometry. Rapid Commun. Mass Spectrom. 24, 801–806.

Fiehn, O., Kopka, J., Dörmann, P., Altmann, T., Trethewey, R.N., Willmitzer, L., 2000. Metabolite profiling for plant functional genomics. Nat. Biotechnol. 18, 1157–1161.

Fuller, C.W., Middendorf, L.R., Benner, S.A., Church, G.M., Harris, T., Huang, X., et al., 2009. The challenges of sequencing by synthesis. Nat. Biotechnol. 27, 1013–1023.

Glas, A.M., Floore, A., Delahaye, L.J., Witteveen, A.T., Pover, R.C., Bakx, N., et al., 2006. Converting a breast cancer microarray signature into a high-throughput diagnostic test. BMC Genomics 7, 278.

Grada, A., Weinbrecht, K., 2013. Next-generation sequencing: methodology and application. J. Invest. Dermatol. 133 (8), e11.

Gregory, B.D., Yazaki, J., Ecker, J.R., 2008. Utilizing tiling microarrays for whole-genome analysis in plants. Plant J. 53, 636–644.

Ho, A., Murphy, M., Wilson, S., Atlas, S.R., Edwards, J.S., 2011. Sequencing by ligation variation with endonuclease V digestion and deoxyinosine-containing query oligonucleotides. BMC Genomics 12, 598.

Hollywood, K., Brison, D.R., Goodacre, R., 2006. Metabolomics: current technologies and future trends. Proteomics 6, 4716–4723.

Huang, M.-Z., Yuan, C.-H., Cheng, S.-C., Cho, Y.-T., Shiea, J., 2010. Ambient ionization mass spectrometry. Annu. Rev. Anal. Chem. 3, 43–65.

Hui, K., Feng, Z.-P., 2013. Efficient experimental design and analysis of real-time PCR assays. Channels 7, 160.

Jiang, K., 2013. Competition intensifies over DNA-based tests for prenatal diagnoses. Nature Med. 19, 381.

Kebarle, P., Verkerk, U.H., 2009. Electrospray: from ions in solution to ions in the gas phase, what we know now. Mass Spectrom. Rev. 28, 898–917.

Kind, T., Tolstikov, V., Fiehn, O., Weiss, R.H., 2007. A comprehensive urinary metabolomic approach for identifying kidney cancer. Anal. Biochem. 363, 185–195.

Koulman, A., Lane, G.A., Harrison, S.J., Volmer, D.A., 2009. From differentiating metabolites to biomarkers. Anal. Bioanal. Chem. 394, 663–670.

Krishnan, P., Kruger, N.J., Ratcliffe, R.G., 2005. Metabolite fingerprinting and profiling in plants using NMR. J. Exp. Bot. 56, 255–265.

Kumar, R., Sharma, A., Tiwari, R.K., 2012. Application of microarray in breast cancer: an overview. J. Pharm. Bioallied. Sci. 4, 21.

Lee, S.H., Woo, H.M., Jung, B.H., Lee, J., Kwon, O.S., Pyo, H.S., et al., 2007. Metabolomic approach to evaluate the toxicological effects of nonylphenol with rat urine. Anal. Chem. 79, 6102–6110.

Lee, W., Lockhart, A.C., Kim, R.B., Rothenberg, M.L., 2005. Cancer pharmacogenomics: powerful tools in cancer chemotherapy and drug development. Oncologist 10, 104–111.

Levy, P.A., 2010. An overview of newborn screening. J. Dev. Behav. Pediatr. 31, 622–631.

Lokhov, P., Dashtiev, M., Bondartsov, L., Lisitsa, A., Moshkovskii, S., Archakov, A., 2010. Metabolic fingerprinting of blood plasma from patients with prostate cancer. Biochemistry (Moscow) Suppl. Ser. B.: Biomed. Chem. 4, 37–41.

Lokhov, P.G., Dashtiev, M.I., Bondartsov, L.V., Lisitsa, A.V., Moshkovskii, S.A., Archakov, A.I., 2009. Metabolic fingerprinting of blood plasma for patients with prostate cancer. Biomed. Khim. 55, 247–254.

Lynch, T.J., Bell, D.W., Sordella, R., Gurubhagavatula, S., Okimoto, R.A., Brannigan, B.W., et al., 2004. Activating mutations in the epidermal growth factor receptor underlying responsiveness of non-small-cell lung cancer to gefitinib. N. Engl. J. Med. 350, 2129–2139.

Ma, Q., Lu, A.Y., 2011. Pharmacogenetics, pharmacogenomics, and individualized medicine. Pharmacol. Rev. 63, 437–459.

March, R.E., Todd, J.F., 2005. Quadrupole ion trap mass spectrometry. John Wiley & Sons, Hoboken.

Metzker, M.L., 2009. Sequencing technologies—the next generation. Nat. Rev. Genet. 11, 31–46.

Minoche, A.E., Dohm, J.C., Himmelbauer, H., 2011. Evaluation of genomic high-throughput sequencing data generated on Illumina HiSeq and genome analyzer systems. Genome. Biol. 12, R112.

Mockler, T.C., Ecker, J.R., 2005. Applications of DNA tiling arrays for whole-genome analysis. Genomics 85, 1–15.

Nelson, D., Cox, M., Lehninger, A., 2008. Principles of biochemistry. W.H. Freeman and Company, New York.

Newberger, D.S., 2000. Down syndrome: prenatal risk assessment and diagnosis. Am. Fam. Physician 62, 825–832, 837–838.

Nicholson, J.K., Lindon, J.C., Holmes, E., 1999. 'Metabonomics': understanding the metabolic responses of living systems to pathophysiological stimuli via multivariate statistical analysis of biological NMR spectroscopic data. Xenobiotica 29, 1181–1189.

Nishiumi, S., Shinohara, M., Ikeda, A., Yoshie, T., Hatano, N., Kakuyama, S., Mizuno, S., et al., 2010. Serum metabolomics as a novel diagnostic approach for pancreatic cancer. Metabolomics 6, 518–528.

Nordström, A., Lewensohn, R., 2010. Metabolomics: moving to the clinic. J. Neuroimmune. Pharm. 5, 4–17.

Novais, R., Thorstenson, Y., 2011. The evolution of Pyrosequencing® for microbiology: from genes to genomes. J. Microbiol. Meth. 86, 1–7.

Nováková, L., Matysová, L., Solich, P., 2006. Advantages of application of UPLC in pharmaceutical analysis. Talanta 68, 908–918.

Oliver, S.G., Winson, M.K., Kell, D.B., Baganz, F., 1998. Systematic functional analysis of the yeast genome. Trends Biotechnol. 16, 373–378.

Orešič, M., 2009. Metabolomics, a novel tool for studies of nutrition, metabolism and lipid dysfunction. Nutr. Metab. Cardiovas. 19, 816–824.

Pareek, C.S., Smoczynski, R., Tretyn, A., 2011. Sequencing technologies and genome sequencing. J. Appl. Genet. 52, 413–435.

Pereira, D.M., Correia-da-Silva, G., Valentão, P., Teixeira, N., Andrade, P.B., 2012a. Marine metabolomics in cancer chemotherapy. In: Debmalya, B. (Ed.), OMICS: Biomedical perspectives and applications. CRC Press, Boca Raton, pp, 379–400.

Pereira, D.M., Valentão, P., Ferreres, F., Andrade, P.B., 2010. Metabolomic analysis of natural products. In: Tzanavaras, P., Zacharis, C. (Eds.), Reviews in pharmaceutical and biomedical analysis. Bentham, pp. 1–19.

Pereira, D.M., Vinholes, J., de Pinho, P.G., Valentão, P., Mouga, T., Teixeira, N., Andrade, P.B., 2012b. A gas chromatography–mass spectrometry multi-target method for the simultaneous analysis of three classes of metabolites in marine organisms. Talanta 100, 391–400.

Pray, L., 2008. Eukaryotic genome complexity. Nat. Edu. 1, 96. Cambridge, UK.

Raffan, E., Semple, R.K., 2011. Next generation sequencing—implications for clinical practice. Brit. Med. Bull. 99, 53–71.

Rehm, H.L., Bale, S.J., Bayrak-Toydemir, P., Berg, J.S., Brown, K.K., Deignan, J.L., et al., 2013. ACMG clinical laboratory standards for next-generation sequencing. Genet. Med. 15, 733–747.

Rezzi, S., Ramadan, Z., Fay, L.B., Kochhar, S., 2007. Nutritional metabonomics: applications and perspectives. J. Proteom. Res. 6, 513–525.

Robertson, D.G., Reily, M.D., Baker, J.D., 2007. Metabonomics in pharmaceutical discovery and development. J. Proteom. Res. 6, 526–539.

Rubin, E., Reisner, H.M., 2009. Essentials of Rubin's pathology. Lippincott Williams & Wilkins, Baltimore.

Sabunciyan, S., Aryee, M.J., Irizarry, R.A., Rongione, M., Webster, M.J., Kaufman, W.E., et al., 2012. Genome-wide DNA methylation scan in major depressive disorder. PloS One 7. e34451.

Salek, R.M., Maguire, M.L., Bentley, E., Rubtsov, D.V., Hough, T., Cheeseman, M., et al., 2007. A metabolomic comparison of urinary changes in type 2 diabetes in mouse, rat, and human. Physiol. Genomics 29, 99–108.

Sanger, F., Nicklen, S., Coulson, A.R., 1977. DNA sequencing with chain-terminating inhibitors. Proc. Natl. Acad. Sci. USA 74, 5463–5467.

Schulze, A., Downward, J., 2001. Navigating gene expression using microarrays—a technology review. Nat. Cell. Biol. 3, E190–E195.

Siest, G., Jeannesson, E., Visvikis-Siest, S., 2007. Enzymes and pharmacogenetics of cardiovascular drugs. Clin. Chim. Acta. 381, 26–31.

Siest, G., Marteau, J.-B., Maumus, S., Berrahmoune, H., Jeannesson, E., Samara, A., et al., 2005. Pharmacogenomics and cardiovascular drugs: need for integrated biological system with phenotypes and proteomic markers. Eur. J. Pharmacol. 527, 1–22.

Sinclair, A.J., Viant, M.R., Ball, A.K., Burdon, M.A., Walker, E.A., Stewart, P.M., et al., 2010. NMR-based metabolomic analysis of cerebrospinal fluid and serum in neurological diseases—a diagnostic tool? NMR. BioMed. 23, 123–132.

Stoehlmacher, J., Lenz, H.-J., 2003. Implications of genetic testing in the management of colorectal cancer. Am. J. Pharmacogenom. 3, 73–88.

Suhre, K., Meisinger, C., Döring, A., Altmaier, E., Belcredi, P., Gieger, C., et al., 2010. Metabolic footprint of diabetes: a multiplatform metabolomics study in an epidemiological setting. PloS One 5, e13953.

Sumner, L.W., Mendes, P., Dixon, R.A., 2003. Plant metabolomics: large-scale phytochemistry in the functional genomics era. Phytochemistry 62, 817–836.

Swartz, M.E., 2005. UPLC™: an introduction and review. J. Liq. Chromatogr. Rel. Technol. 28, 1253–1263.

Tan, Y., Yin, P., Tang, L., Xing, W., Huang, Q., Cao, D., et al., 2012. Metabolomics study of stepwise hepatocarcinogenesis from the model rats to patients: potential biomarkers effective for small hepatocellular carcinoma diagnosis. Mol. Cell. Proteomics 11 (M111), 010694.

Villas-Bôas, S.G., Noel, S., Lane, G.A., Attwood, G., Cookson, A., 2006. Extracellular metabolomics: a metabolic footprinting approach to assess fiber degradation in complex media. Anal. Biochem. 349, 297–305.

Vuckovic, D., 2012. Current trends and challenges in sample preparation for global metabolomics using liquid chromatography–mass spectrometry. Anal. Bioanal. Chem. 403, 1523–1548.

Wang, B., Chen, D., Chen, Y., Hu, Z., Cao, M., Xie, Q., et al., 2012. Metabonomic profiles discriminate hepatocellular carcinoma from liver cirrhosis by ultraperformance liquid chromatography–mass spectrometry. J. Proteome Res. 11, 1217–1227.

Wang, X., Zhang, A., Sun, H., 2013. Power of metabolomics in diagnosis and biomarker discovery of hepatocellular carcinoma. Hepatology 57, 2072–2077.

Watson, J.T., Sparkman, O.D., 2007. Introduction to mass spectrometry: instrumentation, applications, and strategies for data interpretation. John Wiley & Sons, Chichester, NY.

Weckwerth, W., 2003. Metabolomics in systems biology. Annu. Rev. Plant Biol. 54, 669–689.

Wen, H., Yoo, S.S., Kang, J., Kim, H.G., Park, J.-S., Jeong, S., et al., 2010. A new NMR-based metabolomics approach for the diagnosis of biliary tract cancer. J. Hepatol. 52, 228–233.

Whiteaker, J.R., Zhang, H., Eng, J.K., Fang, R., Piening, B.D., Feng, L.-C., et al., 2007. Head-to-head comparison of serum fractionation techniques. J. Proteom. Res. 6, 828–836.

Wong, M.L., Medrano, J.F., 2005. Real-time PCR for mRNA quantitation. Biotechniques 39, 75.

Wood, A.J., Evans, W.E., McLeod, H.L., 2003. Pharmacogenomics—drug disposition, drug targets, and side effects. N. Engl. J. Med. 348, 538–549.

Wu, H., Xue, R., Lu, C., Deng, C., Liu, T., Zeng, H., et al., 2009. Metabolomics study for diagnostic model of oesophageal cancer using gas chromatography/mass spectrometry. J. Chromatogr. B. 877, 3111–3117.

Zhang, A., Sun, H., Wang, X., 2013. Power of metabolomics in biomarker discovery and mining mechanisms of obesity. Obes. Rev. 14, 344–349.

Zhou, J., Thompson, D.K., 2002. Challenges in applying microarrays to environmental studies. Curr. Opin. Biotech. 13, 204–207.

Chapter 2.1.3

Potency Analysis of Cellular Therapies: The Role of Molecular Assays

David F. Stroncek, Ping Jin, Ena Wang, Jiaqiang Ren, Luciano Castellio, Marianna Sabatino and Francesco M. Marincola

Cellular therapies are making a major contribution to the emerging field of biologic therapy. The possibilities for the clinical application of new cellular therapy products are expanding rapidly as is their clinical promise. The diversity and effectiveness of cellular therapies that are now available has encouraged the development of new clinical applications and improved the quality of life of patients. These therapies include adoptive immune therapy utilizing enriched or in vitro manipulated autologous or allogeneic immune cells to treat cancer and viral infections (Brentjens et al., 2011, 2013; Grupp et al., 2013; Kalos et al., 2011; Kochenderfer et al., 2010, 2012, 2013; Leen et al., 2006; Porter et al., 2011; Powell et al., 2006); β islet cell transplantation (Shapiro et al., 2000), hematopoietic stem cells (HSCs) for transplantation; HSC therapy for cardiac ischemia (Kinnaird et al., 2004; Nanjundappa et al., 2007), gene therapy (Ariga, 2006; Le and Ringden, 2006; Corsten and Shah, 2008), and mesenchymal stem cell (MSC) therapy (Caplan, 2007). The ability to produce large quantities of biological products with predictable quality and quantifiable potency is critical to this field.

The complexity of cellular therapies is increasing. For example, the initial adoptive immune therapy protocols to treat cancer once only involved the administration of autologous tumor infiltrating leukocytes (TILs) (Rosenberg et al., 1986) or leukocyte activated killer cells (LAK) (Rosenberg et al., 1987). Now adoptive immune therapy protocols are combination therapies that include high-dose chemotherapy, the administration of in vitro activated and primed TIL, and autologous HSCs (Wrzesinski et al., 2007). Immunosuppressive chemotherapy depletes the patient's naturally occurring repertoire of lymphocytes, including T regulatory cells and myeloid-derived suppressor cells (MDSCs). The lack of T regulatory cells and MDSCs and increased levels of cytokines, including interleukin-7 (IL-7), that are associated with leukopenia allow for the rapid and marked in vivo expansion of TIL administered with HSCs (Wrzesinski et al., 2007).

Adoptive cellular therapy with genetically engineered T cells is becoming an important cancer therapy. This therapy involves the genetic engineering of autologous T cells to express high-affinity T-cell receptors (TCRs) or chimeric antigen receptors (CARs) that are reactive with tumor antigens or cancer testis antigens. T cells engineered to express TCRs reactive with melanoma antigens MAGE1 and gp100 are being used to treat metastatic melanoma (Morgan et al., 2006), and those expressing TCRs reactive with the cancer testis antigen NY-ESO-1 are being used to treat melanoma and sarcoma (Robbins et al., 2011). Autologous T cells engineered to express CAR which consist of the single chain fragment variable (scFv) of an antibody directed to a tumor antigen, the TCR CD3 zeta chain, and a costimulator molecule such as CD28 or 4-1BB ligand are being used to treat patients with cancer, leukemia, and lymphoma. CAR T cells specific for CD19 antigen have shown promising results in treating patients with acute lymphocytic leukemia (ALL), chronic lymphocytic leukemia, and B-cell lymphoma (Brentjens et al., 2011, 2013; Grupp et al., 2013; Kalos et al., 2011; Kochenderfer et al., 2010, 2012, 2013; Porter et al., 2011). The methods used to product T cells expressing CARs and high-affinity TCRs are similar. They require T-cell collection by apheresis from the blood of the patient to be treated, T-cell stimulation, transduction with a retro- or lentiviral vector, and expansion (Tumaini et al., 2013). The effectiveness of these adoptive T-cell therapies is dependent on in vivo proliferation and persistence of the genetically engineered T cells (Fry and Mackall, 2013).

Similarly, HSC transplants have become more complex. While HSC transplants have been used successfully to treat leukemia for more than 35 years, this therapy has been constantly evolving. These changes involve tailoring and optimizing its efficacy by using HSC transplantation in combination with immune therapy to treat leukemia. Either manipulated or unmanipulated lymphocytes from HSC transplant donors are often administered to recipients following transplantation to

Principles of Translational Science in Medicine. http://dx.doi.org/10.1016/B978-0-12-800687-0.00004-9

prevent or treat disease relapse and enhance immune recovery (Fowler et al., 2006; Montero et al., 2006). T cells specific for viruses such as cytomegalovirus (CMV), Epstein–Barr virus (EBV) and parvovirus are being expanded and used to prevent or treat infections following HSC transplantation (Gerdemann et al., 2013; Leen et al., 2013).

Regenerative medicine is becoming an important new part of cellular therapy. Bone marrow stromal cells (BMSCs), which are also known as mesenchymal stromal cells or MSCs, are now being widely used in tissue engineering and regenerative medicine applications. BMSCs are found in the marrow where they support hematopoiesis. They can also differentiate into osteoblasts, chondrocytes, and adipocytes. BMSCs can be collected from the marrow but must be expanded ex vivo in order to obtain sufficient quantities for effective therapy. When BMSCs are administered with an appropriate scaffold, they form bone, and they are being used to repair bone (Bruder et al., 1998; Marcacci et al., 2007; Quarto et al., 2001). BMSCs also secrete a number of cytokines, chemokines, and growth factors that confer onto BMSCs angiogenic, anti-inflammatory, and immune modulatory properties (Elman et al., 2014; Nuschke, 2014; Stagg and Galipeau, 2013). These characteristics have led to clinical trials of BMSCs for the treatment of left ventricular failure due to ischemic heart disease (Zhou et al., 2009), inflammatory bowel disease (Forbes et al., 2014), and acute graft versus host disease following HSC transplantation (Le et al., 2008; Ringden et al., 2006). The quality of BMSC products is dependent on a number of factors, including the degree of expansion, composition of the culture media and media supplements, and age of the donor. There is also considerable interdonor variability in BMSCs (Ren et al., 2013).

Pluripotent embryonic and induced pluripotent stem cells (iPSCs) hold even more potential for regenerative medicine. Since iPSCs can be produced from easily obtained material such as skin fibroblasts or CD34+ cells isolated from the blood (Merling et al., 2013), intense focus has been placed on using iPSCs for regenerative therapies. However, the production of iPSCs is complex requiring reprograming with four to five factors (Takahashi and Yamanaka, 2006), initial expansion on feeder cells, followed by expansion on extracellular matrix and prolonged culture. The characteristics of the iPSCs can be affected by the starting cellular material, the reprograming methods, and the culture conditions. For most clinical applications, iPSCs will likely be differentiated before they are administered. While methods for the differentiation of iPSCs into many cell types have been described, the differentiation is often not complete and is dependent on the starting material used for the production of the iPSCs.

Many of these clinical cellular therapy products require cell mobilization, collection, subset isolation, in vitro or in vivo stimulation, and culture of cells over a period of several days. The production of some cellular therapies involves serial isolation steps and multiple stimulation and/or culturing steps and gene transfer. Cellular therapy product manufacture is further complicated by donor or patient genetic and physiological heterogeneity. The final product is often markedly different from the starting material. Because of the complex nature of producing cellular therapy products and the clinical importance of the final products, most institutes conducting cellular therapy have developed specialized good manufacturing product (GMP) laboratories devoted to the production of these therapeutic agents. The goal of these cell processing laboratories is to produce cellular therapy products that provide the desired clinical effect without resulting in adverse effects. These specialized laboratories ensure that an adequate dose of cells is provided to each patient, each product meets release specifications, and lot-to-lot variation is minimized. In order to produce consistently high-quality products, quality assurance has become a critical part of cellular therapy laboratories.

All cellular therapy products must be demonstrated to be safe, pure, potent, stable, and effective for human use. Objective standards based on clinical trial and manufacturing data should be established to evaluate safety and quality characteristics of clinical products during production and at the time of lot release. Also known as product specifications, these standards are intended to ensure that cellular products consistently meet regulatory and industry requirements for sterility, safety, purity, identity, and potency. Tests to measure and evaluate these parameters are performed at critical steps in the manufacturing process (in-process testing) and at the end of production prior to the release of the product for clinical use (lot release testing). The results of in-process and lot release assays should fall within specified ranges and meet predetermined acceptance criteria before the product can be released for human use. In-process testing and lot release testing are important for assuring individual product quality as well as lot-to-lot consistency. For cellular therapies, these assays include tests of sterility, including mycoplasma, viability and identity, and assessment of product potency. One of the most important aspects of assessing the quality of cellular therapy products is to ensure that all products meet established minimal levels or ranges of potency and that potency levels are consistent across manufacturing lots.

POTENCY TESTING

Potency testing involves the quantitative measure of biological activity of a product. The biological activity describes the ability of a product to achieve a defined biological effect. Potency testing is the quantitative measure of a biological activity that is linked to relevant biological properties of a product. The biological activity measured should be closely related to the product's intended biological effect, and ideally it should be related to the product's clinical response (Cellular, Tissue, and Gene Therapies Advisory Committee, 2006; Committee for Medicinal Products for Human Use, 2006; International Conference on Harmonization of Technical Requirements for Registration of Pharmaceuticals for Human Use, 1999).

Potency assessments are meant to measure a cellular therapy product's critical biological activity within a complex mixture by quantifying the product's activity in a biological system. Measurement of the potency of a product is not the same as measuring clinical efficacy, but rather a means to control product consistency. Generally, potency testing is performed at the time of product lot release and across all production lots.

Since potency assays for cellular products usually take a considerable amount of time to develop, generally, the development of potency assays is progressive. The development of potency assays usually begins during preclinical and early clinical development. Development starts with identifying the critical biological activity of the product and formulation of an approach to potency determination. A potency assay should be validated prior to phase III clinical trials (Cellular, Tissue, and Gene Therapies Advisory Committee, 2006; Committee for Medicinal Products for Human Use, 2006).

COMPLEXITIES ASSOCIATED WITH POTENCY TESTING OF CELLULAR THERAPIES

Potency testing of cellular therapies is particularly challenging for several reasons (Table 2.5). First, since most cellular therapies are patient-specific, there is usually a limited quantity of suitable source material and, therefore, a limited amount of final ready-to-administer biologic material to use for lot release and potency testing. The starting materials for most cellular therapies are cells collected from human subjects. The subjects may be the persons being treated (autologous products) or living donors (allogeneic products). For both situations the quantity of starting material that can be collected is limited, and consequently, the amount of material produced is limited. As a result an entire production lot of a cellular therapy is usually administered to a single patient, and the use of large quantities of the product for lot release testing may adversely affect the dose and clinical effectiveness of the product. This limitation on the quantity of material available prevents the use of some assays and/or limits the number of analytes that can be tested.

Second, the time to test the product is limited since cellular therapy products must be tested at the time production is complete but prior to being released for clinical use. This schedule is particularly problematic for cellular therapies since the potency of most living cells is affected by prolonged storage at physiological temperatures. In fact, some products must be administered within hours upon production completion. In addition, handling affects the potency of some products.

Most potency assays require reference preparations with an established potency that are used as assay standards (Committee for Medicinal Products for Human Use, 2006). The limited availability of reference standards complicates potency testing for cellular therapies. Often "in-house" reference standards must be developed. When reference standards are commercially available, they may be expensive.

Finally, cellular therapy products typically show a large degree of lot-to-lot variability. Product variability is due in part to inherent variability in the starting cells or tissues. Donor genetic factors likely contribute to differences in potency of the final cellular therapy product. Genetic polymorphisms in cytokines, growth factors, and their receptors affect the cellular immune response (Hollegaard and Bidwell, 2006; Howell and Rose-Zerilli, 2006; Howell et al., 2001; McCarron et al., 2002). It is likely that these polymorphisms affect the response of cells to cytokine and growth factor stimulation in vitro and the behavior of cells during culture. Epigenetic changes may also be important. The same types of cells obtained from different donors at different time points and under different physiological conditions could vary significantly due to genetic heterogeneities, epigenetic differences, or transcription regulation diversities.

TABLE 2.5 Challenges associated with potency testing of cellular therapies
Limited quantity of final product to test
Limited time to perform lot release testing usually
Limited stability of most cellular therapy products
Limited availability of reference standards
Generally very high variability among lots

TABLE 2.6 Factors contributing to the complex nature of cellular therapies

Variations may exist in the starting cellular material.
Multiple biological products may be used in the manufacturing process.
Multiple steps can be involved in the manufacturing process.
Clinical effectiveness may be dependent on multiple cellular functions.

FACTORS AFFECTING THE POTENCY OF CELLULAR THERAPIES

Despite the difficulties associated with potency testing of cellular therapies, potency is particularly important for these products since the complexities associated with their production can result in considerable differences in potency among different lots of the same product (Table 2.6). These differences are related to the multiple steps required to produce most cellular therapies, variations in starting materials, limited stability of the final product, complex mechanisms of action of the product, and genetic differences among individuals donating the starting cells.

Advanced cellular therapies may incorporate multiple components. For example, cellular products used for cancer vaccines may require more than one peptide to educate immune cells in vitro, followed by cytokine stimulation. Genetically engineered T cells require gene transfer followed by ex vivo cytokine-stimulated expansion. A manipulated lymphocyte component prepared from an HSC transplant donor may involve isolating and recombining multiple different types of cells. Regenerative therapies require the reprograming of somatic cells to produce iPSCs followed by ex vivo expansion and differentiation. The multiple cell types present in many cellular therapies have the potential to interfere with one another or to act synergistically.

Many cellular therapies are subject to extensive manipulation, including manufacturing processes such as cytokine, growth factor, or antigen stimulation; culture; expansion; and treatment with vectors or toxins. For these products, slight variations in the starting cellular material, reagents, processing methods, or culture conditions may result in significant variation in the final product, leading to heterogeneous clinical outcomes of the same therapies. Variability in starting materials is particularly problematic. Since many cellular therapies make use of autologous cells as starting material, these cells are often affected by the patient's disease or treatments.

Finally, the in vivo function of most cellular therapies is dependent on multiple factors in the host environment. HSCs must travel to specific sites, expand, and differentiate into several mature cell types. Immune therapies must migrate from the site of administration, interact with tumor or other immune cells, and respond to stimuli and/or stimulate other cells. iPSCs and iPSC-derived cells must differentiate into the desired cell type but not create unwanted or abnormal cells in vivo.

MEASURING POTENCY OF CELLULAR THERAPIES

Potency can be tested in a number of ways including in vivo and in vitro systems (Table 2.7). Testing potency using in vivo animal models is generally preferred over in vitro test systems, since animal model assays have the ability to directly measure a product's functional activity. However, existing animal models may not be relevant, and new animal models may be difficult to develop (Committee for Medicinal Products for Human Use, 2006). In addition, the results of in vivo tests are often variable and difficult to reproduce. Furthermore, these assays usually take a considerable amount of time to complete, making it difficult to use these assays for routine lot release testing. Many in vivo assays are best suited for use in product development, as an in-process control, or to evaluate the potential effect of changes in the manufacturing process or materials (Cellular, Tissue, and Gene Therapies Advisory Committee, 2006).

In vitro assays involve the measurement of biochemical or physiological responses at the cellular level (Committee for Medicinal Products for Human Use, 2006). The in vitro measurement of cell surface markers, activation markers, secretion of factors, or protein expression does not directly measure the function of a cellular product; however, these assays have been used as surrogates for potency. When an in vitro assay is used as a surrogate for potency, a correlation should be demonstrated between the assay results and the intended biological activity. Typical in vitro assays used as surrogates for potency testing include ELISA, ELISPOT, flow cytometry, proteomic analysis, and cytotoxicity assays.

When the mechanism of action of a cellular therapy can be attributed to the expression of specific cell surface antigens, the measurement of antigens by flow cytometry can be used as an in vitro potency assay. In fact, the measurement of biomarkers by flow cytometry is often used as a surrogate measure of cell potency. Flow cytometry is useful due to the large

TABLE 2.7 Assays used for potency testing

In vivo assays
• Animal models
In vitro assays
• Cell-based assay systems
• ELISA
• Flow cytometry
• ELISPOT
• Proteomics
• Quantitative real-time PCR

number of reagents and assays that are available as well as the relatively quick turnaround time. It can be used to measure the expression of cell surface markers, viability, and the production of cytokines. Extensive analysis of cell surface markers using flow cytometry has been used to assess cellular therapies, but the maximum number of markers that can be analyzed is limited by the availability of specific antibodies, instrumental detection limits, and final product quantity. In addition, the markers that might be most useful may not be known.

In vitro cell function assays have also been used to measure cell potency. Cytotoxicity assays are sometimes used to reflect the function of adoptive immune therapies. Cytokine release by stimulated cells can also be used to measure cell function. However, cell function assays have many limitations. While they may be able to detect differences in relevant biological activity, these assays are typically highly specialized for each cell type, labor intensive, and require highly skilled staff. Different types of cells and cell subsets require completely different types of cell function technologies. Many cells require the measurement of multiple functions to adequately assess potency. Furthermore, the function(s) that best predict cell potency may not be known. In fact, for many cellular therapies, all aspects that contribute to in vivo activity are not completely understood. In addition to these limitations, many cell processing laboratories working with cellular therapies in phase I and II clinical trials prepare several different types of cellular therapies. It is possible, but may not always be feasible, for a centralized cell processing laboratory to perform several different types of cell function assays.

Cell counts and viability measurements are often performed on cellular therapies. However, since these assays do not measure a relevant biological activity, they are not potency measures (Cellular, Tissue, and Gene Therapies Advisory Committee, 2006; Committee for Medicinal Products for Human Use, 2006).

GENE EXPRESSION ARRAYS FOR POTENCY TESTING

Measurements of the expression of genes related to a specific cellular activity or function could be used as an in vitro biomarker of potency. Quantitative real-time PCR assays are useful tools for assessing the expression of individual genes in order to assess the activity of immune cells. The measurement of changes in interferon gamma transcription by quantitative real-time PCR has been used to as a marker for T-cell activation following stimulation with a recall antigen (Kammula et al., 2000; Panelli et al., 2002; Provenzano et al., 2002, 2003). Quantitative real-time PCR has recently been used to measure the production of mRNA encoding interferon-γ, IL-2, IL-4, and IL-10 by stimulated T cells (Provenzano et al., 2003). Quantitative real-time PCR assays are also available to assess angiogenesis, apoptosis, cell cycle, insulin signaling pathways, cytokines and receptors, nitric oxide signaling pathways, and JAK/STAT signaling pathways.

While using quantitative real-time PCR to measure the expression of single genes or groups of genes is helpful in assessing cell function, the complete assessment of the function of cellular therapies requires the measurement of a broad range of gene transcripts, especially when the mechanisms responsible for effective therapy are not thoroughly understood. The analysis of cells using gene expression microarrays allows the simultaneous assessment of the expression of thousands of genes. One practical advantage of gene expression microarray assays over other analytical assays is that very few cells are needed. Enough RNA can be isolated from 1×10^4 to 1×10^6 cells for analysis with a 17,500 gene cDNA expression microarray (Wang et al., 2000).

Microarrays with 40,000 genes or oligonucleotide probes have been used clinically to characterize lymphomas (Dave et al., 2006), prostate cancer (Halvorsen et al., 2005), ovarian cancer (Wang et al., 2005), small cell lung cancer (Taniwaki et al., 2006), melanoma (Basil et al., 2006), and many other cancers. We have used cDNA gene expression microarrays with 17,500–40,000 genes to investigate the immunologic changes associated with high-dose IL-2 therapy for renal cell carcinoma (Panelli et al., 2004) and imiquimod (a TLR-7 ligand) therapy for basal cell carcinoma (Panelli et al., 2007). We have also used cDNA microarrays to assess the effects of IL-10 on NK cells (Mocellin et al., 2004; Nagorsen et al., 2005; Stroncek et al., 2005) and several different types of interferon on lipopolysaccharide (LPS)-stimulated mononuclear cells, the in vitro response of mononuclear cells to IL-2 (Jin et al., 2006), and the molecular basis of cutaneous wound healing (Deonarine et al., 2007).

While gene expression microarrays have been widely used to assess changes in cells in response to stimuli, or to classify different types of cancers, they have been used only to a limited extent to assess cell potency. However, since gene expression microarrays simultaneously measure the expression of thousands of genes, they capture a snapshot of all possible gene expression signatures that are associated with cellular function and hence could be a very important tool for assessing the potency of cellular therapies. The comprehensive nature of gene expression microarray analysis makes them ideal for measuring both expected and unexpected cell functions. This is particularly important for the analysis of cells with complex and multiple critical functions such as dendritic cells (DCs), cytotoxic T cells, embryonic stem cells, HSCs, and BMSCs.

In addition to assessing potency, gene expression microarrays can also assess other important aspects of cellular therapy products such as stability, purity, comparability, maturation, and differentiation status. Since microarrays can detect the activation of apoptosis pathways that signal the onset of cell death, they have the potential to provide useful information concerning the effects of storage or manipulation on cell viability. The assessment of the expression of apoptosis genes is likely to be much more sensitive in assessing cell viability than dye exclusion assays or the flow cytometric measurement of Fas or annexin. Gene expression profiles can also detect the subpopulation of cells and therefore provide information concerning cell purity.

There are some limitations concerning the use of gene expression microarrays for potency testing. Gene expression microarray analysis involves multiple steps, including RNA isolation, amplification, fluorescent labeling, hybridization, and data analysis. It is impossible at the current technology stage to complete the whole procedure within a few hours, so these global expression microarrays cannot yet be used for lot release testing. However, if global microarrays can identify specific sets of genes whose expression is associated with potency, tailored chips, or quantitative real-time PCR kits that assess only specific "potency genes" could be developed and used for lot release testing. In addition, a recently developed platform allows the measurement of gene expression without amplification (Beard et al., 2013). This platform can measure the expression of more than 90 genes in a single assay. Next-generation DNA/RNA sequencing platforms allow for RNA sequencing and the accurate measurement of the expression of the entire transcriptome.

POTENTIAL APPLICATIONS OF GENE EXPRESSION PROFILING FOR POTENCY TESTING

Predicting the Confluence of Human Embryonic Kidney 293 Cells

Gene expression microarrays have been demonstrated to be useful for some cell therapy applications. They can be used to predict the quality of cells used to manufacture biologic products. Human embryonic kidney (HEK) 293 cells are often used to manufacture products such as adenoviral gene therapy vectors and vaccines (Han et al., 2006). These cells can be grown in bioreactors, tissue culture flasks, and roller bottles. However, when HEK 293 cells grow to form a confluent monolayer, their phenotype changes, as does the quality of the vector or vaccine produced by these cells. Cell confluence can be readily assessed by visual inspection of cells grown in flasks and roller bottles, but for cells grown in bioreactors, the assessment of confluence by visual inspection is not always possible. Gene expression profiling has been used to identify genes whose expression predicts cell confluence (Han et al., 2006). HEK 293 cells that have been grown to 90% confluence have a unique gene expression signature compared to those grown to 40% confluence. A set of 37 of these signature genes is able to predict that quality and confluence of HEK 293 cells. While this use of gene expression profiling does not represent a potency assay, it demonstrates the potential of the use of gene expression profile assays.

Cell Differentiation Status Analysis of Embryonic Stem Cells

Human embryonic stem cells (hESC) have the potential to be useful for a number of clinical applications. Since cultured hESC may undergo spontaneous differentiation, it is important to determine if cultured hESC have maintained their stem cell qualities or if they have begun to acquire properties of more differentiated cells. Gene expression profiling may be useful for assessing cultured hESC. Gene expression profiling has been used to identify genes that are uniquely expressed by hESC (Player et al., 2006). Player et al. have found that 1,715 genes were differentially expressed between hESC and differentiated embryonic cells (Player et al., 2006). The analysis of the expression of genes that are expressed by hESC but not by differentiated cells is likely to be useful in determining if cells in culture have maintained their embryonic stem cell characteristics.

Embryonic stem cells must be differentiated before they can be used clinically. One of the first steps in the differentiation of hESC into mature cells and tissues for clinical use is the production of embryoid bodies (hEB). The production of embryiod bodies involves the aggregation of embryonic cells but the prevention of separation of cells into germ lines by plating them onto a nonpermissive substrate. After these hEB are isolated, they can be induced to generate several different types of cells including hematopoietic, neuronal, myogenic, and cardiac muscle cells. A comparison of gene expression profiles of hESC and hEB has found that the expression of several genes was downregulated and several were upregulated, including 194 whose expression was more than threefold greater in hEB (Bhattacharya et al., 2005). This unique set of genes should also be useful in assessing hESC differentiation.

Potency Testing of Hematopoietic Stem Cells

HSCs are widely used for several clinical applications, and better potency assays for these therapies are needed. Potency assays for HSC products used for transplantation should measure the ability of the product to reconstitute bone marrow hematopoietic cells and peripheral blood cells in the transplant recipient. The potency assay should reflect the period of time that neutrophil, platelet, and red blood cell counts return to and remain above specified levels independent of transfusion therapy. In other words, if the potency assay indicates that a product meets minimum criteria, the therapy should result in at least minimum acceptable neutrophil, platelet, and RBC counts in the recipient for a minimum specified duration of time.

Liquid culture of long-term culture initiating cells (LTC-IC) and the repopulation of marrow in nonobese diabetic (NOD)/severe combined immunodeficiency (SCID) mice assays are considered to be the best measures of the quantity and quality of HSCs. However, these assays require several weeks to complete, highly specialized reagents, and highly trained staff. As a result, these assays have seldom, if ever, been used as potency assays.

The measurement of myeloid, erythroid, and mixed-colony formation in methylcellulose culture systems has been the standard method for assessing bone marrow and peripheral blood stem cell (PBSC) concentrates, but they have been used mainly as in-process controls. The measurement of colony formation in methyl cellulose is an effective biological assay that directly measures a relevant function of HSCs; however, these assays take approximately 14 days to complete, and consequently, they cannot be used as potency assays.

Traditionally, total nucleated cells counts were used to assess the potency of bone marrow and are still used as a measure of potency of Umbilical cord blood (UCB) components prepared for transplantation. Regulations suggest that UCB components contain $\geq 90 \times 10^7$ total nucleated cells including nucleated RBC and that $\geq 85\%$ of nucleated cells are viable (Health Resources and Services Administration, 2006). However, the measurement of CD34+ cells by flow cytometry has become the universal assay for measuring the potency of HSC products collected by apheresis from subjects treated with hematopoietic growth factors. The number of CD34+ cells in an HSC product can be measured within a few hours using anti-CD34 and flow cytometry, and this assay is well suited for lot release testing. Generally, a dose of 1×10^6 CD34+ G-CSF-mobilized PBSCs is considered adequate for an autologous transplant and $3–5 \times 10^6$ for an allogeneic transplant. UCB components must contain $\geq 1.25 \times 10^6$ viable CD34+ cells.

While CD34 antigen expression is widely used as a measure of potency of HPCs collected from the peripheral blood, HSCs expressing CD34 antigen do not represent a homogenous population. Several distinct subpopulations or phenotypes of CD34+ cells have been described (Tjonnfjord et al., 1994). Some subpopulations are more primitive, whereas others are more likely to differentiate into myeloid cells, erythroid cells, or megakaryocytes.

Despite the heterogeneity of CD34+ cells, the measurement of CD34+ cells has been an effective measurement of potency of PBSC concentrates collected by apheresis. The reason is likely that PBSC components are relatively similar in that almost all PBSC components are collected from subjects given granulocyte colony-stimulating factor (G-CSF) alone or in combination with chemotherapy. However, the sources of HSCs and types of mobilizing agents used for transplantation are changing. UCB components are being used in place of PBSC concentrates and marrow for unrelated donor HSC transplantation. A new stem cell mobilizing agent, plerixafor, is being used with G-CSF to mobilize stem cells for autologous transplants (Flomenberg et al., 2005) and will likely soon be used for allogeneic donor transplants (Burroughs et al., 2005; Larochelle et al., 2006). CD34+ cells from both UCB and plerixafor-mobilized PBSC concentrates differ from those found in G-CSF-mobilized PBSC concentrates, and the quantity of CD34+ required for a successful transplant from some of these types of products will likely differ from the quantity required for a successful G-CSF-mobilized PBSC transplant.

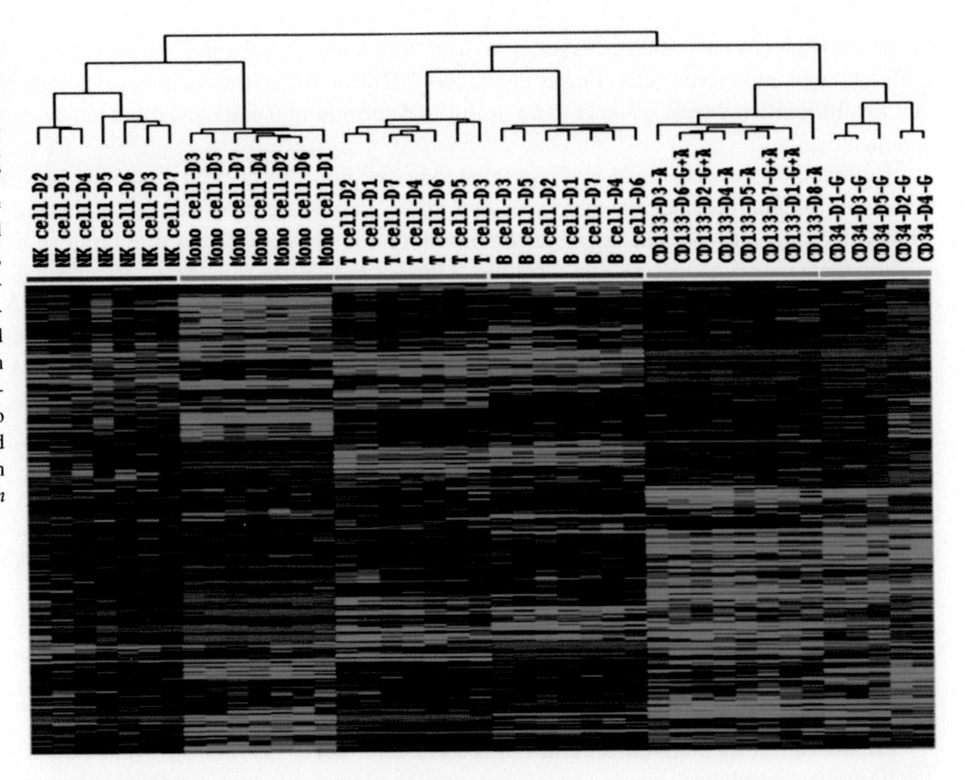

FIGURE 2.7 **Gene expression profiles of hematopoietic stem cells (HSCs) and peripheral blood leukocytes (PBLs).** cDNA was isolated from T cells, B cells, monocytes, and NK cells from seven subjects, G-CSF (G)-mobilized CD34+ cells from five subjects, AMD3100 (A)-mobilized CD133+ cells from four subjects, and AMD3100 plus G-CSF (A+G)-mobilized CD133+ cells from four subjects. cDNA expression was analyzed using an expression microarray with 17,500 cDNA. Unsupervised hierarchical clustering of Eisen was used to analyze the 11,023 genes that remained after filtering (cDNA expressed in ≥ 80% of samples). *(Adapted from Jin et al., 2008.)*

Plerixafor mobilizes stem cells by a different mechanism than G-CSF. Plerixafor is a CXCR4 antagonist, and it mobilizes HSCs within 6 hours by disrupting the binding of stem cell CXCR4 with SDF-1, CXCL12, on marrow osteoblasts (Nervi et al., 2006). In contrast, G-CSF mobilizes stem cells indirectly by downregulating the expression of SDF-1 on marrow osteoblasts and by releasing neutrophil and monocyte proteolytic enzymes including neutrophil elastase, cathepsin G, and matrix metalloproteinase-9 that degrade important HPC trafficking and adhesion molecules c-kit, VCAM-1, CXCR4, and SDF-1 (Nervi et al., 2006). Because of the differences in mechanisms of mobilization between plerixafor and G-CSF, plerixafor mobilizes a CD34+ cell population with a greater long-term marrow repopulating capacity and with a different phenotype than G-CSF (Larochelle et al., 2006).

The potency of UCB CD34+ cells also differs from that of G-CSF-mobilized peripheral blood CD34+ cells. The potency of CD34+ cells from UCB as measured by the ability to repopulate NOD SICD mice is greater than the potency of CD34+ cells from bone marrow or G-CSF-mobilized PBSCs (Bhatia et al., 1997; Broxmeyer, 2005; Broxmeyer et al., 2006). In addition, UCB CD34+ cells show increased proliferative capacity compared to bone marrow and G-CSF-mobilized PBSC CD34+ cells in methylcellulose culture (Broxmeyer et al., 1989, 1992, 2006).

Since the potency of CD34+ cells is dependent on the number and subtypes of CD34+ cells and since the sources of HSCs used in transplantation is increasing, new potency assays are needed. A preliminary comparison of CD34+ cells mobilized by G-CSF and G-CSF plus plerixafor has found that these two types of HSCs can be differentiated by gene expression profiling (Figure 2.7) (Fruehauf et al., 2006; Jin et al., 2008). These two types of HSCs were mobilized with two different protocols, but also two different monoclonal antibodies were used to isolate the HSCs, CD34, and CD133 (Jin et al., 2008). It is not certain if the differences noted in the HSCs were due to the mobilization protocols or the antibody used for HSC isolation. In additional studies, CD34+ cells isolated from rhesus macaques given G-CSF or plerixafor were analyzed by global gene expression analysis. We compared to G-CSF-mobilized CD34+ cells; plerixafor-mobilized CD34+ cells were enriched for B-cell, T-cell, and mast cell genes, and G-CSF-mobilized CD34+ cells were enriched for neutrophil and monocyte genes (Donahue et al., 2009).

Potency Testing of Dendritic Cells

DCs are potent professional antigen-presenting cells capable of capturing and processing antigens in order to present peptides to prime T cells (Schmitt et al., 2007). They express both HLA class I and class II molecules and present peptides to CD4+ and CD8+ T cells. They also express costimulatory molecules such as CD80, CD86, CD40, ICAM-1, and LFA-3. For immune therapy, DCs can be generated from PBMCs after GM-CSF and IL-4 stimulation in vitro, or they can be generated by coculturing in vitro with irradiated tumor cells or virus-infected cells, proteins, or peptides. Mature DCs are administered to patients to stimulate cytotoxic T cells in vivo. Immunotherapies with DCs are being used to treat melanoma, renal cell carcinoma, prostate cancer, and leukemia (Schmitt et al., 2007).

Since few DCs are present in the blood, they must be produced from other types of cells. DCs for clinical therapies produced from CD34+ cells are known as plasmacytoid DCs, and those produced from circulating mononuclear cells are known as myeloid-derived DCs. Either mature or immature DCs can be produced. Immature DCs express lower levels of HLA class II antigens and lower levels of costimulatory molecules but higher levels of Fc and mannose receptors. The ability of immature DCs to phagocytosize and process antigens is better than that of mature DCs, but mature DCs present antigens better than immature DCs. While the function of mature and immature DCs differs, it is not possible with standard analytic assays to precisely distinguish the degree of maturation of DCs.

The potency of DCs can be tested by assessing the ability of DCs loaded with antigen to stimulate autologous T cells (Hinz et al., 2006). However, this is difficult because of the low percentage of T cells in most patients that are responsive to tumor antigens. One alternative to overcome the low number of autologous T cells is to generate and expand T-cell clones that respond to specific antigens. Even so, only T cells with the same HLA restriction elements and antigen specificity could be used in a DC potency assay. For example, HLA-A*0201 T-cell clones specific to a melanoma antigen such as Mart I would not be useful for testing DCs prepared from subjects with other HLA types such as HLA-A*03 or other antigens such as CMV pp65. Consequently, separate clones must be developed for each antigen and HLA restriction being studied.

The potency of DCs can be assessed by using test peptides from recall antigens that are able to stimulate memory T-cell responses (Hinz et al., 2006). These antigens include HLA-restricted tetanus toxin, influenza virus, and EBV antigens, since most people have been immunized against these antigens. However, assays using recall antigens do not directly test DCs' ability to present tumor-associated antigens and efficacy to stimulate tumor-specific T cells. So these assays cannot be used as a lot release test for DCs used for cancer therapy, although testing the ability of DCs to present recall antigens and stimulate T cells is useful as an in-process control.

The measurement of DC costimulatory activity has been used to measure the potency of DCs. Costimulation plays a critical role in the induction of antigen-specific immunity. One method to measure costimulation is the mixed lymphocyte culture reaction that is based on the stimulation of responder cells with replication-competent allogeneic DC stimulator cells. However, it is not known to what degree alloreactivity and costimulation contribute to T-cell stimulation.

Alternatively, gene expression profiling is likely to be useful in assessing the potency of DCs used for clinical therapies. It has been used to characterize the differentiation of monocytes into macrophages and their polarization to macrophages with a type 1 or type 2 phenotype (Martinez et al., 2006). It has also been used to characterize the response of monocytes to LPS and cytokine stimulation (Nagorsen et al., 2005; Stroncek et al., 2005). Preliminary data in our laboratory have also found that gene expression profiling can distinguish monocytes from immature DCs and immature DCs from mature DCs. The ability of gene expression microarrays to assess cells globally may allow them to determine the potency of DCs by evaluating unstimulated cells or cells that have been stimulated with a recall antigen. However, genes whose expression reflects DC maturation as well as specific DC functions must be identified before gene expression profiling can be used as a potency assay for DCs.

One important use of a potency assay is comparability testing, which is an essential part of cellular therapies. While it is desirable to maintain the same manufacturing methods and to use the same materials throughout a clinical trial or as a product moves from early phase to late phase clinical trials and licensure, change is inevitable. There are many reasons that the manufacturing process or reagents and instruments used in manufacturing may change. The supplier of reagents used in the manufacturing process may change, the methods used to produce a reagent may change, or they may stop manufacturing the reagent entirely and a new reagent supplier must be used. Devices used to collect the starting material and containers and bioreactors used for cell culture may also change. In addition, as cellular therapies move from early phase to late phase clinical trials and to licensure, the manufacturing process must be scaled up and possibly out, and this almost always requires some changes to the manufacturing process. It is essential to show that the products manufactured using a new reagent, container, or instrument are comparable to the existing products. Potency assays can be used to show that products are comparable. We have shown that even if the potency of a product is not entirely understood, global gene expression analysis can be used to show that the products are comparable.

We have used gene expressing profiling to show that immature DCs produced from two different types of starting materials were equivalent. The production of clinical cellular therapies often involves multiple centers. For example, cells such as PBMCs are collected from a donor or patient at one center and then shipped to a specialized cellular therapy laboratory where they are processed to produce cells such as DCs. When the processing is complete, the product is shipped to the site where it is administered to a specific patient. The centralized cell processing laboratory may be hundreds or thousands of miles from the collection center, and it may take up to 48 hours after the collection is complete for the starting cellular material to reach the cell processing laboratory.

We wished to determine if immature DCs produced from PBMCs that were stored at 4°C for 48 hours differed from those produced from fresh PBMCs (Shin et al., 2008). Immature DCs were produced from fresh PBMCs and PBMCs from the same donors that were stored for 48 hours. Global gene expression analysis was used to analyze the fresh and stored PBMCs, monocytes isolated from the PBMCs, and immature DCs manufactured from the PBMCs. Hierarchical clustering analysis separated the fresh and stored PBMCs and monocytes into separate groups, but immature DCs prepared from fresh and stored PBMCs were in the same cluster (Figure 2.8). Comparison of genes differentially expressed by fresh compared to stored products (paired *t*-tests, $p < 0.005$) found 273 genes that differed between fresh and stored PBMCs, 711 between elutriated monocytes prepared from fresh and stored PBMCs, but only three differed between immature DCs prepared from elutriated monocytes isolated from fresh PBMCs and immature DCs prepared from elutriated monocytes isolated from stored PBMCs (Shin et al., 2008). These results showed that immature DCs made from PBMCs stored 48 hours did not differ from those prepared from fresh PBMCs and demonstrated that gene expression profiling can be used to show equivalence of DC products manufactured using different methods.

We have also taken a novel approach to identify candidate markers for potency testing of DCs. We assessed the variability of genes expressed by DCs and identified the least and most variable genes in order to determine if they might make good candidates as genes for potency testing (Castiello et al., 2013b). In these studies we manufactured DCs from peripheral blood monocytes using methods and reagents used to manufacture DCs in our clinical GMP manufacturing facility. The monocytes were cultured for 3 days with IL-4 and GM-CSF to produce immature DCs, and they were then cultured for 1 day with LPS and IFN-γ to produce mature DCs. We first assessed the variability of DCs and sources of variability to determine if lot-to-lot differences in DCs might be significant enough to contribute to differences in clinical outcome.

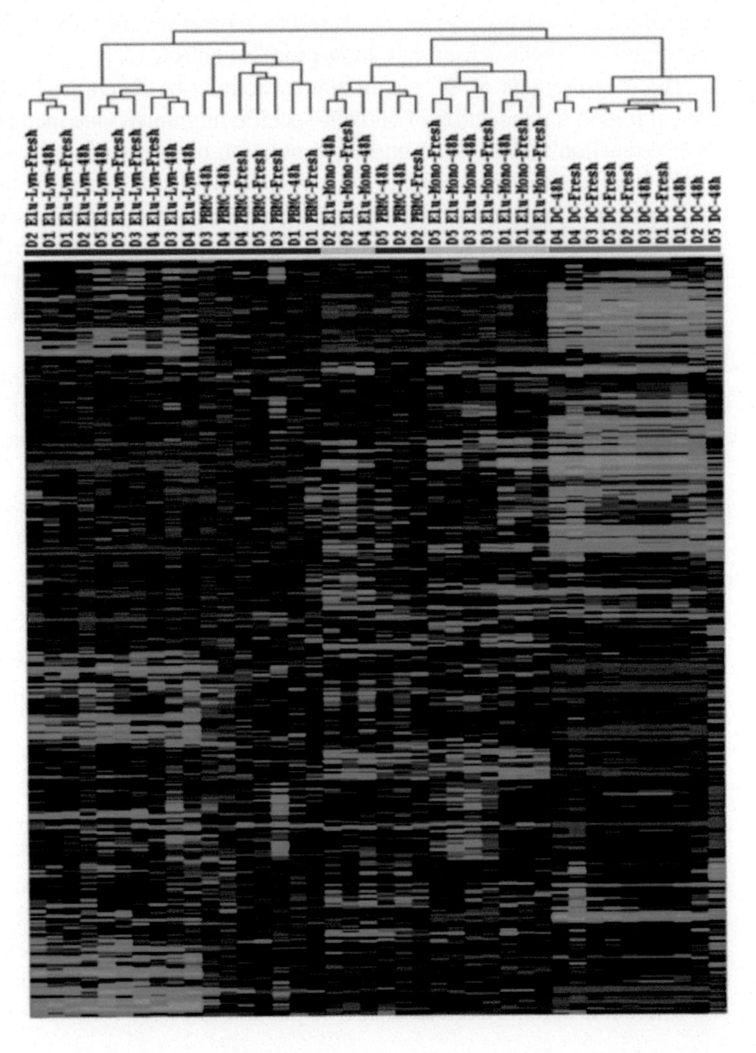

FIGURE 2.8 **Gene expression profiling of fresh and PBMC products stored 48 hours, and elutriated lymphocytes, elutriated monocytes, and immature dendritic cells (DCs) prepared from fresh and stored PBMCs.** Products from five healthy donors were analyzed. The 8,661 genes that remained after filtering (expressed in 80% of samples) were analyzed by unsupervised hierarchical clustering of Eisen. PBMC products are indicated by a blue bar, elutriated lymphocytes by a red bar (Elu-Lym), elutriated monocytes by a yellow bar (Elu-Mono), and immature DCs by a green bar (DC). *(Adapted from Shin et al., 2008.)*

To assess intra-donor variability, we manufactured mature DCs from monocytes collected from the same donor on five different occasions. For the assessment of manufacturing variability, DCs were made five different times using the monocytes obtained from a single apheresis collection. To assess inter-donor variability, we manufactured DCs from nine different donors. The interclass correlation coefficient (ICC) of genes from each sample was calculated to assess the degree that each of these sources of variability effected mature DCs. We found that manufacturing variability was less than intra-donor variability and inter-donor variability was the greatest. These results show variability in the starting material; both intra- and inter-donor variability affect the final DC product, and these factors were more important than manufacturing variability.

We next looked for quality markers for mature DCs (Castiello et al., 2013b). We hypothesized that good markers for assessing the quality of manufactured DCs would be genes whose expression significantly differed by a large degree among monocytes, immature DCs, and mature DCs. Using the global gene expression data, we identified all genes whose average expression differed among mature DCs and monocytes by > fivefold and whose difference in expression was highly significant; false discovery rate (FDR) < 0.005 and $p < 0.001$. Similarly, we identified all genes whose average expression differed among mature DCs and immature DCs by > 5, FDR < 0.005, and $p < 0.001$. Using these criteria, we identified 1,186 genes whose expression differed among mature DCs and monocytes and 647 whose expression differed among mature DCs and immature DCs. Among these two sets of genes, 323 were common to both sets. Among these genes, we selected 291 with good assay reproducibility. We then assessed the variability of each of these 291 genes among mature DCs in order to identify those with the greatest index of variability, which would likely make good markers for identity and potency testing, and those with the least index of variability, which may be good potency markers. Among the 29 least variable genes were AIM2, FEM1C, APOL1, NUB1, IMAZ, DRAM1, AK4, IFI27, WARS, PSME2, and ICAM1 (CD54), all of which are functionally important immune-related genes (Table 2.8). Among the 29 most variable genes were the functionally important DC genes CD80, CCL1, CCRL1, and CD70 (Table 2.8). Many groups that manufacture DCs use the expression of CD80 as measured by flow cytometry as lot release criteria for mDCs. We are currently working to validate these genes by comparing their expression in mDCs manufactured for clinical therapy and correlating their expression with clinical outcomes.

TABLE 2.8 Dendritic cell genes with the least and greatest index of variability

Least Index of Variability		Greatest Index of Variability	
brain and acute leukemia, cytoplasmic	BAALC	basic helix-loop-helix family, member e22	BHLHE22
apolipoprotein L, 1	APOL1	dynein, axonemal, heavy chain 10	DNAH10
tumor necrosis factor, alpha-induced protein 6	TNFAIP6	neuregulin 2	NRG2
phospholipase A1 member A	PLA1A	annexin A3	ANXA3
Predicted: hypothetical LOC100131733	LOC100131733	endothelin 1	EDN1
hyaluronan and proteoglycan link protein 3	HAPLN3	ladinin 1	LAD1
absent in melanoma 2	AIM2	S100 calcium-binding protein A14	S100A14
tryptophanyl-tRNA synthetase	WARS	cingulin-like 1 (CGNL1), mRNA	CGNL1
MYC-associated zinc finger protein	MAZ	dystrophin	DMD
sorcin	SRI	chemokine (C-C motif) ligand 1	CCL1
myosin, heavy chain 11, smooth muscle	MYH11	serine/threonine-protein kinase NIM1	MGC42105
fem-1 homolog c (*C. elegans*)	FEM1C	CD70 molecule	CD70
insulin-induced gene 2	INSIG2	secreted frizzled-related protein 4	SFRP4
mucolipin 2	MCOLN2	immunoglobulin superfamily, member 3	IGSF3
DNA damage-regulated autophagy modulator 1	DRAM1	neuregulin 2	NRG2

Continued

TABLE 2.8 Dendritic cell genes with the least and greatest index of variability—cont'd

Least Index of Variability		Greatest Index of Variability	
adenylate kinase 4	AK4	guanylate cyclase 2F, retinal	GUCY2F
Serpin peptidase inhibitor, clade G	SERPING1	chemokine (C-C motif) receptor-like 1	CCRL1
AT rich interactive domain 5B (MRF1-like)	ARID5B	CD80 molecule	CD80
negative regulator of ubiquitin-like proteins 1	NUB1	LIM and calponin homology domains 1	LIMCH1
neuralized homolog 3 (*Drosophila*) pseudogene	NEURL3	lymphotoxin alpha (TNF superfamily, member 1)	LTA
interferon induced with helicase C domain 1	IFIH1	erythrocyte membrane protein band 4.1 like 4A	EPB41L4A
BTG family, member 3	BTG3	regulating synaptic membrane exocytosis 2	RIMS2
Intercellular adhesion molecule 1	ICAM1	tissue factor pathway inhibitor 2	TFPI2
interferon, alpha-inducible protein 27	IFI27	synaptopodin 2	SYNPO2
interleukin 7	IL7	cytochrome P450, family 3, subfamily A, polypeptide 4	CYP3A4

Cultured CD4+ Cells

One critical aspect of HSC transplants involving HLA-matched siblings is donor T-cell responses. Donor T-cell responses directed toward recipient cells promote HSC engraftment but can cause GVHD, while donor T-cell responses directed toward leukemia or lymphoma cells can prevent post-transplant disease relapse. Considerable effort has been spent on manipulating HSC grafts in order to minimize donor T-cell-mediated GVHD and maximizing T-cell-mediated antitumor effects. Fowler and colleagues have found that following allogeneic HSC transplantation, the infusion of CD4+ donor T cells that had been cultured with anti-CD3/anti-CD28-coated beads and IL-4 in the presence of the mTor pathway inhibitor rapamycin results in a low incidence of both GVHD and disease relapse in patients with B-cell lymphoma (Fowler et al., 2013). While these so-called T-Rapa cells have shown promising clinical results, their mechanism of action is not certain, and consequently, potency markers have yet to be identified. These cells have been difficult to analyze since rapamycin inhibits many aspects of cell metabolism, which limits the effectiveness of the ELISA and flow cytometry assays, which are standard for assessing T-cell function. We used global gene expression analysis to search for the critical biological functions of T-Rapa cells manufactured over 6 and 12 days, T-Rapa$_6$ and T-Rapa$_{12}$ cells, both of which have been tested in clinical trials. Using conventional analytic methods, cytokine release and surface marker expression, we found that T-Rapa$_6$ and T-Rapa$_{12}$ cells had a similar phenotype. To further compare the potency of these two types of cells, we analyzed both types of cells and input CD4+ cells using global gene expression analysis. When compared to CD4+ cells, 6,641 genes were differentially expressed in T-Rapa$_6$ cells and 6,147 in T-Rapa$_{12}$ cells (Castiello et al., 2013a). The genes that were differentially expressed in the T-Rapa$_6$ and T-Rapa$_{12}$ cells were similar in terms of gene families that were upregulated (cell cycle, stress response, glucose metabolism, and DNA metabolism) and downregulated (inflammatory response, apoptosis, and transcriptional regulation). However, when T-Rapa$_6$ and T-Rapa$_{12}$ cells were compared, 1,994 genes were differentially expressed. These results show that the T-Rapa$_6$ and T-Rapa$_{12}$ cells have similar yet distinct properties, and they may have slightly different clinical effects. Conventional analyses failed to reveal these differences. The clinical trial of T-Rapa$_{12}$ cells has been completed (Fowler et al., 2013), but the trial of T-Rapa$_6$ cells is ongoing, so it is not yet certain if these molecular differences result in difference in in vivo function.

Bone Marrow Stromal Cell

Potency testing is challenging for BMSCs because they are used for many diverse purposes, and the critical properties of BMSCs that contribute most to their clinical effectiveness likely differ among applications. For example, when BMSCs are used for treating peripheral vascular disease or left ventricular failure due to coronary artery disease, the angiogenic and anti-inflammatory properties of BMSCs are likely to be important, but when BMSCs are used to treat steroid-resistant acute GVHD, their immune modulatory functions are likely to be most important. However, it would be desirable to have potency markers that could be used for all applications of BMSCs.

We hypothesized that markers of BMSC senescence would be good for potency testing. When BMSCs undergo prolonged culture, their proliferation slows and finally stops; the cells become senescent. Senescence is also associated with changes in morphology, phenotype, and function. In addition to becoming less proliferative, senescent BMSCs lose their ability to form bone and cartilage, and their immune modulatory properties change (Ren et al., 2013). We believed that gene expression changes associated with senescence could be used as a measure of the quality of BMSCs.

BMSCs from marrow aspirates of seven healthy subjects were isolated and cultured until they became senescent (Ren et al., 2013). The expression of the senescence-associated marker beta galactosidase, stromal cell surface markers, and global gene expression was assessed on BMSCs from serial passages measured on cells from passages 2 through senescence. A gene expression signature that was associated with senescence was identified, and changes in these senescence-associated genes begin before changes in proliferation, colony formation, beta gal expression, and cell surface marker expression. Changes in the BMSC genes expression transcriptome occurred with every passage. However, a change from an early passage transcriptional profile to a later passage transcription profile occurred between passages 4 through 8, while changes in cell proliferation, surface marker expression, and senescence-associated Beta-Gal staining occurred at later passages.

We also used a least-angle regression algorithm to identify the minimum-sized set of genes whose expression could be used to calculate BMSC replicative age as expressed as a percent of maximum time in culture or time to senescence. We found that a set of 24 genes could predict the replicative age of BMSCs (Table 2.9). We have used these genes to assess

TABLE 2.9 Twenty-four genes and coefficients use to calculate a predicted replicative age of expanded BMSCs*

UG Cluster	Symbol	Coefficient
Hs.559718	AK5	0.007
Hs.439145	SCN9A	0.007
Hs.535845	RUNX2	–0.003
Hs.124638	TMEM90B	–0.011
Hs.656071	ADAMTS9	–0.001
Hs.269764	BACH2	–0.014
Hs.646614	KLF8	–0.001
Hs.244940	RDH10	0.038
Hs.483238	ARHGAP29	–0.026
Hs.1407	EDN2	0.008
Hs.131933	PLBD1	–0.068
Hs.369265	IRAK3	–0.009
Hs.148741	RNF144B	0.005
Hs.525093	NDFIP2	0.003
Hs.124299	FAM167A	0.009
Hs.183114	ARHGAP28	0.002
Hs.596680	EYA4	–0.005
Hs.652230	TM7SF4	–0.001
Hs.89640	TEK	0.032
Hs.377894	GCA	–0.017
Hs.439145	SCN9A	0.02
Hs.681802	FLJ00254 protein	0.019
Hs.492974	WISP1	–0.001
Hs.122055	C7orf31	–0.006

**The expression of the genes was measured using a microarray platform, and the predicated replicative age was calculated using the formula: $\sum c_i x_i + 0.143$, where c_i and x_i are the coefficient and gene expression for the ith gene, respectively.*
Adapted from Ren et al. (2013).

clinical BMSCs from 12 production lots manufactured by our clinical cellular and gene therapy laboratory. Unsupervised hierarchical clustering analysis separated the 12 lots into 2 groups; one cluster with 8 lots, which has the lowest predicted replicative ages and one cluster with 4 samples with the greatest predicted replicative age. Interestingly, the 8 samples in the low replicative age cluster all met lot-release criteria, whereas only 1 of 4 samples in the high replicative age cluster met lot-release criteria. Two lots failed due to slow proliferation of the primary culture, and one lot failed due to high expression of CD34 antigen. These results show that the predicted replicative age of BMSCs as measured using the expression of a 24-gene set is useful for assessing BMSC quality.

MicroRNAs AS POTENCY ASSAYS

MicroRNAs (miRNAs) are likely to be important indicators of HSC, MSC, embryonic cells, and immune cell potency. MiRNAs are an abundant class of endogenous non-protein-coding small RNAs of 19–23 nucleotides, which are derived from pre-miRNA of 60–120 nucleotides. Mature miRNAs negatively regulate gene expression at the post-transcriptional level. They reduce the levels of target transcripts as well as the amount of protein encoded. So far, 541 human miRNAs have been identified (Wellcome Trust Sanger Institute, 2007). In general, miRNAs are phylogenetically conserved and, therefore, have conserved and defined post-transcription inhibition function. Some miRNAs are expressed throughout an organism, but most are developmentally expressed or are tissue-specific.

MiRNAs play an important role in many cellular development and metabolic processes, including developmental timing, signal transduction, tissue differentiation, and cell maintenance. Most miRNAs are tissue-specific. For example, the expression of miR-1 is restricted to the heart (Lagos-Quintana et al., 2002) and miR-223 to granulocytes and macrophages (Chen et al., 2004). Recently, miRNA have been found to have a role in stem cell self-renewal and differentiation. Several different miRNAs are involved with the differentiation of hematopoietic progenitor cells. MiR-155 is important in preventing the differentiation of CD34+ cells toward myeloid and erythroid cells (Georgantas et al., 2007). In addition, miR-221 and miR-222 prevent the differentiation of HSCs into erythroid progenitors (Felli et al., 2005). MiR-181 is involved in the control of lymphopoiesis (Chen et al., 2004).

MiRNA seems ideally suited for distinguishing primitive from committed hematopoietic, embryonic, mesenchymal, and other stem cells as well as different types of lymphocytes and mononuclear phagocytes. However, they have not been evaluated to determine if they would be useful in this capacity. MiRNA profiles of mononuclear phagocytes and DCs have not been studied extensively, but if miRNA profiles differ between immature and mature DCs, they may be useful in assessing the potency of DCs produced in vitro.

The high throughput analysis of miRNAs requires at least 10 times greater quantities of cells than gene expression profiling, since miRNAs contribute only about 1% of a cell's total mRNA. MiRNA amplification methods have not yet been fully validated and, hence, are not considered reliable. However, targeted miRNA analysis requires a relatively small number of cells, 1×10^6.

An advantage of miRNA expression profiling compared to gene expression profiling is that miRNA expression profiling requires smaller arrays and chips, which make it possible to analyze multiple samples on the same slides containing subarrays. While gene expression cDNA microarrays contain 10,000–35,000 probes, the number of miRNA currently identified is only in the hundreds.

We have shown that miRNA expression can be used to differentiate HSCs from PBMCs and different types of HSCs (Jin et al., 2008). We used miRNA and gene expression profiling to compare plerixafor-mobilized CD133+ cells, G-CSF-mobilized CD34+ cells, and peripheral blood leukocytes (PBLs). Hierarchical clustering of miRNAs separated HSCs from PBLs (Figure 2.9). MiRNAs upregulated in all HSCs included hematopoiesis-associated miRNA: miR-126, miR-10a, miR-221, and miR-17-92 cluster. MiRNAs upregulated in PBLs included miR-142-3p, -218, -21, and -379. Hierarchical clustering analysis of miRNA expression separated the plerixafor-mobilized CD133+ cells from G-CSF-mobilized CD34+ cells. Gene expression analysis of the HSCs and PBLs also naturally segregated samples according to mobilization and isolation protocol and cell differentiation status (Jin et al., 2008).

CONCLUSIONS

As more and more new cellular therapies are being developed and used to treat an increasing variety of diseases and patients, potency testing is becoming a critical and required part of the production of cellular therapies. Existing assays, such as function, flow cytometry, and ELISA, are important but limited by the number of factors analyzed. Gene and miRNA expression microarrays have the potential to become important in potency testing. They are well suited for the assessment of the potency of cellular therapies in phase I and II clinical trials. As data are collected during clinical trials,

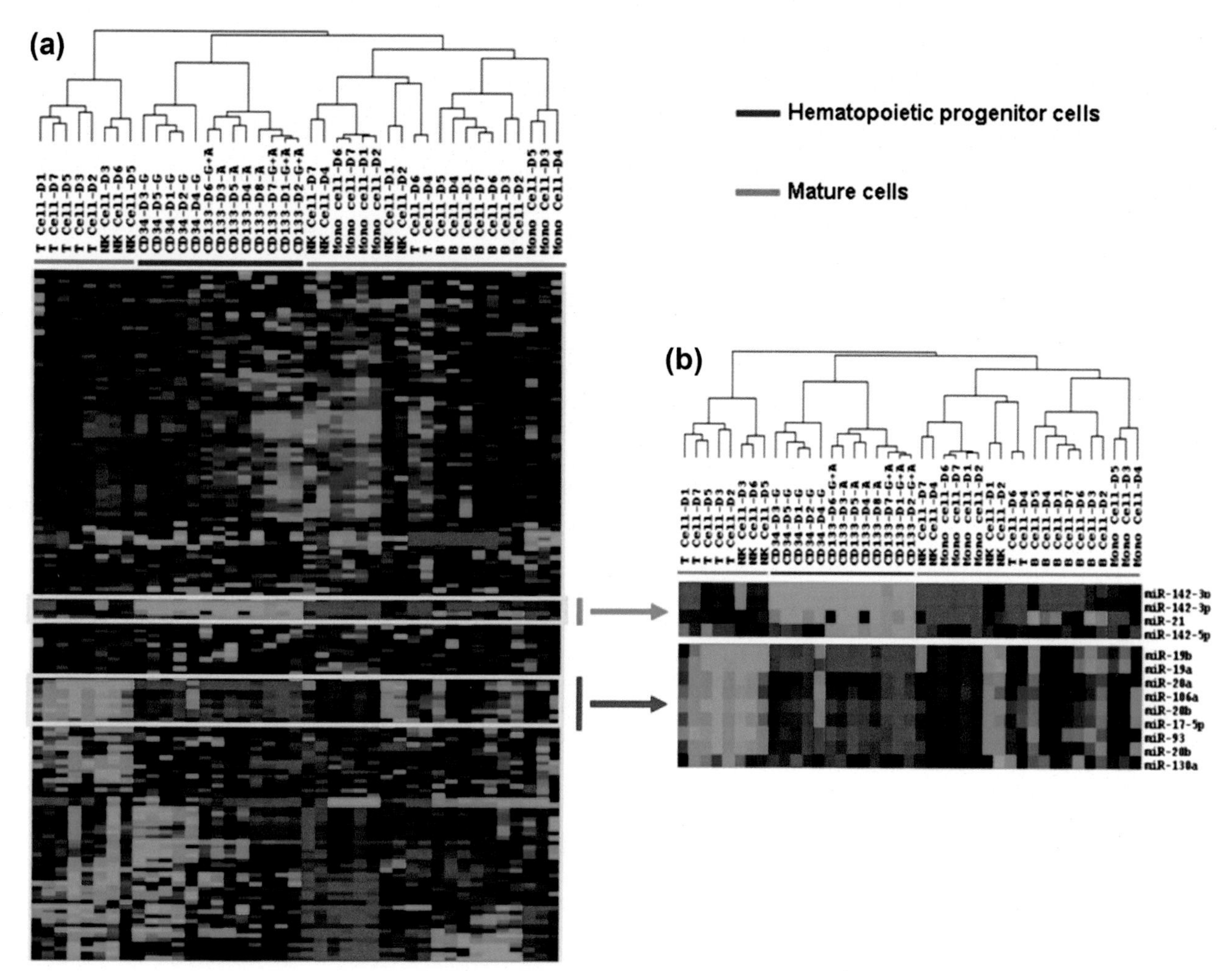

FIGURE 2.9 **MicroRNA (miRNA) expression profiles of hematopoietic stem cells (HSCs) and peripheral blood leukocytes (PBLs).** RNA was isolated from T cells, B cells, monocytes, and NK cells from seven subjects, G-CSF-mobilized CD34+ cells from five subjects, AMD3100 (a)-mobilized CD133+ cells from four subjects, and AMD3100 plus G-CSF (A+G)-mobilized CD133+ cells from four subjects, and miRNA expression was analyzed using an expression array with 457 human miRNAs. (a) Unsupervised hierarchical clustering of Eisen was used to analyze the 148 miRNAs that remained after filtering (miRNA expressed in ≥ 80% of samples). (b) Signature miRNAs whose expression was markedly upregulated in HSCs or PBLs. *(Adapted from Jin et al., 2008.)*

the results of analysis with the gene and miRNA expression microarrays should be compared with the results of traditional function assays and genes whose expression is associated with critical biological function identified and used to develop assays to rapidly measure the expression of genes associated with cell potency.

REFERENCES

Ariga, T., 2006. Gene therapy for primary immunodeficiency diseases: recent progress and misgivings. Curr. Pharm. Des. 12, 549–556.

Basil, C.F., Zhao, Y., Zavaglia, K., Jin, P., Panelli, M.C., Voiculescu, S., et al., 2006. Common cancer biomarkers. Cancer Res. 66, 2953–2961.

Beard, R.E., Abate-Daga, D., Rosati, S.F., Zheng, Z., Wunderlich, J.R., Rosenberg, S.A., et al., 2013. Gene expression profiling using nanostring digital RNA counting to identify potential target antigens for melanoma immunotherapy. Clin. Cancer Res. 19, 4941–4950.

Bhatia, M., Wang, J.C., Kapp, U., Bonnet, D., Dick, J.E., 1997. Purification of primitive human hematopoietic cells capable of repopulating immune-deficient mice. Proc. Natl. Acad. Sci. USA 94, 5320–5325.

Bhattacharya, B., Cai, J., Luo, Y., Miura, T., Mejido, J., Brimble, S.N., et al., 2005. Comparison of the gene expression profile of undifferentiated human embryonic stem cell lines and differentiating embryoid bodies. BMC. Dev. Biol. 5, 22.

Brentjens, R.J., Davila, M.L., Riviere, I., Park, J., Wang, X., Cowell, L.G., et al., 2013. CD19-targeted T cells rapidly induce molecular remissions in adults with chemotherapy-refractory acute lymphoblastic leukemia. Sci. Transl. Med. 5, 177ra38.

Brentjens, R.J., Riviere, I., Park, J.H., Davila, M.L., Wang, X., Stefanski, J., et al., 2011. Safety and persistence of adoptively transferred autologous CD19-targeted T cells in patients with relapsed or chemotherapy refractory B-cell leukemias. Blood 118, 4817–4828.

Broxmeyer, H.E., 2005. Biology of cord blood cells and future prospects for enhanced clinical benefit. Cytotherapy 7, 209–218.

Broxmeyer, H.E., Douglas, G.W., Hangoc, G., Cooper, S., Bard, J., English, D., et al., 1989. Human umbilical cord blood as a potential source of transplantable hematopoietic stem/progenitor cells. Proc. Natl. Acad. Sci. USA 86, 3828–3832.

Broxmeyer, H.E., Hangoc, G., Cooper, S., Ribeiro, R.C., Graves, V., Yoder, M., et al., 1992. Growth characteristics and expansion of human umbilical cord blood and estimation of its potential for transplantation in adults. Proc. Natl. Acad. Sci. USA 89, 4109–4113.

Broxmeyer, H.E., Srour, E., Orschell, C., Ingram, D.A., Cooper, S., Plett, P.A., et al., 2006. Cord blood stem and progenitor cells. Methods Enzymol. 419, 439–473.

Bruder, S.P., Kurth, A.A., Shea, M., Hayes, W.C., Jaiswal, N., Kadiyala, S., 1998. Bone regeneration by implantation of purified, culture-expanded human mesenchymal stem cells. J. Orthop. Res. 16, 155–162.

Burroughs, L., Mielcarek, M., Little, M.T., Bridger, G., Macfarland, R., Fricker, S., LaBrecque, J., et al., 2005. Durable engraftment of AMD3100-mobilized autologous and allogeneic peripheral-blood mononuclear cells in a canine transplantation model. Blood 106, 4002–4008.

Caplan, A.I., 2007. Adult mesenchymal stem cells for tissue engineering versus regenerative medicine. J. Cell. Physiol. 213, 341–347.

Castiello, L., Mossoba, M., Viterbo, A., Sabatino, M., Fellowes, V., Foley, J.E., et al., 2013a. Differential gene expression profile of first-generation and second-generation rapamycin-resistant allogeneic T cells. Cytotherapy 15, 598–609.

Castiello, L., Sabatino, M., Zhao, Y., Tumaini, B., Ren, J., Ping, J., et al., 2013b. Quality controls in cellular immunotherapies: rapid assessment of clinical grade dendritic cells by gene expression profiling. Mol. Ther. 21, 476–484.

Cellular, Tissue, and Gene Therapies Advisory Committee, 2006. Potency measurements for cell and gene therapy products. Cellular, Tissue, and gene Therapies Advisory Committee Meeting #41. 2–9–2006. U.S. Food and Drug Administration. http://www.fda.gov/ohrms/dockets/ac/06/briefing/2006–4205B1_1.htm (accessed 24.06.14.).

Chen, C.Z., Li, L., Lodish, H.F., Bartel, D.P., 2004. MicroRNAs modulate hematopoietic lineage differentiation. Science 303, 83–86.

Committee for Medicinal Products for Human Use (CHMP), 2006. Guideline on potency testing of cell based immunotherapy medicinal products for the treatment of cancer. 10–11–2006. http://www.ema.europa.eu/docs/en_GB/document_library/Scientific_guideline/2009/09/WC500003814.pdf

Corsten, M.F., Shah, K., 2008. Therapeutic stem cells for cancer treatment: hopes and hurdles in tactical warfare. Lancet Oncol. 9, 376–384.

Dave, S.S., Fu, K., Wright, G.W., Lam, L.T., Kluin, P., Boerma, E.J., et al., 2006. Molecular diagnosis of Burkitt's lymphoma. N. Engl. J. Med. 354, 2431–2442.

Deonarine, K., Panelli, M.C., Stashower, M.E., Jin, P., Smith, K., Slade, H.B., et al., 2007. Gene expression profiling of cutaneous wound healing . J. Transl. Med. 5, 11.

Donahue, R.E., Jin, P., Bonifacino, A.C., Metzger, M.E., Ren, J., Wang, E., et al., 2009. Plerixafor (AMD3100) and granulocyte colony-stimulating factor (G-CSF) mobilize different CD34+ cell populations based on global gene and microRNA expression signatures. Blood 114, 2530–2541.

Elman, J.S., Li, M., Wang, F., Gimble, J.M., Parekkadan, B., 2014. A comparison of adipose and bone marrow-derived mesenchymal stromal cell secreted factors in the treatment of systemic inflammation. J. Inflamm. (Lond) 11, 1.

Felli, N., Fontana, L., Pelosi, E., Botta, R., Bonci, D., Facchiano, F., et al., 2005. MicroRNAs 221 and 222 inhibit normal erythropoiesis and erythroleukemic cell growth via kit receptor down-modulation. Proc. Natl. Acad. Sci. USA 102, 18081–18086.

Flomenberg, N., Devine, S.M., Dipersio, J.F., Liesveld, J.L., McCarty, J.M., Rowley, S.D., et al., 2005. The use of AMD3100 plus G-CSF for autologous hematopoietic progenitor cell mobilization is superior to G-CSF alone. Blood 106, 1867–1874.

Forbes, G.M., Sturm, M.J., Leong, R.W., Sparrow, M.P., Segarajasingam, D., Cummins, A.G., et al., 2014. A phase 2 study of allogeneic mesenchymal stromal cells for luminal Crohn's disease refractory to biologic therapy. Clin. Gastroenterol. Hepatol. 12, 64–71.

Fowler, D.H., Mossoba, M.E., Steinberg, S.M., Halverson, D.C., Stroncek, D., Khuu, H.M., et al., 2013. Phase 2 clinical trial of rapamycin-resistant donor CD4+ Th2/Th1 (T-Rapa) cells after low-intensity allogeneic hematopoietic cell transplantation. Blood 121, 2864–2874.

Fowler, D.H., Odom, J., Steinberg, S.M., Chow, C.K., Foley, J., Kogan, Y., et al., 2006. Phase I clinical trial of costimulated, IL-4 polarized donor CD4+ T cells as augmentation of allogeneic hematopoietic cell transplantation. Biol. Blood Marrow Transplant. 12, 1150–1160.

Fruehauf, S., Seeger, T., Maier, P., Li, L., Weinhardt, S., Laufs, S., et al., 2006. The CXCR4 antagonist AMD3100 releases a subset of G-CSF-primed peripheral blood progenitor cells with specific gene expression characteristics. Exp. Hematol. 34, 1052–1059.

Fry, T.J., Mackall, C.L., 2013. T-cell adoptive immunotherapy for acute lymphoblastic leukemia. Hematology. Am. Soc. Hematol. Educ. Program 2013, 348–353.

Georgantas III, R.W., Hildreth, R., Morisot, S., Alder, J., Liu, C.G., Heimfeld, S., et al., 2007. CD34+ hematopoietic stem-progenitor cell microRNA expression and function: a circuit diagram of differentiation control. Proc. Natl. Acad. Sci. USA 104, 2750–2755.

Gerdemann, U., Katari, U.L., Papadopoulou, A., Keirnan, J.M., Craddock, J.A., Liu, H., et al., 2013. Safety and clinical efficacy of rapidly-generated trivirus-directed T cells as treatment for adenovirus, EBV, and CMV infections after allogeneic hematopoietic stem cell. Transplant. Mol. Ther. 21, 2113–2121.

Grupp, S.A., Kalos, M., Barrett, D., Aplenc, R., Porter, D.L., Rheingold, S.R., et al., 2013. Chimeric antigen receptor-modified T cells for acute lymphoid leukemia. N. Engl. J. Med. 368, 1509–1518.

Halvorsen, O.J., Oyan, A.M., Bo, T.H., Olsen, S., Rostad, K., Haukaas, S.A., et al., 2005. Gene expression profiles in prostate cancer: association with patient subgroups and tumour differentiation. Int. J. Oncol. 26, 329–336.

Han, J., Farnsworth, R.L., Tiwari, J.L., Tian, J., Lee, H., Ikonomi, P., et al., 2006. Quality prediction of cell substrate using gene expression profiling. Genomics 87, 552–559.

Health Resources and Services Administration, 2006. Draft guidance for industry; minimally manipulated, unrelated, allogeneic placenta/umbilical cord blood intended for hematopoietic reconstitution in patients with hematological malignancies. 12–1–2006. http://www.fda.gov/downloads/Biologics-BloodVaccines/GuidanceComplianceRegulatoryInformation/Guidances/Blood/UCM187144.pdf.

Hinz, T., Buchholz, C.J., van der Stappen, T., Cichutek, K., Kalinke, U., 2006. Manufacturing and quality control of cell-based tumor vaccines: a scientific and a regulatory perspective. J. ImmunoTher. 29, 472–476.

Hollegaard, M.V., Bidwell, J.L., 2006. Cytokine gene polymorphism in human disease: on-line databases, Supplement 3. Genes Immun. 7, 269–276.

Howell, W.M., Rose-Zerilli, M.J., 2006. Interleukin-10 polymorphisms, cancer susceptibility and prognosis. Fam. Cancer 5, 143–149.

Howell, W.M., Turner, S.J., Bateman, A.C., Theaker, J.M., 2001. IL-10 promoter polymorphisms influence tumour development in cutaneous malignant melanoma. Genes Immun. 2, 25–31.

International Conference on Harmonization of Technical Requirements for Registration of Pharmaceuticals for Human Use, 1999. Specifications: test procedures and acceptance criteria for biotechnological/biological products. Q6B. 3–10–1999. http://www.fda.gov/downloads/Drugs/Guidance ComplianceRegulatoryInformation/Guidances/UCM073488.pdf.

Jin, P., Wang, E., Provenzano, M., Deola, S., Selleri, S., Ren, J., et al., 2006. Molecular signatures induced by interleukin-2 on peripheral blood mononuclear cells and T cell subsets. J. Transl. Med. 4, 26.

Jin, P., Wang, E., Ren, J., Childs, R., Shin, J.W., Khuu, H., et al., 2008. Differentiation of two types of mobilized peripheral blood stem cells by microRNA and cDNA expression analysis. J. Transl. Med. 6, 39.

Kalos, M., Levine, B.L., Porter, D.L., Katz, S., Grupp, S.A., Bagg, A., et al., 2011. T cells with chimeric antigen receptors have potent antitumor effects and can establish memory in patients with advanced leukemia. Sci. Transl. Med. 3, 95ra73.

Kammula, U.S., Marincola, F.M., Rosenberg, S.A., 2000. Real-time quantitative polymerase chain reaction assessment of immune reactivity in melanoma patients after tumor peptide vaccination. J. Natl. Cancer Inst. 92, 1336–1344.

Kinnaird, T., Stabile, E., Burnett, M.S., Shou, M., Lee, C.W., Barr, S., et al., 2004. Local delivery of marrow-derived stromal cells augments collateral perfusion through paracrine mechanisms. Circulation 109, 1543–1549.

Kochenderfer, J.N., Dudley, M.E., Carpenter, R.O., Kassim, S.H., Rose, J.J., Telford, W.G., et al., 2013. Donor-derived CD19-targeted T cells cause regression of malignancy persisting after allogeneic hematopoietic stem cell transplantation. Blood 122, 4129–4139.

Kochenderfer, J.N., Dudley, M.E., Feldman, S.A., Wilson, W.H., Spaner, D.E., Maric, I., et al., 2012. B-cell depletion and remissions of malignancy along with cytokine-associated toxicity in a clinical trial of anti-CD19 chimeric-antigen-receptor-transduced T cells. Blood 119, 2709–2720.

Kochenderfer, J.N., Wilson, W.H., Janik, J.E., Dudley, M.E., Stetler-Stevenson, M., Feldman, S.A., et al., 2010. Eradication of B-lineage cells and regression of lymphoma in a patient treated with autologous T cells genetically engineered to recognize CD19. Blood 116, 4099–4102.

Lagos-Quintana, M., Rauhut, R., Yalcin, A., Meyer, J., Lendeckel, W., Tuschl, T., 2002. Identification of tissue-specific microRNAs from mouse. Curr. Biol. 12, 735–739.

Larochelle, A., Krouse, A., Metzger, M., Orlic, D., Donahue, R.E., Fricker, S., et al., 2006. AMD3100 mobilizes hematopoietic stem cells with long-term repopulating capacity in nonhuman primates. Blood 107, 3772–3778.

Le, B.K., Frassoni, F., Ball, L., Locatelli, F., Roelofs, H., Lewis, I., et al., 2008. Mesenchymal stem cells for treatment of steroid-resistant, severe, acute graft-versus-host disease: a phase II study. Lancet 371, 1579–1586.

Le, B.K., Ringden, O., 2006. Mesenchymal stem cells: properties and role in clinical bone marrow transplantation. Curr. Opin. Immunol. 18, 586–591.

Leen, A.M., Bollard, C.M., Mendizabal, A.M., Shpall, E.J., Szabolcs, P., Antin, J.H., et al., 2013. Multicenter study of banked third-party virus-specific T cells to treat severe viral infections after hematopoietic stem cell transplantation. Blood 121, 5113–5123.

Leen, A.M., Myers, G.D., Sili, U., Huls, M.H., Weiss, H., Leung, K.S., et al., 2006. Monoculture-derived T lymphocytes specific for multiple viruses expand and produce clinically relevant effects in immunocompromised individuals. Nat. Med. 12, 1160–1166.

Marcacci, M., Kon, E., Moukhachev, V., Lavroukov, A., Kutepov, S., Quarto, R., et al., 2007. Stem cells associated with macroporous bioceramics for long bone repair: 6- to 7-year outcome of a pilot clinical study. Tissue Eng. 13, 947–955.

Martinez, F.O., Gordon, S., Locati, M., Mantovani, A., 2006. Transcriptional profiling of the human monocyte-to-macrophage differentiation and polarization: new molecules and patterns of gene expression. J. Immunol. 177, 7303–7311.

McCarron, S.L., Edwards, S., Evans, P.R., Gibbs, R., Dearnaley, D.P., Dowe, A., et al., 2002. Influence of cytokine gene polymorphisms on the development of prostate cancer. Cancer Res. 62, 3369–3372.

Merling, R.K., Sweeney, C.L., Choi, U., De Ravin, S.S., Myers, T.G., Otaizo-Carrasquero, F., et al., 2013. Transgene-free iPSCs generated from small volume peripheral blood non-mobilized CD34+ cells. Blood 121, e98–107.

Mocellin, S., Panelli, M., Wang, E., Rossi, C.R., Pilati, P., Nitti, D., et al., 2004. IL-10 stimulatory effects on human NK cells explored by gene profile analysis. Genes Immun. 5, 621–630.

Montero, A., Savani, B.N., Shenoy, A., Read, E.J., Carter, C.S., Leitman, S.F., et al., 2006. T cell depleted peripheral blood stem cell allotransplantation with T cell add back for patients with hematological malignancies: effect of chronic GVHD on outcome. Biol. Blood Marrow Transplant. 12, 1318–1325.

Morgan, R.A., Dudley, M.E., Wunderlich, J.R., Hughes, M.S., Yang, J.C., Sherry, R.M., et al., 2006. Cancer regression in patients after transfer of genetically engineered lymphocytes. Science 314, 126–129.

Nagorsen, D., Deola, S., Smith, K., Wang, E., Monsurro, V., Zanovello, P., et al., 2005. Polarized monocyte response to cytokine stimulation. Genome Biol. 6, R15.

Nanjundappa, A., Raza, J.A., Dieter, R.S., Mandapaka, S., Cascio, W.E., 2007. Cell transplantation for treatment of left-ventricular dysfunction due to ischemic heart failure: from bench to bedside. Expert Rev. Cardiovasc. Ther. 5, 125–131.

Nervi, B., Link, D.C., Dipersio, J.F., 2006. Cytokines and hematopoietic stem cell mobilization. J. Cell Biochem. 99, 690–705.

Nuschke, A., 2014. Activity of mesenchymal stem cells in therapies for chronic skin wound healing. Organogenesis 10, 29–37.

Panelli, M.C., Stashower, M.E., Slade, H.B., Smith, K., Norwood, C., Abati, A., et al., 2007. Sequential gene profiling of basal cell carcinomas treated with imiquimod in a placebo-controlled study defines the requirements for tissue rejection. Genome Biol. 8, R8.

Panelli, M.C., Wang, E., Monsurro, V., Marincola, F.M., 2002. The role of quantitative PCR for the immune monitoring of cancer patients. Expert Opin. Biol. Ther. 2, 557–564.

Panelli, M.C., White, R., Foster, M., Martin, B., Wang, E., Smith, K., et al., 2004. Forecasting the cytokine storm following systemic interleukin (IL)-2 administration. J. Transl. Med. 2, 17.

Player, A., Wang, Y., Bhattacharya, B., Rao, M., Puri, R.K., Kawasaki, E.S., 2006. Comparisons between transcriptional regulation and RNA expression in human embryonic stem cell lines. Stem Cells Dev. 15, 315–323.

Porter, D.L., Levine, B.L., Kalos, M., Bagg, A., June, C.H., 2011. Chimeric antigen receptor-modified T cells in chronic lymphoid leukemia. N. Engl. J. Med. 365, 725–733.

Powell Jr, D.J., Dudley, M.E., Hogan, K.A., Wunderlich, J.R., Rosenberg, S.A., 2006. Adoptive transfer of vaccine-induced peripheral blood mononuclear cells to patients with metastatic melanoma following lymphodepletion. J. Immunol. 177, 6527–6539.

Provenzano, M., Mocellin, S., Bettinotti, M., Preuss, J., Monsurro, V., Marincola, F.M., et al., 2002. Identification of immune dominant cytomegalovirus epitopes using quantitative real-time polymerase chain reactions to measure interferon-gamma production by peptide-stimulated peripheral blood mononuclear cells. J. Immunother. 25, 342–351.

Provenzano, M., Mocellin, S., Bonginelli, P., Nagorsen, D., Kwon, S.W., Stroncek, D., 2003. Ex vivo screening for immunodominant viral epitopes by quantitative real time polymerase chain reaction (qRT-PCR). J. Transl. Med. 1, 12.

Quarto, R., Mastrogiacomo, M., Cancedda, R., Kutepov, S.M., Mukhachev, V., Lavroukov, A., et al., 2001. Repair of large bone defects with the use of autologous bone marrow stromal cells. N. Engl. J. Med. 344, 385–386.

Ren, J., Stroncek, D.F., Zhao, Y., Jin, P., Castiello, L., Civini, S., et al., 2013. Intra-subject variability in human bone marrow stromal cell (BMSC) replicative senescence: molecular changes associated with BMSC senescence. Stem Cell Res. 11, 1060–1073.

Ringden, O., Uzunel, M., Rasmusson, I., Remberger, M., Sundberg, B., Lonnies, H., et al., 2006. Mesenchymal stem cells for treatment of therapy-resistant graft-versus-host disease. Transplantation 81, 1390–1397.

Robbins, P.F., Morgan, R.A., Feldman, S.A., Yang, J.C., Sherry, R.M., Dudley, M.E., et al., 2011. Tumor regression in patients with metastatic synovial cell sarcoma and melanoma using genetically engineered lymphocytes reactive with NY-ESO-1. J. Clin. Oncol. 29, 917–924.

Rosenberg, S.A., Lotze, M.T., Muul, L.M., Chang, A.E., Avis, F.P., Leitman, S., et al., 1987. A progress report on the treatment of 157 patients with advanced cancer using lymphokine-activated killer cells and interleukin-2 or high-dose interleukin-2 alone. N. Engl. J. Med. 316, 889–897.

Rosenberg, S.A., Spiess, P., Lafreniere, R., 1986. A new approach to the adoptive immunotherapy of cancer with tumor-infiltrating lymphocytes. Science 233, 1318–1321.

Schmitt, A., Hus, I., Schmitt, M., 2007. Dendritic cell vaccines for leukemia patients. Expert. Rev. Anticancer Ther. 7, 275–283.

Shapiro, A.M., Lakey, J.R., Ryan, E.A., Korbutt, G.S., Toth, E., Warnock, G.L., et al., 2000. Islet transplantation in seven patients with type 1 diabetes mellitus using a glucocorticoid-free immunosuppressive regimen. N. Engl. J. Med. 343, 230–238.

Shin, J.W., Jin, P., Fan, Y., Slezak, S., vid-Ocampo, V., Khuu, H.M., et al., 2008. Evaluation of gene expression profiles of immature dendritic cells prepared from peripheral blood mononuclear cells. Transfusion 48, 647–657.

Stagg, J., Galipeau, J., 2013. Mechanisms of immune modulation by mesenchymal stromal cells and clinical translation. Curr. Mol. Med. 13, 856–867.

Stroncek, D.F., Basil, C., Nagorsen, D., Deola, S., Arico, E., Smith, K., et al., 2005. Delayed polarization of mononuclear phagocyte transcriptional program by type I interferon isoforms. J. Transl. Med. 3, 24.

Takahashi, K., Yamanaka, S., 2006. Induction of pluripotent stem cells from mouse embryonic and adult fibroblast cultures by defined factors. Cell 126, 663–676.

Taniwaki, M., Daigo, Y., Ishikawa, N., Takano, A., Tsunoda, T., Yasui, W., et al., 2006. Gene expression profiles of small-cell lung cancers: molecular signatures of lung cancer. Int. J. Oncol. 29, 567–575.

Tjonnfjord, G.E., Steen, R., Evensen, S.A., Thorsby, E., Egeland, T., 1994. Characterization of CD34+ peripheral blood cells from healthy adults mobilized by recombinant human granulocyte colony-stimulating factor. Blood 84, 2795–2801.

Tumaini, B., Lee, D.W., Lin, T., Castiello, L., Stroncek, D.F., Mackall, C., et al., 2013. Simplified process for the production of anti-CD19-CAR-engineered T cells. Cytotherapy 15, 1406–1415.

Wang, E., Miller, L.D., Ohnmacht, G.A., Liu, E.T., Marincola, F.M., 2000. High-fidelity mRNA amplification for gene profiling. Nat. Biotechnol. 18, 457–459.

Wang, E., Ngalame, Y., Panelli, M.C., Nguyen-Jackson, H., Deavers, M., Mueller, P., et al., 2005. Peritoneal and subperitoneal stroma may facilitate regional spread of ovarian cancer. Clin. Cancer Res. 11, 113–122.

Wellcome Trust Sanger Institute, 2007. miRBase. http://microrna.sanger.ac.uk/sequences/index.shtml, release 9.1.2007 (accessed 24.06.14.).

Wrzesinski, C., Paulos, C.M., Gattinoni, L., Palmer, D.C., Kaiser, A., Yu, Z., et al., 2007. Hematopoietic stem cells promote the expansion and function of adoptively transferred antitumor CD8 T cells. J. Clin. Invest. 117, 492–501.

Zhou, Y., Wang, S., Yu, Z., Hoyt Jr., R.F., Sachdev, V., Vincent, P., et al., 2009. Direct injection of autologous mesenchymal stromal cells improves myocardial function. Biochem. Biophys. Res. Commun. 390, 902–907.

Chapter 2.1.4

Translational Pharmacogenetics to Support Pharmacogenetically Driven Clinical Decision Making

Julia Stingl

INTRODUCTION

Despite more than 50 years of pharmacogenetic research showing the impact of inherited variability in the reaction to drugs in so many examples, this knowledge is not being used in a broad way for improving individual drug treatment. In patient care, diagnostics of genetic variants may lead to improved drug safety and efficacy, and, at the time, several efforts, from both clinicians and the drug regulatory agencies, intend to spread out the knowledge on pharmacogenetic diagnostics and validate the benefit for patients.

The current status of companion diagnostics in the treatment of diseases shows a tremendous development in the field of cancer diagnostics with obligatory tests that lead to stratified treatment with substances specifically acting on molecular substructures. More than 150 drug labels have been extended with information on pharmacogenetics or other biomarkers by the Food and Drug Administration (FDA) and more than 70 by the European Medicines Agency (EMA) at this time (see PharmGKB website). Regulatory guidance is provided for the inclusion of pharmacogenetic diagnostics in clinical drug development and in pharmacovigilance in a proactive strategy by the EMA and FDA, incorporating human diversity into clinical drug trials and further developing the clinical use of pharmacogenetic diagnostics.

PHARMACOGENETICS AS A TOOL FOR IMPROVING INDIVIDUAL DRUG THERAPY

Drug safety and efficacy vary considerably among patients, and few molecular or clinical factors exist that might help to better predict an individual's response to a given drug. Many patients will experience the desired drug effect, some may suffer from well-known adverse drug reactions (ADRs), others may experience no effects, and very rarely a patient will die from severe side effects. It is currently difficult for physicians to prescribe the optimal drug in the optimal dose for each patient because prediction of a patient's response to any one specific drug is rarely possible. The absolute number of adverse drug effect or adverse drug reaction (ADR)-related deaths per year is estimated to be around 100,000 in the United States, and estimations from population-based studies show that ADRs account for approximately 3% of all deaths in the general population (Shepherd et al., 2012; Wester et al., 2008). For instance, according to a study on the incidence of adverse drug effects in hospitalized patients, about 6.7% of all hospitalizations are due to severe adverse drug effects, and 0.37% of these cases end with death (Lazarou et al., 1998; Pirmohamed et al., 2004). Adverse drug effects are the fifth leading cause of death in the United States, directly after coronary heart disease, cancer, stroke, and lung diseases, and even before accidents, diabetes mellitus, and pneumonia (Spear et al., 2001). The ADR-related costs are estimated between $1.4 and $4 billion per year in the United States (Lazarou et al., 1998). For Europe, similar data have been published (Schneeweiss et al., 2002); for example, a recent study investigated the burden of ADR through a prospective analysis of 18,820 patients aged >16 years, admitted during a 6-month period to two large general hospitals in Merseyside, UK (Pirmohamed et al., 2004). The study found that ADR was prevalent in 6.5% of cases, with the ADR directly leading to the admission in 80% of cases. The overall fatality was 0.15%.

To understand pharmacogenetic mechanisms and why individual patients respond differently to drugs, it is helpful to envisage the progress of a particular drug from its administration to its observed effect. The effect of a drug will depend first on its systemic concentration and its concentration at the drug target. The systemic concentration of a drug depends on

Principles of Translational Science in Medicine. http://dx.doi.org/10.1016/B978-0-12-800687-0.00005-0

several pharmacokinetic factors, commonly referred to by the acronym ADME (drug absorption, distribution, metabolism, and elimination). It has long been known that pharmacogenetic factors may influence the pharmacokinetics of a drug, and therefore a doctor might take this into account when determining dosage for a given patient. In many cases drug concentration at its target will represent a mere function of the systemic concentration. Active transport processes, however, may influence local target concentrations; for example, the blood–brain barrier is an important determinant of drug concentration in the cerebrospinal fluid. It has become increasingly clear that hereditary variances in drug metabolizing enzymes (DME) and drug transporters can exert considerable influence on drug concentrations and drug exposure.

The second crucial determinant of an observed drug effect is the response of the drug target to a given drug concentration. Differences in activity or gene expression sometimes caused by hereditary variants have been discovered and characterized in receptors, drug transporters, ion channels, lipoproteins, coagulation factors, and many other factors involved in immune response, cell development, cell cycle control, and other functions; these differences significantly influence the manifestation and course of diseases. At the same time many of these polymorphic structures are targets of specific drugs and can thus potentially influence the effect that a specific drug concentration will exert at the drug target.

TYPES OF DRUG THERAPIES THAT MIGHT PROFIT FROM PHARMACOGENETIC DIAGNOSTICS

Not all drug therapies are principally suited for genotype-based optimization concepts. One problem is that results from genetic testing in general are generated during the first or second day after blood sampling. Thus, drug therapies in which quick decisions are demanded probably will not profit from genotype-based dose adjustments. Further, drug therapies with a large therapeutic range do not justify the effort and costs required for pretherapeutic genotyping. In contrast, drug therapies for which the dosage has to be carefully chosen and slowly increased are prone to individual optimization by means such as genotyping. For example, treatment of heart failure with beta blockers demands careful dosing at low dosages; thus, more in-depth knowledge of genetically caused differences in drug metabolism would be of great interest. The use of beta blockers for treatment of hypertension, in contrast, can be monitored by measuring the blood pressure and thus probably will not further profit from genetic diagnostics.

Indications for genotyping in patient care are emerging and more and more in use starting in the field of oncology but also spreading out to other conditions in general medicine. For some drug therapies, when certain genotypes lead to an altered risk of a severe adverse drug reaction, or when a direct improvement of efficacy can be expected by individualized dosage, pharmacogenetic diagnostics should be performed before the start of drug therapy. This might, for example, be the case for antidepressant drug therapy with tricyclic antidepressants (which is largely affected by the CYP2D6 polymorphism), anticoagulant drug therapy and polymorphism of CYP2C9, as well as codeine treatment and CYP2D6.

In anticoagulant drug treatment, bleeding complications occur mostly at the beginning of treatment, and, thus, genotyping for CYP2C9 polymorphisms will make most sense before or during the start of therapy. In other drug therapies, genotyping might be used as a further diagnostic tool for explaining therapeutic failure or adverse drug effects or for explaining abnormally high or low plasma levels at therapeutic drug monitoring. This is the case, for example, in general psychiatric drug treatment. Because many antidepressant and antipsychotic drugs are metabolized by the polymorphic enzyme CYP2D6, genotyping of this enzyme is a common diagnostic tool after adverse drug effects have occurred or after therapeutic failure.

Thus, pharmacogenetic implementation guidelines also comprise tools for deciding at which time point of which drug therapies genetic tests might be useful.

THE STATUS OF TRANSLATIONAL PHARMACOGENETICS IN VARIOUS DRUG THERAPY FIELDS

Depression

The pharmacotherapy of depression, one of the major psychiatric disorders, is characterized by long-running drug therapy, a relatively narrow therapeutic index, and poor predictability of individual response. About 30% of all depression patients do not respond sufficiently to the first antidepressant drug that is given to them (Bauer et al., 2002). Failure to respond to antidepressant drug therapy as well as intolerable side effects not only results in personal suffering for individuals and their families but also imposes considerable costs on society. At present, it is not possible to reliably predict an individual's response probability before onset of a drug treatment.

Many drugs used in psychiatry such as antidepressants, antipsychotics, and mood stabilizers are extensively metabolized by polymorphic DME, with CYP2D6 being involved in the metabolism of approximately half of the commonly prescribed psychotropic drugs (Mulder et al., 2007). Since differences in plasma concentration due to individual variability in drug clearance often vary by 10-fold and even higher, the possibility of rationally justified pharmacogenetic dose adjustments is given in reflection of pharmacokinetic data such as oral drug clearance (Hicks et al., 2013; Swen et al., 2011; Stingl et al, 2013).

Inadequate drug exposure may constitute a risk due to nonresponse or toxicity depending on the therapeutic range of the drug. In some cases, imprecise dosing may have little or no clinical consequences because of a wide therapeutic range. In antidepressant drug treatment, poor metabolizers (PM) for CYP2D6 have been associated with longer time requirements to find the appropriate drug and with more frequent drug switches (Bijl et al., 2008). There are also hints that poor metabolizers suffer more frequently from adverse drug reactions than extensive metabolizers and register longer hospital stays, whereas ultrarapid metabolizers (UM) have a higher risk of therapeutic failure (Bijl et al., 2008; Chou et al., 2000; Ruano et al., 2013).

The pharmacokinetic-based dose-adjustment approach is now being followed by issuing clinical recommendations for specific drug–genotype pairs (Clinical Pharmacogenetics Implementation Consortium, CPIC) (Hicks et al., 2013; Relling and Klein, 2011). Such evidence-based dosing guidelines are now available for 27 drugs, and made accessible in the PharmGKB database (Caudle et al., 2014). Of all genes involved in these evaluations, CYP2C19 and CYP2D6 have been shown to be the most important DME in dose adjustments of psychotropic drugs.

Individuals to whom tricyclic antidepressants are prescribed could benefit from CYP2D6 genotyping if the dose is adjusted for the group of PM and UM of CYP2D6. Within the group of selective serotonin reuptake inhibitors (SSRI), cytochrome (CYP) inhibition poses a problem for drug interaction, but there is growing evidence that assessment of cytochrome P450 polymorphisms may not be clinically useful for guiding SSRI therapy.

For mirtazapine, it could be shown that the CYP2D6 genotype had a significant influence on the variability in the plasma concentration; however, when UM was compared to extensive metabolizers (EM), the magnitude of concentration differences was only moderate (Kirchheiner et al., 2004b).

CYP2D6 is responsible for the transformation of venlafaxine into the equipotent O-desmethyl-venlafaxine (Fukuda et al., 2000, 1999; Otton et al., 1996; Veefkind et al., 2000). However, a higher risk for cardiotoxic events might exist in PM, as cases of severe arrhythmia have been reported in four patients treated with venlafaxine who all were PM according to CYP2D6 (Lessard et al., 1999).

As has been shown in several studies, differences in pharmacokinetic parameters caused by genetic polymorphisms impact the outcome and the risk of adverse drug reactions of antidepressants. In a German study evaluating the effect of the CYP2D6 genotype on adverse effects and nonresponse during treatment with CYP2D6-dependent antidepressants, PM and UM were significantly overrepresented as compared to the control population; they were four times as prevalent in the group of patients suffering from adverse drug reactions and five times as prevalent in the nonresponders (Rau et al., 2004). In a study published in the same year, Grasmader and colleagues did not find any influence of the CYP2D6 and CYP2C19 genotypes on antidepressant drug response, although the incidence of relevant side effects tended to be higher in PM of CYP2D6 (Grasmader et al., 2004). Furthermore, a prospective one-year clinical study of 100 psychiatric inpatients suggested a trend toward longer duration of hospital stay and higher treatment costs in UM and PM of CYP2D6 (Kawanishi et al., 2004). Figure 2.10 expresses the differences in mean oral clearance among carriers of none, one, two, and more active CYP2D6 genes as percentages of the dose adjustments for antidepressants.

In conclusion, for treatment with antidepressants, there is considerable evidence that CYP2D6 and, to a lesser extent, CYP2C19 polymorphisms affect the pharmacokinetics of several antidepressants and possibly the therapeutic outcome and adverse drug effects. The usefulness of genotyping procedures in depressed patients, however, has not been confirmed in prospective clinical trials; therefore, translational pharmacogenetic diagnostics is limited to a few hospitals, and to patients who experienced adverse drug effects or did not respond to standard treatment regimens.

Cardiovascular Disease

Cardiovascular diseases are the most common cause of morbidity and mortality in developed countries. The following discussion addresses the clinical implications resulting from pharmacogenetic research for some of the drugs used in the therapy of cardiovascular diseases, like statins, oral anticoagulants, beta blockers, and sartans.

Beta Blockers

For metoprolol, one of the most often-prescribed beta blockers, the role of CYP2D6 genetic polymorphisms in its pharmacokinetics seems to be well established. CYP2D6 catalyzes O-demethylation and, even more specifically, α-hydroxylation of the drug (Lennard et al., 1982a). Not only metoprolol plasma concentrations but also effects on heart rate correlated

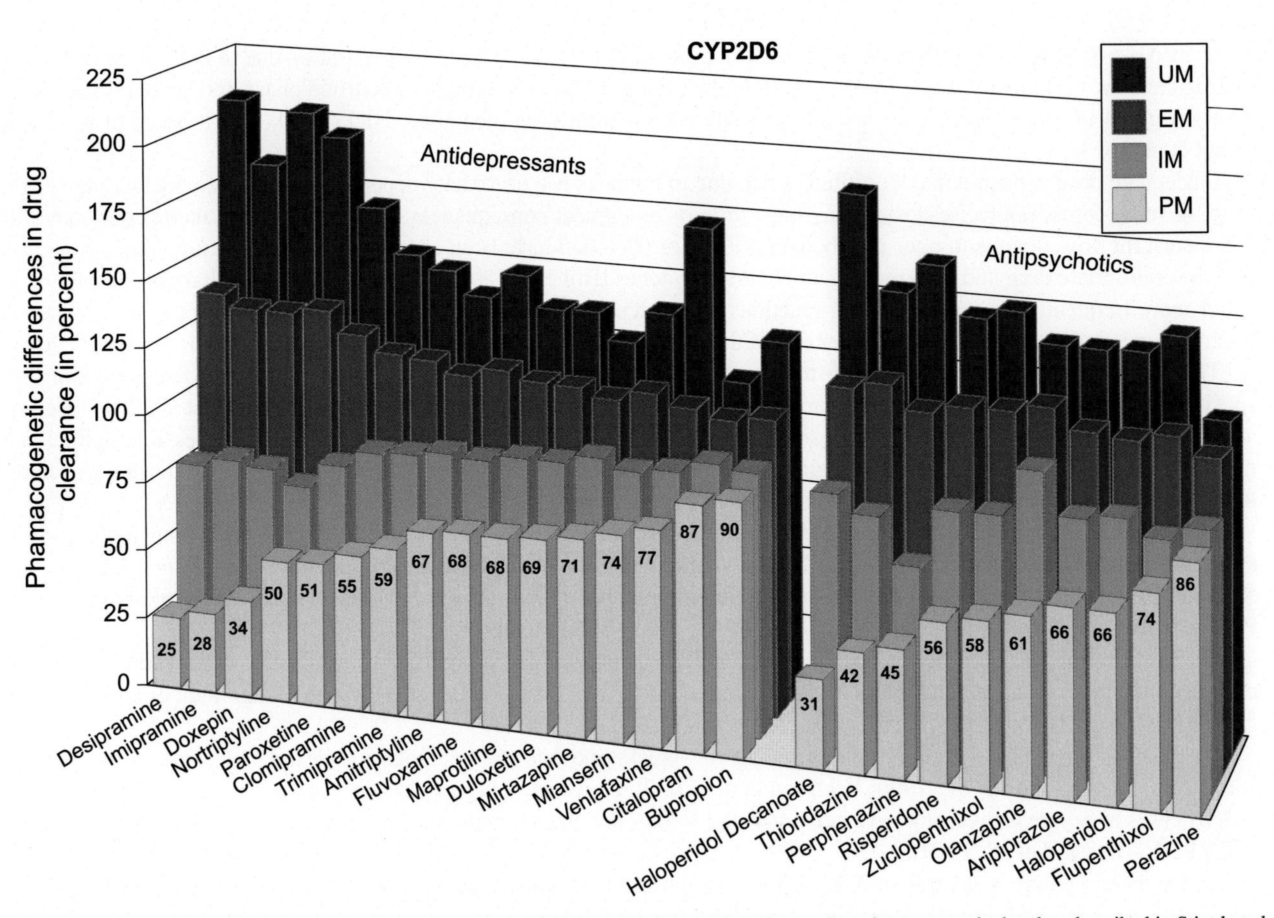

FIGURE 2.10 CYP2D6 PM (light blue), IM (medium blue), EM (blue), UM (dark blue). Dose adaptations were calculated as described in Stingl et al (2013). Dose adaptations are based on an average dose of 100% and are aimed for the Caucasian population. Data from studies in Asiatic or African or other populations were not incorporated in this figure. *(With permission taken from Stingl et al, 2013. CYP2D6 genotype-dependent quantitative changes in pharmacokinetics of antidepressant and antipsychotic drugs expressed as percent dose adaptations.)*

significantly with the CYP2D6 metabolic phenotype (Lennard et al., 1982b). In a study with 281 individuals, heart rate response differed significantly by CYP2D6 phenotype with PM and IM showing greater reduction (Hamadeh et al., 2014). In UM carrying the CYP2D6 gene duplication, total clearance of metoprolol was about 100% higher compared with the EM as the reference group (367 vs 168 L/h) (Kirchheiner et al., 2004a). The reduction of exercise-induced heart rate by metoprolol in the UM group was only about half of that observed in the EM.

On the basis of the considerable impact of the CYP2D6 polymorphism on the disposition of CYP2D6 substrates, it has often been suggested that CYP2D6 PM are more susceptible to adverse effects than EM at standard doses of metoprolol (Rau et al., 2002; Wuttke et al., 2002; Zineh et al., 2004) which is the same as for antidepressive and antipsychotic drugs. For instance, the CYP2D6 PM genotype was overrepresented among 24 patients with severe metoprolol-associated adverse effects in a retrospective study (Wuttke et al., 2002). However, metoprolol is well tolerated in the majority of patients with cardiovascular diseases irrespective of CYP2D6 genotype (Hamadeh et al., 2014). The effect of beta blockers on the quality of life of hypertensive patients has not been extensively studied. The wide therapeutic range of metoprolol may explain why it is well tolerated in the majority of CYP2D6 PM and IM, despite several-fold higher plasma concentrations.

Another beta blocker, the racemate carvedilol, which has been approved as an adjunctive therapy in the treatment of heart failure, is known to be stereoselectively metabolized by cytochrome P450 enzymes (Neugebauer and Neubert, 1991). CYP2D6 polymorphism has been shown to alter the stereoselective disposition of carvedilol with PM demonstrating an impaired clearance of the R-enantiomer; thus, the PM were affected by a more pronounced $\alpha 1$-blockade, which might outweigh the beneficial $\beta 1$-blocking effects (Zhou and Wood, 1995). Hence, CYP2D6 genotyping might predict therapeutic outcomes in cardiac insufficient patients treated with carvedilol. However, inhibition of CYP2D6 metabolism by administration of fluoxetine in EM led to significant changes in plasma pharmacokinetics in favor of the R-enantiomer, without any effect on blood pressure and heart rate, which casts doubt on the clinical significance of the CYP2D6 genotype for treatment with carvedilol in antihypertensive therapy (Graff et al., 2001).

In summary, CYP2D6 genotyping might be beneficial, if at all, for long-term treatment with metoprolol in indications such as heart failure or in post-myocardial infarction patients when no surrogate parameter such as blood pressure is available to predict long-term efficacy. However, pharmacogenetic diagnostics is not generally done in patients using beta blockers, and at present it does not seem likely to become a general tool for therapeutic improvement.

Vitamin K Antagonists

The S-enantiomers of all three vitamin K antagonists—acenocoumarol, phenprocoumon, and warfarin—are substrates of CYP2C9. Warfarin is mainly prescribed in North America and Asia, whereas acenocoumarol and phenprocoumon are more commonly used in Europe. Both CYP2C9 alleles—CYP2C9*2 and *3—have a substantial effect on the intrinsic clearance of (S)-warfarin (Loebstein et al., 2001; Scordo et al., 2002; Takahashi et al., 1998). For acenocoumarol, heterozygous carriers of CYP2C9*3 had only 40% of the clearance measured in CYP2C9*1/*1 carriers, whereas differences in the international normalized ratio (INR) between these two groups resulted in a mean daily dose that was 70% that of wild-type patients in *1/*3 carriers (Tassies et al., 2002; Thijssen et al., 2001; Thijssen and Ritzen, 2003; Verstuyft et al., 2001; Visser et al., 2004a). Many clinical studies reported a higher risk of adverse drug effects, especially bleeding complications in patients at the beginning of treatment with oral anticoagulants (Visser et al., 2004b).

The vitamin K antagonistic effects of the coumarin derivatives are mediated by inhibition of the vitamin K epoxide reductase (Rost et al., 2004). The dosage of vitamin K antagonists required to achieve the target-level anticoagulation largely depends on the activity of the vitamin K epoxide reductase (VKOR). About 25% of variability in warfarin dose was shown to be explained by the VKORC1 haplotypes, with one single functional single nucleotide polymorphism (SNP) being responsible for the differences in gene expression (Geisen et al., 2005; Rieder et al., 2005). The mean daily dose requirement of phenprocoumon differs largely in dependence of the G-1639A polymorphism of VKORC1 with GG carriers, requiring more than twice the mean dose compared to AA carriers (Reitsma et al., 2005). In correspondence with the lower dose requirement in AA carriers, those individuals had a higher risk of bleeding even when the INR values were stable. For acenocoumarol, it has been shown that the 1639AA genotype is the greatest risk factor for overanticoagulation with an odds ratio of 10.5 (3.3–34.1), and that it predisposes to larger INR variations (Osman et al., 2006, Quteineh et al., 2005).

A lot of discussion worldwide on the implementation of pharmacogenetic diagnostics to guide anticoagulant drug treatment is ongoing, and guidelines exist both recommending and not recommending the incorporation of genotype information into therapy, mostly for warfarin (Johnson et al., 2011). Conflicting results of recent large clinical trials keep the discussion on the true clinical utility of genotyping ongoing. The European Pharmacogenetics of AntiCoagulant Therapy (EU-PACT) study reported that the time to receive therapeutic range during the initial weeks of anticoagulation was better in the genotype dose adjusted study arm than in the comparison group during warfarin initiation (Pirmohamed et al., 2013). The Clarification of Optimal Anticoagulation through Genetics trial (COAG), which also was a multicenter randomized trial evaluating the effect of genotype-guided dosing on time within therapeutic range, did not detect a difference between the pharmacogenetic versus clinical algorithm guided arm (Kimmel et al., 2013).

The FDA has recommended genotyping for CYP2C9 and VCORC1 prior to onset of the anticoagulant therapy, in addition to dosing according to INR, in the drug labeling of warfarin, rendering the vitamin K antagonists one of the first drugs in which pharmacogenetic factors have been completely translated into drug therapy.

Statins and Proton Pump Inhibitors

Statins (3-hydroxy-3-methylglutaryl coenzyme A reductase inhibitors) are generally regarded to be safe and have become the most commonly prescribed agents for the treatment of dyslipidemia, to reduce cholesterol and cardiovascular events. The most common (occurring in 1%–5% of exposed subjects) side effect is skeletal myopathy (Buettner et al., 2012). Statin-related myopathy includes weak symptoms of myalgia up to severe muscle damage and kidney injury (rhabdomyolysis). Patient risk factors for statin-induced myopathy include high statin exposure, drug–drug interactions, high statin dose, female sex and low body mass index, intense physical exercise, endocrine abnormalities such as hypothyroidism, and ethnicity.

The transporter OATP1B1 (encoded by the SLCO1B1 gene) mediates the hepatic uptake of statins, and therefore, the activity of the transporter influences bioavailability of statins to a large degree. Interaction studies with cyclosporine, a strong inhibitor of CYP3A4 and OATP1B1, have shown three- to eightfold increases of the area under the curve for simvastatin acid, thereby markedly increasing the risk for statin-related muscle damage (Niemi et al., 2011).

SLCO1B1 polymorphisms clearly impact the pharmacokinetics of simvastatin and, to a lesser degree, the pharmacokinetics of other statins (Niemi, 2010). For simvastatin, the evidence linking myopathy to rs4149056 in SLCO1B1 is strongest with data from randomized trials and clinical practice–based cohorts. Because the association of rs4149056 with myopathy is less compelling for other statins, the CPIC guideline focuses on recommendations for simvastatin (Wilke et al., 2012). At lower simvastatin doses (e.g., 40 mg daily), a modest risk for myopathy in carriers of the C allele of rs4149056 is expected, with a relative risk of myopathy of 2.6 per copy of the C allele at rs4149056. Thus, utility of routine CK surveillance may be recommended to clinicians. If lower doses of simvastatin are not sufficient, physicians may consider an alternative statin based on (i) potency differences, (ii) kinetic differences, (iii) comedications, (iv) hepatic function, (v) renal function, and (vi) relevant comorbidities (Wilke et al., 2012). The risk is much higher for the 80-mg simvastatin dose (OR 4.5 for the TC genotype, ~20.0 for the CC genotype). Nonetheless, the guideline states that simvastatin-related muscle toxicity can still occur in the absence of rs4149056 (Wilke et al., 2012).

The purpose of the CPIC guideline for simvastatin was to provide interpretive guidance when the SLCO1B1 genotype was already available. Thus, it is not argued directly in this guideline that SLCO1B1 genotyping is absolutely necessary. However, guidance is provided for clinicians treating patients with simvastatin in the event that SLCO1B1 genotype is available during routine care (Wilke et al., 2012).

Proton pump inhibitors (PPI), such as omeprazole, esomeprazole, lansoprazole, pantoprazole, or rabeprazole, are commonly prescribed in combination with antibiotics for *Helicobacter pylori* eradication and for patients with peptic ulcer as well as gastroduodenal reflux disease. They undergo extensive presystemic biotransformation in the liver with the involvement of genetically polymorphic CYP2C19 (Klotz et al., 2004; Schwab et al., 2004). CYP2C19*2 and CYP2C19*3 are the most common nonfunctional alleles; they are responsible for the majority of PM phenotypes of CYP2C19 (de Morais et al., 1994).

Andersson and colleagues observed that in PM of mephenytoin (a model substrate of CYP2C19), the area under the curve (AUC) for omeprazole, lansoprazole, and pantoprazole at a steady state was fivefold higher compared with EM, indicating that approximately 80% of the dose for all three PPI is metabolized by CYP2C19 (Andersson et al., 1998). For different PPI, drug exposures, which were defined with AUC values, were 3–13 times higher in PM and about 2–4 times higher in heterozygous EM as compared to homozygous EM of CYP2C19 (Klotz et al., 2004). At the same time, the elevation of intragastric pH as a pharmacodynamic response to PPI can be shown to be directly related to the respective AUC, and a much higher pH can be monitored for 24 hours following the administration of PPIs in PM than in EM (Klotz et al., 2004).

In a systematic review of pharmacogenetic studies with PPI, Chong and Ensom evaluated the effects of CYP2C19 genetic polymorphism on the clinical outcomes, that is, the effects of the *Helicobacter pylori* eradication rates on therapy with these drugs (Chong et al., 2003). The results of most studies supported the hypothesis that eradication rates vary with CYP2C19 genotype and PMs have a significantly better PPI efficacy.

Furuta and colleagues showed that the eradication rates of *Helicobacter pylori* on the triple therapy comprising PPI, clarithromycin, and amoxicillin was 72.7%, 92.1%, and 97.8% in CYP2C19 EM, IM, and PM, respectively. The authors concluded that the CYP2C19 genotype seems to be one of the important factors associated with the cure rates of *Helicobacter pylori* infections on the PPI-based therapy (Furuta et al., 2001). Moreover, the same authors conducted a prospective clinical study showing in a randomized prospective way that CYP2C19 genotyping provides a cost-effective benefit in *Helicobacter pylori* eradication treatment (Furuta et al., 2007).

The role of CYP2C19 polymorphism in the eradication of *Helicobacter pylori* was also studied in the Caucasian population. However, in contrast to the Asian population, the prevalence of PM of CYP2C19 in Caucasians is much lower—about 2.8% in Caucasians versus 21.3% in Japanese (Wedlund, 2000). We can show significant differences in the efficacy of lansoprazole-based quadruple therapy among Caucasian patients carrying wild-type alleles, one CYP2CIP defect allele, and two CYP2C19 defect alleles, with cure rates of 80.2%, 97.8%, and 100%, respectively (Schwab et al., 2004).

Treatment of gastroesophageal reflux disease (GERD) demands long-term application of PPI. Two retrospective analyses in Japanese patients showed that the healing rate of GERD using lansoprazole was 20%–40% higher in PM or IM compared to EM of CYP2C19, and that PM even benefit in the prevention of relapse of GERD (Furuta et al., 2002; Kawamura et al., 2007; Kawamura et al., 2003).

In summary, according to the results of numerous clinical studies, CYP2C19 polymorphisms are an important factor affecting the pharmacokinetics of most PPI, the value of intragastric pH, and the eradication rates of *Helicobacter pylori.* Therefore, genotyping for CYP2C19 polymorphisms, especially in Asian populations, characterized by a high prevalence of defect CYP2C19 alleles, should be generally applied. The natural consequence of this approach could be higher dosages for extensive metabolizers and/or adjusted treatment regimens. Only prospective controlled clinical trials can show whether the patients might benefit from such an individually tailored therapy.

Pain Treatment

One large field in which the CYP2C9 polymorphism plays a role is the treatment of pain and inflammation with nonsteroidal anti-inflammatory drugs (NSAID). At least 16 different registered NSAIDs are currently known to be at least partially metabolized via CYP2C9. They include aceclofenac, acetylsalicylic acid, azapropazone, celecoxib, diclofenac, flurbiprofen, ibuprofen, indomethacin, lornoxicam, mefenamic acid, meloxicam, naproxen, phenylbutazone, piroxicam, suprofen, and tenoxicam (Rendic, 2002). There were significant inter-genotypic differences in the pharmacokinetics of celecoxib, flurbiprofen, ibuprofen, and tenoxicam that could be translated into dose recommendations based on CYP2C9 genotype. Celecoxib was one of the first drugs for which the manufacturers' drug information recommends caution when administering celecoxib to "poor metabolizers of CYP2C9 substrates," as they might have abnormally high plasma levels (Celebrex® drug information, Searle Pfizer, Chicago, USA). In contrast, the relative contribution of CYP2C9 to the pharmacokinetics of diclofenac was found to be independent of CYP2C9 polymorphisms in several clinical studies (Brenner et al., 2003; Dorado et al., 2003; Kirchheiner et al., 2003; Yasar et al., 2001).

Codeine is converted to morphine via CYP2D6. After several cases of severe or fatal intoxication with codeine in children, the FDA and the EMA restricted the use of codeine and issued a warning about the issue that variable drug metabolism capacity may lead to dangerous intoxication with morphine. According to the EMA, codeine is now contraindicated in all pediatric patients (0–18 years of age) who undergo tonsillectomy or adenoidectomy for obstructive sleep apnea syndrome and in patients of any age who are known to be CYP2D6 UM (up to approximately 10% of Caucasians are CYP2D6 UM, but prevalence differs according to racial and ethnic group), due to an increased risk of developing serious and life-threatening adverse reactions, and in women who are breastfeeding (http://www.ema.europa.eu/).

Malignant Diseases

Most anticancer drugs are characterized by a very narrow therapeutic index and severe consequences of over- or underdosing in the form of, respectively, life-threatening adverse drug reactions or treatment failure.

Tamoxifen

The clinical benefit of the antiestrogen tamoxifen for the treatment of estrogen receptor (ER) positive breast cancer has been evident for more than three decades, but 30%–50% of patients undergoing adjuvant tamoxifen therapy relapse or die. There are several lines of evidence which suggest that the metabolites 4-hydroxy tamoxifen (4-OH-TAM) and 4-hydroxy-N-desmethyl tamoxifen (endoxifen) are the active therapeutic moieties because, compared to the parent drug, these two metabolites have up to 100-fold higher potency in terms of binding to ER (Desta et al., 2004). CYP2D6 is one of the key enzymes for the formation of 4-OH-TAM and endoxifen (Coller et al., 2002), and CYP2D6 PMs have been shown to have very low endoxifen plasma levels (Borges et al., 2006) and nonfavorable clinical outcomes (Goetz et al., 2007). It is of note that between 15% and 20% of European patient populations carry genetic CYP2D6 variants associated with a pronounced impairment in the formation of antiestrogenic tamoxifen metabolites. Tamoxifen-treated patients carrying the CYP2D6 alleles *4, *5, *10, and *41 associated with impaired formation of antiestrogenic metabolites had significantly more recurrences, shorter relapse-free periods (HR = 2.24; 95% CI, 1.16 to 4.33; P = .02), and lower rates of event-free survival (HR = 1.89; 95% CI, 1.10 to 3.25; P = .02) compared to carriers of functional alleles. Moreover, in 2006 Sim and colleagues reported the first genetic evidence for the CYP2C19*17 polymorphism as a putative supplementary biomarker for the classification of patients with favorable treatment outcome (Sim et al., 2006). Patients with the CYP2C19 high enzyme activity promoter variant *17 had a more favorable clinical outcome (HR = .45; 95% CI, .21 to .92; P = .03) than carriers of *1, *2, and *3 alleles. CYP2C19 contributes to tamoxifen metabolism toward the antiestrogenic metabolite 4-OH-TAM with in vitro activities similar to CYP2D6 (Coller et al., 2002; Desta et al., 2004). These studies support the feasibility of treatment outcome prediction for tamoxifen based on the patients' genetic constitution.

Altogether these findings are particularly important in light of the current debate on the effectiveness of tamoxifen for postmenopausal women with hormone receptor positive breast cancer. Because the definition of patient groups with nonfavorable and favorable tamoxifen treatment appears to be feasible through a priori genetic assessment, and given the long-term use of tamoxifen as a safe and effective antihormonal treatment, it is possible to use the results of the pharmacogenetic evidence to refine the choice (Brauch et al., 2013). The Royal Dutch Pharmacists Association—Pharmacogenetics Working Group has issued therapeutic recommendations for tamoxifen based on CYP2D6 genotypes (Swen et al., 2011). For PM and IM genotypes, they recommend considering using aromatase inhibitors for postmenopausal women due to increased risk for relapse of breast cancer with tamoxifen. For IM genotypes, they recommend avoiding concomitant CYP2D6 inhibitor use and/or sequencing of hormonal therapy in prospective clinical trials (Swen et al., 2011).

Thiopurines

Thiopurines, such as 6-mercaptopurine and thioguanine, are largely used in the treatment of acute leukemias, whereas autoimmune diseases and chronic inflammatory bowel diseases are common fields of application for azathioprine (Teml et al., 2007). The cytosolic enzymes thiopurine S-methyltransferase (TPMT) and xanthin oxidase (XO) are the predominant catabolic enzymes in the metabolism of thiopurines. TPMT-dependent methylation is critical in white blood cells, leading to an enhanced cytotoxic effect in patients with low TPMT activity. In 1980, Weinshilboum and Sladek first reported on a trimodal frequency distribution of TPMT activity with about 10% IM and 1 out of 300 homozygous for the trait with very low or undetectable TPMT activity (Weinshilboum and Sladek, 1980). To date, 24 mutant alleles responsible for variation in TPMT enzyme activity have been described, and most of these alleles are characterized by one or more nonsynonymous SNP, resulting in decrease or loss of enzyme activity (Schaeffeler et al., 2006). In several independent studies, the TPMT genotype showed excellent concordance with the TPMT phenotype; the most comprehensive analysis, conducted by Schaeffeler and colleagues, included 1,214 healthy Caucasian blood donors (Schaeffeler et al., 2004). The overall concordance rate between the TPMT phenotype and genotype was 98.4%, and specificity, sensitivity, and the positive and negative predictive power of the genotyping test were estimated to be higher than 90%. In addition, patients with TPMT deficiency treated with standard doses of AZA or 6-MP are at high risk of developing severe myelosuppression, independent of the underlying disease, within a few weeks after commencing drug therapy (e.g., childhood leukemia or inflammatory bowel disease) (Evans et al., 1991; Sebbag et al., 2000). This can be explained by an 8- to 15-fold increase in 6-thioguanine nucleotide (6-TGN) levels in red blood cells (RBC) compared to wild-type patients, subsequently leading to an exaggerated cytotoxic effect. Moreover, an increased risk of hematotoxicity in TPMT heterozygous subjects receiving standard thiopurine therapy is also well known from patients with childhood acute lymphocytic leukemia (ALL) and rheumatologic diseases (Black et al., 1998; Relling et al., 1999). Thus, dose adjustment, at least in TPMT-deficient patients, is required, with an initial dose reduction to 10%–15% of the standard dose as exemplarily shown for patients with ALL and Crohn's disease (Evans et al., 1991; Kaskas et al., 2003). Moreover, it was also demonstrated that the TPMT phenotype or genotype influences the effectiveness of therapy. Low TPMT activity has been associated with higher 6-TGN levels and improved survival, while high TPMT activity has been associated with lower 6-TGN concentrations and an increased relapse risk (Lennard et al. 1990). The German ALL BFM Study Group recently reported on the association of the TPMT genotype and minimal residual disease (MRD) in 810 children with childhood ALL (Stanulla et al., 2005), indicating a significant impact of the TPMT genotype on MRD after administration of 6-MP in the early course of childhood ALL. This may offer a rationale for genotype-based adaptation of 6-MP dosing in the early course of childhood ALL, provided that the described observations translate into improved long-term outcomes.

In the CPIC guideline for Thiopurine Methyltransferase Genotype and Thiopurine Dosing: 2013 Update, the literature published between June 2010 and November 2012 was reviewed, and no new evidence that would change the original guidelines was identified (Relling et al., 2013). Therefore, the previously issued dosing recommendations remain clinically relevant. For patients with one nonfunctional TPMT allele, it is recommended to start with reduced doses of thioguanine, and for patients with malignancy and two nonfunctional alleles, it is recommended to adjust dose based on degree of myelosuppression and disease-specific guidelines. In addition, it is recommended to consider alternative non-thiopurine immunosuppressant therapy for patients with nonmalignant conditions and two nonfunctional alleles (Relling et al., 2011).

For 6-MP the FDA has recently included in the product label respective pharmacogenetic data considering the impact of the polymorphic TPMT on severe toxicity as well as the availability of genotypic and phenotypic testing (Maitland et al., 2006).

5-Fluorouracil

Following administration, 5-FU undergoes complex anabolic and catabolic biotransformations that play an eminent role for both antitumor activity and toxicity. Although the effectivity of 5-FU depends on its bioactivation, resulting in 5-fluoronucleotides that interfere with normal DNA and RNA function preferentially in cancer cells, more than 80% of a given dose is rapidly metabolized by dihydropyrimidine dehydrogenase (DPD). This rate-limiting enzyme of pyrimidine catabolism converts 5-FU to the inactive 5-fluoro-5,6-dihydrouracil (Heggie et al., 1987). Diasio and colleagues were the first to suggest the potential role of DPD as a factor determining 5-FU toxicity; they showed complete deficiency of DPD enzyme activity in a patient with the rare familial metabolic disorder pyrimidinemia who had developed severe neurotoxicity during 5-FU treatment (Diasio et al., 1988). Later studies also showed a correlation between lymphocyte DPD activity and 5-FU clearance (Etienne et al., 1994; Soong and Diasio, 2005).

Other polymorphic candidate genes considered as potential factors for 5-FU toxicity include the thymidylate synthase (TYMS), which is strongly inhibited by 5-FU, as well as methylenetetrahydrofolate reductase (MTHFR), which forms the reduced folate cofactor essential for TYMS inhibition (Robien et al., 2005).

A large study by Schwab and colleagues investigated markers of 5-FU toxicity in a group of patients that is so far the largest of all similar studies but also rather diverse (Schwab et al., 2008). The experimental design focused on an examination of only a few sequence variants in three selected genes (*DPYD*, the gene that encodes DPD enzyme; *TYMS*; and methylene tetrahydrofolate reductase [*MTHFR*]), coupled with nongenetic factors (sex and diet) previously suggested to be associated with severe toxicity from 5-FU.

The CPIC Dosing Guideline for fluoropyrimidines (i.e., 5-fluorouracil, capecitabine, or tegafur) recommends an alternative drug for patients who are homozygous for DPYD nonfunctional variants—*2A (rs3918290), *13 (rs55886062), and rs67376798 A (on the positive chromosomal strand)—as these patients are typically DPD deficient (Caudle et al., 2013). This guideline recommends a 50% reduction in starting dose for heterozygous patients with intermediate activity. The CPIC authors recommend that the *DPYD*4*, **5*, **6* and **9A* alleles be categorized as "normal" activity, in part based on the recent publication (Offer et al., 2014). However, the guideline states that available evidence does not clearly indicate a degree of dose reduction needed to prevent fluoropyrimidine-related toxicities.

Irinotecan

Recently, much attention has been paid to the importance of pharmacogenetic polymorphisms in uridine diphosphate glucuronosyltransferase (UGT) and its role in toxicity and therapeutic response in cancer patients undergoing treatment with the potent antitumor agent irinotecan. SN-38, the active metabolite of irinotecan acting as topoisomerase I inhibitor, is glucuronidated to its inactive form by UGT1A1, the UGT isoform, which is also responsible for the glucuronidation of bilirubin. Several variant alleles leading to reduced enzymatic UGT1A1 activity have been identified in the UGT1A1 gene. These polymorphisms become manifest as unconjugated hyperbilirubinemias in the form of the Crigler–Najjar or Gilbert's syndromes; in addition, in patients treated with irinotecan, they result in reduced SN-38 glucuronidation rates (Van Kuilenburg et al., 2000). For UGT1A1 genetic variations, major interethnic differences have been shown; whereas the UGT1A1*28 promoter polymorphism (TA repeat in the promoter) is quite common in the Caucasian population, UGT1A1*6 and UGT1A1*27 situated in the coding region of UGT1A1 are found mainly in Asians (Sai et al., 2004).

The impact of UGT1A1 polymorphisms on irinotecan toxicity, which is characterized by severe diarrhea and neutropenia, has been studied in several retrospective as well as prospective studies. It can be shown that the presence of common UGT1A1 mutant alleles, even in heterozygous carriers, significantly changed the disposition of irinotecan, causing severe toxicity in patients (Araki et al., 2006; Innocenti et al., 2001: Sai et al., 2004). Finally, given the cumulative evidence, the FDA has approved a new labeling for irinotecan in favor of UGT1A1*28 genotyping, which reflects the prevalence of the UGT1A1*28 allele in Caucasian Americans. A reduced initial dose of irinotecan has been recommended for homozygous carriers of this variant allele, as they are at increased risk of neutropenia (see the FDA's website at http://www.fda.gov). The Royal Dutch Pharmacists Association—Pharmacogenetics Working Group has evaluated therapeutic dose recommendations for irinotecan based on UGT1A1 genotype (Swen et al., 2011). They recommend reducing the dose for *28 homozygous patients receiving more than 250 mg/m^2 (Swen et al., 2011).

TRANSLATIONAL PHARMACOGENETICS AND THE NEED FOR CLINICAL STUDIES TO SUPPORT PHARMACOGENETICALLY DRIVEN PRESCRIBING

The use of pharmacogenetic markers as a diagnostic tool to improve and individualize drug therapy depends on the existing evidence of the clinical impact of the respective variant. In addition to functional characterization of a polymorphism on the molecular level, further clinical studies that use prospective methods to characterize the clinical consequences in different situations in drug therapy are necessary. The lack of clinical data often hampers the application of pharmacogenetic analyses as a diagnostic tool and therefore impairs the translational process of pharmacogenetic research.

A variety of parameters are of interest when validating the usefulness of a pharmacogenetic variant as a biomarker for therapeutic outcome prediction. Data should be available characterizing the impact of a variant on pharmacokinetics or the direct pharmacological effects of drugs that are affected by the pharmacogenetic variant. In addition, surrogate markers for drug response or so-called intermediate phenotypes should be studied with regard to the genetic variant. Finally, data on complex phenotypes such as drug response or toxicity are of main interest. Beyond this, data on long-term parameters such as hospitalization time, quality of life, disability, or survival are warranted in order to estimate the cost–benefit ratio of a pharmacogenetic test.

Different types of study designs are—and should be—commonly used in order to support pharmacogenetically driven prescribing. Consideration of the most valid rational study designs for different situations in drug therapy may lead to a more effective translation of pharmacogenetics into clinical practice.

Figure 2.11 shows the flow of pharmacogenetic clinical studies necessary for translation of pharmacogenetic findings into clinical practice. Beyond the functional molecular characterization of a given variant, several clinical studies should confirm the impact of the findings in different clinical situations. One study type that is quite suitable for pharmacogenetic research is the panel design. This design is usually applied to study the effect of hepatic or renal impairment on a certain drug therapy, and results from those phase I studies are mostly directly involved in drug labeling. Thus, if information in the drug label is seen as the ultimate translation of pharmacogenetics into clinical practice, the use of panel studies certainly is a cost-saving and effective way to achieve this. Figure 2.12 depicts the pharmacogenetic panel design.

In this design, volunteers are included into the study after genotyping and, as much as possible, participants are selected with the aim of getting equal sample sizes for each genotype group to be studied. Preselecting genotypes enables researchers to study a sufficient number of individuals, even those with the more rare genotypes, without having to include large numbers of carriers of the common genotypes. Of course, prior to such volunteer studies, a genotype selection screening will have to take place, and this will also reveal the population frequencies of the genotypes of interest. The aim of such studies is to find out in a very short time and with high statistical power the impact of a genetic variant on pharmacokinetics or pharmacodynamic parameters, which are, for the most part, surrogate parameters of intermediate phenotypes, when studying healthy volunteers.

Cohort studies of all patients receiving a certain drug treatment coupled with subsequent analysis of the differences in therapeutic outcomes related to genotype are among the most common study designs in pharmacogenetic research. This

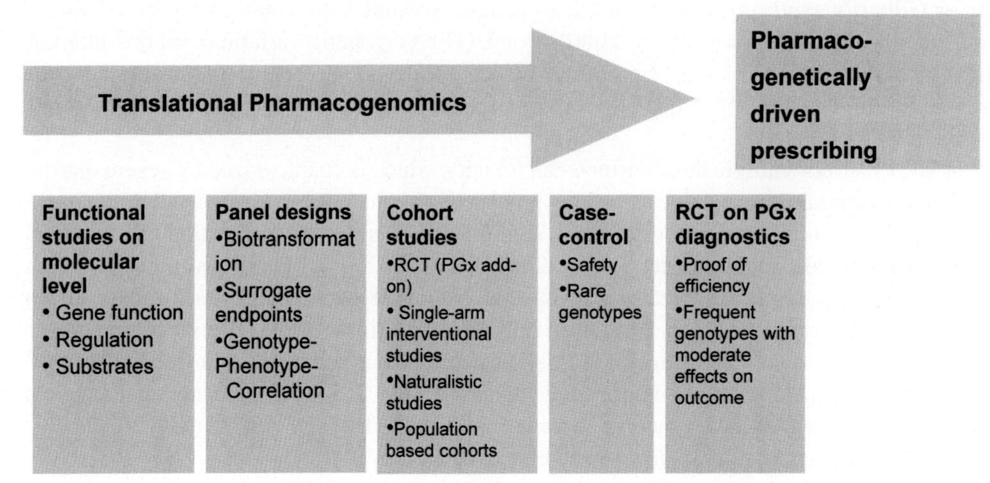

FIGURE 2.11 The flow of pharmacogenetic studies necessary for translation of pharmacogenetics to clinical application is depicted. After preclinical characterization of the respective variant, several rational clinical study designs might help to provide the evidence needed to support pharmacogenetically driven prescribing.

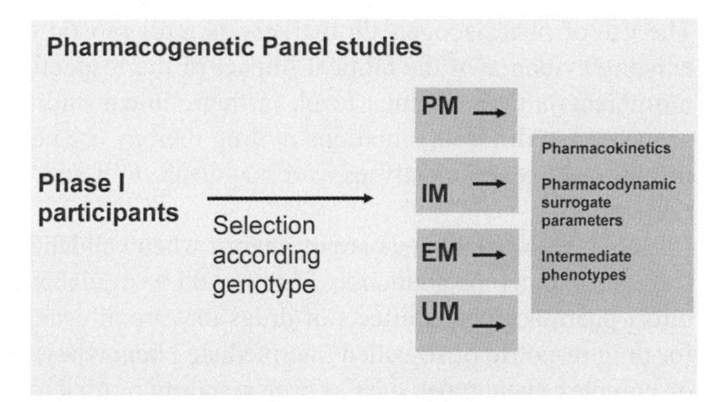

FIGURE 2.12 The principle of the panel study design for pharmacogenetic studies is depicted. Similar to phase I studies in drug development, a panel of healthy volunteers are selected due to their genotypes and tested for pharmacokinetic or pharmacodynamic parameters. A small sample size of homozygous and heterozygous carriers of genetic variants generates maximum statistical power to prove genotype effects.

approach usually requires a relatively large sample size but is feasible as a pharmacogenetic add-on study in single-armed or randomized controlled trials. In addition, cohort studies can also be designed in a naturalistic manner, which has the advantage that a lack of intervention more realistically depicts the true clinical situation. If informed consent for genotyping can be obtained from all patients included in the clinical trial, there is relatively little selection bias, and the data are representative for treatment with the drug in question. The true relative risk of nonresponse or adverse effects based on genetic variation, as well as the health and economic value of genotyping, can be estimated with reasonable validity from these kinds of cohort studies.

However, one main problem with the use of data from cohort studies for translating pharmacogenetics into clinical practice is the possibility of false-positive results. The lack of independent replications, small sample sizes—which make the studies heavily underpowered—and the bias of nonreporting negative study designs hamper the usefulness of the results of these trials. Recently, a structured methodology for developing pharmacogenetic predictive tests from data from association studies was proposed (Weiss et al., 2008). The first step is to identify a set of genes that replicate an association with a drug-response phenotype across several clinical trials in a single ethnic group. The second step is to try to replicate the interaction model in an independent clinical trial population and explain a significant portion of the model's sensitivity and specificity. Once replication is achieved, the third step is to define the functional variants in the genes identified. The last step is to develop a predictive model and validate the hypothetical pharmacogenetic test (Weiss et al., 2008). Figure 2.13 depicts the design of pharmacogenetic cohort studies.

In the pharmacogenetic analysis of rare adverse events, a case-control design study in which patients with a certain clinical outcome (e.g., a specific type of adverse event or nonresponse) are compared with appropriate controls who do not have such an outcome is often the only existing feasible approach. In this study design, the so-called odds ratios as indicators of the strength of genotype effects can be calculated. A case-control study may often be an appropriate clinical research method in safety-related pharmacogenetic research, but it also bears the risk of several types of bias (Rothman and Greenland, 1998). Despite these limitations, a formal and correctly performed case-control study may often be the best and only existing feasible study design because prospective cohort or controlled trials would require inclusion of 100,000 or more subjects if adverse events with frequencies of 1:5,000 or less are to be analyzed. Such studies will rarely be possible.

A randomized controlled clinical trial is generally considered the most valid design for interventions and diagnostic tests (Patsopoulos et al., 2005), and it provides the most stringent proof-of-concept available for pharmacogenetics-based optimization of drug therapy. Such valid studies on pharmacogenetic diagnostics are relatively expensive and usually involve multicenter trials. The design of a controlled pharmacogenetic prospective trial is usually suited for common drug therapies and for rather common genetic variants. At the time of this writing, several international multicenter trials on prospective evaluation of genotype-specific dosing—such as treatment with oral anticoagulants, tamoxifen, abacavir, isoniazid, proton pump inhibitors, and azathioprine—are ongoing in several fields of drug therapy. The results of these trials will finally give the information needed on cost and clinical benefit of pharmacogenetic diagnostics in the specific fields of drug therapy.

In conclusion, in order to achieve translation of pharmacogenetic data into clinical practice, data from a variety of clinical studies in different fields of drug therapy are required, and large cooperative approaches are necessary to obtain the critical mass of evidence needed to convince clinicians and health-care providers to use and pay for pharmacogenetic diagnostics as a tool to improve and individualize patient treatment.

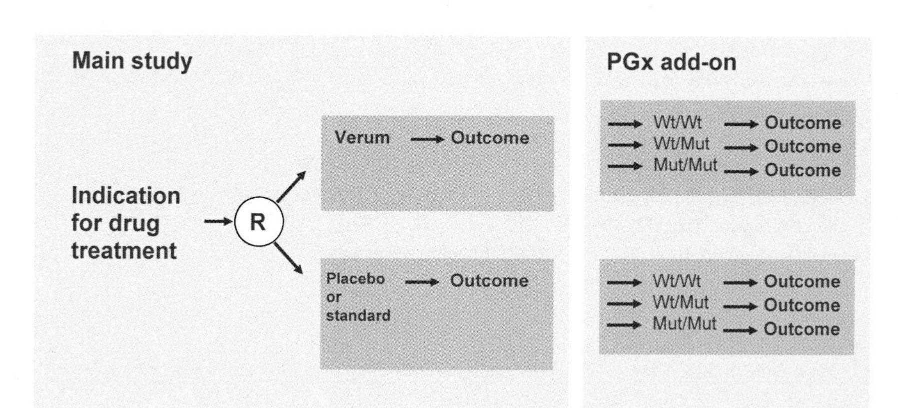

FIGURE 2.13 The design of cohort studies with pharmacogenetic add-ons is depicted. In a two-armed trial, the role of pharmacogenetic variants can be tested in one drug therapy and can be compared in another arm, which is often therapy as usual. Data on outcome and toxicity can be compared between the two groups, and the number needed to genotype can be assessed to quantify the importance of the genetic variant. Wt: wildtype; Mut: mutant

REFERENCES

Andersson, T., Holmberg, J., Rohss, K., Walan, A., 1998. Pharmacokinetics and effect on caffeine metabolism of the proton pump inhibitors, omeprazole, lansoprazole, and pantoprazole. Br. J. Clin. Pharmacol. 45, 369–375.

Araki, K., Fujita, K., Ando, Y., Nagashima, F., Yamamoto, W., Endo, H., et al., 2006. Pharmacogenetic impact of polymorphisms in the coding region of the UGT1A1 gene on SN-38 glucuronidation in Japanese patients with cancer. Cancer Sci. 97, 1255–1259.

Bauer, M., Whybrow, P.C., Angst, J., Versiani, M., Moller, H.J., 2002. World Federation of Societies of Biological Psychiatry (WFSBP) Guidelines for Biological Treatment of Unipolar Depressive Disorders, Part 1: Acute and continuation treatment of major depressive disorder. World. J. Biol. Psychiatry 3, 5–43.

Bijl, M.J., Visser, L.E., Hofman, A., Vulto, A.G., van Gelder, T., Stricker, B.H., et al., 2008. Influence of the CYP2D6*4 polymorphism on dose, switching and discontinuation of antidepressants. Br. J. Clin. Pharmacol. 65, 558–564.

Black, A.J., McLeod, H.L., Capell, H.A., Powrie, R.H., Matowe, L.K., Pritchard, S.C., et al., 1998. Thiopurine methyltransferase genotype predicts therapy-limiting severe toxicity from azathioprine. Ann. Intern. Med. 129, 716–718.

Borges, S., Desta, Z., Li, L., Skaar, T.C., Ward, B.A., Nguyen, A., et al., 2006. Quantitative effect of CYP2D6 genotype and inhibitors on tamoxifen metabolism: implication for optimization of breast cancer treatment. Clin. Pharmacol. Ther. 80, 61–74.

Brauch, H., Schroth, W., Goetz, M.P., Murdter, T.E., Winter, S., Ingle, J.N., et al., 2013. Tamoxifen use in postmenopausal breast cancer: CYP2D6 matters. J. Clin. Oncol. 31, 176–180.

Brenner, S.S., Herrlinger, C., Dilger, K., Murdter, T.E., Hofmann, U., Marx, C., et al., 2003. Influence of age and cytochrome P450 2C9 genotype on the steady-state disposition of diclofenac and celecoxib. Clin. Pharmacokinet. 42, 283–292.

Buettner, C., Rippberger, M.J., Smith, J.K., Leveille, S.G., Davis, R.B., Mittleman, M.A., 2012. Statin use and musculoskeletal pain among adults with and without arthritis. Am. J. Med. 125, 176–182.

Caudle, K.E., Klein, T.E., Hoffman, J.M., Muller, D.J., Whirl-Carrillo, M., Gong, L., et al., 2014. Incorporation of pharmacogenomics into routine clinical practice: the Clinical Pharmacogenetics Implementation Consortium (CPIC) guideline development process. Curr. Drug Metab. 15, 209–217.

Caudle, K.E., Thorn, C.F., Klein, T.E., Swen, J.J., McLeod, H.L., Diasio, R.B., et al., 2013. Clinical Pharmacogenetics Implementation Consortium guidelines for dihydropyrimidine dehydrogenase genotype and fluoropyrimidine dosing. Clin. Pharmacol. Ther. 94, 640–645.

Chong, S.A., Tan, E.C., Tan, C.H., Mythily, Chan, Y.H., 2003. Polymorphisms of dopamine receptors and tardive dyskinesia among Chinese patients with schizophrenia. Am J. Med. Genet. B. 116, 51–54.

Chou, W.H., Yan, F.X., de Leon, J., Barnhill, J., Rogers, T., Cronin, M., et al., 2000. Extension of a pilot study: impact from the cytochrome P450 2D6 polymorphism on outcome and costs associated with severe mental illness. J. Clin. Psychopharmacol. 20, 246–251.

Coller, J.K., Krebsfaenger, N., Klein, K., Endrizzi, K., Wolbold, R., Lang, T., et al., 2002. The influence of CYP2B6, CYP2C9 and CYP2D6 genotypes on the formation of the potent antioestrogen Z-4-hydroxy-tamoxifen in human liver. Br. J. Clin. Pharmacol. 54, 157–167.

de Morais, S.M., Wilkinson, G.R., Blaisdell, J., Nakamura, K., Meyer, U.A., Goldstein, J.A., 1994. The major genetic defect responsible for the polymorphism of S-mephenytoin metabolism in humans. J. Biol. Chem. 269, 15419–15422.

Desta, Z., Ward, B.A., Soukhova, N.V., Flockhart, D.A., 2004. Comprehensive evaluation of tamoxifen sequential biotransformation by the human cytochrome P450 system in vitro: prominent roles for CYP3A and CYP2D6. J. Pharmacol. Exp. Ther. 310, 1062–1075.

Diasio, R.B., Beavers, T.L., Carpenter, J.T., 1988. Familial deficiency of dihydropyrimidine dehydrogenase. Biochemical basis for familial pyrimidinemia and severe 5-fluorouracil-induced toxicity. J. Clin. Invest. 81, 47–51.

Dorado, P., Berecz, R., Norberto, M.J., Yasar, U., Dahl, M.L., Llerena, A., 2003. CYP2C9 genotypes and diclofenac metabolism in Spanish healthy volunteers. Eur. J. Clin. Pharmacol. 59, 221–225.

Etienne, M.C., Lagrange, J.L., Dassonville, O., Fleming, R., Thyss, A., Renee, N., et al., 1994. Population study of dihydropyrimidine dehydrogenase in cancer patients. J. Clin. Oncol. 12, 2248–2253.

Evans, W.E., Rodman, J.H., Relling, M.V., Crom, W.R., Rivera, G.K., Pratt, C.B., et al., 1991. Concept of maximum tolerated systemic exposure and its application to phase I-II studies of anticancer drugs. Med. Pediatr. Oncol. 19, 153–159.

Fukuda, T., Nishida, Y., Zhou, Q., Yamamoto, I., Kondo, S., Azuma, J., 2000. The impact of the CYP2D6 and CYP2C19 genotypes on venlafaxine pharmacokinetics in a Japanese population. Eur. J. Clin. Pharmacol. 56, 175–180.

Fukuda, T., Yamamoto, I., Nishida, Y., Zhou, Q., Ohno, M., Takada, K., et al., 1999. Effect of the CYP2D6*10 genotype on venlafaxine pharmacokinetics in healthy adult volunteers. Br. J. Clin. Pharmacol. 47, 450–453.

Furuta, T., Shirai, N., Kodaira, M., Sugimoto, M., Nogaki, A., Kuriyama, S., et al., 2007. Pharmacogenomics-based tailored versus standard therapeutic regimen for eradication of H. pylori. Clin. Pharmacol. Ther. 81, 521–528.

Furuta, T., Shirai, N., Takashima, M., Xiao, F., Hanai, H., Nakagawa, K., et al., 2001. Effects of genotypic differences in CYP2C19 status on cure rates for *Helicobacter pylori* infection by dual therapy with rabeprazole plus amoxicillin. Pharmacogenetics 11, 341–348.

Furuta, T., Shirai, N., Watanabe, F., Honda, S., Takeuchi, K., Iida, T., et al., 2002. Effect of cytochrome P4502C19 genotypic differences on cure rates for gastroesophageal reflux disease by lansoprazole. Clin. Pharmacol. Ther. 72, 453–460.

Geisen, C., Watzka, M., Sittinger, K., Steffens, M., Daugela, L., Seifried, E., et al., 2005. VKORC1 haplotypes and their impact on the inter-individual and inter-ethnical variability of oral anticoagulation. Thromb. Haemost. 94, 773–779.

Goetz, M.P., Knox, S.K., Suman, V.J., Rae, J.M., Safgren, S.L., Ames, M.M., et al., 2007. The impact of cytochrome P450 2D6 metabolism in women receiving adjuvant tamoxifen. Breast Cancer Res. Treat. 101, 113–121.

Graff, D.W., Williamson, K.M., Pieper, J.A., Carson, S.W., Adams Jr., K.F., Cascio, W.E., et al., 2001. Effect of fluoxetine on carvedilol pharmacokinetics, CYP2D6 activity, and autonomic balance in heart failure patients. J. Clin. Pharmacol. 41, 97–106.

Grasmader, K., Verwohlt, P.L., Rietschel, M., Dragicevic, A., Muller, M., Hiemke, C., et al., 2004. Impact of polymorphisms of cytochrome-P450 isoenzymes 2C9, 2C19 and 2D6 on plasma concentrations and clinical effects of antidepressants in a naturalistic clinical setting. Eur. J. Clin. Pharmacol. 60, 329–336.

Hamadeh, I.S., Langaee, T.Y., Dwivedi, R., Garcia, S., Burkley, B.M., Skaar, T.C., et al., 2014. Impact of CYP2D6 polymorphisms on clinical efficacy and tolerability of metoprolol tartrate. Clin. Pharmacol. Ther. 96, 175–181.

Heggie, G.D., Sommadossi, J.P., Cross, D.S., Huster, W.J., Diasio, R.B., 1987. Clinical pharmacokinetics of 5-fluorouracil and its metabolites in plasma, urine, and bile. Cancer Res. 47, 2203–2206.

Hicks, J.K., Swen, J.J., Thorn, C.F., Sangkuhl, K., Kharasch, E.D., Ellingrod, V.L., et al., 2013. Clinical Pharmacogenetics Implementation Consortium guideline for CYP2D6 and CYP2C19 genotypes and dosing of tricyclic antidepressants. Clin. Pharmacol. Ther. 93, 402–408.

Innocenti, F., Iyer, L., Ratain, M.J., 2001. Pharmacogenetics of anticancer agents: lessons from amonafide and irinotecan. Drug Metab. Dispos. 29, 596–600.

Johnson, J.A., Gong, L., Whirl-Carrillo, M., Gage, B.F., Scott, S.A., Stein, C.M., et al., 2011. Clinical Pharmacogenetics Implementation Consortium Guidelines for CYP2C9 and VKORC1 genotypes and warfarin dosing. Clin. Pharmacol. Ther. 90, 625–629.

Kaskas, B.A., Louis, E., Hindorf, U., Schaeffeler, E., Deflandre, J., Graepler, F., et al., 2003. Safe treatment of thiopurine S-methyltransferase deficient Crohn's disease patients with azathioprine. Gut 52, 140–142.

Kawamura, M., Ohara, S., Koike, T., Iijima, K., Suzuki, H., Kayaba, S., et al., 2007. Cytochrome P450 2C19 polymorphism influences the preventive effect of lansoprazole on the recurrence of erosive reflux esophagitis. J. Gastroenterol. Hepatol. 22, 222–226.

Kawamura, M., Ohara, S., Koike, T., Iijima, K., Suzuki, J., Kayaba, S., et al., 2003. The effects of lansoprazole on erosive reflux oesophagitis are influenced by CYP2C19 polymorphism. Aliment. Pharmacol. Ther. 17, 965–973.

Kawanishi, C., Lundgren, S., Agren, H., Bertilsson, L., 2004. Increased incidence of CYP2D6 gene duplication in patients with persistent mood disorders: ultrarapid metabolism of antidepressants as a cause of nonresponse. A pilot study. Eur. J. Clin. Pharmacol. 59, 803–807.

Kimmel, S.E., French, B., Kasner, S.E., Johnson, J.A., Anderson, J.L., Gage, B.F., et al., 2013. A pharmacogenetic versus a clinical algorithm for warfarin dosing. N. Engl. J. Med. 369, 2283–2293.

Kirchheiner, J., Heesch, C., Bauer, S., Meisel, C., Seringer, A., Goldammer, M., et al., 2004a. Impact of the ultrarapid metabolizer genotype of cytochrome P450 2D6 on metoprolol pharmacokinetics and pharmacodynamics. Clin. Pharmacol. Ther. 76, 302–312.

Kirchheiner, J., Meineke, I., Steinbach, N., Meisel, C., Roots, I., Brockmoller, J., 2003. Pharmacokinetics of diclofenac and inhibition of cyclooxygenases 1 and 2: no relationship to the CYP2C9 genetic polymorphism in humans. Br. J. Clin. Pharmacol. 55, 51–61.

Kirchheiner, J., Nickchen, K., Bauer, M., Wong, M.L., Licinio, J., Roots, I., et al., 2004b. Pharmacogenetics of antidepressants and antipsychotics: the contribution of allelic variations to the phenotype of drug response. Mol. Psychiatry 9, 442–473.

Klotz, U., Schwab, M., Treiber, G., 2004. CYP2C19 polymorphism and proton pump inhibitors. Basic Clin. Pharmacol. Toxicol. 95, 2–8.

Lazarou, J., Pomeranz, B.H., Corey, P.N., 1998. Incidence of adverse drug reactions in hospitalized patients: a meta-analysis of prospective studies. JAMA 279, 1200–1205.

Lennard, L., Lilleyman, J.S., Van Loon, J., Weinshilboum, R.M., 1990. Genetic variation in response to 6-mercaptopurine for childhood acute lymphoblastic leukaemia. Lancet 336, 225–229.

Lennard, M.S., Silas, J.H., Freestone, S., Ramsay, L.E., Tucker, G.T., Woods, H.F., 1982a. Oxidation phenotype—a major determinant of metoprolol metabolism and response. N. Engl. J. Med. 307, 1558–1560.

Lennard, M.S., Silas, J.H., Freestone, S., Trevethick, J., 1982b. Defective metabolism of metoprolol in poor hydroxylators of debrisoquine. Br. J. Clin. Pharmacol. 14, 301–303.

Lessard, E., Yessine, M., Hamelin, B., O'Hara, G., LeBlanc, J., Turgeon, J., 1999. Influence of CYP2D6 activity on the disposition and cardiovascular toxicity of the antidepressant agent venlafaxine in humans. Pharmacogenetics 9, 435–443.

Loebstein, R., Yonath, H., Peleg, D., Almog, S., Rotenberg, M., Lubetsky, A., et al., 2001. Interindividual variability in sensitivity to warfarin—nature or nurture? Clin. Pharmacol. Ther. 70, 159–164.

Maitland, M.L., Vasisht, K., Ratain, M.J., 2006. TPMT, UGT1A1 and DPYD: genotyping to ensure safer cancer therapy? Trends Pharmacol. Sci. 27, 432–437.

Mulder, H., Heerdink, E.R., van Iersel, E.E., Wilmink, F.W., Egberts, A.C., 2007. Prevalence of patients using drugs metabolized by cytochrome P450 2D6 in different populations: a cross-sectional study. Ann. Pharmacother. 41, 408–413.

Neugebauer, G., Neubert, P., 1991. Metabolism of carvedilol in man. Eur. J. Drug Metab. Pharmacokinet. 16, 257–260.

Niemi, M., 2010. Transporter pharmacogenetics and statin toxicity. Clin. Pharmacol. Ther. 87, 130–133.

Niemi, M., Pasanen, M.K., Neuvonen, P.J., 2011. Organic anion transporting polypeptide 1B1: a genetically polymorphic transporter of major importance for hepatic drug uptake. Pharmacol. Rev. 63, 157–181.

Offer, S.M., Fossum, C.C., Wegner, N.J., Stuflesser, A.J., Butterfield, G.L., Diasio, R.B., 2014. Comparative functional analysis of DPYD variants of potential clinical relevance to dihydropyrimidine dehydrogenase activity. Cancer Res. 74, 2545–2554.

Osman, A., Enstrom, C., Arbring, K., Soderkvist, P., Lindahl, T.L., 2006. Main haplotypes and mutational analysis of vitamin K epoxide reductase (VKORC1) in a Swedish population: a retrospective analysis of case records. J. Thromb. Haemost. 4, 1723–1729.

Otton, S.V., Ball, S.E., Cheung, S.W., Inaba, T., Rudolph, R.L., Sellers, E.M., 1996. Venlafaxine oxidation in vitro is catalysed by CYP2D6. Br. J. Clin. Pharmacol. 41, 149–156.

Patsopoulos, N.A., Analatos, A.A., Ioannidis, J.P., 2005. Relative citation impact of various study designs in the health sciences. JAMA 293, 2362–2366.

Pirmohamed, M., Burnside, G., Eriksson, N., Jorgensen, A.L., Toh, C.H., Nicholson, T., et al., 2013. A randomized trial of genotype-guided dosing of warfarin. N. Engl. J. Med. 369, 2294–2303.

Pirmohamed, M., James, S., Meakin, S., Green, C., Scott, A.K., Walley, T.J., et al., 2004. Adverse drug reactions as cause of admission to hospital: prospective analysis of 18 820 patients. BMJ 329, 15–19.

Quteineh, L., Verstuyft, C., Descot, C., Dubert, L., Robert, A., Jaillon, P., et al., 2005. Vitamin K epoxide reductase (VKORC1) genetic polymorphism is associated to oral anticoagulant overdose. Thromb. Haemost. 94, 690–691.

Rau, T., Heide, R., Bergmann, K., Wuttke, H., Werner, U., Feifel, N., et al., 2002. Effect of the CYP2D6 genotype on metoprolol metabolism persists during long-term treatment. Pharmacogenetics 12, 465–472.

Rau, T., Wohlleben, G., Wuttke, H., Thuerauf, N., Lunkenheimer, J., Lanczik, M., et al., 2004. CYP2D6 genotype: impact on adverse effects and nonresponse during treatment with antidepressants—a pilot study. Clin. Pharmacol. Ther. 75, 386–393.

Reitsma, P.H., van der Heijden, J.F., Groot, A.P., Rosendaal, F.R., Buller, H.R., 2005. A C1173T dimorphism in the VKORC1 gene determines coumarin sensitivity and bleeding risk. PLoS Med. 2, e312.

Relling, M.V., Gardner, E.E., Sandborn, W.J., Schmiegelow, K., Pui, C.H., Yee, S.W., et al., 2011. Clinical Pharmacogenetics Implementation Consortium guidelines for thiopurine methyltransferase genotype and thiopurine dosing. Clin. Pharmacol. Ther. 89, 387–391.

Relling, M.V., Gardner, E.E., Sandborn, W.J., Schmiegelow, K., Pui, C.H., Yee, S.W., et al., 2013. Clinical Pharmacogenetics Implementation Consortium guidelines for thiopurine methyltransferase genotype and thiopurine dosing: 2013 update. Clin. Pharmacol. Ther. 93, 324–325.

Relling, M.V., Hancock, M.L., Rivera, G.K., Sandlund, J.T., Ribeiro, R.C., Krynetski, et al., 1999. Mercaptopurine therapy intolerance and heterozygosity at the thiopurine S-methyltransferase gene locus. J. Natl. Cancer Inst. 91, 2001–2008.

Relling, M.V., Klein, T.E., 2011. CPIC: Clinical Pharmacogenetics Implementation Consortium of the Pharmacogenomics Research Network. Clin. Pharmacol. Ther. 89, 464–467.

Rendic, S., 2002. Summary of information on human CYP enzymes: human P450 metabolism data. Drug Metab. Rev. 34, 83–448.

Rieder, M.J., Reiner, A.P., Gage, B.F., Nickerson, D.A., Eby, C.S., McLeod, H.L., et al., 2005. Effect of VKORC1 haplotypes on transcriptional regulation and warfarin dose. N. Engl. J. Med. 352, 2285–2293.

Robien, K., Boynton, A., Ulrich, C.M., 2005. Pharmacogenetics of folate-related drug targets in cancer treatment. Pharmacogenomics 6, 673–689.

Rost, S., Fregin, A., Ivaskevicius, V., Conzelmann, E., Hortnagel, K., Pelz, H.J., et al., 2004. Mutations in VKORC1 cause warfarin resistance and multiple coagulation factor deficiency type 2. Nature 427, 537–541.

Rothman, K., Greenland, S., 1998. Modern Epidemiology. Lippincott Williams & Wilkins, Philadelphia.

Ruano, G., Szarek, B.L., Villagra, D., Gorowski, K., Kocherla, M., Seip, R.L., et al., 2013. Length of psychiatric hospitalization is correlated with CYP2D6 functional status in inpatients with major depressive disorder. Biomark. Med. 7, 429–439.

Sai, K., Saeki, M., Saito, Y., Ozawa, S., Katori, N., Jinno, H., et al., 2004. UGT1A1 haplotypes associated with reduced glucuronidation and increased serum bilirubin in irinotecan-administered Japanese patients with cancer. Clin. Pharmacol. Ther. 75, 501–515.

Schaeffeler, E., Eichelbaum, M., Reinisch, W., Zanger, U.M., Schwab, M., 2006. Three novel thiopurine S-methyltransferase allelic variants (TPMT*20, *21, *22)—association with decreased enzyme function. Hum. Mutat. 27, 976.

Schaeffeler, E., Fischer, C., Brockmeier, D., Wernet, D., Moerike, K., Eichelbaum, M., et al., 2004. Comprehensive analysis of thiopurine S-methyltransferase phenotype-genotype correlation in a large population of German-Caucasians and identification of novel TPMT variants. Pharmacogenetics 14, 407–417.

Schneeweiss, S., Hasford, J., Gottler, M., Hoffmann, A., Riethling, A.K., Avorn, J., 2002. Admissions caused by adverse drug events to internal medicine and emergency departments in hospitals: a longitudinal population-based study. Eur. J. Clin. Pharmacol. 58, 285–291.

Schwab, M., Schaeffeler, E., Klotz, U., Treiber, G., 2004. CYP2C19 polymorphism is a major predictor of treatment failure in white patients by use of lansoprazole-based quadruple therapy for eradication of Helicobacter pylori. Clin. Pharmacol. Ther. 76, 201–209.

Schwab, M., Zanger, U.M., Marx, C., Schaeffeler, E., Klein, K., Dippon, J., et al., 2008. Role of genetic and nongenetic factors for fluorouracil treatment-related severe toxicity: a prospective clinical trial by the German 5-FU Toxicity Study Group. J. Clin. Oncol. 26, 2131–2138.

Scordo, M.G., Pengo, V., Spina, E., Dahl, M.L., Gusella, M., Padrini, R., 2002. Influence of CYP2C9 and CYP2C19 genetic polymorphisms on warfarin maintenance dose and metabolic clearance. Clin. Pharmacol. Ther. 72, 702–710.

Sebbag, L., Boucher, P., Davelu, P., Boissonnat, P., Champsaur, G., Ninet, J., et al., 2000. Thiopurine S-methyltransferase gene polymorphism is predictive of azathioprine-induced myelosuppression in heart transplant recipients. Transplantation 69, 1524–1527.

Shepherd, G., Mohorn, P., Yacoub, K., May, D.W., 2012. Adverse drug reaction deaths reported in United States vital statistics, 1999–2006. Ann. Pharmacother. 46, 169–175.

Sim, S.C., Risinger, C., Dahl, M.L., Aklillu, E., Christensen, M., Bertilsson, L., et al., 2006. A common novel CYP2C19 gene variant causes ultrarapid drug metabolism relevant for the drug response to proton pump inhibitors and antidepressants. Clin. Pharmacol. Ther. 79, 103–113.

Soong, R., Diasio, R.B., 2005. Advances and challenges in fluoropyrimidine pharmacogenomics and pharmacogenetics. Pharmacogenomics 6, 835–847.

Spear, B.B., Heath-Chiozzi, M., Huff, J., 2001. Clinical application of pharmacogenetics. Trends Mol. Med. 7, 201–204.

Stanulla, M., Schaeffeler, E., Flohr, T., Cario, G., Schrauder, A., Zimmermann, M., et al., 2005. Thiopurine methyltransferase (TPMT) genotype and early treatment response to mercaptopurine in childhood acute lymphoblastic leukemia. JAMA 293, 1485–1489.

Stingl, J.C., Brockmoller, J., Viviani, R., 2013. Genetic variability of drug-metabolizing enzymes: the dual impact on psychiatric therapy and regulation of brain function. Mol. Psychiatry 18, 273–287.

Swen, J.J., Nijenhuis, M., de Boer, A., Grandia, L., Maitland-van der Zee, A.H., Mulder, H., et al., 2011. Pharmacogenetics: from bench to byte—an update of guidelines. Clin. Pharmacol. Ther. 89, 662–673.

Takahashi, H., Kashima, T., Nomizo, Y., Muramoto, N., Shimizu, T., Nasu, K., et al., 1998. Metabolism of warfarin enantiomers in Japanese patients with heart disease having different CYP2C9 and CYP2C19 genotypes. Clin. Pharmacol. Ther. 63, 519–528.

Tassies, D., Freire, C., Pijoan, J., Maragall, S., Monteagudo, J., Ordinas, A., et al., 2002. Pharmacogenetics of acenocoumarol: cytochrome P450 CYP2C9 polymorphisms influence dose requirements and stability of anticoagulation. Haematologica 87, 1185–1191.

Teml, A., Schaeffeler, E., Herrlinger, K.R., Klotz, U., Schwab, M., 2007. Thiopurine treatment in inflammatory bowel disease: clinical pharmacology and implication of pharmacogenetically guided dosing. Clin. Pharmacokinet. 46, 187–208.

Thijssen, H.H., Drittij, M.J., Vervoort, L.M., de Vries-Hanje, J.C., 2001. Altered pharmacokinetics of R- and S-acenocoumarol in a subject heterozygous for CYP2C9*3. Clin. Pharmacol. Ther. 70, 292–298.

Thijssen, H.H., Ritzen, B., 2003. Acenocoumarol pharmacokinetics in relation to cytochrome P450 2C9 genotype. Clin. Pharmacol. Ther. 74, 61–68.

Van Kuilenburg, A.B., Van Lenthe, H., Tromp, A., Veltman, P.C., Van Gennip, A.H., 2000. Pitfalls in the diagnosis of patients with a partial dihydropyrimidine dehydrogenase deficiency. Clin. Chem. 46, 9–17.

Veefkind, A.H., Haffmans, P.M., Hoencamp, E., 2000. Venlafaxine serum levels and CYP2D6 genotype. Ther. Drug Monit. 22, 202–208.

Verstuyft, C., Morin, S., Robert, A., Loriot, M.A., Beaune, P., Jaillon, P., et al., 2001. Early acenocoumarol overanticoagulation among cytochrome P450 2C9 poor metabolizers. Pharmacogenetics 11, 735–737.

Visser, L.E., van Schaik, R.H., van Vliet, M.M., Trienekens, P.H., De Smet, P.A., Vulto, A.G., et al., 2004a. The risk of bleeding complications in patients with cytochrome P450 CYP2C9*2 or CYP2C9*3 alleles on acenocoumarol or phenprocoumon. Thromb. Haemost. 92, 61–66.

Visser, L.E., van Vliet, M., van Schaik, R.H., Kasbergen, A.A., De Smet, P.A., Vulto, A.G., et al., 2004b. The risk of overanticoagulation in patients with cytochrome P450 2C9*2 and CYP2C9*3 alleles on acenocoumarol or phenprocoumon. Pharmacogenetics 14, 27–33.

Wedlund, P.J., 2000. The CYP2C19 enzyme polymorphism. Pharmacology 61, 174–183.

Weinshilboum, R.M., Sladek, S.L., 1980. Mercaptopurine pharmacogenetics: monogenic inheritance of erythrocyte thiopurine methyltransferase activity. Am. J. Hum. Genet. 32, 651–662.

Weiss, S.T., McLeod, H.L., Flockhart, D.A., Dolan, M.E., Benowitz, N.L., Johnson, J.A., et al., 2008. Creating and evaluating genetic tests predictive of drug response. Nat. Rev. Drug Discov. 7, 568–574.

Wester, K., Jonsson, A.K., Spigset, O., Druid, H., Hagg, S., 2008. Incidence of fatal adverse drug reactions: a population based study. Br. J. Clin. Pharmacol. 65, 573–579.

Wilke, R.A., Ramsey, L.B., Johnson, S.G., Maxwell, W.D., McLeod, H.L., Voora, D., et al., 2012. The clinical pharmacogenomics implementation consortium: CPIC guideline for SLCO1B1 and simvastatin-induced myopathy. Clin. Pharmacol. Ther. 92, 112–117.

Wuttke, H., Rau, T., Heide, R., Bergmann, K., Bohm, M., Weil, J., et al., 2002. Increased frequency of cytochrome P450 2D6 poor metabolizers among patients with metoprolol-associated adverse effects. Clin. Pharmacol. Ther. 72, 429–437.

Yasar, Ü., Eliasson, E., Forslund-Bergengren, C., Tybring, G., Lind, M., Dahl, M.L., et al., 2001. The role of CYP2C9 genotype in the metabolism of diclofenac in vivo and in vitro. Pharmacol. Toxicol. 89, 106.

Zhou, H.H., Wood, A.J., 1995. Stereoselective disposition of carvedilol is determined by CYP2D6. Clin. Pharmacol. Ther. 57, 518–524.

Zineh, I., Beitelshees, A.L., Gaedigk, A., Walker, J.R., Pauly, D.F., Eberst, K., et al., 2004. Pharmacokinetics and CYP2D6 genotypes do not predict metoprolol adverse events or efficacy in hypertension. Clin. Pharmacol. Ther. 76, 536–544.

Chapter 2.1.5

Tissue Biobanks

W. Peeters, F.L. Moll, R.J. Guzman, D.P.V. de Kleijn and G. Pasterkamp

INTRODUCTION

Understanding the etiology and pathogenesis of human diseases, in order to improve early diagnosis and to allow personalized treatment, is a major priority for biomedical research. Diseases are initiated by multiple different stimuli, and progress may vary widely among affected individuals. In addition, humans respond heterogeneously when it comes to therapeutic efficacy. The determinants of the wide variations in clinical presentation and response to treatment are multifactorial in origin. Besides many other factors, the knowledge of individuals' molecular content of the affected tissue, the genetic profile, and the immunological and neurological responses are also important hallmarks in the pathogenesis of disease. The pharmaceutical industry is still struggling with disease heterogeneity and in search of biomarkers that will help to understand the underlying pathogenesis of disease and its different presentation in subgroups of patients; the purpose is to develop innovative diagnostic modalities and allow personalized medical treatments.

Biobanks include collections of tissue specimens to examine the pathological condition of diseases, including the cellular and molecular behavior of the disease, as well as the genetic predisposition. Translational biobanks also include (clinical) patient data to investigate the pathological conditions at the molecular and cellular level in relation to patient characteristics and clinical outcome. These initiatives are critical to bridge the gap between bench and bedside. Therefore, translational biobank studies have gained a central position in the search for the personalized medicinal strategy by integrating the different biological networks in a systems biology approach (Hellings et al., 2007b; Oosterhuis, Coebergh, and van Veen, 2003).

Collaborations between different medical specialties, such as surgery, pathology, and epidemiology, have inspired the initiation of tissue biobanks. The first biobank studies were based on cross-sectional designs. Characteristics of tissues obtained from diseased patients were associated with those from healthy individuals. Prospective biobank studies have gained more interest, since they provide insights into the pathogenesis of diseases over time. Upcoming research technologies and bioinformatics have consolidated the position of biobanks because they provide knowledge of systems biology and allow high-throughput screening of DNA, plasma, and tissue samples (Aggarwal and Lee, 2003; Traylor et al., 2012).

Vascular tissue biobanking research was dominated by cross-sectional studies in the beginning, which hampered research to gain insight into the progression of the disease due to the descriptive design. Over the last decade, large longitudinal prospective vascular tissue biobanks have been initiated including atherosclerotic specimens to investigate biomarkers and genetic "common variants" to determine individuals' risk for acute cardiovascular events. In this chapter we briefly discuss the position of cross-sectional designed biobanks and review the advantages, principles, and position of prospective tissue biobanking in translational research. Atherosclerotic tissue biobanking is highlighted as an example of a well-established initiative concerning a multifactorial disease that demands a multifactorial approach (Hellings et al., 2007b; Hoefer et al., 2013).

PRINCIPLES AND TYPES OF TISSUE BIOBANKS: PROS AND CONS

Biorepositories have increasingly been initiated over the last decade, contributing to numerous fundamental research studies from different medical specialties. For many years epidemiologic studies have been initiated to study potential risk factors for cardiovascular diseases. The Framingham Heart Study, founded in 1948, is the oldest multigenerational epidemiologic and biobank study on a global scale, collecting blood samples and clinical data (Dawber, Meadors, and Moore, Jr., 1951). The consequences of traditional risk factors on heart disease such as smoking, gender, and age have been explored extensively in this study, and outcomes have been implemented in guidelines for clinical practice, patient stratification, and preventive measures (Lloyd-Jones et al., 2004). However, traditional risk factors can, to some extent, predict cardiovascular outcome on a population level but lack power for individual risk assessment to identify those patients who require

Principles of Translational Science in Medicine. http://dx.doi.org/10.1016/B978-0-12-800687-0.00006-2

aggressive treatment and accurate follow-up to prevent devastating atherosclerotic events. Blood biomarker studies also suggest but do not prove the underlying pathological mechanisms of vascular diseases in the tissue. Biopsies and tissue specimens hide specific pathological information that may reveal potential mechanisms and help to elucidate the etiology and pathogenic stages of the disease process. Knowledge of disease pathogenesis facilitates the development of innovative medical therapies and new diagnostic modalities, and it enables monitoring of the efficacy of risk factor management or medical therapeutic regimes (van Lammeren et al., 2014).

Biobanks can be subdivided, based on types of biological samples, such as DNA, plasma, and tissue samples; or differ in study design, concerning cross-sectional, translational, retrospective, or prospective. Translational tissue biobanks are archived collections of human tissue samples in combination with associated individual information and clinical data. These collections are essential resources to understand the function of biomarkers including genes and proteins and cells to explore the biological network in which they are operating (Asslaber and Zatloukal, 2007). Many cross-sectional studies have been performed to unravel the pathophysiology of diseases; however, these are descriptive and lack power to examine the natural course of the disease over time. The most common biobanks include plasma or DNA samples with clinical and epidemiological baseline and follow-up data (Asslaber and Zatloukal, 2007). Plasma samples enable identification of systemic markers representing progression or regression of the disease. In the field of cardiovascular disease, plasma markers serve as surrogate markers reflecting the stage of the disease and individuals' risk for acute events. The most investigated plasma marker is C-reactive protein (CRP), which has been observed being associated with fatal and nonfatal coronary artery disease (van Wijk et al., 2013). Plasma markers may also serve as surrogate targets for measuring treatment efficacy (Barter et al., 2007; Danesh et al., 2004; Tsimikas, Willerson, and Ridker, 2006). More recently, gene profiles have been increasingly considered as valuable determinants for risk prediction and fingerprints of the underlying pathophysiology. Over the past few years, large genomewide studies have been conducted to investigate "common variants" related with cardiovascular events for individual risk assessment (Dichgans et al., 2014; Traylor et al., 2012).

As mentioned earlier, tissue biobanks have been increasingly initiated in the past decade. In the field of oncology, tissue biopsies are recognized as the gold standards for disease stage determination and for diagnostic and prognostic purposes (Pesch et al., 2014). Moreover, the choice for therapeutic intervention is based on the specimens' morphology and radiologic imaging observations. The search for a bioinformatics-based systems biology approach that integrates patient characteristics, tissue phenotypes, and genetic profiling for individual risk assessment and therapeutic target discoveries is common in the oncology research field. In other words, the tissue represents the stage of the disease and reflects patients' prognosis. Although atherosclerosis is the major killer worldwide, biobank studies remain relatively rare in the field of cardiovascular disease. Translational vascular biobank studies contribute to a minority of cardiovascular research, despite their major potential to bridge the gap between bench and bedside, leading to the development of innovative interventions.

Tissue biobanking can be a promising research initiative to unravel pathologies in the field of oncology and cardiovascular medicine, but often suffers from inherent drawbacks. First, the majority of tissue biobanks are cross-sectional, whereas large tissue collections included in prospective studies are still rare. Cross-sectional studies are descriptive and do not elucidate the pathology in relation to time. Measurements of risk management efficacy and the effect of therapeutic interventions on histology and biomarker levels also do not benefit with cross-sectional study designs (van Lammeren et al., 2014). Second, tissue biobank studies include a selective patient population, and control patients are difficult to include for obvious ethical reasons. Since patients may reveal large matches in risk profile and medication use, it can be more difficult to correct for confounding and to extrapolate the study outcome to a nondiseased population when biomarker value is suspected. Third, tissues are often dissected in an end stage of the disease, which makes it difficult to draw inferences regarding causality. Tissue samples, except for plasma or urine, are collected at the time the patient needs to be operated on, due to life-threatening progression of the disease. Alternatives for tissue collections from a relatively healthy population are scarce and difficult to arrange due to ethical reasons. Fourth, when studies on tissue biobanks are generated in a single center, indications to operate, referral patterns, pre- and postoperative medicinal policies, and many other factors may hamper the definition of a study cohort. Extrapolation of the study outcome would be impeded, and external validity may therefore be difficult to assess. Finally, long-term follow-up of large cohorts is essential to examine risk management or drug efficacy. In addition, very large numbers of specimens are inevitable to examine genetic "common variants" in relation to clinical outcome or tissue composition. This requires strong international collaborations and a very well-developed infrastructure to keep data clean and consistent.

Based on this, the determinants of a successful tissue biobank study are as follows: (1) A longitudinal prospective study design. Prospective biobank studies provide better opportunities to study the predictive value of biomarkers for the onset, progression, or recurrence of diseases over time and to investigate the influence of risk management or drug efficacy over time at biomarker level (van Lammeren et al., 2014). Prospective studies are essential to get insight in etiological and pathophysiological conditions. (2) Accurate clinical patient data documentation at entry of the study (baseline) and during follow-up. Specifically, symptoms reflecting onset and timing of the disease should be clearly documented. Risk factors, family history, and detailed monitoring of medication use including dosage and time of prescription are of great value, in order to reveal potential confounders. As mentioned earlier, a well-developed infrastructure is inevitable for the inclusion of a large population and the accurate collection of individual medical data and complete follow-up. The inclusion rate and labor to obtain the follow-up data with a low "cost to follow up" number are major challenges. (3) Tissue examination at immunohistochemical, genetic, and protein level. Concise protocols including extraction of RNA and protein are a must. Standardized protocols for quantitative assessment of protein and RNA content as well as degradation are necessary for future comparisons and improve the reproducibility. (4) Professional data and tissue management. Modern technical equipment and detailed, well-organized documentation are essential for comprehensive biobank research. The freezers encompass the full value of the biobank. Poor freezer management and inaccurate sample labeling may be devastating (Haga and Beskow, 2008). Professionalized centralized storage of biobanks within an academic center should therefore be considered. (5) A good and well-described informed consent. Clear and accurately defined informed consents are the basis of confidential tissue biobank research and enlarge the opportunities of biobanks. Unforeseen ethical issues about secondary sample use, data sharing, and assignments of ownership and intellectual properties can be avoided when they have been well and clearly documented in advance of patients' agreement of participation in the study (Haga and Beskow, 2008; Oosterhuis, Coebergh, and van Veen, 2003).

To summarize, highly developed and well-organized infrastructures, including accurate tissue storage and patient data documentation, and well-defined longitudinal prospective study designs are essential for the development of valuable biorepositories, contributing to progressive translational research and the development of individual risk prediction models and clinical innovations.

DEVELOPMENTS IN VASCULAR BIOBANKING RESEARCH AND CLINICAL RELEVANCE

Previous oncology and vascular biobank studies demonstrated that tissue could be considered as a fingerprint representing the stage, (vulnerable) condition, and progression of the local disease. On the other hand, systemic plasma studies are less specific and may be influenced by systemic pathologies or other factors such as dietary and medication use. Progression in the development of innovative technologies (genomics, proteomics, metabolomics, microarrays) has enabled high-throughput tissue research to search for biomarkers that are involved in etiology and progression of diseases (Bleijerveld et al., 2013; Chung et al., 2007; Hollywood, Brison, and Goodacre, 2006; Kullo and Cooper, 2010).

As mentioned earlier, prospective designed translational biobank studies play a fundamental role in bridging the gap between basic research and the clinical setting. Particularly in vascular translational research, the clinical application of fundamental research observations was hampered by the cross-sectional designed studies.

In 2002, the world largest prospective longitudinal atherosclerotic tissue biobank (Athero-Express) was initiated in Utrecht (the Netherlands), including specimens from patients undergoing carotid and femoral endarterectomy. The primary aim of the biobank project was to identify predictive biomarkers for systemic cardiovascular events and determine individuals' risk for acute atherosclerotic manifestation. The combination of predictive biomarkers could serve as a fingerprint to determine an individual's risk to suffer from a cardiovascular event.

In the following sections we focus on atherosclerosis and associated tissue biobanks as an example how systematic vascular tissue collections and extensive tissue screening may facilitate the research field toward biomarkers and therapeutic targets to monitor, prevent, and treat this disease. Thereby, we highlight some promising observations and discuss actual challenges in the field of cardiovascular biobanking.

Atherosclerosis

Atherosclerotic disease is a systemic disease, which is initiated early in life and presents with clinical symptoms generally after 50 years of age. It is still the most important cause of morbidity and mortality in the Western world, the East European countries, Asia, and South America, making it a worldwide health care problem. Due to the growth of the aging population and the increasing incidence of diabetes and kidney failure, the population with atherosclerotic disease is expected to increase tremendously. Despite advances in risk factor management contributing to plaque-stabilizing effects (van Lammeren et al., 2014), cardiovascular events contribute to the majority of death worldwide. Each year, 12 million patients die because of myocardial infarction or cerebrovascular accidents (Lopez et al., 2006; Mathers, Boerma, and Ma Fat, 2009). Based on the high mortality rates, there is a strong need to unravel the underlying pathologic biomarkers and genetic "common variants" to identify patients at risk for future cardiovascular events and to develop therapeutic interventions and diagnostic modalities (Bis et al., 2011; Dichgans et al., 2014; Traylor et al., 2012).

Atherosclerotic postmortem tissue biobanks have made a significant contribution to the understanding of plaque destabilization features associated with plaque rupture, subsequent intraluminal thrombosis, and arterial occlusion, which clinically presents as myocardial and cerebral infarction or vascular death (Davies, 2000; Davies et al., 1993; Verhoeven et al., 2005). The atherosclerotic plaque that has a high likelihood of rupturing is called the "vulnerable plaque." A vulnerable (unstable) plaque is characterized by an increased number of macrophages, increased expression of tissue factor, reduced levels of smooth muscle cells, a lipid core that occupies a high proportion of the overall plaque volume, and a thin fibrous cap lying over the lipid core that can easily rupture, resulting in thrombus formation (Davies, 2000).

These pathological observations have made a significant contribution to the understanding of the mechanisms of plaque rupture and luminal thrombosis. However, conclusions regarding causality and diagnostic value of the described vulnerable plaque characteristics cannot be made, since these plaque features have been described by cross-sectional studies. Prospective studies are necessary to elucidate the natural course of vulnerable plaque progression to identify the vulnerable patient who is at risk of suffering from plaque rupture. Biomarkers causally related with atherosclerotic manifestations may serve as drug targets and imaging markers to identify those advanced plaques that are prone to rupture. However, despite the lack of evidence that classical vulnerable plaque characteristics have an acceptable predictive value, these plaque features have been implemented in clinical imaging studies as a surrogate definition of the vulnerable plaque or assess plaque progression (Amirbekian et al., 2007; Polonsky et al., 2014; Schaar et al., 2007; van 't Klooster et al., 2014; Wallis de Vries et al., 2009). More recent large genomewide population studies revealed genetic profiles, the so-called common variants that are related with acute cardiovascular events (Bis et al., 2011; Dichgans et al., 2014; Ikram et al., 2009; Traylor et al., 2012). The genetic profiles elucidate new unexplored pathologic pathways, and in combination with general risk factors and tissue biomarkers, these items may hold additional predictive value for individual risk assessment.

Clinical Value of Vascular Biobanks and Diagnostic Imaging

Interest in identifying the vulnerable atherosclerotic plaque has led to the rapid development of new innovative imaging techniques. Tissue biobanks have a unique position in determination of plaque characteristics for imaging studies. For instance, the presence of atheroma is studied using Raman spectroscopy, magnetic resonance imaging (MRI), or intravascular ultrasound (IVUS) elastrography or spectroscopic intravascular photoacoustics (sIVPA) (Cascón-Pérez et al., 2013; Jansen et al., 2014; Nissen and Yock, 2001; Puppini et al., 2006; Schaar et al., 2007). Thrombus and lipid-rich plaques can also be detected by angioscopy, which allows visualization of the plaque surface by color detection (Ohtani et al., 2006). High macrophage density can be detected with thermography that registers temperature rise in the arterial wall and positron emission tomography (PET) (Casscells et al., 1996; Nahrendorf et al., 2008). Biomarkers may be detected by MRI or photon emission CT 15070807-based plaque imaging (Jaffer and Weissleder, 2005; Kietselaer et al., 2004; Sadat et al., 2014).

However, since these images of the vulnerable plaque are based on cross-sectional observations in a selected patient population, prospective tissue biobank studies have been regarded as being essential in knowing which characteristics contribute to plaque vulnerability (Vink and Pasterkamp, 2003). Due the multifactorial nature of atherosclerosis, it is too simplistic to assume that a single histological feature or systemic biomarker would suffice to identify the vulnerable plaque that will rupture and lead to a clinical event. To investigate histopathological characteristics, tissue biomarkers, and genetic profiles in relation to plaque rupture leading to clinical ischemic events, the Athero-Express biobank study was initiated in 2002. This longitudinal biobank, including atherosclerotic specimens from carotid and femoral arteries, introduced a new concept and investigated the association between local histological and molecular information in the atherosclerotic plaque with systemic clinical outcome to identify the vulnerable patient who is at risk for acute systemic ischemic events due to plaque rupture (Hellings et al., 2007b; Verhoeven et al., 2004).

The combination of prognostic atherosclerotic tissue biobank studies and upcoming technologies and bioinformatics provides a unique format to clarify the underlying pathology and natural course of cardiovascular disease and validate drug and risk management efficacy. These initiatives are an important step towards the improvement and development of diagnostic imaging modalities, therapeutic interventions, and accurate risk models with respect to personalized medicine (Hellings et al., 2012; Kullo and Cooper, 2010).

Vascular Biobank Example: The Athero-Express Biobank

The Dutch Athero-Express study is a vascular tissue biobank including atherosclerotic carotid tissue and associated clinical and personal data. Athero-Express was established in 2002 through a collaboration of the cardiology and vascular surgery departments from the University Medical Center in Utrecht and the St. Antonius Hospital in Nieuwegein. The tissue biobank has a longitudinal prospective study design with 3-year postoperative follow-up. The main objective is to determine the predictive value of local histological plaque characteristics, molecular biomarkers, and genetic profiles as determinants for local restenosis or future systemic cardiovascular events, such as myocardial infarction, stroke, or peripheral intervention. The concept is based on the knowledge that atherosclerosis is a systemic disease and that local atherosclerotic vulnerable plaque biomarkers reflect the vulnerable state of lesions through the entire arterial system. Previous studies demonstrated that plaque features correlated between different arterial segments (Vink et al., 2001).

Before the follow-up data were analyzed, the magnitude of the biobank revealed several determinants that influence the plaque phenotype. Cross-sectional studies provided the first observations from the Athero-Express biobank and demonstrated differences in plaque phenotype from symptomatic patients compared to asymptomatic patients. Lesions from symptomatic patients (TIA, stroke) demonstrated predominantly atheromatous plaque phenotype in comparison with asymptomatic lesions, which showed a more fibrous phenotype (Verhoeven et al., 2005). Subsequent studies demonstrated that atherosclerotic lesions from women are associated with a more stable plaque phenotype, indicating that women may benefit less from a carotid endarterectomy (Hellings et al., 2007a). Other parameters that strongly influence plaque characteristics are age, medication use, and time between the last clinical event and surgery (Peeters et al., 2009; van Oostrom et al., 2005).

In addition, Hellings and colleagues demonstrated that after a duration of 5 years, the phenotype of restenotic lesions is comparable to the plaque phenotype of primary symptomatic lesions (Hellings et al., 2008a). They demonstrated in another prognostic study that morphological carotid plaque characteristics, concerning macrophage infiltration and lipid core size, are associated with less restenosis during 4 years of follow-up (Hellings et al., 2008b). This is a remarkable observation, since the presence of these morphological characteristics is presumed to be a high-risk marker for local restenosis following endarterectomy. Another morphology study demonstrated that except for the vessel density within atherosclerotic lesions and plaque hemorrhage, general vulnerable carotid plaque characteristics do not hold predictive value to predict adverse systemic cardiovascular events (Hellings et al., 2010). Other prospective biomarker studies demonstrated that local plaque markers are associated with future atherosclerotic manifestations (de Kleijn et al., 2010; Peeters et al., 2011a; Peeters et al., 2011b). The composition of these predictive plaque markers may serve as a fingerprint to identify vulnerable patients and discriminate cardiovascular risk between individuals.

CHALLENGES FOR FUTURE BIOBANKS

The development of new diagnostic modalities and pharmaceutical drugs would also benefit from the discovery of new biomarkers and ultimately enable identification and treatment of patients who are prone to cardiovascular events. Tissue biobanks fulfill a pivotal role in this first step toward personalized medicine (Meyer and Ginsburg, 2002). However, the following challenges remain to improve the position and outcome of prospective tissue biobanks. (1) Validation: It is essential to validate predictive targets of interest in other populations, which is time-consuming and goes hand in hand with very high costs. (2) Incorporating gene studies: Individual genetic profiles may also play a role in disease development and progression. Genetic studies may reveal unexplored new biological pathways contributing to plaque vulnerability, or serve as predictive markers for adverse atherosclerotic events. (3) Because atherosclerotic manifestations are increasing worldwide: International collaboration in longitudinal prospective tissue biobanking projects is fundamental, since genes and environmental and behavioral conditions differ between races and populations on different continents. This will strongly improve the knowledge of atherosclerosis but requires an exceptionally well-defined study design and accurate protocols as well as good laboratory technical equipment and outstanding bioinformatics to assure high-quality research (Vaught and Lockhart, 2012).

SUMMARY

Tissue biobanking has gained a central position in the field of oncology research, and translational studies that include plasma samples and patient data are being executed on a regular basis. Serological protein expression often reflects a disease state and may serve as a surrogate measure for therapeutic efficacy. On the other hand, well-designed prospective translational tissue biorepositories are relatively rare in cardiovascular research, but gain a central role in investigating the natural course of atherosclerosis and examining risk management or drug efficacy at the tissue level. Moreover, genome studies may reveal new pathologic pathways that can directly be investigated at the biomarker level within the vascular specimens. Genetic profiles may additionally serve as predictive risk markers.

In this chapter we discussed the advantages of tissue biobanking and looked at the dissection and storage of atherosclerotic plaque as an example. Principles and developments of atherosclerotic tissue biobank research and their position in cross-sectional and prognostic translational research and the opportunities with respect to systems biology and personal medicine were discussed.

The Athero-Express study is a large-scale longitudinal tissue biobank collecting vascular arterial specimens, to investigate biomarkers for individual risk prediction. It is a unique example of the pivotal position of tissue biobanks in the field of translational vascular research, due to its prospective longitudinal design, the use of innovative technologies, and its good infrastructure. Several studies have demonstrated prognostic targets related to systemic cardiovascular manifestations of atherosclerotic disease.

With respect to future perspectives, prospective translational biobanks with long-term follow-up are the ultimate bioresources for investigating the natural course of disease progression and testing drug and risk management efficacy. Biomarker analyses and genomewide studies of large numbers of samples in combination with patient data and clinical outcome data will allow the study design to go far beyond the descriptive level of cross-sectional studies. Validation studies are inevitable for bringing the observations to the clinical setting, and international collaborations are fundamental and should be pursued. This makes the construction of tissue biobanks an increasingly interesting and challenging activity.

REFERENCES

Aggarwal, K., Lee, K.H., 2003. Functional genomics and proteomics as a foundation for systems biology. Brief Funct. Genomic. Proteomic. 2, 175–184.

Amirbekian, V., Lipinski, M.J., Briley-Saebo, K.C., Amirbekian, S., Aguinaldo, J.G., Weinreb, D.B., et al., 2007. Detecting and assessing macrophages in vivo to evaluate atherosclerosis noninvasively using molecular MRI. Proc. Natl. Acad. Sci. USA 104, 961–966.

Asslaber, M., Zatloukal, K., 2007. Biobanks: transnational, European and global networks. Brief. Funct. Genomic. Proteomic. 6, 193–201.

Barter, P., Gotto, A.M., LaRosa, J.C., Maroni, J., Szarek, M., Grundy, S.M., et al., 2007. HDL cholesterol, very low levels of LDL cholesterol, and cardiovascular events. N. Engl. J. Med. 357, 1301–1310.

Bis, J.C., Kavousi, M., Franceschini, N., Isaacs, A., Abecasis, G.R., Schminke, U., et al., 2011. CARDIoGRAM Consortium. Meta-analysis of genome-wide association studies from the CHARGE consortium identifies common variants associated with carotid intima media thickness and plaque. Nat. Genet. 43, 940–947.

Bleijerveld, O.B., Zhang, Y.N., Beldar, S., Hoefer, I.E., Sze, S.K., Pasterkamp, G., et al., 2013. Proteomics of plaques and novel sources of potential biomarkers for atherosclerosis. Proteomics. Clin. Appl. 7, 490–503.

Cascón-Pérez, J.D., de la Torre-Hernández, J.M., Ruiz-Abellón, M.C., Martínez-Pascual, M., Mármol-Lozano, R., López-Candel, J., et al., 2013. Characteristics of culprit atheromatous plaques obtained in vivo by intravascular ultrasound radiofrequency analysis: results from the CULPLAC study. Am. Heart J. 165, 400–407.

Casscells, W., Hathorn, B., David, M., Krabach, T., Vaughn, W.K., McAllister, H.A., et al., 1996. Thermal detection of cellular infiltrates in living atherosclerotic plaques: possible implications for plaque rupture and thrombosis. Lancet 347, 1447–1451.

Chung, C.H., Levy, S., Chaurand, P., Carbone, D.P., 2007. Genomics and proteomics: emerging technologies in clinical cancer research. Crit. Rev. Oncol. Hematol. 6, 1–25.

Danesh, J., Wheeler, J.G., Hirschfield, G.M., Eda, S., Eiriksdottir, G., Rumley, A., et al., 2004. C-reactive protein and other circulating markers of inflammation in the prediction of coronary heart disease. N. Engl. J. Med. 350, 1387–1397.

Davies, M.J., 2000. The pathophysiology of acute coronary syndromes. Heart 83, 361–366.

Davies, M.J., Richardson, P.D., Woolf, N., Katz, D.R., Mann, J., 1993. Risk of thrombosis in human atherosclerotic plaques: role of extracellular lipid, macrophage, and smooth muscle cell content. Br. Heart J. 69, 377–381.

Dawber, T.R., Meadors, G.F., Moore Jr., F.E., 1951. Epidemiological approaches to heart disease: the Framingham Study. Am. J. Public Health Nations Health 41, 279–281.

de Kleijn, D.P., Moll, F.L., Hellings, W.E., Ozsarlak-Sozer, G., de Bruin, P., Doevendans, P.A., et al., 2010. Local atherosclerotic plaques are a source of prognostic biomarkers for adverse cardiovascular events. Arterioscler. Thromb. Vasc. Biol. 30, 612–619.

Dichgans, M., Malik, R., König, I.R., Rosand, J., Clarke, R., Gretarsdottir, S., et al., 2014. METASTROKE Consortium; CARDIoGRAM Consortium; C4D Consortium; International Stroke Genetics Consortium. Shared genetic susceptibility to ischemic stroke and coronary artery disease: a genome-wide analysis of common variants. Stroke 45, 24–36.

Haga, S.B., Beskow, L.M., 2008. Ethical, legal, and social implications of biobanks for genetics research. Adv. Genet. 60, 505–544.

Hellings, W.E., Moll, F.L., de Kleijn, D.P., Pasterkamp, G., 2012. 10-years experience with the Athero-Express study. Cardiovasc. Diagn. Ther. 2, 63–73.

Hellings, W.E., Moll, F.L., de Vries, J.P., de Bruin, P., de Kleijn, D.P., Pasterkamp, G., 2008a. Histological characterization of restenotic carotid plaques in relation to recurrence interval and clinical presentation: a cohort study. Stroke 39, 1029–1032.

Hellings, W.E., Moll, F.L., de Vries, J.P., Ackerstaff, R.G., Seldenrijk, K.A., Met, R., et al., 2008b. Atherosclerotic plaque composition and occurrence of restenosis after carotid endarterectomy. JAMA 299, 547–554.

Hellings, W.E., Pasterkamp, G., Verhoeven, B.A., de Kleijn, D.P., de Vries, J.P., Seldenrijk, K.A., et al., 2007a. Gender-associated differences in plaque phenotype of patients undergoing carotid endarterectomy. J. Vasc. Surg. 45, 289–296.

Hellings, W.E., Peeters, W., Moll, F.L., Pasterkamp, G., 2007b. From vulnerable plaque to vulnerable patient: the search for biomarkers of plaque destabilization. Trends Cardiovasc. Med. 17, 162–171.

Hellings, W.E., Peeters, W., Moll, F.L., Piers, S.R., van Setten, J., Van der Spek, P.J., et al., 2010. Composition of carotid atherosclerotic plaque is associated with cardiovascular outcome: a prognostic study. Circulation 121, 1941–1950.

Hoefer, I.E., Sels, J.W., Jukema, J.W., Bergheanu, S., Biessen, E., McClellan, E., et al., 2013. Circulating cells as predictors of secondary manifestations of cardiovascular disease: design of the Circulating Cells study. Clin. Res. Cardiol. 102, 847–856.

Hollywood, K., Brison, D.R., Goodacre, R., 2006. Metabolomics: current technologies and future trends. Proteomics 6, 4716–4723.

Ikram, M.A., Seshadri, S., Bis, J.C., Fornage, M., DeStefano, A.L., Aulchenko, Y.S., et al., 2009. Genomewide association studies of stroke. N. Engl. J. Med. 360, 1718–1728.

Jaffer, F.A., Weissleder, R., 2005. Molecular imaging in the clinical arena. JAMA 293, 855–862.

Jansen, K., van der Steen, A.F., Wu, M., van Beusekom, H.M., Springeling, G., Li, X., et al., 2014. Spectroscopic intravascular photoacoustic imaging of lipids in atherosclerosis. J. Biomed. Opt. 9, 026006.

Kietselaer, B.L., Reutelingsperger, C.P., Heidendal, G.A., Daemen, M.J., Mess, W.H., Hofstra, L., et al., 2004. Noninvasive detection of plaque instability with use of radiolabeled annexin A5 in patients with carotid-artery atherosclerosis. N. Engl. J. Med. 350, 1472–1473.

Kullo, I.J., Cooper, L.T., 2010. Early identification of cardiovascular risk using genomics and proteomics. Nat. Rev. Cardiol. 7, 309–317.

Lloyd-Jones, D.M., Wilson, P.W., Larson, M.G., Beiser, A., Leip, E.P., D'Agostino, R.B., et al., 2004. Framingham risk score and prediction of lifetime risk for coronary heart disease. Am. J. Cardiol. 94, 20–24.

Lopez, A.D., Mathers, C.D., Ezzati, M., Jamison, D.T., Murray, C.J., 2006. Global and regional burden of disease and risk factors, 2001: systematic analysis of population health data. Lancet 367, 1747–1757.

Mathers, C.D., Boerma, T., Ma Fat, D., 2009. Global and regional causes of death. Br. Med. Bull. 92, 7–32.

Meyer, J.M., Ginsburg, G.S., 2002. The path to personalized medicine. Curr. Opin. Chem. Biol. 6, 434–438.

Nahrendorf, M., Zhang, H., Hembrador, S., Panizzi, P., Sosnovik, D.E., Aikawa, E., et al., 2008. Nanoparticle PET-CT imaging of macrophages in inflammatory atherosclerosis. Circulation 117, 379–387.

Nissen, S.E., Yock, P., 2001. Intravascular ultrasound: novel pathophysiological insights and current clinical applications. Circulation 103, 604–616.

Ohtani, T., Ueda, Y., Mizote, I., Oyabu, J., Okada, K., Hirayama, A., et al., 2006. Number of yellow plaques detected in a coronary artery is associated with future risk of acute coronary syndrome: detection of vulnerable patients by angioscopy. J. Am. Coll. Cardiol. 47, 2194–2200.

Oosterhuis, J.W., Coebergh, J.W., van Veen, E.B., 2003. Tumour banks: well-guarded treasures in the interest of patients. Nat. Rev. Cancer 3, 73–77.

Peeters, W., de Kleijn, D.P., Vink, A., van de Weg, S., Schoneveld, A.H., Sze, S.K., et al., 2011a. Adipocyte fatty acid binding protein in atherosclerotic plaques is associated with local vulnerability and is predictive for the occurrence of adverse cardiovascular events. Eur. Heart J. 32, 1758–1768.

Peeters, W., Hellings, W.E., de Kleijn, D.P., de Vries, J.P., Moll, F.L., Vink, A., et al., 2009. Carotid atherosclerotic plaques stabilize after stroke: insights into the natural process of atherosclerotic plaque stabilization. Arterioscler. Thromb. Vasc. Biol. 29, 128–133.

Peeters, W., Moll, F.L., Vink, A., van der Spek, P.J., de Kleijn, D.P., de Vries, J.P., et al., 2011b. Collagenase matrix metalloproteinase-8 expressed in atherosclerotic carotid plaques is associated with systemic cardiovascular outcome. Eur. Heart J. 32, 2314–2325.

Pesch, B., Brüning, T., Johnen, G., Casjens, S., Bonberg, N., Taeger, D., et al., 2014. Biomarker research with prospective study designs for the early detection of cancer. Biochim. Biophys. Acta. 1844, 874–883.

Polonsky, T.S., Liu, K., Tian, L., Carr, J., Carroll, T.J., Berry, J., et al., 2014. High-risk plaque in the superficial femoral artery of people with peripheral artery disease: prevalence and associated clinical characteristics. Atherosclerosis 237, 169–176.

Puppini, G., Furlan, F., Cirota, N., Veraldi, G., Piubello, Q., Montemezzi, S., et al., 2006. Characterisation of carotid atherosclerotic plaque: comparison between magnetic resonance imaging and histology. Radiol. Med. (Torino) 111, 921–930.

Sadat, U., Jaffer, F.A., van Zandvoort, M.A., Nicholls, S.J., Ribatti, D., Gillard, J.H., 2014. Inflammation and neovascularization intertwined in atherosclerosis: imaging of structural and molecular imaging targets. Circulation 130, 786–794.

Schaar, J.A., Mastik, F., Regar, E., den Uil, C.A., Gijsen, F.J., Wentzel, J.J., et al., 2007. Current diagnostic modalities for vulnerable plaque detection. Curr. Pharm. Des. 13, 995–1001.

Traylor, M., Farrall, M., Holliday, E.G., Sudlow, C., Hopewell, J.C., Cheng, Y.C., et al., 2012. Australian Stroke Genetics Collaborative, Wellcome Trust Case Control Consortium 2 (WTCCC2), International Stroke Genetics Consortium. Genetic risk factors for ischaemic stroke and its subtypes (the METASTROKE collaboration): a meta-analysis of genome-wide association studies. Lancet Neurol. 11, 951–962.

Tsimikas, S., Willerson, J.T., Ridker, P.M., 2006. C-reactive protein and other emerging blood biomarkers to optimize risk stratification of vulnerable patients. J. Am. Coll. Cardiol. 47 (8 Suppl.), C19–C31.

van Lammeren, G.W., den Ruijter, H.M., Vrijenhoek, J.E., van der Laan, S.W., Velema, E., de Vries, J.P., et al., 2014. Time-dependent changes in atherosclerotic plaque composition in patients undergoing carotid surgery. Circulation 129, 2269–2276.

van Oostrom, O., Velema, E., Schoneveld, A.H., de Vries, J.P., de Bruin, P., Seldenrijk, C.A., et al., 2005. Age-related changes in plaque composition: a study in patients suffering from carotid artery stenosis. Cardiovasc. Pathol. 14, 126–134.

van 't Klooster, R., Truijman, M.T., van Dijk, A.C., Schreuder, F.H., Kooi, M.E., van der Lugt, A., et al., 2014. Visualization of local changes in vessel wall morphology and plaque progression in serial carotid artery magnetic resonance imaging. Stroke 45, 160–163.

van Wijk, D.F., Boekholdt, S.M., Wareham, N.J., Ahmadi-Abhari, S., Kastelein, J.J., Stroes, E.S., et al., 2013. C-reactive protein, fatal and nonfatal coronary artery disease, stroke, and peripheral artery disease in the prospective EPIC-Norfolk cohort study. Arterioscler. Thromb. Vasc. Biol. 33, 2888–2894.

Vaught, J., Lockhart, N.C., 2012. The evolution of biobanking best practices. Clin. Chim. Acta. 413, 1569–1575.

Verhoeven, B., Hellings, W.E., Moll, F.L., de Vries, J.P., de Kleijn, D.P., de Bruin, P., et al., 2005. Carotid atherosclerotic plaques in patients with transient ischemic attacks and stroke have unstable characteristics compared with plaques in asymptomatic and amaurosis fugax patients. J. Vasc. Surg. 42, 1075–1081.

Verhoeven, B.A., Velema, E., Schoneveld, A.H., de Vries, J.P., de Bruin, P., Seldenrijk, C.A., et al., 2004. Athero-Express: differential atherosclerotic plaque expression of mRNA and protein in relation to cardiovascular events and patient characteristics. Rationale and Design. Eur. J. Epidemiol. 19, 1127–1133.

Vink, A., Pasterkamp, G., 2003. Atherosclerotic plaques: how vulnerable is the definition of "the vulnerable plaque"? J. Interv. Cardiol. 16, 115–122.

Vink, A., Schoneveld, A.H., Richard, W., de Kleijn, D.P., Falk, E., Borst, C., et al., 2001. Plaque burden, arterial remodeling and plaque vulnerability: determined by systemic factors? J. Am. Coll. Cardiol. 38, 718–723.

Wallis de Vries, B.M., Hillebrands, J.L., van Dam, G.M., Tio, R.A., de Jong, J.S., Slart, R.H., et al., 2009. Images in cardiovascular medicine. Multispectral near-infrared fluorescence molecular imaging of matrix metalloproteinases in a human carotid plaque using a matrix-degrading metalloproteinase-sensitive activatable fluorescent probe. Circulation 119, 534–536.

Chapter 2.1.6

Animal Models: Value and Translational Potency

Philipp Mergenthaler and Andreas Meisel

WHAT IS THE VALUE OF ANIMAL MODELS? PATHOPHYSIOLOGICAL CONCEPTS

The majority of translational research relies on preclinical animal models. However, given an incredible number of examples of failed translation, that is, phase II or phase III clinical trials, which were not able to reproduce the beneficial effect of preclinical findings (O'Collins et al., 2006; Perrin, 2014; Prinz, Schlange, and Asadullah, 2011), the translational value of animal models has been questioned. In particular, rodent models have been accused of falsely modeling human disease conditions.

Nonetheless, many animal models are geared to replicate pathophysiological conditions found in patients. An ideal animal model of a human disease is characterized by similarities between both in terms of (1) pathophysiology, (2) phenotypical and histopathological characteristics, (3) predictive biomarkers for course or prognosis, (4) response to therapies, and (5) drug safety or toxicity (Perrin, 2014; Prabhakar, 2012).

Four types of animal models are used in preclinical research: (1) disease induction models, (2) xenograft animal models, (3) inbred strains, and (4) transgenic models (Prabhakar, 2012). The rodent stroke model of middle cerebral artery occlusion is a typical disease induction model. Xenografting, or transplantation of organs or tissues from one species into another, is often used in cancer research. "Humanized" mice are another example of xenograft models (see later). Inbred animals are genetically homogeneous, allowing investigation of pathobiology with small sample sizes (Prabhakar, 2012). Through the use of molecular biology methods, specific genes are either deleted (knock-out), mutated, or overexpressed in transgenic animals, mainly mice. Often these models are combined; e.g., disease induction models in transgenic mice are often used to investigate the contribution of specific genes in diseases.

Rodent models of cerebral ischemia are good examples of animal models that replicate human pathophysiology well (Astrup et al., 1977; Heiss, 2011). The ischemic penumbra is defined as the area surrounding the core of the ischemic lesion. While physiological cascades are compromised, this area of brain tissue can potentially be rescued by medical intervention. This concept was first described in animal models (Astrup et al., 1977; Heiss, 2011) and has since been found to be relevant for human stroke pathophysiology (Dirnagl, Iadecola, and Moskowitz, 1999; Donnan et al., 2008; Mergenthaler, Dirnagl, and Meisel, 2004; Mergenthaler and Meisel, 2012). The same has been found true for the concept of stroke-induced immunodepression. While the pathophysiological concept has initially been described in animal models (Chamorro et al., 2012; Prass et al., 2003), clinical trials have been able to replicate this concept in human stroke pathophysiology (Chamorro et al., 2012; Harms et al., 2008; Mergenthaler and Meisel, 2012), albeit therapeutic protocols making use of this concept are still under development (Mergenthaler and Meisel, 2012).

Likewise, animal models of cancer, and in particular genetically engineered mouse models, have significantly contributed to the understanding of tumor biology and cancer pathophysiology. In particular, advances in genetic engineering have allowed modeling the manifold genetic defects underlying many forms of cancer (Cheon and Orsulic, 2011). Likewise, the concept that several mutations in the genome might be required for tumor development as well as prototypic oncogenes has been established by the use of mouse models (Cheon and Orsulic, 2011). However, similar to the situation in stroke (Dirnagl and Fisher, 2012), mouse models in preclinical cancer research have yet to prove their translational capacity (Cheon and Orsulic, 2011).

WHAT IS A GOOD ANIMAL MODEL FOR TRANSLATIONAL RESEARCH?

It is clear that there is no single ideal animal model of human disease conditions. Likewise, the design of preclinical experimental studies at present offers substantial room for improvement. While this topic has recently received significant

Principles of Translational Science in Medicine. http://dx.doi.org/10.1016/B978-0-12-800687-0.00007-4

attention, many of the proposed remedies for the "translational roadblock" have yet to prove themselves in translational studies and the design of clinical trials. Among others, considering the complex characteristics of the animal models as well as of the human disease state is essential when selecting an appropriate model for preclinical studies. Three aspects are often not considered in preclinical studies: the heterogeneous nature of disease, the presence of comorbidities, and appropriate outcome measures (Mergenthaler and Meisel, 2012).

Several approaches to improve translation from animals to the clinic have been suggested. Before clinical trials are started, preclinical investigations should be performed in multiple experimental settings involving different small and large animals modeling different disease states, including the characterization of the optimal therapeutic window, optimal administration routes and schemes, as well as dose–response curves (Xiong, Mahmood, and Chopp, 2013). Furthermore, preclinical studies need to reflect the clinical scenarios. Importantly, these include relevant treatment windows and outcome parameters. For example, drug administration at onset or even before injury, as performed in many preclinical studies investigating disease mechanisms, is of minor relevance for therapy.

Most preclinical research in stroke or traumatic brain injury (TBI) suffers from short-term studies demonstrating treatment effects 1 – 7 days after the event (Xiong, Mahmood, and Chopp, 2013). Investigations on long-term outcome weeks to months after injury are still scarce. On the contrary, primary endpoints of clinical phase III trials have to focus on relevant long-term outcome measures.

Disease modeling focused on pathophysiological research is invariably an oversimplification of the clinical situation. For example, stroke patients often suffer from a variety of other diseases such as hypertension, diabetes mellitus, or chronic obstructive pulmonary disease, which are commonly not modeled. Beyond the comorbidities patients have before stroke onset, patients are often affected by several post-stroke complications, such as infection or depression, which are also usually either not modeled or not considered. The same holds true for other disease models such as TBI. Moreover, stroke patients receive a myriad of treatments including medication and general care such as nursing and physiotherapy, among others. Although stroke unit care is efficient without any doubt, we do not know which single pieces of treatment are of relevance. Nevertheless, modeling of care is probably one prerequisite in successful translation of treatment strategies of complex disorders such as stroke (Mergenthaler and Meisel, 2012).

Modeling Comorbidities

Most investigators disregard the fact that most patients are not young or middle-aged males without any comorbidities (Howells et al., 2010; Sena et al., 2010). One fundamental criticism of animal research is that most models do not consider age (Howells et al., 2010), which is one of the most relevant cofactors of outcome for most noncommunicable disorders (Howells et al., 2010; Lozano et al., 2012). However, young to middle-aged inbred rodents of one gender and of homogeneous genetic backgrounds are typically used for preclinical animal studies. Ideally, preclinical animal studies should use animal populations of mixed gender, advanced age, and with various comorbidities, such as diabetes mellitus, hyperlipidemia, hypertension, obesity, or other risk factors that are relevant for the respective human disease. Such an approach would model the human etiology of most diseases more closely. In many cases, such models are readily available (Howells et al., 2010). In addition, experimental animal populations should be increasingly complex as a therapeutic intervention advances in the translational pipeline (Figure 2.14). The concept of establishing a framework as well as funding schemes to enable such preclinical randomized controlled trials (pRCTs) has been suggested in many medical disciplines including cancer (Cheon and Orsulic, 2011) and stroke (Bath, Macleod, and Green, 2009; Dirnagl and Fisher, 2012; Mergenthaler and Meisel, 2012).

Modeling Care of Patients

Many successful therapeutic strategies rely on "intensified care" of (critically ill) patients in the acute phase of the disease on dedicated and highly specialized hospital wards. Acute care is usually complex and committed to optimize physiological parameters. Including such a strategy in preclinical modeling would aid in better modeling clinical care of patients as well as its associated complications.

In cerebral ischemia, stroke units are prepared to treat the clinical condition as well as potential complications (Donnan et al., 2008). Infections have largely been neglected in preclinical stroke research (Meisel and Meisel, 2011; Meisel et al., 2005), although they heavily influence stroke outcome (Mergenthaler and Meisel, 2012; Westendorp et al., 2011). Preventive antibacterial treatment not only prevents infections but also improves survival and neurological outcome after experimental stroke compared with placebo treatment (Meisel et al., 2004); recent phase IIb trials have successfully proven this experimental concept (Chamorro et al., 2005; Harms et al., 2008; Schwarz et al., 2008) by demonstrating that prevention of infection is effective in stroke patients (van de Beek et al., 2009). Thus, basic research findings and preclinical modeling preceded the development of this new treatment approach (Mergenthaler and Meisel, 2012).

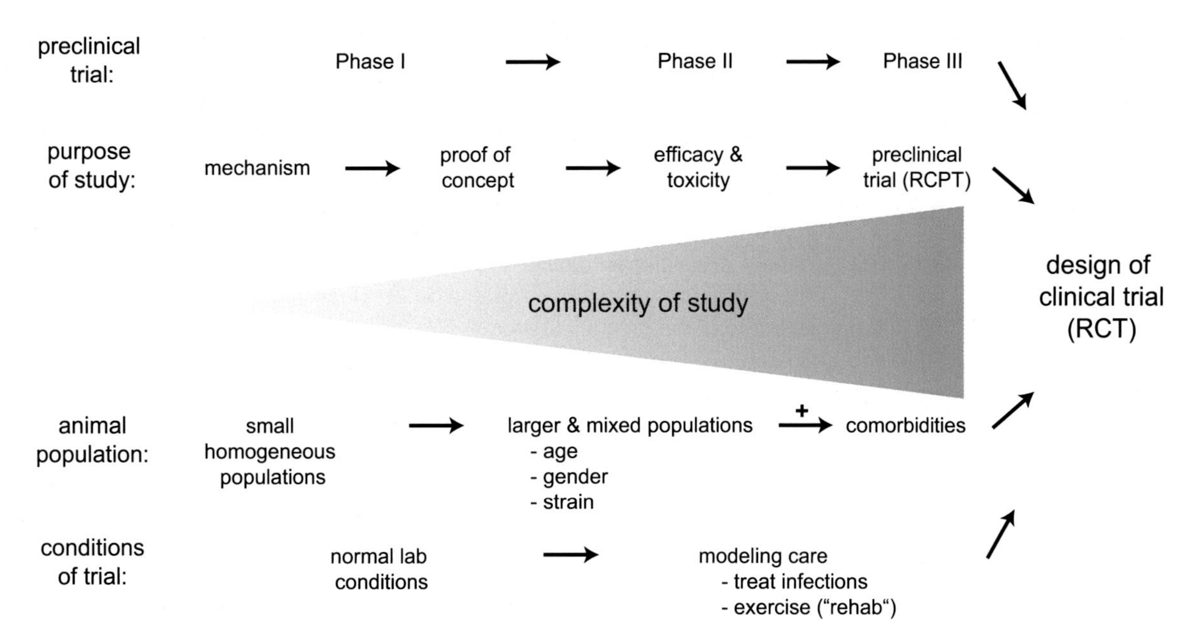

FIGURE 2.14 The preclinical trial phases of translational research. As therapeutic agents or concepts advance in development, the experimental setting increases in complexity. It ranges from small cohorts to investigate novel mechanisms to large mixed populations with (multiple) comorbidities and additional modeling of stroke care. The final stage of preclinical development is the conduct of a randomized-controlled preclinical trial (RCPT), ideally in a stroke unit setting. Randomized clinical trials (RCT) commence after this process has been completed and is based on evidence gained in preclinical testing. *(Reproduced with permission from Mergenthaler P and Meisel A. (2012). Do stroke models model stroke?* Dis. Model Mech. *5, 718–725.)*

A novel approach to preclinical research would include modeling the acute, subacute, and chronic phase of disease. Clinical and empirical evidence indicate that intensified and specialized treatments are beneficial for long-term outcome. Thus, taking "care" of patients should be reflected in future preclinical trials. In summary, preclinical trials as the foundation for future clinical trials should include large and complex cohorts of animals, and include gender-mixed, aged animals from different strains, ideally with different comorbidities, and model care of (hospitalized) patients. Furthermore, complex long-term outcome analyses should be performed to evaluate the success of a novel therapeutic concept or pharmacological agent (Figure 2.14).

WHAT IS THE TRANSLATIONAL VALUE OF ANIMAL MODELS?

Recurrent failure to translate promising treatment strategies in animal models into the clinic has challenged the value of animal research for predicting the effectiveness of treatment strategies in humans. Thus, animal models of human disorders are more and more condemned, have been considered meaningless or at best as imprecise for the human setting, and all medical areas employ models that have advantages or limitations. At least, animal models are used successfully to define basic pharmacokinetic properties as well as to investigate safety and toxicity issues (McGonigle and Ruggeri, 2014).

One example for this approach is the following. The devastating neurodegenerative disorder amyotrophic lateral sclerosis (ALS) is characterized by a progressive degeneration of motor neurons leading to a generalized paralysis, respiratory insufficiency, and death usually within 3 – 5 years. Stem cell transplantation has emerged as a promising approach for ALS patients. Rather than motor neuron replacement, current approaches consider mesenchymal or neural stem cells as supporters for motor neurons, delaying neurodegeneration. Although some ALS models suggest that stem-cell-based approaches might delay motor neuron degeneration, current strategy in the field is rather not proving efficacy than demonstrating safety in preclinical models aiming at quick "translation" to the clinical setting investigating efficacy in patients. The main argument for this approach is the rather poor understanding of the ALS pathobiology (Thomsen et al., 2014). However, whether or not this safety-focused approach in translation is successful remains to be demonstrated.

Even preclinical studies aiming at toxicity analysis might fail in predicting safety for humans. For example, the immunomodulatory humanized agonistic anti-CD28 monoclonal antibody TGN1412, which was developed for autoimmune disorders such as multiple sclerosis or rheumatoid arthritis, was tested successfully for safety in various animal models including mice. However, in the first in man (phase I) trial, TGN1412 caused a severe systemic inflammatory response syndrome due to a "cytokine storm," resulting in a disastrous outcome with a multi-organ failure for the study participants, despite the fact that the dose used was 500 times lower than the dose found to be safe in animal studies (Suntharalingam et al., 2006).

Drug discovery begins with target identification and validation and proceeds with identification and development of candidate therapeutic agents. At each step of this process, which often requires more than 12 years, animal models are needed (Whiteside, Pomonis, and Kennedy, 2013). However, only 15% of novel drugs successfully tested in animal models pass early clinical trials, and approximately half of them surviving phase III become finally approved by the regulatory authorities for clinical practice (Ledford, 2011).

Extrapolation of preclinical findings into the clinical settings might also depend on the substances under investigation. For example, animal models mimicking airway susceptibility in different lung disorders have been demonstrated to be predictive for the human situation for anesthetic drugs like halothane, isoflurane, propofol, and ketamine, but not lidocaine, morphine, or muscle relaxants. Among others, variability between species in different receptor distributions and drug affinities might account for the different predictability of the preclinical models (Habre and Petak, 2013).

Animal models of human tumors are considered as indispensable for drug discovery and development. The commonly used ectopic and orthotopic xenografts models, primary human tumorgraft models, genetically engineered models, or various multistage carcinogen-induced models all have different strengths and weaknesses (Cheon and Orsulic, 2011; Heyer et al., 2010). These models should be used as sophisticated biological tools at specific stages of drug development in a hierarchical manner of increasingly complex modeling (Figure 2.14) of the diversity of human cancers (Ruggeri, Camp, and Miknyoczki, 2014).

One approach to test the predictive power of animal models is to conduct reverse-translational studies investigating known effective treatment strategies of human disorders in appropriate animal models. Temozolomide is a good example of a successful forward and reverse translational approach for the treatment of glioblastoma. A systematic review and meta-analysis of temozolomide in animal models of glioblastoma predicted clinical efficacy. This treatment is effective in reducing tumor volume and improving survival clinically as well as in experimental models of malignant glioma. The reported efficacy for treatment has not significantly changed after publication (Hirst et al., 2013) of the seminal phase III temozolomide trial demonstrating efficacy in glioblastoma patients (Stupp et al., 2005), although evidence suggests a publication bias overemphasizing its therapeutic efficacy (Hirst et al., 2013).

Genetic mouse models of Huntington's disease (HD) should help to identify and prioritize the most promising treatment strategies to be tested in clinical trials (Menalled and Brunner, 2014). Many neural circuits affected by Huntington's disease are evolutionarily conserved. More than a dozen genetic mouse models express a mutation similar to that responsible in HD with many variations in CAG length of the Huntington gene. These models mimic the human genetic insult with different phenotypic aspects of HD (Menalled and Brunner, 2014).

Numerous transgenic or surgically induced pig models of neurodegenerative disorders have been established in order to develop cell-replacement strategies. Defining the optimal cell dose and immunosuppression protocols and testing new cell delivery devices were prerequisites for designing human clinical trial protocols in neurodegenerative disorders such as ALS, stroke, spinal cord and traumatic brain injury, Huntington's disease, Alzheimer's disease, and Parkinson's disease. In contrast to other animal models, fully or partially MHC-matched pig strains model the human situation, thereby better modeling host versus graft and graft versus host reactions of cell and tissue replacement strategies (Dolezalova et al., 2014).

In neuropathic pain research, the effect size of successful pain treatment is almost twice in animal models as in clinical trials. Correspondingly, the number needed to treat (NNT), which reflects the number of individuals that must be treated in order to see one successful treatment outcome, is almost half in animal compared to clinical pain trials. Among others, placebo effects in clinical trials, which are absent in animal research, are significant confounders. Effect sizes of at least 60% pain relief in animal models are required to predict clinical efficacy (Whiteside, Pomonis, and Kennedy, 2013).

Psychiatric disease is not directly translatable to animal models. For example, even transgenic mouse models of neuropsychiatric disorders cannot fully represent the broad spectrum of symptoms, including confusion or suicidal thoughts. However, these models serve to explore psychiatric disorders by unraveling disturbances of neural circuits underlying disease-relevant phenotypes, in particular how environmental and (epi-)genetic factors interact to shape behavioral phenotype and predispositions to psychiatric disorders (Donaldson and Hen, 2015). Traditionally, in psychiatric animal models, abnormal animal behavior was created, phenotypically resembling the aspects of mental disorders. Reverse translation using knowledge about the mechanisms of human disorders has been used to identify and develop animals that have the molecular and cellular abnormalities found in these diseases (Malkesman et al., 2009). For example, depression has been modeled in mice having point mutations in the mitochondrial DNA polymerase (Kasahara et al., 2006) and glutamate receptor 6 knockout mice have a high face and predictive validity for mania (Shaltiel et al., 2008).

"Lost in translation" has become a very popular paraphrase for the obstacles encountered in translational research. Three reasons for the "Lost in Translation Problem" have been suggested. First, small differences in the models might lead to vast differences in the results, which has been attributed to the chaotic behavior of the models and termed the "butterfly effect." Second, the effect size is decreasing from biochemical models over cell and tissue cultures to animal experiments to human studies, which seems to be unexpected according to the "princess and the pea" story. Finally, the "two cultures" of preclinical and clinical research are different (Ergorul and Levin, 2013; Mergenthaler and Meisel, 2012).

REMEDIES FOR FAILED TRANSLATION: IMPROVING PRECLINICAL RESEARCH

Improving Models

In order to improve the quality of translational biomedicine, it has been suggested to make the process of preclinical research more like clinical research. Among them, applying similar rules used by regulatory agencies for clinical trial has been suggested also for preclinical studies. Using methods such as systematic reviews and meta-analyses has become more and more popular in animal research to identify robust treatment effects. Commonly accepted "futility" and "stopping" rules in clinical research become increasingly accepted in preclinical research. These approaches have been demonstrated to improve the predictive value of animal research (Perel et al., 2007).

An ideal animal model will meet all the following three criteria: face validity, predictive validity, and construct validity. Face validity refers to the phenomenological similarity between the model and its corresponding disorder. Predictive validity refers to the ability of the model to have comparable biomarkers and treatment responses as the human disorder. Construct validity reflects the degree to which a model measures what it claims to be measuring (Willner, 1986).

In order to improve construct validity, it has been proposed that therapeutic interventions should be tested in animal models of central nervous system (CNS) disorders under conditions of greater environmental enrichment. One limitation of current research is that most animal studies are performed under caging conditions with sedentary, unstimulated animals having unlimited access to food. Enriched environments stimulating sensory system, cognition, and physical exercise have been demonstrated to affect outcome significantly (McOmish, Burrows, and Hannan, 2014).

In order to improve translational power, the use of more humanized models has been suggested (Ergorul and Levin, 2013). Immunodeficient mice that have been engrafted with human primary hematopoietic cells and tissues generating a functional human immune system in these mice are well-established examples of humanized mice. These models have been successfully used to investigate infectious diseases, autoimmune disorders, and tumors (Shultz et al., 2012).

Recent exciting findings in stem cell biology open the door to novel approaches in disease modeling. Terminally differentiated human somatic cells may be reprogrammed to a pluripotent stem cell (iPSC) state in order to then differentiate these cells into any cell type of interest (Lee and Studer, 2010). These developments might revolutionize investigations of human disorders, in particular those affecting the CNS (Philips, Rothstein, and Pouladi, 2014). Accessible cells from patients, e.g., fibroblasts from skin or monocytes from blood, might be used to generate iPSCs. These cells might be differentiated in specific neuronal subpopulations, e.g., striatal medium spiny neurons (Philips, Rothstein, and Pouladi, 2014), which are affected in brains of patients suffering from Huntington's disease. Obviously, these cells are not directly accessible, neither for study disease mechanisms nor for specific treatment. Using iPSC technology and refined genomic editing tools correction of mutations is feasible, and specific treatment is conceivable (Kaye and Finkbeiner, 2013).

However, cell-based models cannot reflect the complexity of an organism. For example, investigating systemic effects of local disease, such as post-stroke pneumonia, requires animal models (Prass et al., 2003) to complement mechanistic cellular modeling. Another example is the blood–brain barrier (BBB), a highly selective permeability barrier separating the blood from the brain extracellular fluid. Although sophisticated in vitro models of BBB have been developed in the last decade, drug transport across the BBB and brain-specific drug delivery strategies remain challenging for development of successful treatment strategies (Bicker et al., 2014). Enzymes usually cannot pass the BBB. However, local enzyme replacement therapy in the brain by intrathecal application is a promising strategy for the treatment of patients with metabolic disorders caused by the absence or malfunction enzymes involved in cerebral metabolism. For example, repeated injections of a recombinant enzyme into the spinal fluid (intrathecal) corrects enzyme deficiency and normalizes lysosomal storage in a canine model of mucopolysaccharidosis (Dickson and Chen, 2011).

Improving Rigor of Preclinical Studies

The lack of reproducibility of preclinical studies and the failure of translation to the clinic have attracted attention in the last few years (Howells, Sena, and Macleod, 2014; Ioannidis, 2005; Macleod et al., 2014; Perrin, 2014; Prinz, Schlange, and Asadullah, 2011). One important reason is the publication bias toward reporting positive results due to difficulties or missing incentives in publishing negative results (Dirnagl and Lauritzen, 2010; Dwan et al., 2013). Moreover, experimental design, including statistics, has been challenged as a quality problem in preclinical trials. For example, definition and declaration of statistical approaches and endpoint measures need to be performed before preclinical trials are finally analyzed or even started (Dirnagl and Lauritzen, 2011). Whereas clinical trial registries are widely accepted as good clinical research practice, preclinical trial registries are rather uncommon and need to be established. Thereby, post-hoc analyses generating hypotheses in an exploratory manner can be clearly distinguished from a primary hypothesis that has been tested in a confirmatory approach. *A priori* power calculations and sample size considerations, randomized assignment to groups, and blinding for treatment groups are further important issues well established in clinical but not preclinical research.

Finally, it has been suggested that bringing the rigor and quality of study design expected in clinical trials to preclinical trials will improve translational success (Dirnagl and Fisher, 2012; Macleod et al., 2014). This includes better knowledge about the drug before starting a preclinical trial. For example, pharmacokinetics might be different between mutants and wild-type mice (Menalled et al., 2010). Confirmation of research findings includes replication of preclinical research in independent laboratories (Figure 2.15). Using different models will increase robustness of the observed findings in treatment effects (Menalled and Brunner, 2014).

Endpoint measures are of great importance in preclinical research as well as in clinical research and should therefore follow endpoints used in clinical research as close as possible. For example, Huntington's disease is characterized not only by motor symptoms but also by cognitive and psychiatric symptoms appearing years before the loss of motor control. These complaints often have a large impact on the quality of life. Although survival also is an important outcome measure in clinical trials, caution is required when translating preclinical into clinical findings. In contrast to animals even in preclinical research, survival in patients depends not only on the specific intervention under investigation but also on general care as well as ethical and religious issues leading to end-of-life decisions.

Specific suggestions for improving the predictiveness of preclinical stroke research have been oriented on accepted standards of clinical research (Bath, Macleod, and Green, 2009; Macleod et al., 2009; Mergenthaler and Meisel, 2012). In order to improve internal validity, good clinical research avoids any kind of bias, in particular selection bias (biased allocation to treatment groups), performance (biased care of treatment groups apart from intervention under study), assessment (biased rating due to knowledge of treatment assignment), and attrition (biased handling of protocol violation and loss in follow-up).

Preclinical research in the final stages of translation into clinical trials should follow the guidelines of clinical research by (1) improving internal validity by predefined inclusion/exclusion criteria and primary endpoint(s), randomization, blinding for treatment allocation, and outcome assessment intention-to-treat analysis; (2) improving external validity by studying pathophysiology and treatment strategies in animals of both sexes, old age, and with comorbidities, disease-related appropriate dosing, and treatment windows for the drug under investigation; (3) replicating pivotal findings; (4) publishing negative as well as positive results; (5) focusing on long-term functional outcome; and (6) using meta-analyses of preclinical studies; (7) establishing registries of preclinical studies; and (8) establishing international multicenter phase III preclinical trials (Dirnagl and Endres, 2014; van der Worp et al., 2010). Moreover, preclinical trials need a standardized and "humanized" modeling of general as well as disease-specific patient care (Mergenthaler and Meisel, 2012).

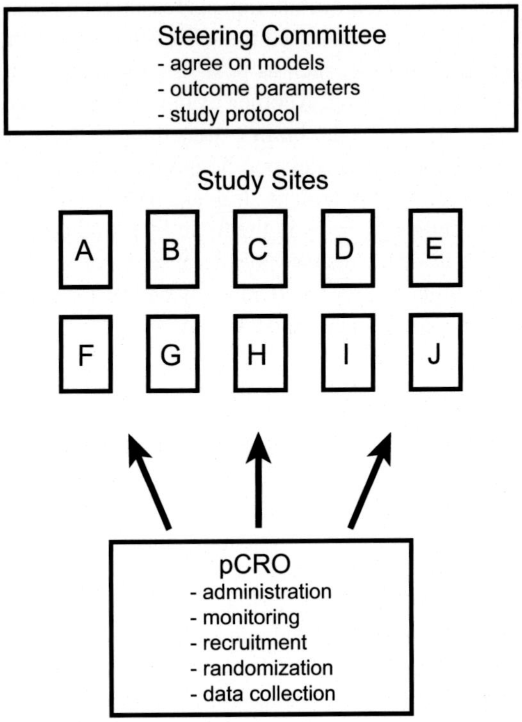

FIGURE 2.15 Modeled after randomized controlled clinical trials (RCTs), the final stage of preclinical testing is to conduct a randomized controlled preclinical trial (RCPT). A steering committee agrees on the intervention to be tested and all related aspects (e.g., models, outcome parameters, etc.). All administrative matters are centrally organized by a preclinical research organization (pCRO) and include objective criteria for the recruitment of study sites, the modes of randomization, collection of the data from the study sites, and central monitoring of all aspects of the trial. Ideally, all study sites are capable of performing the same experiments (i.e., they have access to the same models and equipment). All aspects of the RCPT are monitored by an independent organization. *(Reproduced with permission from Mergenthaler P and Meisel A. (2012). Do stroke models model stroke?* Dis Model Mech. *5, 718–725.)*

SUMMARY

In summary, many well-defined animal models for human disease are employed in modern preclinical and pathophysiology-driven research. However, the scientific community across all fields of modern biomedicine has become aware of weaknesses in current preclinical animal modeling. Here, we have outlined several strategies that have already been set into action to overcome the translational gap that is common to all current preclinical modeling of human disease.

REFERENCES

Astrup, J., Symon, L., Branston, N.M., Lassen, N.A., 1977. Cortical evoked potential and extracellular k+ and h+ at critical levels of brain ischemia. Stroke 8, 51–57.

Bath, P.M., Macleod, M.R., Green, A.R., 2009. Emulating multicentre clinical stroke trials: a new paradigm for studying novel interventions in experimental models of stroke. Int. J. Stroke 4, 471–479.

Bicker, J., Alves, G., Fortuna, A., Falcao, A., 2014. Blood-brain barrier models and their relevance for a successful development of CNS drug delivery systems: a review. Eur. J. Pharm. Biopharm. 87, 409–432.

Chamorro, A., Horcajada, J.P., Obach, V., Vargas, M., Revilla, M., Torres, F., et al., 2005. The early systemic prophylaxis of infection after stroke study: a randomized clinical trial. Stroke 36, 1495–1500.

Chamorro, A., Meisel, A., Planas, A.M., Urra, X., van de Beek, D., Veltkamp, R., 2012. The immunology of acute stroke. Nat. Rev. Neurol. 8, 401–410.

Cheon, D.J., Orsulic, S., 2011. Mouse models of cancer. Annu. Rev. Pathol. 6, 95–119.

Dickson, P.I., Chen, A.H., 2011. Intrathecal enzyme replacement therapy for mucopolysaccharidosis I: translating success in animal models to patients. Curr. Pharm. Biotechnol. 12, 946–955.

Dirnagl, U., Endres, M., 2014. Found in translation: preclinical stroke research predicts human pathophysiology, clinical phenotypes, and therapeutic outcomes. Stroke 45, 1510–1518.

Dirnagl, U., Fisher, M., 2012. International, multicenter randomized preclinical trials in translational stroke research: it's time to act. J. Cereb. Blood Flow Metab. 32, 933–935.

Dirnagl, U., Iadecola, C., Moskowitz, M.A., 1999. Pathobiology of ischaemic stroke: an integrated view. Trends Neurosci. 22, 391–397.

Dirnagl, U., Lauritzen, M., 2010. Fighting publication bias: introducing the negative results section. J. Cereb. Blood Flow Metab. 30, 1263–1264.

Dirnagl, U., Lauritzen, M., 2011. Improving the quality of biomedical research: guidelines for reporting experiments involving animals. J. Cereb. Blood Flow Metab. 31, 989–990.

Dolezalova, D., Hruska-Plochan, M., Bjarkam, C.R., Sorensen, J.C., Cunningham, M., Weingarten, D., et al., 2014. Pig models of neurodegenerative disorders: utilization in cell replacement-based preclinical safety and efficacy studies. J. Comp. Neurol. 522, 2784–2801.

Donaldson, Z.R., Hen, R., 2015. From psychiatric disorders to animal models: a bidirectional and dimensional approach. Biol. Psychiatry. 77, 15–21.

Donnan, G.A., Fisher, M., Macleod, M., Davis, S.M., 2008. Stroke. Lancet 371, 1612–1623.

Dwan, K., Gamble, C., Williamson, P.R., Kirkham, J.J., Group, Reporting Bias, 2013. Systematic review of the empirical evidence of study publication bias and outcome reporting bias—an updated review. PLoS One 8, e66844.

Ergorul, C., Levin, L.A., 2013. Solving the lost in translation problem: improving the effectiveness of translational research. Curr. Opin. Pharmacol. 13, 108–114.

Habre, W., Petak, F., 2013. Anaesthesia management of patients with airway susceptibilities: what have we learnt from animal models? Eur. J. Anaesthesiol. 30, 519–528.

Harms, H., Prass, K., Meisel, C., Klehmet, J., Rogge, W., Drenckhahn, C., et al., 2008. Preventive antibacterial therapy in acute ischemic stroke: a randomized controlled trial. PLoS One 3, e2158.

Heiss, W.D., 2011. The ischemic penumbra: correlates in imaging and implications for treatment of ischemic stroke. The Johann Jacob Wepfer Award 2011. Cerebrovasc. Dis. 32, 307–320.

Heyer, J., Kwong, L.N., Lowe, S.W., Chin, L., 2010. Non-germline genetically engineered mouse models for translational cancer research. Nat. Rev. Cancer 10, 470–480.

Hirst, T.C., Vesterinen, H.M., Sena, E.S., Egan, K.J., Macleod, M.R., Whittle, I.R., 2013. Systematic review and meta-analysis of temozolomide in animal models of glioma: was clinical efficacy predicted? Br. J. Cancer 108, 64–71.

Howells, D.W., Porritt, M.J., Rewell, S.S., O'Collins, V., Sena, E.S., van der Worp, H.B., et al., 2010. Different strokes for different folks: the rich diversity of animal models of focal cerebral ischemia. J. Cereb. Blood Flow Metab. 30, 1412–1431.

Howells, D.W., Sena, E.S., Macleod, M.R., 2014. Bringing rigour to translational medicine. Nat. Rev. Neurol. 10, 37–43.

Ioannidis, J.P., 2005. Why most published research findings are false. PLoS Med. 2, e124.

Kasahara, T., Kubota, M., Miyauchi, T., Noda, Y., Mouri, A., Nabeshima, T., et al., 2006. Mice with neuron-specific accumulation of mitochondrial DNA mutations show mood disorder-like phenotypes. Mol. Psychiatry 11, 577–593, 523.

Kaye, J.A., Finkbeiner, S., 2013. Modeling Huntington's disease with induced pluripotent stem cells. Mol. Cell Neurosci. 56, 50–64.

Ledford, H., 2011. Translational research: 4 ways to fix the clinical trial. Nature 477, 526–528.

Lee, G., Studer, L., 2010. Induced pluripotent stem cell technology for the study of human disease. Nat. Methods 7, 25–27.

Lozano, R., Naghavi, M., Foreman, K., Lim, S., Shibuya, K., Aboyans, V., et al., 2012. Global and regional mortality from 235 causes of death for 20 age groups in 1990 and 2010: a systematic analysis for the global burden of disease study 2010. Lancet 380, 2095–2128.

Macleod, M.R., Fisher, M., O'Collins, V., Sena, E.S., Dirnagl, U., Bath, P.M., et al., 2009. Good laboratory practice: preventing introduction of bias at the bench. Stroke 40, e50–52.
Macleod, M.R., Michie, S., Roberts, I., Dirnagl, U., Chalmers, I., Ioannidis, J.P., et al., 2014. Biomedical research: increasing value, reducing waste. Lancet 383, 101–104.
Malkesman, O., Austin, D.R., Chen, G., Manji, H.K., 2009. Reverse translational strategies for developing animal models of bipolar disorder. Dis. Model Mech. 2, 238–245.
McGonigle, P., Ruggeri, B., 2014. Animal models of human disease: challenges in enabling translation. Biochem. Pharmacol. 87, 162–171.
McOmish, C.E., Burrows, E.L., Hannan, A.J., 2014. Identifying novel interventional strategies for psychiatric disorders: integrating genomics, 'enviromics' and gene-environment interactions in valid preclinical models. Br. J. Pharmacol. 171, 4719–4728.
Meisel, C., Meisel, A., 2011. Suppressing immunosuppression after stroke. N. Engl. J. Med. 365, 2134–2136.
Meisel, C., Prass, K., Braun, J., Victorov, I., Wolf, T., Megow, D., et al., 2004. Preventive antibacterial treatment improves the general medical and neurological outcome in a mouse model of stroke. Stroke 35, 2–6.
Meisel, C., Schwab, J.M., Prass, K., Meisel, A., Dirnagl, U., 2005. Central nervous system injury-induced immune deficiency syndrome. Nat. Rev. Neurosci. 6, 775–786.
Menalled, L., Brunner, D., 2014. Animal models of Huntington's disease for translation to the clinic: best practices. Mov. Disord. 29, 1375–1390.
Menalled, L.B., Patry, M., Ragland, N., Lowden, P.A., Goodman, J., Minnich, J., et al., 2010. Comprehensive behavioral testing in the r6/2 mouse model of Huntington's disease shows no benefit from CoQ10 or minocycline. PLoS One 5, e9793.
Mergenthaler, P., Dirnagl, U., Meisel, A., 2004. Pathophysiology of stroke: lessons from animal models. Metab. Brain Dis. 19, 151–167.
Mergenthaler, P., Meisel, A., 2012. Do stroke models model stroke? Dis. Model Mech. 5, 718–725.
O'Collins, V.E., Macleod, M.R., Donnan, G.A., Horky, L.L., van der Worp, B.H., Howells, D.W., 2006. 1,026 experimental treatments in acute stroke. Ann. Neurol. 59, 467–477.
Perel, P., Roberts, I., Sena, E., Wheble, P., Briscoe, C., Sandercock, P., et al., 2007. Comparison of treatment effects between animal experiments and clinical trials: systematic review. BMJ 334, 197.
Perrin, S., 2014. Preclinical research: make mouse studies work. Nature 507, 423–425.
Philips, T., Rothstein, J.D., Pouladi, M.A., 2014. Preclinical models: needed in translation? A pro/con debate. Mov. Disord. 29, 1391–1396.
Prabhakar, S., 2012. Translational research challenges: finding the right animal models. J. Investig. Med. 60, 1141–1146.
Prass, K., Meisel, C., Hoflich, C., Braun, J., Halle, E., Wolf, T., et al., 2003. Stroke-induced immunodeficiency promotes spontaneous bacterial infections and is mediated by sympathetic activation reversal by poststroke T helper cell type 1-like immunostimulation. J. Exp. Med. 198, 725–736.
Prinz, F., Schlange, T., Asadullah, K., 2011. Believe it or not: how much can we rely on published data on potential drug targets? Nat. Rev. Drug Discov. 10, 712.
Ruggeri, B.A., Camp, F., Miknyoczki, S., 2014. Animal models of disease: pre-clinical animal models of cancer and their applications and utility in drug discovery. Biochem. Pharmacol. 87, 150–161.
Schwarz, S., Al-Shajlawi, F., Sick, C., Meairs, S., Hennerici, M.G., 2008. Effects of prophylactic antibiotic therapy with mezlocillin plus sulbactam on the incidence and height of fever after severe acute ischemic stroke: the Mannheim infection in stroke study (miss). Stroke 39, 1220–1227.
Sena, E.S., van der Worp, H.B., Bath, P.M., Howells, D.W., Macleod, M.R., 2010. Publication bias in reports of animal stroke studies leads to major overstatement of efficacy. PLoS Biol. 8, e1000344.
Shaltiel, G., Maeng, S., Malkesman, O., Pearson, B., Schloesser, R.J., Tragon, T., et al., 2008. Evidence for the involvement of the kainate receptor subunit GluR6 (GRIK2) in mediating behavioral displays related to behavioral symptoms of mania. Mol. Psychiatry 13, 858–872.
Shultz, L.D., Brehm, M.A., Garcia-Martinez, J.V., Greiner, D.L., 2012. Humanized mice for immune system investigation: progress, promise and challenges. Nat. Rev. Immunol. 12, 786–798.
Stupp, R., Mason, W.P., van den Bent, M.J., Weller, M., Fisher, B., Taphoorn, M.J., et al., 2005. Radiotherapy plus concomitant and adjuvant temozolomide for glioblastoma. N. Engl. J. Med. 352, 987–996.
Suntharalingam, G., Perry, M.R., Ward, S., Brett, S.J., Castello-Cortes, A., Brunner, M.D., et al., 2006. Cytokine storm in a phase 1 trial of the anti-CD28 monoclonal antibody TGN1412. N. Engl. J. Med. 355, 1018–1028.
Thomsen, G.M., Gowing, G., Svendsen, S., Svendsen, C.N., 2014. The past, present and future of stem cell clinical trials for ALS. Exp. Neurol. 262, 127–137.
van de Beek, D., Wijdicks, E.F., Vermeij, F.H., de Haan, R.J., Prins, J.M., Spanjaard, L., et al., 2009. Preventive antibiotics for infections in acute stroke: a systematic review and meta-analysis. Arch. Neurol. 66, 1076–1081.
van der Worp, H.B., Howells, D.W., Sena, E.S., Porritt, M.J., Rewell, S., O'Collins, V., et al., 2010. Can animal models of disease reliably inform human studies? PLoS Med. 7, e1000245.
Westendorp, W.F., Nederkoorn, P.J., Vermeij, J.D., Dijkgraaf, M.G., van de Beek, D., 2011. Post-stroke infection: a systematic review and meta-analysis. BMC. Neurol. 11, 110.
Whiteside, G.T., Pomonis, J.D., Kennedy, J.D., 2013. An industry perspective on the role and utility of animal models of pain in drug discovery. Neurosci. Lett. 557 (Pt A), 65–72.
Willner, P., 1986. Validation criteria for animal models of human mental disorders: learned helplessness as a paradigm case. Prog. Neuropsychopharmacol. Biol. Psychiatry 10, 677–690.
Xiong, Y., Mahmood, A., Chopp, M., 2013. Animal models of traumatic brain injury. Nat. Rev. Neurosci. 14, 128–142.

Chapter 2.1.7

Localization Technologies and Immunoassays: Promises and Benefits for Molecular Medicine

Estelle Marrer, Frank Dieterle and Jacky Vonderscher

INTRODUCTION

After the identification of biomarkers, which involves technologies allowing whole-genome, proteome, or metabolome scans, the candidate biomarkers (which will realistically number fewer than 20) have to be validated using localization techniques and immunoassays.[1] These technologies are generally more suitable for measuring biomarkers with a high throughput and at low costs. Validation is a critical step to ensure a smooth translation to the next level—clinical application and routine use. Key steps of this validation involve the following:

- Validation of a scientific hypothesis
  - Validation of genomics-derived hypotheses on the protein level
  - Cross-species validation
  - Translation into humans
- Technology development
  - Assay development (e.g., antibody generation)
  - Selection of a platform
  - Adaptation of existing methods to specific needs
- Development of the diagnostic test or medical device
  - Development of a diagnostic device (bedside device) or, alternatively, establishment and validation of a central laboratory procedure
  - Generation of qualification data (baseline data, normal ranges, thresholds)

To be eligible, these methods and techniques have to be economical, easy to use, and automatable and must allow simple data capturing and straightforward validation. The previously mentioned methods and techniques, in particular in situ hybridization (ISH), immunohistochemistry (IHC), and immunoassays, are discussed in the following sections.

LOCALIZATION TECHNOLOGIES

Genetics and Genomics

ISH is the method of choice for localizing specific DNA or RNA in tissue. Labeled RNA or DNA probes can be designed and applied to tissues to overlay particular genes and their expression in specific structures of an organ or tissue. For the detection, two techniques exist: (1) the radioactive labeled probes, which are detected by autoradiography, and

1. This chapter is reprinted and modified from the first edition.

Principles of Translational Science in Medicine. http://dx.doi.org/10.1016/B978-0-12-800687-0.00008-6

(2) fluorescent labeling (FISH), which is detected by immunocytochemistry. The classical ISH protocol is composed of the following steps:

- Fixation: The samples, which are usually paraffin-embedded tissue blocks, are fixated onto glass slides.
- Permeabilization: The cells are permeabilized and, in the case of DNA, denaturized.
- Hybridization: The probe is hybridized onto the sample. The hybridization parameters are specific to the experiment (i.e., probe, tissue, or cell type).
- Microscopy: Microscopy techniques are used to visualize the samples.

There are two challenges to overcome: (1) preserving high RNA quality during tissue preprocessing and (2) designing the optimal probe. The different probe types used for ISH are double-stranded DNA, single-stranded DNA, synthetic oligodeoxyribonucleotides for DNA, and single-stranded complementary RNA (riboprobe). Oligonucleotide probes should be relatively small (20–40 base pairs) and specific to go through cell membranes. The labeling of the probes can be direct (fluorescein and rhodamine) or indirect via a specific antibody or labeled binding protein coupled with biotin or digoxigenin (Kher and Bacallao, 2006; Levsky and Singer, 2003).

There is a growing number of examples that use ISH for medical diagnoses. Two of them are discussed in the following text. The first example illustrates the use of an FISH-based test for the selection of patients and choice of the appropriate therapy, Herceptin®. The Food and Drug Administration (FDA) approval of Herceptin® as a frontline therapy in combination with paclitaxel for the treatment of Her2-positive metastatic breast cancer was based on the request to develop a diagnostic test to select the patients who will benefit from the treatment. These patients overexpress the molecular target of Herceptin®: Her2. Herceptin® is indicated for metastatic breast cancer, in which the overexpression of Her2 contributes to aggressive growth and spreading to other parts of the body. This overexpression of Her2 occurs in approximately 25% of breast cancer cases. In a normal context, a healthy breast cell has two copies of the Her2 gene. Her2 breast cancer gets started when breast cells have more than two copies of that gene, causing an overproduction of the Her2 protein. Two diagnostic methods exist to detect this anomaly: the FISH-based test (gene level) and the IHC-based test (protein level). The IHC-based diagnostic test, Herceptest™, developed by DAKO is described in the following section on IHC. The FISH-based test assessing HER2 gene copy number, PathVysion™, was developed by Vysis (now Abott Molecular). The PathVysion Her2 DNA Probe Kit consists of two labeled DNA probes:

1. The Her2 probe is labeled in SpectrumOrange and covers the entire Her2 gene.
2. The CEP 17 probe, labeled in SpectrumGreen, picks up the alpha satellite DNA located at the centromere of chromosome 17. Inclusion of the CEP 17 probe as a normalization gene allows the determination of the relative copy number of Her2.

Biopsies with a Her2 and CEP 17 signal ratio ≥ 2.0 are considered as Her2 abnormal. The Her2 test is also used as a prognostic factor in stage II, node-positive breast cancer patients.

The second example illustrates the importance of "FISH-ing" translocations, specifically the t(14;18)(q32;q21) translocation, involving the bcl-2 gene (B cell leukemia or lymphoma 2) and the IgH (immunoglobulin heavy chain) gene, leading to the overproduction of the BCL2 protein. This specific translocation is observed in about 90% of follicular lymphomas (FLs) (accounting for 20%–25% of all non-Hodgkin's lymphomas), which usually undergo an indolent course. By contrast, the t(11;14) (q13;q32) translocation is a hallmark of mantle-cell lymphoma (MCL), which is a rapidly progressing disease (Zink, Fischer, and Nickerson, 2004).

For treatment strategy, it is absolutely critical to differentiate FL from MCL. In this context, FISH has been demonstrated to perform well. In addition to histopathology, the diagnosis of FL can be based on

- IHC: This detects the overexpression of the BCL2 protein in > 90% of cases but lacks specificity for FL.
- Polymerase chain reaction methods: These demonstrate molecular rearrangements, but the locations of the breakpoints at the BCL2 locus can vary. Therefore, in 10%–15% of the cases, no specific amplification results from these primers.
- Cytogenetics: This is the best way to detect karyotypic changes in tumor samples, but fresh samples are needed for accuracy.
- FISH: This is an alternative that allows researchers to optimize the diagnosis of cytogenetic changes. BCL2 rearrangements can then be observed in 85%–100% of the cases.

The diagnosis, characterization, and classification of specific subtypes of hematologic malignancies are based on cytomorphologic testing, immunophenotyping, cytogenetics, and molecular genetics. The detection of chromosomal translocations is a key aspect of accurate and detailed cancer subtyping, especially with regard to specific translocations, which are characteristic of types and subtypes of leukemias or lymphomas. An accurate diagnosis is essential to the management of most of these neoplasms, also impacting patient prognosis.

Protein Localization

IHC is the counterpart to ISH for proteins. Monoclonal or polyclonal antibodies are used to localize protein targets in a tissue in either a direct or indirect manner. Direct localization implies a labeling of the target antibodies themselves, whereas in indirect localization the target antibodies are detected by labeled secondary antibodies. The detection can be done by fluorescence or enzymatic reaction (phosphatase alkaline or horseradish peroxidase). The classical protocol for IHC involves

- Generation of 4 μm tissue slices
- Detergent treatment of the tissue to break the membranes (e.g., Triton X-100)
- Hybridization
  - The direct method is performed in one step with the labeled antibody. This technique is therefore simple and rapid, but sensitivity can be an issue due to the lack of signal amplification.
  - The indirect method involves a primary antibody raised against the tissue antigen and a second labeled antibody reacting against the animal species in which the primary antibody was raised. As several molecules of the secondary antibody can bind different binding sites of one molecule of the primary antibody, the signal is amplified (Cregger, Berger, and Rimm, 2006).

The first example of an IHC-based diagnostic test is the Her2/neu diagnostic test used for choice of treatment with Herceptin®. Herceptest™ (developed by DAKO) is an IHC test that measures the overexpression of the protein Her2 with, as readouts, rankings of 0, 1+, 2+, and 3+. Herceptin treatment is considered only when a patient scores a +2. In 2003, the testing became a prerequisite for choosing Herceptin® (Egervari, Szollosi, and Nemes, 2007).

Another example is the detection of the c-kit protein in tissue biopsy to support the diagnosis of gastrointestinal stromal tumors (GIST) and decide on patient eligibility for imatinib mesylate therapy (Gleevec, from Novartis Pharmaceuticals Corp.). Imatinib mesylate is indicated for the treatment of kit-positive unresectable and/or metastatic malignant GIST. In 2005, the FDA approved c-Kit pharmDx, a qualitative immunohistochemical kit system developed and commercialized by DakoCytomation to detect c-kit protein (CD117 antigen) expression in tissue biopsies. The use of IHC to detect the c-kit protein figures on the label of Gleevec is for "information only."

IMMUNOASSAYS

Most of the previously mentioned assays rely on biopsies. In the oncology field, tumor biopsies are often taken for diagnosis purposes but, in other medical contexts, biopsy sampling is considered extremely invasive for patients. Consequently, assays for secreted proteins in peripheral biofluids (typically serum, plasma, or urine) are the methods of choice. Antibody-based diagnostic tests are extremely well suited to quantifying proteins and therefore the best proof-of-concept for proteomic and genomic findings.

The application of an existing immunoassay (e.g., a commercially available assay) is the easiest context, allowing high-throughput, cost-effective, and straightforward data analysis and widespread and standardized instrumentations. Yet, the development of immunoassays can be a difficult task and is associated with many different challenges:

- Designing and producing appropriate and price-sensitive antigens and antibodies.
- Guaranteeing the sensitivity to the native protein and not only to the recombinant protein typically used to generate the antibodies.
- Optimization of the assay conditions for a specific antibody–antigene combination to achieve a sensitive assay.
- Multiplexing comes with the additive complexity of individual assays combined with a re-evaluation of the optimal parameters. Often this means compromising and not having the perfect individual assays in a single "pot" (well).

Enzyme-Linked Immunosorbent Assays

The classical enzyme-linked immunosorbent assay (ELISA) is the standard, most widespread technology for protein measurement. It combines the specificity of antibodies with the sensitivity of simple enzyme assays. ELISAs can be used to detect not only the presence of antigens (proteins) by antibodies but also antibodies in biofluids recognized by antigens.

Direct ELISAs

The most basic ELISA is composed of five steps:

1. An unknown amount of antigen (sample) is nonspecifically affixed to a surface via adsorption (usually a polystyrene microtiter plate).
2. All unbound sites are blocked to prevent false positive results (typically with bovine serum albumin [BSA] or casein). The serum proteins block nonspecific adsorption.
3. A specific antibody is applied to the surface so that it can bind to the antigen.
4. This antibody is linked to an enzyme, whose substrate is added in the final step to generate a chromogenic, fluorogenic, or electrochemically detectable signal.
5. In the case of fluorescence ELISAs, any antigen or antibody complexes will fluoresce upon light excitation so that the amount of antigen in the sample can be measured. The quantification is performed with a spectrophotometer, spectrofluorometer, or other optical or electrochemical devices.

There are many different types of ELISAs. The alternatives to the previously described direct assays are indirect ELISAs, sandwich ELISAs, and competitive ELISAs, which are described in the following sections.

Indirect ELISAs

Indirect ELISAs include all of the steps of the direct ELISAs described previously. Additionally, between steps 3 and 4, secondary antibodies, which will bind to the detection antibodies, are added to the wells. These secondary antibodies (and not the primary antibodies) are conjugated to the substrate-specific enzyme. Because several secondary antibodies can bind to one detection antibody, a signal amplification will be obtained. An additional advantage of the use of a secondary antibody is the universality of the enzymatic step: the use of an enzyme-linked antibody binding the Fc region of other antibodies can be included in many different assays.

A major disadvantage of the direct and indirect ELISA methods is the nonspecific immobilization of the antigen onto the surface. This means that any protein in the sample can be immobilized on the surface. The competition for the binding of the proteins to the surface leads to a possible loss of sensitivity.

Sandwich ELISAs

The sandwich ELISA overcomes the problem of competition for surface binding. In this assay, all the steps described in the direct ELISA protocol are maintained. The difference is that an antibody (capture antibody) is coated to the solid support (usually a polystyrene microtiter plate) to capture specifically the antigen of interest within the sample measured. The main advantage of a sandwich ELISA is its ability to deal with sample complexity (impurities) because any present antigen can still be selectively bound.

Competitive ELISAs

The competitive ELISA demonstrates its usefulness for low concentrations of the antigen in samples. After a known amount of antigen is coated to the surface, the sample is preincubated with a known quantified excess of the detection antibody in an additional step. After incubation, the sample is applied to the coated antigen. The excess antibody will be captured by the coated antigen, and the previously formed antigen–antibody complexes will be washed out. As a result, the amount of antigen in the sample is antiproportionally correlated to the amount of antibody bound to the antigen in the well, therefore generating competition.

Diagnostic Immunoassay Devices: High-Throughput Immunoassay Platforms

Diagnostic companies have been working to develop diagnostic immunoassays to fully integrate them into highly automated, user-friendly, high-throughput, standardized instrumentation and technologies, and an increasing number of FDA-validated assays has become available (U.S. Food and Drug Administration, 2007). This instrumentation can be divided into two main categories: on the one hand, high-throughput platforms mainly used in dedicated clinical laboratories and, on the other hand, integrated, easy-to-use bedside devices.

Two types of platforms are widespread: bead-based technologies such as Luminex® and VeraCode® and the electrochemiluminescence-based MesoScale Discovery® platform.

Bead-Based Assays

The principle of Luminex® xMAP® technology is based on capture antibodies conjugated to the surface of color-coded microspheres (beads), which react with specific antigens present in the sample similarly to sandwich assays. Then detection antibodies labeled with a fluorescent reporter molecule bind in proportion to the captured antigen. The quantification is performed by passing the suspended microspheres (beads) through the detection chamber of a flow cytometer. A green laser detects the amount of analytes bound to the beads, and a red laser identifies the color-coded beads (the nature of the target). Up to 100 color codes exist, which allows a theoretical multiplexing of up to 100 simultaneous assays (Luminex, 2007). Another key advantage is the low sample volume needed—only about 10 μl. An alternative comparable platform is the BD Cytometric Bead Array from BD Biosciences (BD Biosciences, 2007).

Bead-Based Arrays

Another available bead-based platform is VeraCode® (Illumina, 2007), which is based on the principle of the sandwich assay. The major difference is the coding of the beads, which for VeraCode® is a holographic binary code instead of a color code, allowing a higher multiplexing capacity (hundreds of targets). In addition, the beads are randomly arrayed in a "groove cell" with a special flow-system. Eight wells can be read in parallel, resulting in a throughput of about 100 samples per hour. Thirty replicates per bead (for one protein), optically measured around 10 times, result in 300 data points, ensuring a high robustness. The principle is applicable to a variety of analytes, including proteins, peptides, nucleic acids, and other ligands.

Electrochemiluminescence-Based Arrays

Electrochemiluminescence-based assays (e.g., the assay by Meso Scale Diagnostics [MSD]) also rely on the sandwich immunoassay. The antibodies are immobilized on planar arrays in microplate format, and the readout is the light emitted by an electrochemiluminescence reaction. Each well is equipped with a working electrode and a counter electrode generating an electrical circuit, which initiates the electrochemical stimulation of the ruthenium-labeled detection antibodies. As a consequence, light is emitted. Different entities can be immobilized, such as antibodies, peptides, proteins, cells or membranes, DNA, or RNA. Different plate formats are available. Multispotting up to 10 spots in a 96-well format and up to 100 spots in a 24-well format allows multiplexing (Meso Scale Discovery, 2007). All of these new technologies have in common the fact that only a few microliters of sample are needed.

Dynamic Arrays

Independently of the platform, multiplexing relies on the fact that the same assay conditions—such as incubation time, reagents, dilutions, and temperature—need to be applied to all markers investigated. One has to find a balance between the loss of sensitivity over single assays and the extent of development. Furthermore, if the quality check of one single analyte fails, the complete multiplex panel needs to be remeasured.

Fluidigm has recently introduced a new concept and device, called a dynamic array, that combines multiformat assays with microfluidics technology (Fluidigm, 2007). The device is composed of integrated fluidics circuits with nanovalves, nanochannels, and nanochambers. In contrast to multiplexing, each assay on the array is controlled individually in its nanochamber, which results in reduced pipeting and mixing of antibodies and antigens, which therefore prevents competition, cross-reactivity, and variability. Most importantly, the dilution, incubation time, and type and amount of reagents are optimized for each assay. In this multiformat system, 48 samples with 18 assays can be measured in duplicates on each card.

Diagnostic Immunoassay Devices: Bedside Devices

So-called bedside devices have been developed to allow nonspecialized laboratory technicians to perform biomarker assays, both at the lab bench and directly in the room of the patient (bedside). Plasma, serum, or urine samples can thereby be directly analyzed in real time. As a consequence, the samples do not need to be shipped to a lab, and the data are immediately available for diagnosis, therefore lowering costs and streamlining the operational processes. In this fast-growing field, a few products have reached the market and have already been implemented in clinics. The first example is the Triage Cardiac Panel (created by Biosite Incorporated, San Diego), which measures three cardiac markers (myoglobin, creatine kinase MB, and troponin I) for the diagnosis of acute myocardial infarction and its differentiation from other cardiac abnormalities (Clark, McPherson, and Buechler, 2002). The device is a combination of three fluorescence immunoassays

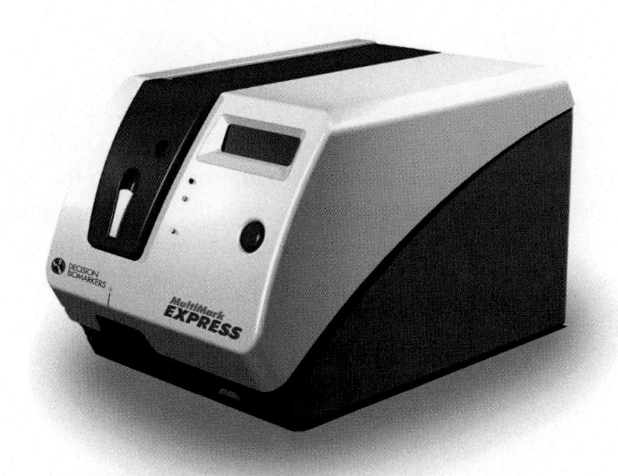

FIGURE 2.16 Depiction of the AVANTRA® Biomarker Workstation *Reprinted by permission from Courtagen Life Sciences, Inc.*

on a disposable chip measured and analyzed by a portable spectrophotometer, the Triage MeterPlus. The steps in a sample measurement are the following:

1. Blood or plasma is added to the sample port.
2. Red blood cells are retained on a filter within the device.
3. The remaining plasma is applied to the surface of a reaction chamber covered with fluorescent antibody conjugates, localized in different zones according to the protein type to be detected.
4. After incubation, the captured cardiac proteins of each zone can be measured by the meter.

The results for all analytes are quantified simultaneously within approximately 15 minutes after addition of the sample to the device. The same device is also used to detect the presence of commonly abused drugs (Triage Tox Drug Screen), and the Triage® Parasite Panel can be used to diagnose intestinal parasitic diseases.

The second example, the AVANTRA® Biomarker Workstation from Courtagen Life Sciences (Woburn, MA; see Figure 2.16), offers, in addition to their standard panels, a fully customized service. The biomarkers of choice can be multiplexed in an optimized assay in a biochip (Figure 2.17). Multiplex panels with up to 20 analytes in six replicates can be measured in approximately 1 hour (Courtagen Life Sciences, 2014).

CASE STUDY: SCREENING OF A BIOMARKER FOR KIDNEY INJURY USING LOCALIZATION AND IMMUNOASSAYS

An illuminating example demonstrating the power of combining localization techniques and immunoassays for the confirmation and validation of biomarkers is the renal safety biomarker Kim-1 (kidney injury molecule-1). Kim-1 is a type 1 membrane protein containing a 6-cystein immunoglobulin-like domain and a mucin domain, suggesting its possible involvement in cell–cell or cell–matrix interactions and extracellular matrix remodeling. The identification of Kim-1 as a biomarker for drug-induced proximal tubular injury was first achieved by an interplay of genomics and proteomics (Ichimura et al., 1998, 2004). Because Kim-1 mRNA and protein are expressed at very low levels in a normal kidney, the increased expression in injured proximal tubules is dramatic, making it an ideal safety biomarker candidate. Furthermore, it is conserved between different species, including humans. Research in IHC and ISH combined with histopathological evaluation highlighted that the expression of the gene and the protein were mainly within the proximal tubules and directly correlated with tubular injury. In clinical practice, it is rare to be able to obtain biopsies because of the invasiveness of the method; therefore, immunoassays using peripheral fluids (urine, blood, or plasma) had to be developed, and the data generated had to be correlated to histological data. The cleaved fragment of Kim-1 is measured in rodent and human urinary-based immunoassays with high sensitivity and specificity (Vaidya et al., 2005). In preclinical studies using nephrotoxicants, such as cisplatin, Kim-1 demonstrates increased expressions at all levels, including gene levels (ISH for localization shown in the top row of Figure 2.19 and reverse transcriptase polymerase chain reaction for quantification shown in the top row of Figure 2.18) and protein levels (IHC for localization demonstrated in the bottom row of Figure 2.19 and immunoassays

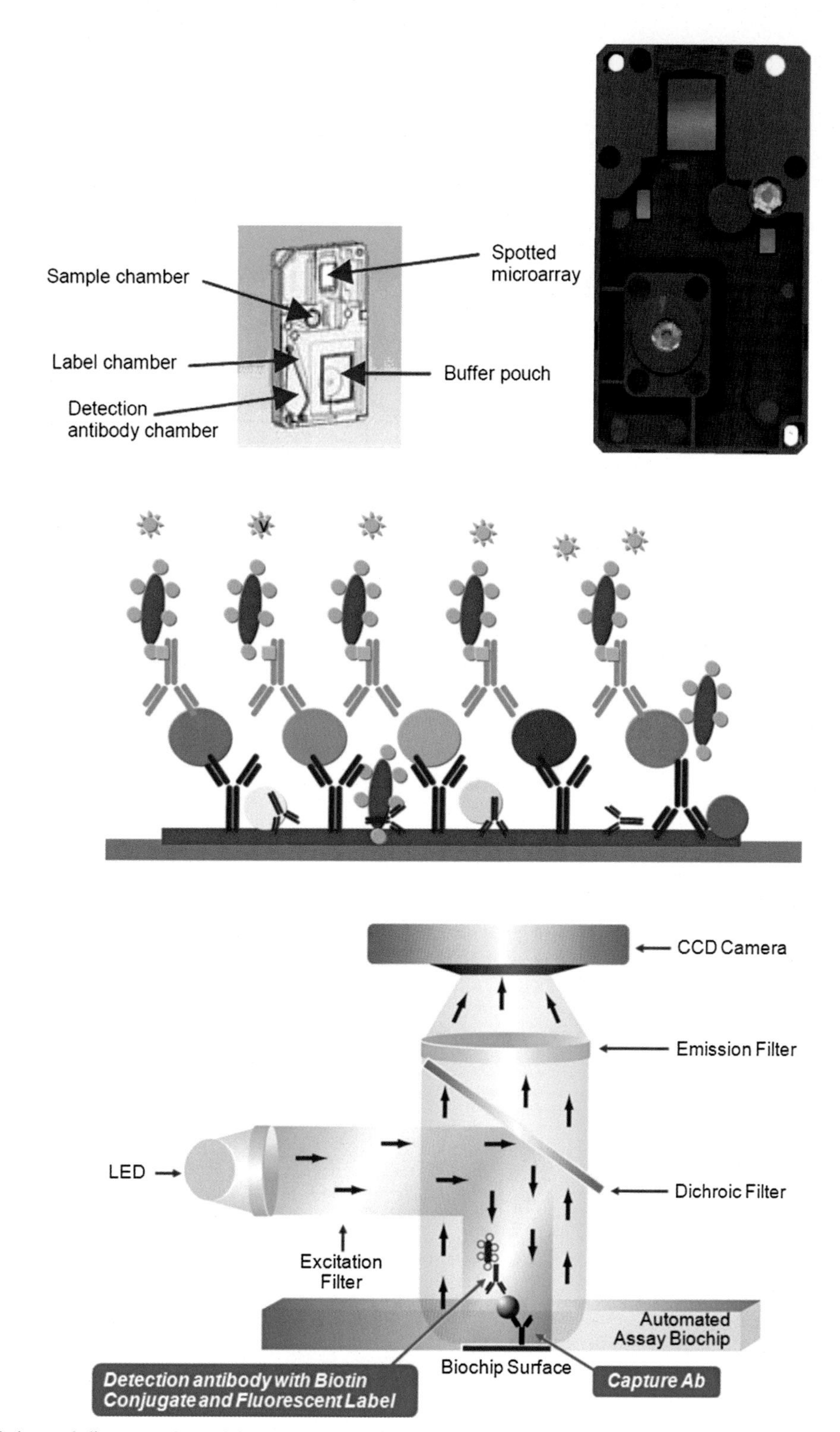

FIGURE 2.17 Depiction and diagrams of a BIOCHIP (top) and schematic diagram of the assay principle (bottom) *Reprinted by permission from Courtagen Life Sciences, Inc.*

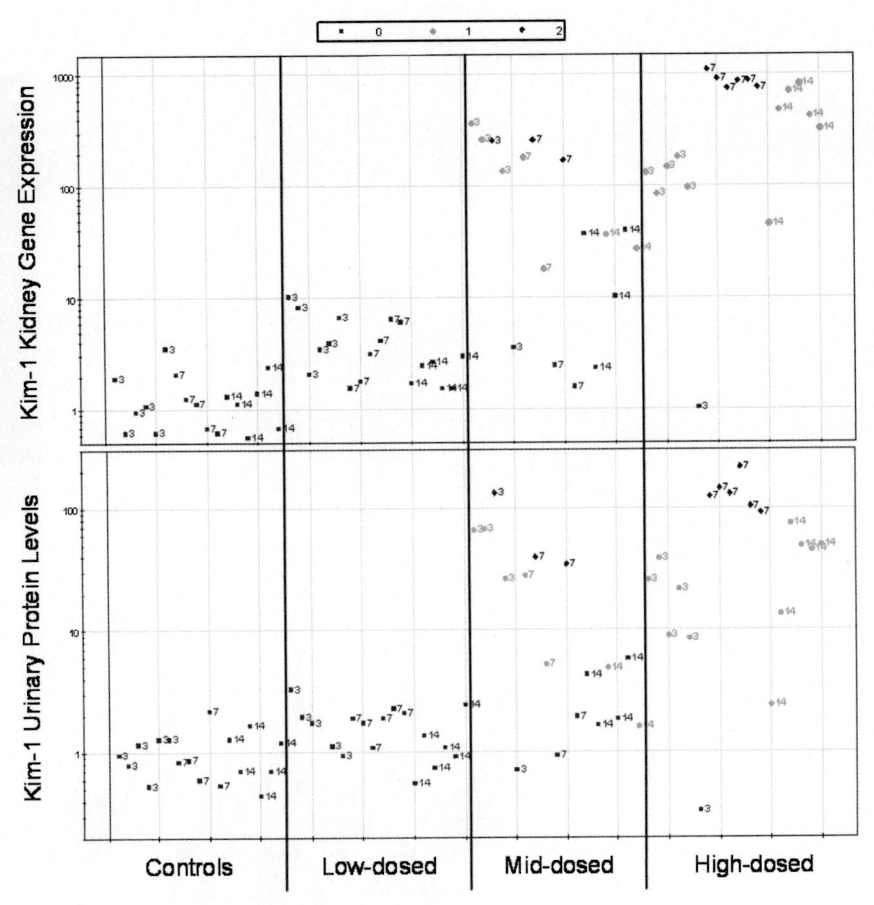

FIGURE 2.18 Graph showing the correlation of Kim-1 with severity grades of histopathology for 72 Han Wistar rats in a single-dose cisplatin study. The top panel shows the Kim-1 gene expression in the kidney, and the bottom panel highlights the urinary Kim-1 protein levels. All values are represented as fold changes versus the average values of time-matched control animals on a logarithmic scale. The animals are ordered by termination time point within each dose group, whereby the values, which are vertically aligned in both panels, belong to the same animal. The animals are labeled by termination time point (3, 7, and 14 days). The symbols and the color represent the histopathology readout for proximal tubular damage (red = no histopathology finding observed, green = grade 1, and blue = grade 2, on a 5-grade scale).

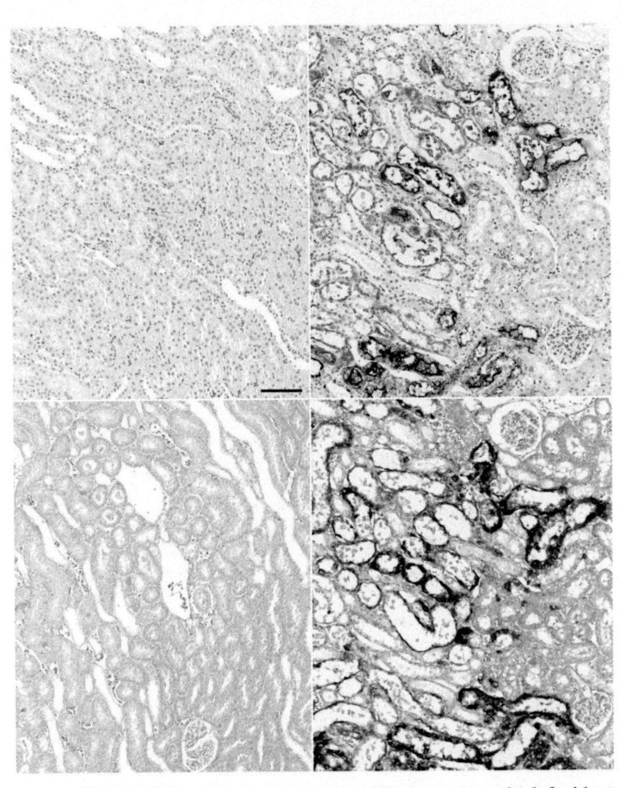

FIGURE 2.19 ISH of Kim-1 in the top row and IHC of Kim-1 in the bottom row. The images on the left side represent a control animal; the images on the right side represent a cisplatin-treated animal. The scale bar is 100 μm.

for quantification demonstrated in the bottom row of Figure 2.18). This biomarker has been demonstrated to encompass the power of current standards, for example, serum creatinine and blood urea nitrogen (BUN). Compared to the used urinary biomarkers, Kim-1 is specifically expressed in the kidney and cochlea after injury. In addition, no interferences with other pathologies have been reported, elevating Kim-1 to the top of the list of kidney injury biomarkers. Its excellence has been shown in numerous publications, in various species, and in different clinical trials (Han et al., 2006; Ichimura et al., 2004; Liangos et al., 2006).

REFERENCES

Biosciences, B.D., 2007. Company website. https://www.bdbiosciences.com/ (accessed 2014).

Clark, T.J., McPherson, P.H., Buechler, K.F., 2002. The triage cardiac panel. Point of Care 1, 42–46.

Courtagen Life Sciences, 2014. Company website. http://www.courtagen.com/company-protein-biomarker-workstaton.htm (accessed 2014).

Cregger, M., Berger, A.J., Rimm, D.L., 2006. Immunohistochemistry and quantitative analysis of protein expression. Arch. Pathol. Lab. Med. 130, 1026–1030.

Egervari, K., Szollosi, Z., Nemes, Z., 2007. Tissue microarray technology in breast cancer HER2 diagnostics. Path Res. Pract. 203, 169–177.

Fluidigm, 2007. Company website. http://www.fluidigm.com (accessed 2014).

Han, W.K., Waikar, S.S., Johnson, A., Curhan, G.C., Devarajan, P., Bonventre, J.V., 2006. Urinary biomarkers for early detection of acute kidney injury. J. Am. Soc. Nephrol. 17, 403A.

Ichimura, T., Bonventre, J.V., Bailly, V., Wei, H., Hession, C.A., Cate, R.L., et al., 1998. Kidney injury molecule-1 (KIM-1), a putative epithelial cell adhesion molecule containing a novel immunoglobulin domain, is up-regulated in renal cells after injury. J. Biol. Chem. 273, 4135–4142.

Ichimura, T., Hung, C.C., Yang, S.A., Stevens, J.L., Bonventre, J.V., 2004. Kidney injury molecule-1: a tissue and urinary biomarker for nephrotoxicant-induced renal injury. Am. J. Physiol. Renal. Physiol. 286, F552–F563.

Illumina, 2007. Illumina VeraCode™ Technology. http://res.illumina.com/documents/products/brochures/brochure_veracode.pdf (accessed 2014).

Kher, R., Bacallao, R.L., 2006. Imaging gene expression. Nephron Exp. Nephrol. 103, e75–e80.

Levsky, J.M., Singer, R.H., 2003. Fluorescence in situ hybridization: past, present and the future. J. Cell. Sci. 116, 2833–2838.

Liangos, O., Han, W.K., Wald, R., Perianayagam, M.C., MacKinnon, R.W., Dolan, N., et al., 2006. Urinary kidney injury molecule-1 level is an early and sensitive marker of acute kidney injury following cardiopulmonary bypass. J. Am. Soc. Nephrol. 17, 403A.

Luminex, 2007. About xMAP® technology. http://www.luminexcorp.com/technology/index.html (accessed 2007).

Meso Scale Discovery, 2007. Technology. http://www.mesoscale.com/CatalogSystemWeb/WebRoot/ (accessed 2014).

U.S. Food and Drug Administration, 2007. Approval letter for ImmnueTech Total IgE System. http://www.accessdata.fda.gov/cdrh_docs/reviews/K032251.pdf (accessed 2014).

Vaidya, V.S., Ramirez, V., Ichimura, T., Bobadilla, N.A., Bonventre, J.V., 2005. Urinary kidney injury molecule-1: a sensitive quantitative biomarker for early detection of tubular injury. Am. J. Phys. Renal. Physiol. 290, F517–F529.

Zink, D., Fischer, A.H., Nickerson, J.A., 2004. Nuclear structure in cancer cells. Nat. Rev. Cancer. 4, 677–687.

Figure 2.

and

to the top of

diff

biosciences.com/

cardiac panel. P

.com

chemistry and

technology in b

(accessed 2014)

, B. Bonventre

C.A., Cate, R.L.,

domain is up-reg

, J.V., 2001. Ki

286, F552–F563.

illumina.com/doc

Nephron Exp. Neph

hybridization: past, pre

McKinnon, R.W., De

pulmonary bypas

luminexcorp.com/tec

mesoscale.com/Catalog

Letter for ImmuneTe

J.A., Bonventre, J.V., 200

Renal Physiol. 290, F517–

structure in cancer cel

Chapter 2.1.8

Biomarkers in the Context of Health Authorities and Consortia

Frank Dieterle, Estelle Marrer and Jacky Vonderscher

THE CRITICAL PATH INITIATIVE

The stagnation of medical product development has been pointed out in numerous contexts, in particular in an impactful report from the U.S. Food and Drug Administration (FDA) titled "Challenge and Opportunity on the Critical Path to New Medical Products" (U.S. Food and Drug Administration, 2004). As a consequence, public and private stakeholders joined forces to identify the gaps in drug development and to find promising ways to make the medical product development process more efficient, predictable, and productive. The consultation process led to the critical path opportunities list (U.S. Food and Drug Administration, 2006a), which proposed 76 specific opportunities, organized under six topics:

- Creating better evaluation tools (development of new biomarkers and disease models to improve clinical trials and medical therapy)
- Streamlining clinical trials (creating innovative and efficient clinical trials and improved clinical endpoints)
- Harnessing bioinformatics (data pooling and simulation models)
- Moving manufacturing into the twenty-first century (manufacturing, scale-up, and quality management)
- Developing products to address urgent public health needs (rapid pathogen identification and predictive disease models)
- Focusing on specific at-risk populations such as children (unlocking innovation in pediatric products)

In 2006 and 2007, reports were released listing the projects initiated, which covered a number of the previously mentioned topics, with a special emphasis on creating better evaluation tools—in particular, biomarkers.

With the release of the critical path opportunities list, the FDA aimed to increase the awareness of the need for collaboration. Most of the listed opportunities imply considerable resources; therefore, single entities (regulatory authorities, single companies, universities, or other government agencies) cannot achieve the stated objectives on their own. Collaboration means openly sharing data to feed the needs of the individual projects within the Critical Path infrastructure. As one of the first projects of the Critical Path Initiative, the C-Path Institute was founded in 2005 as a neutral ground and catalyst for programs. The first project hosted by the C-Path Institute, the Predictive Safety Testing Consortium (PSTC), initially brought 16 pharmaceutical and biotechnical companies together to exchange data and methodologies and led recently to the first ever submission of newly qualified kidney toxicity biomarkers.

NEW TECHNOLOGIES, HEALTH AUTHORITIES, AND REGULATORY DECISION MAKING

Until recently, none of the newly discovered entities for monitoring the side effects of drugs had a clear path to full validation for acceptance by health authorities. In the context of the PSTC, new biomarkers of drug-induced kidney injury were initially negatively affected by skepticism. This skepticism was linked to their lack of considerable historical biological evidence, due to their steady state—even though they had a higher sensitivity than the current clinical standards blood urea nitrogen (BUN) and serum creatinine. In the light of the known weaknesses of the standards, the use of the new biomarkers in preclinical and clinical practice provided a body of evidence that was large enough to facilitate interpretation and knowledge of limitations. The PSTC compiled a package for the seven biomarkers consisting of different proposed contexts of uses and data supporting these intended uses, and submitted them to the European Medicines Agency (EMA) and FDA for regulatory endorsement. Following FDA and joint approval of the seven new renal safety biomarkers (European Medicines

Principles of Translational Science in Medicine. http://dx.doi.org/10.1016/B978-0-12-800687-0.00009-8

Agency, 2008), the use of these biomarkers in clinical practice or in nonclinical and clinical drug development is expected to take off in the near future.

Voluntary Genomics Data Submission and Voluntary Exploratory Data Submission

As a first reaction to the emergence of genomics datasets as potential tools for diagnostic and prognostic purposes, the FDA has defined clear submission processes, which are known as voluntary genomics data submission (VGDS) processes. In March 2005, a guidance (U.S. Food and Drug Administration, 2006b) was issued to foster submissions of genomics data, which, in addition to genetic test data, includes gene expression profiles as panels or algorithm-based biomarkers. This was followed in May 2006 by an FDA/EMEA joint VGDS process, which opened the door to multiparametric qualifications but was still restricted to genomics datasets without achieving any regulatory approval. It quickly became clear that the VGDS guidance was too restrictive for genomics data, and a new frame emerged that included biomarker panels of different kinds, including proteins, metabolites, and imaging types of markers. The VGDS process was subsequently renamed the voluntary exploratory data submission (VXDS) process, to reflect this essential evolution. In addition, the VXDS process allowed health authorities to give regulatory approval for the use of the biomarkers in specific defined contexts.

Current Guidances for Biomarker Qualification and Validation

Subsequent to the first couple of biomarker data submissions under the VXDS guidance, the EMA and the FDA released guidances to applicants with a well-defined biomarker qualification process, incorporating the learnings of the first successful biomarker qualification as well as lessons learned from several ongoing biomarker qualifications.

The guidance of the EMA (European Medicines Agency, 2009) defines the framework and process how biomarkers are reviewed and could finally obtain regulatory approval. The guidance defines the "CHMP Qualification Opinion on the acceptability of a specific use of the proposed method (e.g., use of a novel methodology or an imaging method) in a research and development (R&D) context (nonclinical or clinical studies), based on the assessment of submitted data," which is a 190-day formal review process with the goal of regulatory approval. The guideline also defines a "CHMP Qualification Advice on future protocols and methods for further method development towards qualification, based on the evaluation of the scientific rationale and on preliminary data submitted," which is a useful tool to obtain informal advice at the beginning or during the conduct of biomarker qualification programs. The guidance also proposes the content and formats for presenting data when submitting a biomarker qualification application.

The FDA has released a similar guidance called "Qualification Process for Drug Development Tools" (U.S. Food and Drug Administration, 2014). In contrast to the EMA, the guidance covers not only biomarkers, but also patient-reported outcomes and animal models. Recommended format and content for biomarker qualification submissions are similar to the EMA guidance.

Besides the first successful biomarker qualification for kidney biomarkers by the PSTC, only one more qualification has been endorsed by both the EMA and FDA. In this case, the ILSI/HESI consortium submitted preclinical data for the qualification of two new kidney biomarkers and additional data for a biomarker already submitted by the PSTC and qualified. This shows that the biomarker qualification process can be seen as an incremental process when additional data can extend but also limit the context of use. In addition, the EMA has endorsed a number of biomarkers to enroll patients in a prodromal Alzheimer state into clinical trials on the basis of cerebrospinal fluid (CSF) proteins and positron emission tomography (PET) and NMR imaging. The FDA has not qualified these biomarkers, although the same data were submitted to both agencies.

PRIVATE–PUBLIC PARTNERSHIPS (COOPERATIVE R&D AGREEMENTS)

Collaboration (sharing of research and development data, samples, expertise, and resources) between pharmaceutical companies and federal agencies like the FDA is dictated through what is known as a cooperative research and development agreement (CRADA) contract. By definition, a CRADA is a legal document describing the terms by which one or more federal parties and one or more private parties join together to conduct specified research or development. In particular, the research protocol, the resources each party will contribute to the research or development effort (personnel, services, facilities, equipment, etc.), and the documentation of the financial support are outlined in the contract. The CRADA partner must be able to make significant intellectual and/or technical resource contributions that otherwise would not have been available to the federal agency. Among the issues typically addressed in a CRADA are confidentiality; dissemination of research results so as to protect, as appropriate, proprietary materials; and the intellectual property rights of the federal party and of the CRADA partner.

In the field of biomarkers and drug development, one CRADA has drawn a lot of attention recently: the validation of preclinical biomarkers of safety, which aimed at identifying the process and analysis standards by which preclinical safety biomarkers can be validated for their use in regulatory decision making for new drug therapies. Together with the FDA, Novartis conducted preclinical studies with 10 reference compounds to evaluate the degree to which a panel of plasma or urine biomarkers (genomics and proteins) predicts renal toxicity before the current standards (serum creatinine and BUN), possibly even being prodromal (Novartis, 2008). To increase the evidence about the opportunities and limitations of the kidney biomarker, Novartis shared the data with the PSTC consortium, which was the basis of the first successful regulatory biomarker qualification.

CONSORTIA

The call for consortia was triggered by the intense and broad challenges described in the FDA's critical path opportunities report; consortia were created to combine the expertise, resources, and data of their members. This is especially necessary considering the amount of clinical trial data that need to be reassembled for rare events or qualifying new assays across databases of pooled samples from numerous studies. Because a biomarker is considered applicable for decision with regulatory impact only after it has been shown to be valid or qualified in a large or diverse set of samples and compounds, many pharmaceutical companies decided to participate in such an effort to have access to a critical mass of samples. At the time of this writing, joining consortia is the only way to validate and qualify a biomarker for regulatory purposes, outside the limited context of use for one specific drug, for example, biomarkers for organ toxicity. Also, joining a consortium means accelerating the generation of a dataset, which will support the biomarker qualification. In the context of the creation of a consortium, several points need to be taken into serious consideration:

- Consistency with applicable antitrust regulations
- Fulfillment of legal requirements such as membership, operations, management, and intellectual property
- Clearly stated missions and goals that cannot be achieved by individual entities
- Definition of the parties' rights and obligations (new and pre-existing intellectual property, governance, confidentiality)
- Subscription to common and strict project management ground rules, full commitment to a common goal, and compliance with agreed-upon timelines and plans

The Critical Path Institute's Predictive Safety Testing Consortium

Although the high-level frame for biomarker qualification was set through the VGDS and VXDS processes, the type of data and amount of biological evidence needed to support a qualification as a "known valid biomarker" was not specified in the guidance. This black-and-white approach—either broadly validating a biomarker or keeping it as an exploratory biomarker—has discouraged sponsors. A new opportunity to validate a biomarker for a specific context of use ("fit-for-purpose qualification") was proposed in 2007 by the PSTC in the framework of the first biomarker submission for qualification. This new context was endorsed by the PSTC group working on drug-induced kidney injury markers; these markers were submitted to the FDA and EMEA in July 2007 with claims that the markers in question were either superior or added value to BUN and serum creatinine. The first level of usefulness of these new markers in drug development is expected to be translational between toxicology studies and the first safety trials in healthy subjects and patients. The PSTC is a precompetitive consortium formed under the C-Path Institute (Tucson, Arizona) and was launched in early 2006. The PSTC consists of 19 pharmaceutical companies, one patient organization, and several academic institutions; it focuses on defining or testing a safety biomarker qualification process for markers of drug-induced injuries in different target organs of high interest that often lead to drug development failures (liver, kidney, vascular system, muscles, etc.). While the PSTC initially focused on the preclinical qualification of safety biomarkers for kidney, liver, carcinogenicity and vascular systems, which led to the first successful qualification of kidney biomarkers, the consortium has teamed up with the FNIH to run qualification studies for kidney biomarkers in humans with the goal to extend the FDA- and EMA-approved context of use of the kidney biomarkers as translational tools to bring potentially nephrotoxic drugs into human studies and evaluate their safety and efficacy profile in humans.

The PSTC collaboration model has been very successful, so subsequently additional consortia were founded with the Critical Path Institute as a driver (Critical Path Institute, 2014): the Coalition Against Major Diseases (CAMD) focusing on new tools, standards, and biomarkers for Alzheimer's disease and Parkinson's disease; the Multiple Sclerosis Outcomes Assessments Consortium (MSOAC); the Electronic Patient-Reported Outcomes Consortium (EPRO); the Coalition For Accelerating Standards & Therapies (CFAST); the Polycystic Kidney Disease Outcome Consortium (PKD); the Critical Path to Drug TB Regimen initiative (CPTR); and the Patient Reported Outcomes initiative (PRO).

The Innovative Medicines Initiative in Europe

Although the Critical Path Initiative was mainly driven by American health authorities, government, and federal institutions, the European side, including the European Commission, the EMEA, and the European Federation of Pharmaceutical Industries and Associations (EFPIA), started a comparable program several years later, named the Innovative Medicines Initiative (IMI).

IMI projects (IMI, 2014) are conducted by public–private consortia composed of a variety of stakeholders, in particular small- and medium-sized enterprises (SMEs), academia, research centers, patient groups, public authorities (including regulators), and the research-based pharmaceutical industry. In contrast to the PSTC, public funding is exclusively reserved for public participants and SMEs to operate cost-effectively, whereas big pharma companies will bear their own costs and contribute in kind. This is a unique public–private partnership between the European Community (represented by the European Commission) and the pharmaceutical industry (represented by the EFPIA). IMI has a total budget of €3.3 billion for the period 2014–2024. Although half of the IMI's budget (€1.6 billion) comes from Horizon 2020, the EU's framework program for research and innovation, about €1.4 billion have been committed to the program by EFPIA companies, plus up to €200 million that could be committed by other life science industries or organizations that decide to contribute to IMI as members or associated partners in individual projects. As the biggest public–private partnership in life sciences, IMI will fund innovative projects addressing the causes of delay or bottlenecks in biopharmaceutical R&D.

IMI currently has 46 ongoing projects, with more in the pipeline. A number of projects focus on biomarkers, in particular:

- BioVacSafe: Biomarkers for Enhanced Vaccine Immunosafety
- IMIDIA: Improving beta-cell function and identification of diagnostic biomarkers for treatment monitoring in diabetes
- MARCAR: Biomarkers and molecular tumour classification for nongenotoxic carcinogenesis
- Onco Track: Methods for systematic next-generation oncology biomarker development
- PRECISESADS: Molecular reclassification to find clinically useful biomarkers for systemic autoimmune diseases
- U-BIOPRED: Unbiased biomarkers for the prediction of respiratory disease outcomes
- SAFE-T: Safer and faster evidence-based translation

A number of additional consortia have biomarker identification, qualification, and development as a secondary objective. The most extensive program focusing on biomarkers is the SAFE-T consortium (SAFE-T, 2014). The goal is the development and qualification of biomarkers for liver, kidney, and vascular safety. While the PSTC consortium focuses on the preclinical qualification of safety biomarkers, the SAFE-T consortium focuses on the clinical qualification of biomarkers and currently conducts numerous clinical studies to compile clinical evidence about the potential and limitations of kidney, liver, and vascular safety biomarkers. Since clinical biomarker qualification programs are very resource-intensive, SAFE-T has teamed up with several other initiatives and networks such as the PSTC consortium, the Centre for Drug Safety Science at Liverpool University, and the Spanish DILI registry.

The objectives of the Innovative Medicines Initiative are very similar to the FDA Critical Path Initiative (with the added incentive of financial support), and indeed the IMI intends to complement the FDA Critical Path Initiative rather than compete with it. For example, the PSTC and SAFE-T consortia have started to work together and combine resources for the qualification of safety biomarkers, but also other consortia are looking for partners worldwide, as the qualification of biomarkers is a very resource-intensive task.

The PhRMA Biomarkers Consortium

In the PhRMA Biomarkers Consortium (PhRMA Biomarkers Consortium, 2014) founded in October 2006, U.S. government organizations and private companies are working to identify and validate new biomarkers for use in the prevention and detection of disease. Companies share early, nonproprietary information on biomarkers. This collaboration is supported by the Foundation for the National Institutes of Health (FNIH), the National Institutes of Health (NIH), the FDA, and the Pharmaceutical Research and Manufacturers of America (PhRMA). The first group of projects approved under this consortium will focus on

- Atherosclerosis in Silico Modeling: Deliver a time-dependent panel of short-term biomarkers that predict long-term (clinical) outcomes after statin therapy, as well as markers of residual risk after statin therapy.
- Osteoarthritis Biomarkers Project: Establish predictive validity of disease progression biomarkers and assesses the responsiveness of several imaging and biochemical markers pertinent to knee osteoarthritis using the resources of the NIH Osteoarthritis Initiative.

- CABP/ABSSSI Project: Develop reliable and well-defined endpoints for clinical trials of antibacterial drugs for Acute Bacterial Skin and Skin Structure Infections (ABSSSI) and Community-Acquired Bacterial Pneumonia (CABP).
- Identification and Validation of Markers That Predict Long-Term Beta-Cell Function and Mass: Use systematic methodologic studies of the Meal Tolerance and the Maximum Stimulation Tests to generate an effective protocol for implementation in longitudinal clinical trials of diabetes progression.
- Clinical Evaluation and Qualification of Biomarkers of Acute Kidney Injury: Qualify novel, more sensitive biomarkers of drug-induced acute kidney injury for use in early drug development decisions in collaboration with the PSTC consortium.
- AD Targeted Cerebrospinal Fluid (CSF) Proteomics Project: Qualify a multiplexed panel of known AD CSF-based biomarkers; examine Beta-Site APP Cleaving Enzyme (BACE) levels in CSF; and qualify a mass spectroscopy panel using CSF samples from ADNI.
- ISPY-2 TRIAL: Establish a new paradigm to accelerate the clinical trial process using a novel adaptive, multidrug phase II trial of investigational agents in the neoadjuvant setting in high-risk breast cancer patients. Each patient's tumor is extensively molecularly profiled, and the profile is used to assign a treatment likely to work for that tumor profile.
- AD/Mild Cognitive Impairment (MCI) Placebo Data Analysis Project: Combine placebo data from large industry clinical trials and analyze them to provide better measures of cognition and disease progression
- FDG-PET Lung and Lymphoma Projects (2 projects): Build the case for FDA incorporation of FDG-PET and volumetric CT into outcome measures for lung cancer and lymphoma.

The PhRMA Biomarkers Consortium has already completed a number of projects, and the results and conclusions have been published:

- Sarcopenia Consensus Summit: Generate the first consensus evidence-based definition of sarcopenia (age-related decrease in skeletal muscle mass) to provide specific guidelines for clinical diagnosis and enable future regulatory decisions regarding treatments of this currently unrecognized condition.
- Carotid MRI Reproducibility Project: Establish a standardized carotid MRI protocol and determine the impact of site/platform on reproducibility of the measurements.
- Alzheimer's Disease (AD) Targeted Plasma Proteomics Project: Qualify a multiplexed panel of known AD plasma-based biomarkers using plasma samples from the Alzheimer's Disease Neuroimaging Initiative (ADNI).
- PET Radioligand Project: Develop improved, more sensitive radioligands with higher binding to the peripheral benzodiazepine receptor.
- Adiponectin Project: Determine whether adiponectin has utility as a predictive biomarker of glycemic control.

The International Serious Adverse Event Consortium

The international Serious Adverse Event Consortium (iSAEC) (iSAEC, 2014) is a nonprofit organization composed of leading pharmaceutical companies and academic institutions with scientific and strategic input from the FDA. The mission of the SAEC is to help identify and validate DNA variants useful in predicting the risk of drug-related serious adverse events (SAEs). The working hypothesis of the SAEC is that the differences observed in adverse responses to medication have a genetic basis. The emphasis will be placed on the identification of genetic markers associated with the following disease conditions:

- Hepatotoxicity (DILI)
- Serious skin rash (DISI)
- Acute hypersensitivity syndrome (DRESS)
- Nephrotoxicity (DIRI)
- TdP/PQT effects (DITdP)
- Excessive weight gain (associated with class 2 anti-psychotic medications)
- Jaw osteonecrosis (ONJ)

In contrast to the other consortia, the focus of the iSAEC is only on genetic and epigenetic biomarkers.

REFERENCES

Critical Path Institute, 2014. http://c-path.org/ (accessed 16.08.14.).

European Medicines, 2008. Medicines and emerging science. http://www.ema.europa.eu/ema/index.jsp?curl=pages/special_topics/general/general_content_000339.jsp (accessed 16.08.14.).

European Medicines, 2009. Qualification of novel methodologies for drug development: guidance to applicants. http://www.ema.europa.eu/docs/en_GB/document_library/Regulatory_and_procedural_guideline/2009/10/WC500004201.pdf (accessed 16.08.14.).

IMI, 2014. www.imi.europa.eu (accessed 16.08.14.).

iSAEC, 2014. http://www.saeconsortium.org/ (accessed 16.08.14.).

Novartis, 2008. Novartis and FDA complete two-year cooperative research and development agreement (CRADA) to define and test a process for qualifying preclinical safety biomarkers. http://www.evaluategroup.com/Universal/View.aspx?type=Story&id=152911 (accessed 16.08.14.).

PhRMA Biomarkers Consortium, 2014. http://www.biomarkersconsortium.org (accessed 16.08.14.).

SAFE-T, 2014. www.imi-safe-t.eu/ (accessed 16.08.14.).

U.S. Food and Drug Administration, 2004. Challenge and opportunity on the critical path to new medical products. http://www.fda.gov/scienceresearch/specialtopics/criticalpathinitiative/criticalpathopportunitiesreports/ucm077262.htm (accessed 16.08.14.).

U.S. Food and Drug Administration, 2006a. Critical Path Opportunities List. http://www.fda.gov/downloads/scienceresearch/specialtopics/criticalpathinitiative/criticalpathopportunitiesreports/UCM077258.pdf (accessed 16.08.14.).

U.S. Food and Drug Administration, 2006b. Guidance for industry—pharmacogenomic data submissions. www.fda.gov/downloads/regulatoryinformation/guidances/ucm126957.pdf (accessed 16.08.14.).

U.S. Food and Drug Administration, 2014. Guidance for industry and FDA staff—Qualification process for drug development tools. http://www.fda.gov/downloads/Drugs/GuidanceComplianceRegulatoryInformation/Guidances/UCM230597.pdf (accessed 16.08.14.).

Chapter 2.1.9

Human Studies as a Source of Target Information

Martin Wehling

An important option for identifying and validating key drug targets is the analysis of data from human studies pointing to targets other than the one for which a particular drug has been developed. As is shown in this chapter, this approach is a particularly powerful component in the toolbox for target identification or validation, as those targets are human ones and, thus, validated in the best test system possible—the human body. This approach is also termed *reverse pharmacology* and covers results derived from the use of both small organic molecules and natural compounds contained in traditional ("herbal") medicines.

USING OLD DRUGS FOR NEW PURPOSES: BACLOFEN

The first example is archetypical, in that an unexpected side effect of a particular drug has subsequently been closely looked at and used as a lead observation for the profiling and establishment of a potentially useful target.

Baclofen was developed as a skeletal muscle relaxant and is being used in this capacity, especially in patients with lumbago, spastic disorders (including multiple sclerosis, vertebral trauma with paraplegia, or stroke sequelae), or cramps. It is used orally or even intrathecally. In an entirely different setting, the related pharmacological activity of baclofen (gamma-aminobutyric acid receptor type B agonism, or $GABA_B$ receptor agonism) has been used as a proof-of-principle tool to validate a human target. Reflux esophagitis is very common in Western societies, and its major treatment, acid secretion inhibition by proton pump inhibitors (PPIs; e.g., omeprazole), is effective only in about two-thirds of patients (Coron, Hatlebakk, and Galmiche, 2007; Jones et al., 2006). Thus, mechanisms other than acid damage have been increasingly investigated as potential causes of PPI-resistant disease. Subsequently, transient lower esophageal sphincter relaxations (TLESR) have been found to be important contributing factors in this regard, and researchers have begun to seek out agents to suppress them.

It became obvious that $GABA_B$ receptor agonism seemed to represent an effective principle, at least in animals such as ferrets or dogs. How should this hypothesis be tested in humans? Should researchers wait until a new compound is developed that specifically targets $GABA_B$ receptors in the lower esophageal sphincter to find out whether this mechanism is important in humans as well? In this particular situation, an elegant solution was discovered and consequently tested: baclofen had been used for many years in patients, but no one had done studies on esophageal reflux. Given the medicinal use of this compound, there was no ethical hurdle to test its effects on human esophageal reflux; thus, experiments were performed in both adults and children. Though larger randomized trials show the limitations of effects in adults, effects in children seem promising, even for therapeutic purposes (for a review, see Sifrim and Zerbib, 2012). More important than the use of baclofen in reflux treatment, which will not be feasible at least in adults because of its main muscle relaxant and central nervous effects, is the fact that those studies clearly show that

- $GABA_B$ receptors do exist and function in the lower esophageal sphincter of humans.
- $GABA_B$ receptors in the lower esophageal sphincter can be reached by compounds in humans.
- Activating $GABA_B$ receptors in the lower esophageal sphincter can reduce TLESR in humans.
- Reducing TLESR in humans can reduce esophageal reflux episodes (Figures 2.20 and 2.21), and baclofen may even be used clinically in children.

The reader might think the approach failed because baclofen was not clearly successful in clinical practice in adults, but this reflects a gross misunderstanding of the approach: it was clear from the beginning that baclofen would not become

Principles of Translational Science in Medicine. http://dx.doi.org/10.1016/B978-0-12-800687-0.00010-4

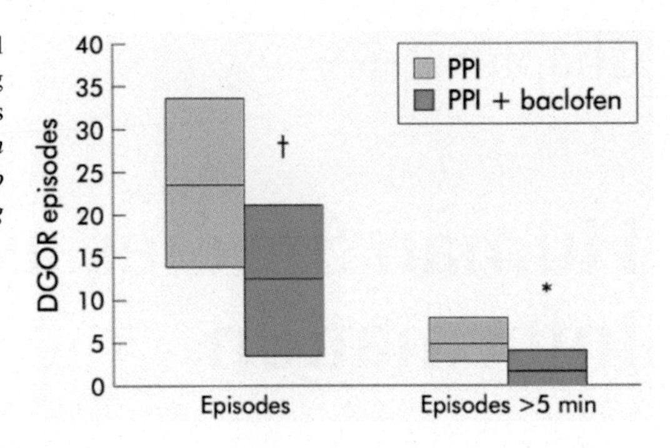

FIGURE 2.20 Number of duodenal reflux episodes (DGOR) and number of duodenal reflux episodes lasting longer than 5 minutes during treatment with omeprazole 20 mg twice daily (PPI) and during omeprazole 20 mg twice daily plus baclofen 20 mg three times daily. * $p < 0.05$; † $p < 0.06$. *(Reproduced from [Effect of the GABA(B) agonist baclofen in patients with symptoms and duodeno-gastro-oesophageal reflux refractory to proton pump inhibitors. Koek et al., 2003,* Gut **52**, *1397–1402] with permission from BMJ Publishing Group Ltd.)*

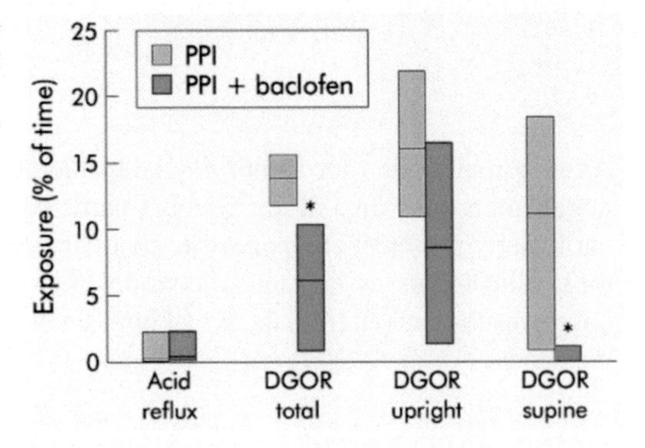

FIGURE 2.21 Acid and duodenal reflux exposure during treatment with omeprazole 20 mg twice daily and during omeprazole 20 mg twice daily (PPI) plus baclofen 20 mg three times daily. PPI, proton pump inhibitor; DGOR, duodeno-gastro-oesophageal reflux. * $p < 0.05$. *(Reproduced from [Effect of the GABA(B) agonist baclofen in patients with symptoms and duodeno-gastro-oesophageal reflux refractory to proton pump inhibitors. Koek et al. 2003,* Gut **52**, *1397–1402] with permission from BMJ Publishing Group Ltd.)*

the new drug to treat TLESR and PPI-resistant reflux. This would be almost impossible, as baclofen was developed to treat spinal cord-related disorders and, thus, was designed to penetrate into this tissue, which is probably not necessary in TLESR treatment. Its pharmacokinetics was chosen to treat neuromuscular disorders, which tend to fluctuate, and a longer half-life may be desirable in TLESR treatment. Baclofen served only as a clinically available test compound for which all prerequisites for human use had been fulfilled because of its clinical use in another human indication. It was, thus, extremely useful for validating the $GABA_B$ receptors in the lower esophageal sphincter as effective drug targets in human disease, and it conceivably raised considerable interest in this target on the part of drug companies. Obviously, baclofen is not ideal for this indication (as it has been designed for something else), but investors in industry and funding bodies in academia gained strong confidence in this target on the grounds of the baclofen experience in human studies. Investing millions of U.S. dollars in the lead optimization process (even the lead identification phase could potentially be reduced or eliminated, as baclofen—depending on intellectual property issues—might be able to serve as a lead compound) is now a relatively safe decision, or at least one that is much less speculative than in situations in which target properties in humans can only be extrapolated from information such as gene screens or animal experiments.

Consequently, several companies are working on this target, and there is good reason to hope that within a reasonable time a $GABA_B$ receptor agonist will reach the market that has been specifically designed for $GABA_B$ receptors in the lower esophageal sphincter in humans and, thus, will be less reminiscent of baclofen's actions elsewhere (muscle relaxation).

This research led to the development of a peripherally restricted, novel $GABA_B$ receptor agonist, lesogaberan, which, unfortunately, did not meet its goal to be clearly superior to PPIs (Shaheen et al., 2013).

Interestingly, the lead compound, baclofen, has meanwhile been studied further and may be used in refractory reflux inducing chronic cough (Xu et al., 2013) or even hiccups (Chang and Lu, 2012).

This example of experiences from human trials reflecting back to early drug discovery and development belongs to a category that is probably best termed *target validation* in humans by a test compound unrelated to an existing drug development program in the particular area.

This means that baclofen as a test compound was not an early but imperfect result of the drug discovery program in the reflux disease area; this category is thus distinct from the category of compounds that are imperfect (e.g., due to the half-life being too short for rationalistic use), novel, and have never been used in humans. It is possible to test such a compound on the grounds of a limited safety package under the guidance of the experimental investigational new drug (eIND) rules by the FDA or EMEA (see Chapter 4.2).

The scientific question behind this example in the context of this book is, Can this concept be generalized and become a standard tool for drug discovery and development? The clear answer is to ask a simple question in all drug discovery projects: if a particular target has been somehow discovered, for example, by chance findings in animals or by genetic approaches, is it possible to test the system in humans using a test compound that can easily be applied to humans because it is being used for other purposes already? If the answer is yes, the next question would be, Can the effects in question be discerned or separated from the main effects for which the compound had been developed? In other words, are toxic effects or pharmacokinetic properties compatible with testing in a different system that may require, for example, prolonged exposures or exposures at higher doses?

The main purpose of this book—the creation of a scientific backbone for translation in medicine—consists of algorithms such as this one that could be reproducibly and transparently used to increase the translational output. The different categories in this context are discussed at the end of this chapter.

SERENDIPITY: SILDENAFIL

The most prominent recent example in the category of serendipity is sildenafil: though originally developed as an anti-anginal or antihypertensive agent, it became obvious during early clinical trials in healthy male volunteers that it had an additional, unexpected effect on erectile function. It would be redundant to summarize the subsequent revolution in this therapeutic area, with congener development, debates about lifestyle drugs, reimbursement, and so on. The main point is the fact that the early clinical trials revealed unexpected results, led to modification of the approval process, and were reflected back into drug discovery. In this case, drug development was mainly restricted to me-too compounds with minor but distinct advantages (toxicity, half-life) over the index compound.

How should we capitalize on this principle? In a scientific context, this is probably the most unreliable and nonsystematic—thus nonscientific—approach to obtaining data for reverse drug development from clinical experiences. The main reason is the fact that this chance finding depends on what is nowadays called serendipity. By definition, it is impossible to establish rules to increase the likelihood of serendipity. Pharmaceutical history is full of examples of serendipitous findings, including the discovery of penicillin and, in particular, psychotropic drugs such as chlorpromazine or imipramine (Ban, 2006). The dependence on serendipity of discoveries in the central nervous system (CNS) area reflects the tremendous translational difficulties in this context due to the lack of appropriate animal models and robust cross-species biomarkers.

Is there no way to utilize serendipity? The only reasoning in this regard is reflective of the fact that the flexibility, open-mindedness, and sensitivity of clinical (and preclinical) researchers toward unexpected findings should be increased. Teaching examples of serendipitous findings might create awareness and inspiration, which could serve as prerequisites to the willingness to sense "strange" results and interpret them correctly. From a negative viewpoint, safety findings and unwanted effects come unexpectedly in many instances. Of course, defense mechanisms may exist that keep researchers from recognizing such shortcomings, which may jeopardize long-term work and big expenditures. Safety issues (see Chapter 5) may represent good examples of attempts to avoid being taken by surprise or becoming a victim of inverse serendipity. If the safety program undertakes to test sensitive and specific biomarkers as early as possible in the developmental program of a drug, surprises will be reduced. On the other hand, the interpretation of safety signals such as increases in liver enzymes (transaminases) requires skills, experience, and open-mindedness, which is key to positive dealing with serendipitous findings.

These examples show that the grid by which serendipitous findings can be detected and thus retained can be tailored and tuned to increase the likelihood of detecting valuable, though unexpected, findings, especially in safety-related issues. This approach can be called *facilitated serendipity* (Figure 2.22).

This includes:

- Heating the pot to increase emanations: In other words, utilizing multifaceted trials and multiple readout strategies; provocation tests; and challenging patients as early as possible as the true target population, rather than healthy volunteers, who may be more robust and thus may not expose toxicity signs.
- Increasing the sensitivity of the sensors: In other words, making use of better, more sensitive biomarkers, especially for safety issues. How can liver transaminases ALAT and ASAT, which were introduced into medical practice more than half a century ago with early utility reports from the 1950s, still represent the standard test biomarkers for liver toxicity assessment?

From a scientific standpoint, however, serendipity is the least structured and controlled area in translational medicine, and it is not very likely that this will ever change.

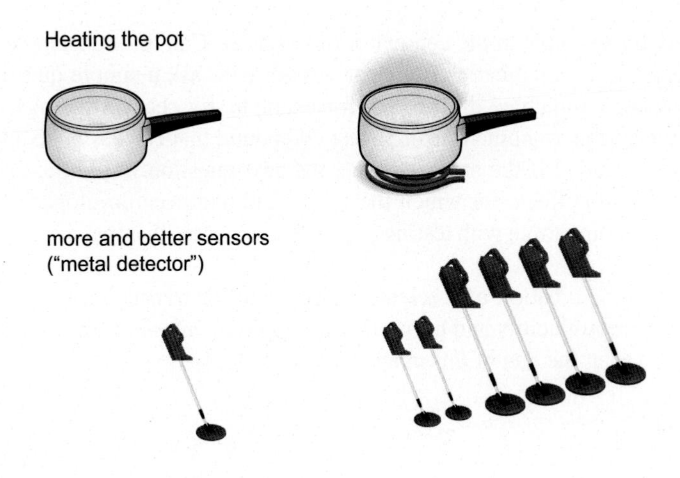

FIGURE 2.22 Facilitated serendipity.

REVERSE PHARMACOLOGY

The third paradigm is a far-reaching, backward-oriented translational drug discovery process. The term *reverse pharmacology* has been utilized especially for this approach. It is based on the fact that many major drug discoveries have been derived from natural products, from which either the drug itself or a lead compound that was refined in the subsequent drug discovery process was isolated.

Historic examples are manifold: they include morphine, which was isolated from poppy plants—specifically from opium, the primary extract—or digoxin, which was isolated from *Digitalis lanata*, woolly foxglove. Even more famous is the use of *Salix alba* extracts containing salicylic acid, which was the starting point for the most successful drug synthesis ever, namely that of acetylsalicylic acid by Felix Hoffmann in 1897. The most impressive development any drug ever had (Bayer aspirin) was simply a retrograde translational approach starting from observations made by the use of natural products. Salicylic acid from salix and other plants had been used for at least 3,500 years by almost all developed cultures; the first known documentation of this is from ancient Egypt and describes a myrtle extract for pain relief. Acetylsalicylic acid was gentler on the stomach than its progenitor and easy to synthesize. In the early twentieth century, acetylsalicylic acid was advertised as an efficient painkiller that did not affect the heart (as opposed to painkillers used at that time). Since then, due to clinical observations, we have discovered that acetylsalicylic acid is a potent platelet inhibitor and saves thousands of lives by protecting from heart attacks. This addition of a new indication has inspired other companies to find other platelet inhibitors that are currently on the market, for example, clopidogrel. In this sense, acetylsalicylic acid also served in the category of retrograde translation discussed at the beginning of this section, in that it became a model or test compound to demonstrate and validate a very important target in humans, namely the platelet in the prophylaxis of heart attacks.

One might think that the approach described in this section is quite outdated, and that modern rationalistic technology for drug development has replaced those natural products with small, smartly designed organic chemicals. This is untrue, and one of the clues to resuscitating successful translation may still come from this angle. The most famous recent example is the market approval of exenatide, a glucagons-like peptide type 1 (GLP-1) analogue. This compound is one of the few innovations in the treatment of the current epidemic of diabetes mellitus, a safe and reliable compound to reduce blood glucose and improve metabolic control in type II diabetics. It was developed synthetically from a compound called exendin-4, which had been isolated from one of the only two venomous lizards in the world, the Gila monster, after wide screens of chemical structures and biological effects of lizard venoms. The key idea to the search for an antidiabetic compound in lizard venom was the "clinical" observation that victims bitten by the lizard may develop pancreatitis, and it was assumed that the venom somehow overstimulated this organ, thus causing its inflammation. Subsequently, researchers discovered that the active ingredient could be isolated, and its blood glucose-lowering potency seemed to result from its similarity with a human hormone, GLP-1, which stimulates insulin incretion from the pancreas. It took another 15 years to bring the related drug to the market, but it all started from observations of the impact of lizard venom on humans, retrograde translation, preclinical design, and clinical development of the drug exenatide. Its success will come from its unique feature of reducing HbA1C as a measure of glucose control and weight reduction in the same place (DeFronzo et al., 2005). Insulin supplementation or sensitization by glitazones is accompanied by weight gain, the last thing one wants to see in obese type II diabetics.

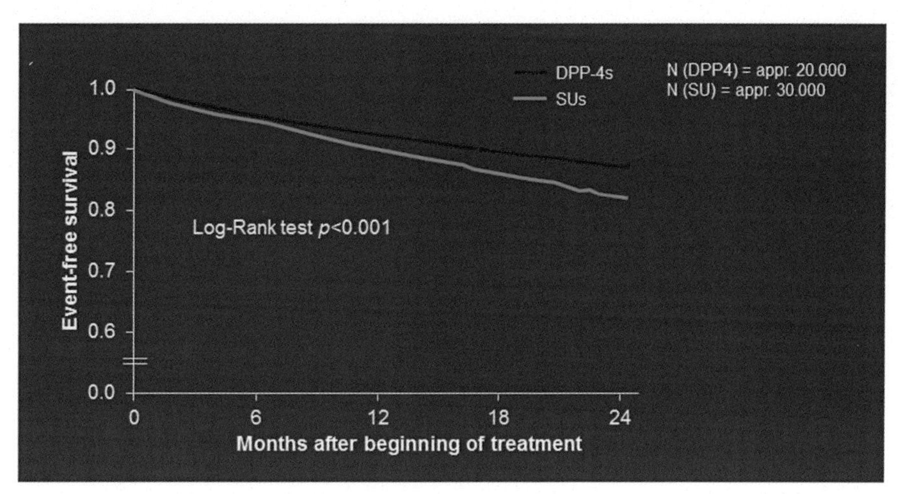

FIGURE 2.23 Kaplan–Meier curves for event-free (macrovascular events) survival over 24 months show significantly fewer endpoints/better event-free survival of patients on DPP IV inhibitors (DPP-4s) versus sulfonylureas (SU); data from an IMS analysis. *(Modified from Rathmann W. et al., 2013. Treatment persistence, hypoglycaemia and clinical outcomes in type 2 diabetes patients with dipeptidyl peptidase-4 inhibitors and sulphonylureas: a primary care database analysis.* Diabetes Obes. Metab. ***15**, 55–61 by kind permission from John Wiley and Sons.)*

The so-called incretin-analogues (or GLP1-analogues) such as exenatide have to be applied parenterally as they are peptidic drugs; attempting to overcome this disadvantage, the development of orally applicable drugs inhibiting the degrading enzyme (dipeptidyl peptidase type IV or DPP IV) for endogenous GLP1 was successfully induced. The resulting DPP IV inhibitors ("gliptins," e.g., sitagliptin) have been introduced into the modern treatment of diabetes mellitus type 2 and represent the most promising principle at present, as they do not induce weight gain or hypoglycemia—the two main downsides of former, insulinotropic therapies. These novel drugs seem to be safe even in the elderly, vulnerable population (Strain et al., 2013), which underscores their innovative potential. The superiority of DPP IV inhibitors over the older sulfonylurea drugs has been underlined by a recent metaanalysis clearly demonstrating higher event-free survival rates for DPP IV inhibitors (Rathmann et al., 2013; see Figure 2.23).

This example demonstrates that facilitated serendipity may not only precipitate directly useful results, but may also open wide avenues of innovation.

The systematic search for drugs from animal venoms has been successfully performed for other areas as well, such as anticoagulation (hirudin from leaches and bivalirudin as a novel derivative or ancrod from snake venoms). Those developments were triggered by the observation that the wounds of victims of animal venoms fail to clot, an easily available, very obvious observation. As animal venoms need to have strong biological effects, at least if they are to be poisonous to humans, systematic screening has been and is being performed by many companies and academic institutions. The main limitation is the fact that venoms are designed for rapid and short action (for defense and hunting purposes), and thus treatment of chronic illnesses might only occasionally be derived. Still, there are some snake venom approaches aiming at a cure for cancer, which seems quite distant from what snake venoms are originally made for.

This shortcoming must not exist for a by-far larger source of organic compounds with potential medical importance, namely those derived from plants. Some plant-derived principles may have begun their career in human treatment also due to initial intoxication experiences with subsequent dose-ranging trials in early humankind, which made them usable, but an even larger asset of natural compounds may act only if used chronically with little or no immediate effects. Such compounds will be much harder to detect than those with immediate actions.

This is actually a fascinating area in which shamanism meets scientific medicine. Undoubtedly, there are medicinal uses of plant extracts, including those cited above. And it is absolutely feasible to assume that there is a lot more to discover from plant ingredients. This approach is being increasingly promoted and advertised by Indian scientists who claim that thousands of years of Ayurveda medicine should have produced useful evidence to be explored by reverse pharmacology. There is evidence for *Rauwolfia* use in hypertension (reserpine has been in use for many years, but it is essentially an outdated drug due to its side effects), or *Boswellia serrata* (incense resin) in rheumatic diseases or inflammatory bowel disease due to boswellic acids. Even Ayurvedic pharmacoepidemiology is proposed as a new research area (Vaidya et al., 2003). There is very wide-reaching, cross-cultural evidence that various plants contain anti-TNFalpha-activity (Ichikawa et al., 2007). TNFalpha is a major mediator in cancer disease, and herbal drugs would certainly raise a lot of interest.

The latter example points to a very scientific way to explore the efficacy of plant extracts and separate true effects from shamanism, which is the biggest threat to the rationalistic utilization of herbal medicines. This transcultural approach

TABLE 2.10 Evidence from human experiences which can guide drug discovery and development (*reverse pharmacology*)*

Approach	Example	Strength	Weakness	Scientific Value
Using "old" test compound developed and already marketed for other indication	Baclofen for treatment of reflux disease	No hurdle to human testing as already used	Side effects to be used as main target may be too weak versus main effect and dosing into useful ranges, thus limited	+
Using newly developed test compound for same indication, which is, however, suboptimal as drug	Examples within proprietary protection	Lower hurdle to human testing if used as eIND, limited toxicology testing, may deliver valuable target validation data in humans	May delay development of better drug if decision could be taken without further target validation	++
Serendipity finding in clinical development or use	Sildenafil	Redirection of development still possible	May be overlooked, especially in regard to safety findings	+++, if discovered and rightly interpreted
Natural compounds from venoms and "toxic" plants	Exenatide, digoxin, morphine	Strong biological activities of venoms if poisonous to humans	Mostly acute actions, long-term effects need distant extrapolation	++++
Natural compounds from plants	Salicylic acid	Long-standing, traditional experiences, cross-cultural independent observations may provide objective evidence	Hard to isolate active ingredient, sometimes multiple active ingredients, standardization of dose, shamanism as threat	+–? for newer approaches

**For categorization, see text.*

depends on the fact that distant cultures would use the same or similar plants for the same diseases if they produce true effects. This means that different societies would independently come to the same uses of plants if their observations of biological effects were objective. This is one of the objectives of the *Journal of Ethnopharmacology*, and systematic comparisons of, for example, Chinese, Indian, and Native American habits in this regard (like the one cited previously) could yield valuable results.

It should be mentioned that all drug developments starting from plant preparations are hampered by the difficulties of isolating one or a few active ingredients, and by the difficulties of standardizing the composition in regard to the active compounds, especially if they are yet unknown. It is assumed that India and China will be leading countries in the exploitation of this route of reverse pharmacology, which shares important translational aspects.

These examples of backward translational activities indicate five principal approaches in this regard, which are summarized in Table 2.10.

Utilization of the side effects of marketed drugs is amenable to systematic, pharmacoepidemiologic approaches and may become an important addendum in the drug discovery toolbox if properly employed. Using novel drug candidates as eINDs is another option. In general, if drug development has been done in an area in which the target is still largely unknown, as has been the case in obesity drug development in many instances, taking suboptimal compounds rapidly into humans can be very helpful, especially if confidence in a new target can be considerably increased by this test compound. However, proprietary problems (target validation can be used by all competitors) and the relatively modest time gain by the limitation of the toxicology package may be offset by the limitation of the respective trial to short exposure times (maximum of 7–10 days).

Serendipity findings are most resistant to scientific approaches. As said previously, increasing the challenges at early human trials and increasing the quality and number of relevant biomarkers in conjunction with open eyes and minds may facilitate serendipitous findings, but this is still unreliable. In particular, such attempts should be maximized for the safety aspects of drug development.

The feedback route starting from natural compounds seems especially viable for strong biological effectors, as seen in venoms or toxic plants (and mushrooms). This route is being systematically exploited and has been successful even in the present day. Exploiting longstanding experiences with nontoxic plants is a cross-cultural task and seems very ambitious given the lack of standardization and multiple ingredient problems inherent to this approach. If the old successes, such as acetylsalicylic acid, could be repeated with new compounds, this area would certainly gain momentum. So far, the majority of attempts have come from researchers in India and China.

REFERENCES

Ban, T.A., 2006. The role of serendipity in drug discovery. Dialogues. Clin. Neurosci. 8, 335–344.

Chang, F.Y., Lu, C.L., 2012. Hiccup: mystery, nature and treatment. J. Neurogastroenterol. Motil. 18, 123–130.

Coron, E., Hatlebakk, J.G., Galmiche, J.P., 2007. Medical therapy of gastroesophageal reflux disease. Curr. Opin. Gastroenterol. 23, 434–439.

DeFronzo, R.A., Ratner, R.E., Han, J., Kim, D.D., Fineman, M.S., Baron, A.D., 2005. Effects of exenatide (exendin-4) on glycemic control and weight over 30 weeks in metformin-treated patients with type 2 diabetes. Diabetes Care 28, 1092–1100.

Ichikawa, H., Nakamura, Y., Kashiwada, Y., Aggarwal, B.B., 2007. Anticancer drugs designed by mother nature: ancient drugs but modern targets. Curr. Pharm. Des. 13, 3400–3416.

Jones, R., Armstrong, D., Malfertheiner, P., Ducrotte, P., 2006. Does the treatment of gastroesophageal reflux disease (GERD) meet patients' needs? A survey based study. Curr. Med. Res. Opin. 22, 657–662.

Koek, G.H., Sifrim, D., Lerut, T., Janssens, J., Tack, J., 2003. Effect of the GABA(B) agonist baclofen in patients with symptoms and duodeno-gastro-oesophageal reflux refractory to proton pump inhibitors. Gut 52, 1397–1402.

Rathmann, W., Kostev, K., Gruenberger, J.B., Dworak, M., Bader, G., Giani, G., 2013. Treatment persistence, hypoglycaemia and clinical outcomes in type 2 diabetes patients with dipeptidyl peptidase-4 inhibitors and sulphonylureas: a primary care database analysis. Diabetes Obes. Metab. 15, 55–61.

Shaheen, N.J., Denison, H., Björck, K., Karlsson, M., Silberg, D.G., 2013. Efficacy and safety of lesogaberan in gastro-oesophageal reflux disease: a randomised controlled trial. Gut 62, 1248–1255.

Sifrim, D., Zerbib, F., 2012. Diagnosis and management of patients with reflux symptoms refractory to proton pump inhibitors. Gut 61, 1340–1354.

Strain, W.D., Lukashevich, V., Kothny, W., Hoellinger, M.J., Paldánius, P.M., 2013. Individualised treatment targets for elderly patients with type 2 diabetes using vildagliptin add-on or lone therapy (INTERVAL): a 24 week, randomised, double-blind, placebo-controlled study. Lancet 382, 409–416.

Vaidya, R.A., Vaidya, A.D., Patwardhan, B., Tillu, G., Rao, Y., 2003. Ayurvedic pharmacoepidemiology: a proposed new discipline. J. Assoc. Physicians India 51, 528.

Xu, X.H., Yang, Z.M., Chen, Q., Yu, L., Liang, S.W., Lv, H.J., Qiu, Z.M., 2013. Therapeutic efficacy of baclofen in refractory gastroesophageal reflux-induced chronic cough. World J. Gastroenterol. 19, 4386–4392.

Chapter 2.2

Target Profiling in Terms of Translatability and Early Translation Planning

Martin Wehling

If a target gains attention as a potential drug action site, it is of paramount importance to estimate its translational relevance and feasibility. Given that 20,000–30,000 human genes have been cloned and that their variable products may add up to more than 200,000, it is obvious that there is no shortage of potential targets. Nowadays the key question is not target identification (in scholarly terms, one could say that by the time of this writing all targets have been identified in the human genome cloning project) but target validation. Thus, the critical assessment of a potential target is the most relevant and crucial assessment in the pharmaceutical industry, and a lot of this has to do with translational issues.

One dimension of this assessment is the so-called druggability, which means the chemical accessibility of a target to intervention with pharmaceutical agents. Enzymes may be easily druggable if they are soluble proteins in many instances, if they can be functionally expressed, and if their quantitative structure activity relationship can be derived; activators are harder to find than inhibitors. In addition, a target may be attacked by a known protein (antibody), but if it is intracellular, this receptor is not druggable. Receptors within the brain that can be attacked only by proteins may be hard to reach. Receptors requiring hydrophilic compounds may lead to water-soluble but poorly absorbed compounds, which would require parenteral therapy, just to name some basic chemical druggability criteria. However, aspects of medicinal chemistry such as these that are very important in the assessment of a target will not be further debated, as this book is focused on translation.

It has become obvious to many in the field that the selection of targets with strong translational validation is key to success. If one would need serial brain slicing to detect the beneficial effect of a drug, for example, against Alzheimer's disease, such a target could lead to success in animal models of the disease, but human testing would rely on entirely different testing methods (e.g., functional testing) that would certainly be closer to the desired clinical effect but may not strictly be predicted from the histological findings in animals.

If—at the time of translational assessment—there are no biomarkers that could bridge the gap from animal to man, the result of the assessment could be to invest in biomarker development, and, if success is sensed in this undertaking, the target would gain enough translational power to be heavily investigated in a full-blown drug R&D project. As such biomarkers would probably benefit various targets, an integrated view of such an investment would be necessary.

This example shows that translational assessment of targets—whether they be drug, device, or diagnostic test targets—needs to be done from day one. It is therefore strongly recommended to establish a translational medicine plan already when a novel target is discussed for the first time. This certainly applies to commercial ventures, but in academia, translational assessment should become a constant challenge that cannot be done too early in any given project. Certainly, this assessment should become an essential part of funding applications in the public domain, and funding agencies should routinely require this. Nowadays, funding applications routinely contain sections on "exploitation" or result dissemination, but this is normally just a short list of standard technology transfer or publication strategies that do not imply particular attention to the translational assessment beforehand.

If we accept that translational profiling of a target is a key asset in the biomedical development processes, and that it should commence very early on, the main question is which dimensions this profiling should have. Obviously, the dimensions are variable depending on the stage of the developmental process.

ESSENTIAL DIMENSIONS OF EARLY TRANSLATIONAL ASSESSMENT

The main dimensions of translational assessment are summarized in Table 2.11.

Even if a novel gene encoding for enzymatic activity, for example, has just been detected, it is of paramount importance to clearly state the area of clinical interest for which a potential drug should be developed. This is important because it is

Principles of Translational Science in Medicine. http://dx.doi.org/10.1016/B978-0-12-800687-0.00011-6

TABLE 2.11 Translational assessment dimensions

Area of clinical interest
Unmet clinical need
Competitors—generic or proprietary—and patent expiry
Cost pressure by reimbursement or insurance mechanisms
Starting evidence
In vitro evidence, including animal genetics
In vivo evidence, including animal genetics (e.g., knock-out, overexpression models)
Animal disease models
Multiple species
Human evidence
Genetics
Model compounds
Clinical trials
Biomarkers for efficacy or safety prediction
Biomarker grading
Biomarker development
Concept for proof-of-mechanism, proof-of-principle, and proof-of-concept testing
Biomarker strategy
Surrogate or endpoint strategy
Personalized medicine aspects
Disease subclassification, responder concentration
Pharmacogenomics
Chemical tractability*
Lead identification
Lead optimization
IP position and patent expiry*

**Although not directly linked to translation, these dimensions are essential contributors to project feasibility.*
(Adapted by permission from MacMillan Publishers Ltd.: *Nature Reviews Drug Discovery*, Wehling, 2009.)

the only way to define the unmet clinical need as the main driver for innovation and—as has become nearly as important in recent years—reimbursement. Obviously, treatment of arterial hypertension is almost ideal; thus, unmet medical need seems low. This may be reflected by the fact that truly innovative antihypertensive agents have rarely been introduced in recent years. And yet, treatment options for many malignant tumors and neurological disorders (e.g., Alzheimer's disease) are still poor, and unmet medical need is huge.

This assessment implies that competitors are identified in the generic and proprietary domains. Patent expiries have a strong negative impact, as expensive development programs cannot be refinanced. Cost pressure by reimbursement or insurance mechanisms needs to be extrapolated to the expected market approval time, which contains an assessment of willingness-to-pay for innovation in the health-care community.

From the scientific, translational view, obtaining the starting evidence for the validation of applicable and translatable targets is the biggest scientific challenge in this regard. In vitro and animal in vivo evidence for potentially useful effects will be collected, multiple species experiments compared, and potential human effects extrapolated. Animal models for normal physiology and—even more important—pathology (disease models) will be tested, and genetically engineered or genetically inbred animals (e.g., spontaneously hypertensive rats) will be utilized. Certainly, at the very early stages, even those animal experiments are not yet available, and translation assessment remains truly speculative.

From the first day of discussion of a novel target, human evidence for its validation is the most relevant source. Human genetics should be searched meticulously for clues to the genetic impact, as experiments by "Mother Nature" may guide translational efforts. Genetic association studies are most important in this context, although their reliability or predictivity may be very limited. A prominent recent example is the polymorphism of leukotriene A4 hydrolase or 5-lipoxygenase-activating protein (called FLAP), which, according to the Icelandic DECODE Consortium, may be relevant to ethnic differences in myocardial infarction incidences (Helgadottir et al., 2006). The contributions of the different constituents of the 5-lipoxigenase pathway to arterial thrombotic diseases have been quantified in a Spanish population in between (Camacho et al., 2012). Related studies have sparked interest in drug development with FLAP or other contributors to this pathway as targets, and this has been investigated by several drug companies. So far, zileuton is the only marketed drug in this context, but was approved for the treatment of asthma. At present, the most advanced novel compound is licofelone, which is profiled for the treatment of osteoarthritis. Clinical applications in the important areas of cardiovascular disease and cancer for which evidence of pathophysiological contributions of this pathway is available are yet missing (Steinhilber and Hofmann, 2014).

Model compounds can be very useful for target validation if they have generated evidence of the importance of a target. As mentioned previously, baclofen used as a muscle relaxant produced effects on TLESR and thus provided strong evidence for $GABA_B$ receptors as potential targets in reflux disease treatment. It should be said here that this point is the main reason for the development of me-too compounds. The first approved drug aiming at a particular target makes translation relatively reliable and safe for its congeners—with the important exception of clinical safety, which can reflect multiple effectors other than the main target, for example, allergies. Still, the reliability of translation for congeners is antagonizing innovation, and increased reliability of translation for unrelated, innovative developments would also be desirable in this regard. In certain cases, however, pharmacokinetic or pharmacodynamic (PK/PD) optimization by second- or third-entry congeners has been considered a successful innovation; take, for example, tiotropium versus ipratropium in chronic obstructive pulmonary disease (COPD) treatment. Tiotropium can be applied once daily rather than four times daily and is more efficacious than ipratropium (Voshaar et al., 2008), but it acts via the same bronchial muscarinergic receptors. Clinical trials as part of developmental processes are often published before market approval and are the main source of human evidence.

Biomarkers for efficacy or safety prediction are the major vehicle for translational processes. As specified in the next chapter, the availability of cross-species, easily accessible, accurately measurable, and validated biomarkers will largely determine the success of a translational program. Biomarker grading or quantitative predictivity assessment is an essential part of translation planning from very early days. The linear strategy of drug development is not successful if such biomarkers are lacking; a prominent example is drug development in psychiatric sciences (such as treatment for schizophrenia) in which animal models are absent or very weak. The lack of models and, thus, cross-species (including human) biomarkers has resulted in clinical empiricism rather than rationalistic drug development. Biomarker development is an important element for improving the validity of available biomarkers, and strategies for biomarker development need to be planned from early stages.

Proof-of-mechanism, proof-of-principle, and proof-of-concept (PoC) testing need to be conceptualized at early stages as well. This includes determining the biomarker strategy (which biomarker to use when) and, at later stages, defining surrogate or clinical endpoints to test efficacy. The proof-of-principle study design will be the key gating trial for translational processes, and early assessment of the feasibility, validity, and size of such a trial is elementary. For grading purposes, early conceptualizations of even the final endpoints in the last phase III trial may be instrumental if multiple projects compete for limited resources. If one project will ultimately have to comprise an expensive and long mortality trial because accepted surrogates (e.g., low-density lipoprotein (LDL)-cholesterol for statin development) do not exist, another project with an accepted surrogate endpoint may be preferable.

Personalized medicine aspects such as disease subclassification, responder concentration, and pharmacogenomics need to be covered as early as possible. This concept addresses the key question of which patients are the most appropriate for a given drug, as classified by disease subtype, stage or severity, and the patient's pharmacogenetic setting. The blockbuster concept is being replaced by a niche buster concept, which exactly reflects this dimension.

As said in the beginning of this section, chemical tractability, lead identification issues, and lead optimization issues are very important for a drug project, but they are not directly linked to translation. In an overall assessment, however, they are essential contributors to project feasibility. The same holds true for the IP position (patent strength, freedom to operate, patent expiry).

A NOVEL TRANSLATABILITY SCORING INSTRUMENT: RISK BALANCING OF PORTFOLIOS AND PROJECT IMPROVEMENT

If translation planning is performed early on and if those major dimensions are graded accordingly, the risk of embarking on development of a given target can be estimated comparatively and competitively. The same holds true for assessment of in-licensing opportunities, and investments should be guided by such objective rules. Developing and validating a scoring system to grade the evidence and facts in a translational planning document should be a future goal of the new scientific approach to translational medicine. A similar scoring system has been developed to grade biomarkers (see Chapter 3), and this would become an essential part of the overall scoring of the translational potential of a given target.

A proposal for translatability scoring is given in Table 2.12. The individual scores are chosen between 1 and 5, multiplied by the weight factor, and then divided by 100. Any score above 4 is indicative of fair to good translatability. If no data are available, no score will be given in this line. Stoppers (such as no evidence or very weak evidence for animal effects at large) will be rated 0 and render the project untranslatable. The biomarker grading score should correspond to the one described in Chapter 3, divided by 10.

This score should show the relative importance of translational parameters and help in the translatability assessment. In the pharmaceutical setting, unmet medical need will certainly be very important, but translation can be successful even without any unmet medical need. Me-too drugs have a very low translational risk, and scores are very high for them. Yet, due to market considerations, this approach may not be feasible although not jeopardized by translational risk.

The same holds true for chemical tractability (translation can be performed even with suboptimal compounds; see eINDs, Chapter 4.2) and IP issues. This chapter describes in more detail the overall assessment of a novel target, whereas the score given here is focused on translational items only. It is obvious that arithmetic scoring alone is insufficient, as there could be stoppers (no effects or weak effects in animal experiments at large) that would not be detected by this score.

Obviously, translational dimensions of target validation will change according to the stage of the developmental procedure. With the important transition from animal to human testing, translation of PK/PD modeling will be a major task. Extrapolation of essentials in PK/PD modeling, such as half-life time, volume of distribution, absorption rate, or route of

TABLE 2.12 Proposal for scoring the translatability of an early project

	Score 1–5	Weight (%)	Σ (score × weight/100)
Starting evidence			
In vitro including animal genetics		2	
In vivo including animal genetics (e.g. knock-out, overexpression models)		3	
Animal disease model		3	
Multiple species		3	
Human evidence			
Genetics		5	
Model compounds		13	
Clinical trials		13	
Biomarkers for efficacy/safety prediction			
Biomarker grading		24	
Biomarker development		13	
Concept for proof-of-mechanism, proof-of-principle, and proof-of-concept testing			
Biomarker strategy		5	
Surrogate/endpoint strategy		8	
Personalized Medicine Aspects			
Disease subclassification, responder concentration		3	
Pharmacogenomics		5	
SUM		**100**	

(Wehling, 2009, by kind permission from MacMillan Publishers Ltd.: *Nature Reviews Drug Discovery*).

elimination from various species to humans, is described elsewhere, but successful translation in this area is a prerequisite for dose-finding studies. Dose-finding studies are the ultimate necessity for designing human trials that are fit to test the target in humans in a meaningful way—meaning that the doses chosen yield the correct exposure to the drug. Thus, this piece of translational extrapolation is an essential part of target validation as a process.

Traditionally, drug metabolism and pharmacokinetics (DMPK) studies in animals, which are the only source of predictive data for the transition to humans and the corresponding early human dose-finding trials (single ascending dose and multiple ascending dose), are not seen as central components of the translational process. This separation, which is variably expressed among different drug companies, is not logical, and DMPK studies should be integrated into translational medicine efforts.

As a drug project approaches phase II, either the target has been validated by congener or model compounds in humans, or the proof-of-principle study has revealed that the compound works in principle. This means that the efficacy assessment is based on biomarkers close to the surrogates or endpoints in patients suffering from the disease to be treated. Because those biomarkers are not good enough to become surrogates or clinical endpoints (see Chapter 3), surprising efficacy failures are still not uncommon after this phase.

At this stage, a human target is almost fully validated and just awaits PoC validation, which is the last phase before the pivotal phase III (market approval) trial. If the underlying concept is right, the compound affecting the target in question will ultimately treat the disease as measured by surrogates or clinical endpoints. The huge phase III trial is then confirmatory in terms of target validation only and mainly covers safety aspects, which are dominated by the unknown (or almost unknown) extra-target activities of a compound, such as liver or muscle toxicity that is not linked to the target receptor. In this sense, target validation is finalized in the PoC stage, and, thus, efficacy translation has been completed successfully if the compound works at this level.

However, it is of paramount importance to make it clear that this efficacy validation of a target as a major task of translational medicine is only half the battle. Toxicology or safety translation is usually seen as target related and consequently neglected; thus, accessible toxic effects represent a minority of potential safety issues (as stated previously). One of the biggest challenges in the drug industry is setting up nonspecific test batteries in animal and early human testing to better predict toxic effects. This translational aspect has just started to yield considerable novel safety biomarkers or strategies for screening, e.g., for novel renal safety biomarkers. Established procedures are mandatory screening for human ether-a-go-go related gene (hERG) channel effects (potential of QT interval prolongation and, thus, proarrhythmic effects) and electrocardiogram (ECG) changes in early human trials. See Figure 2.24.

As the science behind translational medicine progresses, scoring will be only one of the possible approaches to judge objectively on translatability issues, and in the end will hopefully become obsolete. Peer review and consensus processes would then regain weight as peers and reviewers are raised to the expert level; as of the time of this writing, translational experts are rare, and consensus processes still seem to lack objectivity in this area.

The translatability scoring instrument reflects the decreasing risk along the developmental path with earlier projects being necessarily more risky than later ones. Thus, it may be wise to apply it preferably to compare projects at a similar stage of development. Thereby, drug companies, but also public funding agencies or venture capitalists, may risk-balance their portfolios in different strata (e.g. preclinical, early/late human). The average risk curve as depicted in Figure 2.25 shows the progressive decline of risk along the R&D development, with a major value deflection point at human proof-of-concept (PoC).

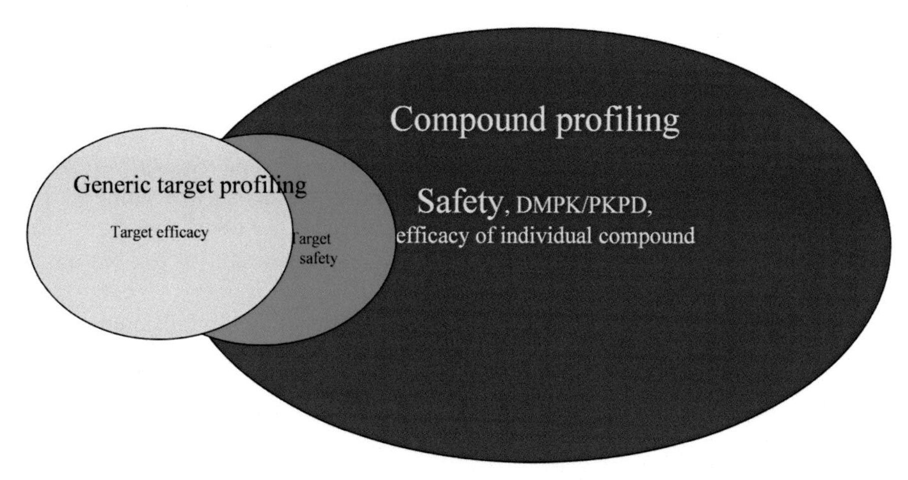

FIGURE 2.24 The target validation process is attached to the overall translational profiling of a compound, with safety concerns being the biggest challenge at present. The generic target validation aspects of receptor-mediated efficacy and receptor-related safety concerns are, in most instances, embedded or close to the translational profiling of a given compound. The generic nature of target validation is important for congener compounds within the program of the company hosting the project and for its competitors as well, if published in the public domain, for example, in clinical trial registries, patent applications, or even journals or books.

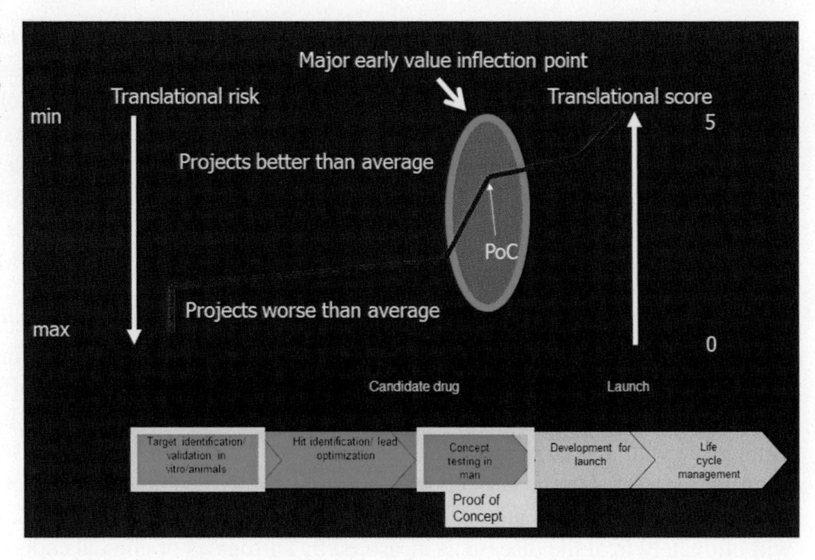

FIGURE 2.25 "Natural" course of translational risk in the pseudolinear drug R&D process. The scores from the translatability assessment instrument (Wehling, 2009) are depicted as right-hand ordinate, the inverse level of risk on the left-hand ordinate. Min: minimal; Max: maximal.

If PoC is positive, the value of a project may increase by one-hundred-fold; thus, any license holder should aim at earning this value increase and make the best possible prediction into the likelihood of success.

This fact mandates not only the translatability assessment beforehand, a speedy rush to PoC, but also the smart use of knowledge on weaknesses of a given project as identified by the scoring approach. If the score is low, e.g., due to the lack of viable and predictive biomarkers, the money should be spent to develop the biomarkers first before frustration emerges from failed PoC.

CASE STUDIES: APPLYING THE NOVEL TRANSLATABILITY SCORING INSTRUMENT TO REAL-LIFE EXPERIENCES

The translatability assessment tool as described in the preceding chapter has been applied and evaluated in retrospective case studies (Wendler and Wehling, 2012). In Table 2.13, the detailed individual score results are given.

Those drugs were assessed starting from a literature search at a fictive date before the phase III trial was commenced. The question to be answered was as follows: which translational risk had been taken at the transition from phase IIb to phase III, and how did it correlate with the outcome in terms of market approval?

One drug, gefitinib, which failed at the first attempt, was assessed twice because new essential biomarker data helped to improve responder concentration.

Dabigatran as treatment for atrial fibrillation was developed at relatively low risk because several features of dabigatran such as safety and the effect on coagulation had already been investigated in earlier studies. Translatability scores were high for the items' model compounds, clinical trials, biomarker grading, and surrogates in Table 2.13. This represents the development of a new therapeutic indication of an already-approved drug, with a much lower risk than the development of a new drug for a new application. The overall translatability score, which is 3.77 (Table 2.13) is, thus, plausible and indicates mean to fair translatability. Subsequently, this drug was approved for this indication. All other drugs with scores in excess of 3.5 were subsequently approved (ipilimumab, varenicline, gefitinib with improved biomarker), whereas those with lower scores failed (vilazodone, latrepirdine, semegacestat, torcetrapib, gefitinib/first attempt).

The gefitinib story is remarkable as an example of improved biomarkers that were discovered only after the first phase III trial had been started. Several studies showed that only patients with activating mutations in the epithelial growth factor receptor (EGFR) responded well to this treatment (e.g., Lynch et al., 2004). With the novel biomarker (activating EGFR mutation) introduced, results were improved by far, and the drug approved. This was reflected in the translatability scoring instruments at the level of "clinical trials, biomarkers, and disease subclassification (= personalization)" as can be seen in increased scores in Table 2.13 after the introduction of this biomarker.

Both drugs (latrepirdine, semegacestat) against Alzheimer's disease (AD) failed. The major implication of the assessment of these two potential AD drugs is the fact that the etiology of AD is not yet understood, and amyloid-ß targeted therapies are likely to attack an epiphenomenon. This gap of knowledge is mainly reflected by the lack of a biomarker placed more proximal in AD etiology. The scores for biomarker-related items are 0 for both compounds in Table 2.13 which could be seen as a stopper as such.

TABLE 2.13 Assessment of translatability for eight drugs

Compound	Dabigatran	Ipilimumab	Gefitinib	Gefitinib*	Vilazodone	Latrepirdine	Semegacestat	Torcetrapib [37]**	Varenicline [37]
Aspect									
Starting evidence									
In vitro data including animal genetics	0.1 [4]	0.06 [38]	0.1 [6,7,39,40]	0.1 [6,7,39,40]	0.08 [41]	0.06 [42–44]	0.1 [12]	0.1	0.1
In vivo data including animal genetics	0.15 [4]	0.06 [45] [46]	0.15 [6,7,14]	0.15 [6,7,14]	0.12 [41,47,48]	0.06 [42,49]	0.15 [12,50]	0.15	0.15
Animal disease models	0.09 [51]	0.06 [1,52]	0.09 [1]	0.09 [1]	0.06 [41,47,48,53]	0.06 [49]	0.03 [18]	0.12	0.15
Data from multiple species	0.15 [4]	0.09 [46]	0.06 [54]	0.06 [54]	0.06 [41,55]	0.03 [42]	0.15 [12]	0.03	0.15
Human evidence									
Genetics	0.05	0.05	0.05	0.25 [8–10,22,28]	0.05	0.05	0.25 [56–59]	0.05	0.05
Model compounds	0.52 [60]	0.52 [61]	0.13 [54]	0.13 [54]	0.65 [62–64]	0.13 [42,43,65,66]	0.13		0.65
Clinical trials	0.52 [31,67]	0.65 [68–74]	0.26 [24,25,75–78]	0.65 [8,9,24,25, 75–81]	0.26 [15,82]	0.13 [15,42,83]	0.39 [50,84,85]	0.26	0.52
Biomarkers for efficacy and safety prediction									
Biomarker grading	0.96 [35,36]	0.96 [5]	1.2 [24,25]	1.2 [8,9,24,25]	0.72 [26,27]	0	0 [23]	0.48	1.2

Continued

TABLE 2.13 Assessment of translatability for eight drugs—cont'd

Compound	Dabigatran	Ipilimumab	Gefitinib	Gefitinib*	Vilazodone	Latrepirdine	Semegacestat	Torcetrapib [37]**	Varenicline [37]
Biomarker development	0.26	0.52 [5]	0.13 [24,25]	0.65 [8–10,22,28]	0.13 [26,27]	0	0 [23]	0.26	0.52
Proof-of-mechanism, proof-of-principle, and proof of concept testing									
Biomarker strategy	0.2 [35,36]	0.2 [5]	0.05 [24,25]	0.25 [8–10,22,28]	0.15 [26,27]	0	0 [23]	0.1	0.25
Surrogate or endpoint strategy	0.4	0.4 [5]	0.24 [24,25]	0.32 [8–10,22,28]	0.24 [26,27]	0	0 [23]	0.16	0.32
Personalized medicine aspects									
Disease subclassification and responder concentration	0.12	0.03	0.03	0.15 [8–10,22,28]	0.03	0.03	0.03	0.09	0.03
Pharmacogenetics	0.25	0.05	0.05	0.25 [8–10,22,28]	0.05	0.05	0.05	0.15	0.05
SUM	**3.77**	**3.65**	**2.54**	**4.25**	**2.6**	**0.60 (0)**	**1.28 (0)**	**1.95**	**4.14**

**After the development of the pivotal biomarker (EGFR mutation status).*
***Numbers in brackets refer to citations in the paper; though not referenced here, this should underline that those scores are based on data in the literature.*
(According to [Wehling, 2009], from Wendler et al., 2012.)

Further studies are needed to analyze the pathogenesis of the disease and to develop suitable biomarkers before any drug should be developed.

The development of torcetrapib was driven by the hypothesis that increasing levels of HDL-cholesterol should be cardioprotective for the patients. This was derived from epidemiological data, for example, from the famous Framingham study that showed a negative correlation between cardiovascular risk and HDL-cholesterol (Gordon et al., 1977). Unfortunately, this biomarker had never been proven to be protective if manipulated by drugs, as opposed to LDL-cholesterol, which is one of the best surrogate parameters. Almost any intervention lowering LDL-cholesterol was beneficial for patients. Such evidence was (and still is) lacking for HDL-cholesterol, which in Table 2.13 got low biomarker scores, rendering the project high to intermediate risk. It finally failed, and yet the insufficiency of the biomarker was not fully acknowledged, but "off-target" effects (increase in blood pressure) blamed for the excess mortality were observed. It soon became clear that the blood pressure increase was not intense enough to increase mortality if a beneficial effect by the target effects would have been present (Nissen et al., 2007).

Despite this clinical failure, which is the ultimate evidence against translational success, a congener drug, dalcetrapib, was taken into phase III. It also failed although this drug, being devoid of off-target effects, at least did not increase mortality. In the scoring system, the score went down from 1.95 for torcetrapib to 1.06 for dalcetrapib as the item's "model compounds, clinical trials, biomarker strategy" turned to 0, and "biomarker strategy, surrogate/endpoint strategy" to only 1 (from 5, resp. 4). Thus, there were stoppers ("0"), so the expensive phase III trial was performed at the highest possible risk. This assessment was done prospectively and, unfortunately, proved to be correct.

With the exception of this last case, the cases presented were retrospective ones; the prospective validation of the tool is still missing. It would probably take 10–15 years given the slow cycle time in drug R&D. However, the results are referenced and plausible. Therefore, it is hoped that this structured approach to translatability assessment would be adapted by the key stakeholders; it is open for modifications regarding the items and the [better] weighing factors, and may be developed to be more specific for different therapeutic areas. In cancer projects as an example, the weight of personalization appears to be much higher than proposed here. Conversely, the uniform score allows for transtherapeutic comparisons that are becoming increasingly important. At least, a common understanding seems to be gaining ground in that "gut-feeling" approaches are outdated, and structured, transparent, reproducible, and citable procedures should be instituted at large.

REFERENCES

Camacho, M., Martinez-Perez, A., Buil, A., Siguero, L., Alcolea, S., López, S., et al., 2012. Genetic determinants of 5-lipoxygenase pathway in a Spanish population and their relationship with cardiovascular risk. Atherosclerosis 224, 129–135.

Gordon, T., Castelli, W.P., Hjortland, M.C., Kannel, W.B., Dawber, T.R., 1977. High-density lipoprotein as a protective factor against coronary heart disease. The Framingham Study. Am. J. Med. 62, 707–714.

Helgadottir, A., Manolescu, A., Helgason, A., Thorleifsson, G., Thorsteinsdottir, U., Gudbjartsson, D.F., et al., 2006. A variant of the gene encoding leukotriene A4 hydrolase confers ethnicity-specific risk of myocardial infarction. Nat. Genet. 38, 68–74.

Lynch, T.J., Bell, D.W., Sordella, R., Gurubhagavatula, S., Okimoto, R.A., Brannigan, B.W., Harris, P.L., Haserlat, S.M., Supko, J.G., Haluska, F.G., et al., 2004. Activating mutations in the epidermal growth factor receptor underlying responsiveness of non-small-cell lung cancer to gefitinib. N. Engl. J. Med. 350, 2129–2139.

Nissen, S.E., Tardif, J.C., Nicholls, S.J., Revkin, J.H., Shear, C.L., Duggan, W.T., et al., 2007. Effect of torcetrapib on the progression of coronary atherosclerosis. N. Engl. J. Med. 356, 1304–1316.

Voshaar, T., Lapidus, R., Maleki-Yazdi, R., Timmer, W., Rubin, E., Lowe, L., et al., 2008. A randomized study of tiotropium Respimat((R)) Soft Mist(TM) Inhaler vs. ipratropium pMDI in COPD. Respir. Med. 102, 32–41.

Steinhilber, D., Hofmann, B., 2014. Recent advances in the search for novel 5-lipoxygenase inhibitors. Basic Clin. Pharmacol. Toxicol. 114, 70–77.

Wehling, M., 2009. Assessing the translatability of drug projects: what needs to be scored to predict success? Nature Rev. Drug Discov. 8, 541–546.

Wendler, A., Wehling, M., 2012. Translatability scoring in drug development: eight case studies. J. Transl. Med. 10, 39.

Chapter 3

Biomarkers

Chapter 3.1

Defining Biomarkers as Very Important Contributors to Translational Science

Martin Wehling

If one defines the aim of drug or medical device development as the prevention, palliation, or cure of diseases, there is an obvious need to measure this impact and relate it to disease stage, types, and severity. This need is the basis for the development of the concept of biomarkers.

As with translational medicine in general, biomarkers are anything but new; they have been used at all times by scientists trying to describe biological systems. Thus, the most generic definition of biomarkers is based on their role as descriptors or measures of biological systems (Table 3.1, definition 1). Thus, anything one can use to describe or measure features of a biological system is a biomarker. Biomarkers may be obvious measures such as blood pressure or heart rate in humans or less obvious measures such as hair color or cell cycle stages, gene polymorphisms, life or death, and ability to fertilize and produce offspring. In biomedical sciences, it is helpful to focus the definition of biomarkers on their operation, as critically selecting what to look at or measure is indispensable if scientific approaches to drug or medical device development are to succeed. Under these operational auspices, the definition is limited in that biomarkers must reflect biological effects induced by or linked to disease and/or therapeutic interventions and must be the main tools for predicting and describing their efficacy and safety (Table 3.1, definition 2; for a further discussion, see Frank and Hargreaves, 2003; Strimbu and Tavel, 2010).

Biomarkers with potential relevance to drug or medical device development are likely to be used if they are easily accessible and tightly linked to disease and its modification by intervention (for classification and grading, see Subchapters 3.2

TABLE 3.1 What is a biomarker? (1) A generic definition; (2) An operational definition for therapeutic development
1) Biomarkers describe or measure biological systems
2) A biomarker relevant to treatment development describes or measures
Disease
Type
Subtype
Severity
Activity
Stage
Impact of intervention on key biological variables
Linked to disease, pathophysiology, or outcome
Linked to safety

Principles of Translational Science in Medicine. http://dx.doi.org/10.1016/B978-0-12-800687-0.00012-8

through 3.4). In this context, serum parameters (e.g., the primary product of an enzyme that is inhibited by the drug); generic markers of disease severity (e.g., high-sensitivity C-reactive protein [CRP] and other markers in inflammatory processes [Schonbeck and Libby, 2004; Pereira et al., 2014]); histology (especially in cancer); and particularly imaging (e.g., plaque morphology in drugs affecting atherosclerosis, liver fat content as a safety biomarker, positron emission tomography [Stahl et al., 2004], or neuroimaging in Alzheimer's disease [Shaw et al., 2007; Leuzy et al., 2014]) are frequently used as biomarkers.

Biomarkers may be as obvious as LDL-cholesterol measurements in the development of lipid-lowering drugs or arterial blood pressure in that of antihypertensive agents, but at preclinical stages and, in particular, at the critical transition phase to human trials, they may not be easily identifiable. Rather, a large program may be devoted to the discovery and validation of novel markers because the existing ones are not good enough. "Omics" approaches (e.g., genomics, metabonomics) may generate huge panels of biomarkers; as stated in Subchapter 2.1 in greater detail, the huge number of measures ("spots") aligning into patterns may create more problems (mainly statistical in nature) than solutions to discovery of reliable drug effects (Bilello, 2005). The validation process of biomarkers resembles the drug development process and is described in Subchapter 3.3.

How should biomarkers be used? Figure 3.1 describes the three key questions to be answered during the drug discovery or development process. They are related to proof-of-mechanism testing (Does the biomarker hit the target?), proof-of-principle testing (Does the biomarker alter the mechanism? In other words, does it change the parameters related to the disease, regardless of whether it can be proved that the disease is beneficially impacted?), and proof-of-concept testing (Does the biomarker affect the disease? In other words, does the drug actually treat the disease and benefit the patient?). All yes-or-no decisions are based on appropriate biomarker measurements. The most obvious biomarker at the first level is a direct product of the receptor or enzymatic activity to be affected by the drug, for example, leukotriene LTB4 for inhibitors of 5-lipoxygenase. A typical biomarker for the second level would be closely linked to the disease; for example, generic inflammatory markers would be used if inflammatory bowel disease treatment was the aim of drug development. Finally, biomarkers for the last stage would directly describe the clinical

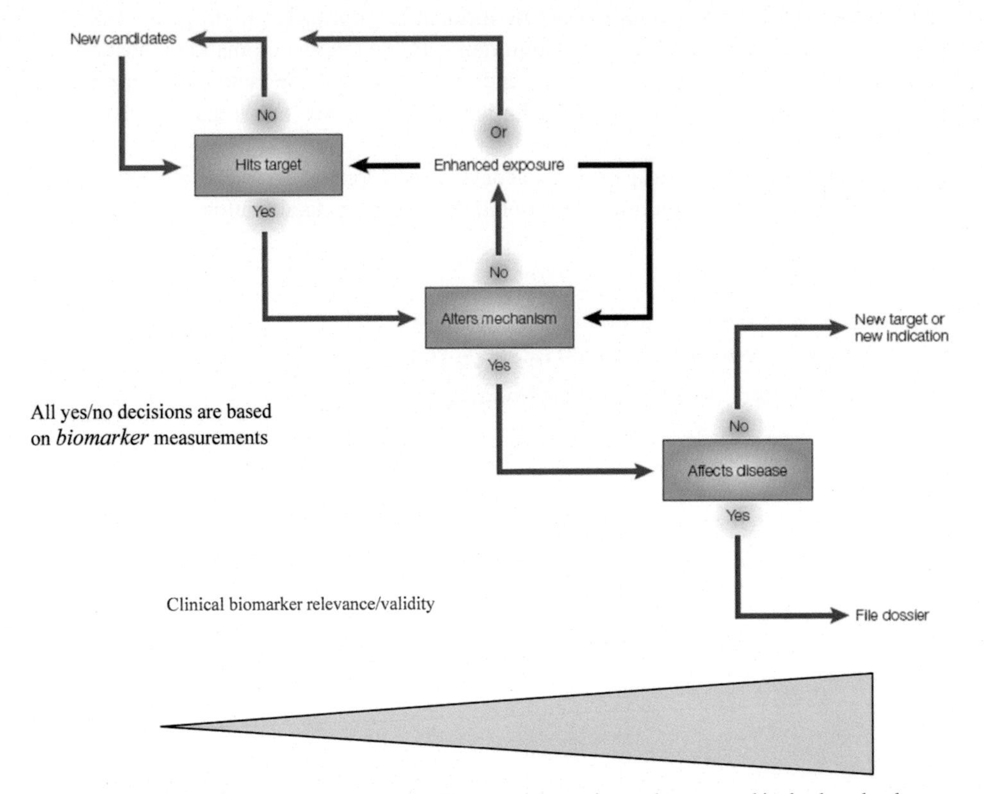

FIGURE 3.1 Impact and remits of biomarkers in drug development. The essential questions to be answered in the drug development process are related to proof-of-mechanism testing (Does the biomarker hit the target?), proof-of-principle testing (Does the biomarker alter the mechanism? In other words, does it change the parameters related to the disease, regardless of whether it can be proved that the disease is beneficially impacted?), and proof-of-concept testing (Does the biomarker affect the disease? In other words, does the drug actually treat the disease as needed?). *(Modified from Frank and Hargreaves, 2003, adapted by permission from Macmillan Publishers Ltd:* Nature Rev. Drug Discov., *2: 566–580, © 2003.)*

endpoints by which disease activity is clinically assessed, for example, biomarkers that describe the clinical symptoms of inflammatory bowel disease (stool frequency and consistency), histology, complications (fistulas, need for surgery), and so on. This example demonstrates that the clinical relevance and validity of biomarkers increase at each stage and that they increasingly resemble or are identical to the clinically used categories for disease grading, staging, or activity assessment. Those criteria represent biomarkers as well. If translation from preclinical to clinical stages is one of the core processes determining the success or failure of drug or medical device developments, then biomarkers are the core of the translational process. The validity of translational projections is closely tied to the predictive value, robustness, reproducibility, and accessibility of the central biomarker(s) on which go-or-no-go decisions are based. The right choice and assessment of biomarkers, therefore, can be considered the major contribution of translational medicine in general. Successful or failed translational processes as described in Chapter 11 have a lot to do with the choice and assessment of biomarkers. Take, for example, the aborted development of torcetrapib (discussed in greater detail in Chapter 11). There has been considerable interest in raising HDL-cholesterol levels, and this CETP-inhibitor was a powerful tool for achieving this effect. Unfortunately and unexpectedly, mortality was increased when torcetrapib was used. Apparently, HDL-cholesterol as such is not a good biomarker for describing and predicting the beneficial intervention in this area. Looking back, it seems that HDL-functionality (as measured by nitrosylation), rather than merely its concentration, should have been determined and would probably have predicted the failure and loss of patient lives. Of course, the nontarget effects of the drug, especially on blood pressure and aldosterone production, have been blamed as the cause of failure, but this has been falsified by testing a congener, dalcetrapib, which also failed (Schwartz et al., 2012).

Thus, the inherent value of a biomarker or a panel of biomarkers is the most precious asset in a translational process. There are areas of biomedical research in which the quality of biomarkers is weak in general, at least when it comes to animal or human translation. In particular, this holds true for psychiatric drug developments, as animal models for complex psychiatric syndromes without clear morphological changes are lacking. This applies to common disorders such as schizophrenia and depression as well.

The opposite is true for antibiotic agents: because the pathogens can be treated in vitro (either directly in cultures or—in the case of viruses—grown in cell cultures), efficacy translation is very predictive even before animal experiments have been conducted. Obviously, efficacy prediction from test tube results into humans is quite good, because the biomarker, microbial growth, is very closely tied to the disease induced (e.g., bacterial angina lacunaris—the bacteria are killed, and the disease is healed). Although in vivo conditions can be different from test tube conditions, especially with regard to target accessibility (e.g., CNS infection, blood-brain barrier), in vivo growth and growth inhibition are generally in line with in vitro results. The main problem remaining is the challenge of safety issues, which are not different from those in other projects, as safety is mostly unrelated to the test tube but entirely dependent on the host organism. Apparently, many novel antibiotic drugs "die" even after market approval (e.g., trovafloxacin, grepafloxacin, and telithromycin in Australia) mostly due to safety concerns. According to Figure 1.4 (Chapter 1), this difference in biomarker quality is reflected in the drug development success rates: 8% versus 16% for CNS versus antimicrobial developments. Cardiovascular projects are claimed to have a similarly positive scoring, but looking at the very rare recent market approvals in this segment, this figure in the plot seems to be overrated.

Accepting the pivotal role of quality biomarkers for successful translation, it is crucial both to develop assessment strategies for biomarkers to define their predictive quality and, if gaps or suboptimal biomarkers become identified, to promote the development of novel biomarkers. The latter is pertinent to almost all projects in which novel targets are addressed. In general, little is known on how those targets are linked to the disease, and whether biomarkers other than the clinical endpoints can be trusted as the basis of decisions on heavy investment into late-stage clinical trials (especially phase III trials, which can cost up to several hundred million U.S. dollars). There are few situations in which surrogates (for a definition of this term, see Subchapter 3.3) or clinical endpoints are easy to obtain and can be utilized early in clinical development. Take, for example, LDL-cholesterol (unlike HDL-cholesterol), which is an accepted surrogate for lipid-lowering agents (that means proof of efficacy can be given by a simple blood test rather than by clinical endpoints such as myocardial infarction or death). Thus, there was no need for the development of biomarkers for early efficacy testing, as this one could be determined in subjects in all clinical trials, including early-stage trials. The situation is quite different in the development of anti-atherosclerotic drugs, which mainly aims at inflammatory processes (e.g., FLAP-inhibitors). The ultimate proof of efficacy will be derived only from huge clinical trials on morbidity (cardiovascular events, e.g., myocardial infarction, stroke) or even mortality. Obviously, biomarkers for earlier clinical stages of development will have to be defined that provide the best evidence that a major endpoint trial is reasonably likely to yield positive results. At present, combined biomarker panels comprising imaging and serum inflammatory parameters are seen as favorites in this regard. This example should show that investing in biomarker development and choice can vary considerably and may excessively increase the risk of a drug

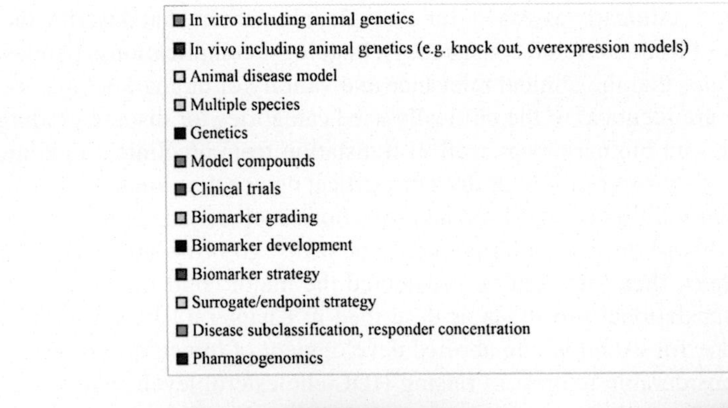

FIGURE 3.2 The relative importance of biomarkers in the translational plot varies depending on the availability of good surrogates that can be utilized early in development (e.g., LDL-cholesterol for lipid-lowering drugs). The pie chart on the left is based on definition 2 from Table 3.1 and represents the average situation requiring intense biomarker work with no surrogate available. The situation depicted in the pie chart on the right is rare.

project. The feasibility check on a given drug project should therefore include the biomarker status—the extent to which biomarkers are available or need to be developed. Development (see Subchapter 3.3) not only costs money but also carries considerable risk of failure, as is the case in all developmental projects.

Figure 3.2 summarizes the central role of biomarkers in drug or medical device projects. The investment depends on the availability of good surrogates, which diminish biomarker work dramatically but are rarely available.

REFERENCES

Bilello, J.A., 2005. The agony and ecstasy of "OMIC" technologies in drug development. Curr. Mol. Med. 5, 39–52.

Frank, R., Hargreaves, R., 2003. Clinical biomarkers in drug discovery and development. Nat. Rev. Drug Discov. 2, 566–580.

Leuzy, A., Zimmer, E.R., Bhat, V., Rosa-Neto, P., Gauthier, S., 2014. Imaging biomarkers for amyloid: a new generation of probes and what lies ahead. Int. Psychogeriatr. 3, 1–5.

Pereira, M.M., Santos, T.P., Aras, R., Couto, R.D., Atta, M.L., Atta, A.M., 2014. Serum levels of cytokines and chemokines associated with cardiovascular disease in Brazilian patients treated with statins for dyslipidemia. Int. Immunopharmacol. 18, 66–70.

Schonbeck, U., Libby, P., 2004. Inflammation, immunity, and HMG-CoA reductase inhibitors: statins as anti-inflammatory agents? Circulation 109 (21 Suppl 1), II18–II26.

Schwartz, G.G., Olsson, A.G., Abt, M., Ballantyne, C.M., Barter, P.J., Brumm, J., et al., 2012. Effects of dalcetrapib in patients with a recent acute coronary syndrome. N. Engl. J. Med. 367, 2089–2099.

Shaw, L.M., Korecka, M., Clark, C.M., Lee, V.M., Trojanowski, J.Q., 2007. Biomarkers of neurodegeneration for diagnosis and monitoring therapeutics. Nat. Rev. Drug Discov. 6, 295–303.

Stahl, A., Wieder, H., Piert, M., Wester, H.J., Senekowitsch-Schmidtke, R., Schwaiger, M., 2004. Positron emission tomography as a tool for translational research in oncology. Mol. Imaging Biol. 6, 214–224.

Strimbu, K., Tavel, J.A., 2010. What are biomarkers? Curr. Opin. HIV AIDS 5, 463–466.

Chapter 3.2

Classes of Biomarkers

Martin Wehling

Because the definition of biomarkers used in this book is very broad, a classification of biomarkers seems desirable. It is obvious that the function and value of a biomarker can be very different; for example, death or life is a very clear and serious biomarker, whereas other biomarkers such as appetite are quite diffuse and hard to standardize and measure. Some biomarkers just describe mechanistic aspects of physiology or pathophysiology, such as blood pressure or heart rate, and—by themselves—are not the ultimate goal of treatment (reduction of cardiovascular events, e.g., myocardial infarction or stroke). However, we know that having an effect on those mechanistic biomarkers is—at least in general—associated with altered clinical outcomes. Other biomarkers that have been thought to be closely linked to disease severity have been disappointingly vague or weak predictors of outcome (e.g., uric acid).

Thus, the inherent value of a biomarker as a translationally relevant predictor is widely variable, and its assessment is an obvious—if not the most important—task in translational medicine. Attempts in this direction have been made, for example, by Rolan, Atkinson, and Lesko (2003), who classified the biomarkers by their mechanistic character (e.g., genotype or phenotype or target occupancy; see Table 3.2).

This approach consists of a seven-point classification based on the location of the biomarker in the sequence of events from underlying subject genotype or phenotype through to clinical measures. This classification has shortcomings; for example, the type 0 biomarker relating to a subject's genotype or phenotype is traditionally considered a covariate rather than a biomarker. Similarly, the type 6 biomarker, a clinical scale, can be regarded as a measurement of a clinical endpoint, and this is not covered by all biomarker definitions.

Previous to Rolan and colleagues' classification, the NIH biomarker definition working group (Biomarkers Definitions Working Group, 2001) simplified the existing biomarker spectrum into only three categories: biomarkers per se, surrogates (which are closely linked to endpoints and can substitute for them), and clinical endpoints (Table 3.3).

The Biomarkers Definitions Working Group (2001) published a paper in *Clinical Pharmacology & Therapeutics*, in which the following explanations are given:

> Biological marker (biomarker): *A characteristic that is objectively measured and evaluated as an indicator of normal biological processes, pathogenic processes, or pharmacologic responses to a therapeutic intervention. Biomarkers may have the greatest value in early efficacy and safety evaluations such as in vitro studies in tissue samples, in vivo studies in animal models, and early-phase*

TABLE 3.2 Proposal for a biomarker classification
Type 0: Genotype or phenotype
Type 1: Concentration
Type 2: Target occupancy
Type 3: Target activation
Type 4: Physiologic measures or laboratory tests
Type 5: Disease processes
Type 6: Clinical scales
(From Rolan, Atkinson, & Lesko 2003, reprinted by permission from Macmillan Publishers Ltd: *Clinical Pharmacology & Therapeutics*, 73: 284–291, © 2003.)

Principles of Translational Science in Medicine. http://dx.doi.org/10.1016/B978-0-12-800687-0.00013-X

TABLE 3.3 Use of biomarkers for decision making: NIH Biomarkers Definitions Working Group Consensus Language

	Definition	Additional Information
Biomarker	A characteristic that is objectively measured and evaluated as an indicator of normal biological processes, pathogenic processes, or pharmacologic responses to a therapeutic intervention	Categories: • Disease relevant (diagnosis, prognosis) • Mechanistic (pharmacological activity)—linked to drug • Safety* • Efficacy**
Surrogate	A characteristic that is intended to substitute for a clinical endpoint, often thought of as a clinical endpoint	• Should predict clinical benefit (or harm or lack of benefit or harm) • Based on epidemiologic, therapeutic, pathophysiologic, or other scientific evidence
Clinical endpoint	A characteristic or variable that reflects how a patient feels, functions, or survives	• Disease characteristics observed in a study or a clinical trial that reflect the effect of a therapeutic intervention

Expected to predict clinical benefit or harm based on epidemiologic, therapeutic, pathophysiologic, or other evidence.
**Note that biomarkers are "evaluated," not "validated."*
(Extracted and modified from Biomarkers Definitions Working Group, 2001.)

clinical trials to establish "proof of concept." Biomarkers have many other valuable applications in disease detection and monitoring of health status. These applications include the following:

- *use as a diagnostic tool …*
- *use as a tool for staging of disease … or classification of the extent of disease …*
- *use as an indicator of disease …*
- *use for prediction and monitoring of clinical response to an intervention …*

Clinical endpoint: *A characteristic or variable that reflects how a patient feels, functions, or survives. Clinical endpoints are distinct measurements or analyses of disease characteristics observed in a study or a clinical trial that reflect the effect of a therapeutic intervention. Clinical endpoints are the most credible characteristics used in the assessment of the benefits and risks of a therapeutic intervention in randomized clinical trials.*

Surrogate endpoint*: A biomarker that is intended to substitute for a clinical endpoint. A surrogate endpoint is expected to predict clinical benefit (or harm or lack of benefit or harm) based on epidemiologic, therapeutic, pathophysiologic, or other scientific evidence. … Although all surrogate endpoints can be considered biomarkers, it is likely that only a few biomarkers will achieve surrogate endpoint status. The term* surrogate endpoint *applies primarily to endpoints in therapeutic intervention trials; however, it may sometimes apply in natural history or epidemiologic studies. It is important to point out that the same biomarkers used as surrogate endpoints in clinical trials are often extended to clinical practice in which disease responses are similarly measured. … The term* surrogate *literally means "to substitute for"; therefore use of the term* surrogate marker *is discouraged because the term suggests that the substitution is for a marker rather than for a clinical endpoint. The use of surrogate endpoints to establish therapeutic efficacy in registration trials is an established concept that has been addressed in regulation that enables the US Food and Drug Administration (FDA) to grant accelerated marketing approval for certain therapeutics (Food and Drug Modernization Act of 1997, 1997). (Reprinted by permission from Macmillan Publishers Ltd:* Clinical Pharmacology & Therapeutics, *69: 89–95, © 2001.)*

A list of examples of potential surrogates is shown here:

- Blood pressure
- Intraocular pressure (glaucoma)
- HbA1c (diabetes mellitus)
- Psychometric testing
- Tumor shrinkage (cancer)
- ACR criteria (rheumatoid arthritis)
- Pain scales (pain)
- LDL-cholesterol
- HIV particle concentration (AIDS)

These surrogates may fulfill the criteria mentioned previously and have been shown to work in practice. It may be surprising that other parameters such as blood glucose (diabetes mellitus) or CRP (inflammatory processes) are not

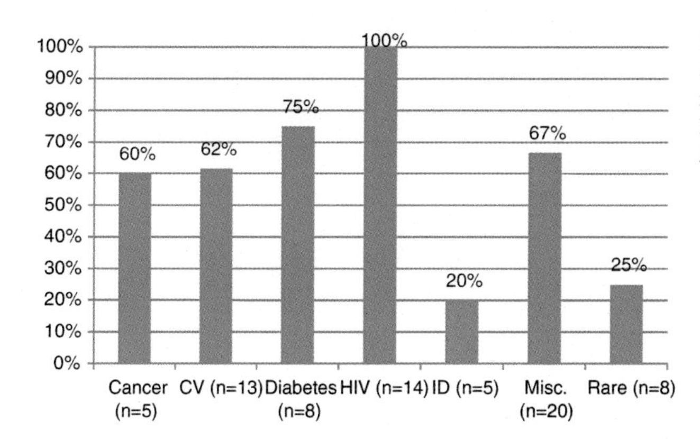

FIGURE 3.3 Percentage of drug recommendations with surrogate acceptability by therapeutic area. *Y* axis: Percent of recommendations using surrogate outcomes. *X* axis: Therapeutic area. CV: cardiovascular disease, HIV: human immunodeficiency virus, ID: infectious disease. *(From Rocchi et al., 2013, Open Access.)*

listed. They lack specificity, are too variable, or simply do not predict hard endpoints such as death or MACE well enough. The most prominent negative example of a former surrogate that caused patients to die was the number of ventricular premature beats (VPBs). In the cardiac arrhythmia suppression trial (CAST), the antiarrhythmic drugs encainide and flecainide did indeed suppress VPBs, but patients died earlier (Echt et al., 1991). This was apparently not a sufficient surrogate, and nowadays we know that the proarrhythmic effects of many antiarrhythmic drugs outweigh by far their benefits on VPBs.

On the other hand, a big leap (one might even say a quantum leap) occurred in AIDS research with the discovery of the close correlation between HIV particle concentration (viral RNA or p24 antigen as direct measures; Ledergerber et al., 2000) in plasma and morbidity or mortality. With the establishment of this correlation in expensive outcome trials, development of HIV drugs has become fast and reliable and has contributed to the tremendous success drugs have brought about in this disease. It is currently possible to develop a new HIV drug within 3–5 years (rather than 8–10 years), as lengthy morbidity or mortality trials are no longer required, and efficacy can be proven in relatively small trials of, for example, 50 participants.

Unfortunately, the acceptance of surrogates across regulatory authorities or health technology assessment (HTA) institutions is not consistent. Rocchi et al. (2013) showed a variable extent of surrogate acceptance for different therapy areas and for different institutions. As can be seen in Figure 3.3, HIV drugs have always been approved on the base of surrogate markers, whereas in cardiovascular disease or diabetes, this was true only in two-thirds of cases.

The acceptance by different institutions varies widely, as shown in Figure 3.4. The Centers for Disease Control and Prevention (CDC) does not accept any surrogates while the authorities EMA and FDA accept them in the majority of examples.

The use of biomarkers in drug development according to these definitions is depicted in Figure 3.5. In this model, it is obvious that biomarkers represent everything but the clinical endpoint and serve more or less as weak surrogates (biomarkers in the narrow sense) or good surrogates for the clinical endpoint. This is the case both for efficacy and safety concerns. The major concern according to this model is the fact that some effects, especially at the safety level, may be missed because the appropriate biomarkers have not been measured (Figure 3.6).

In my opinion and that of others, the classification of the Biomarkers Definitions Working Group is too simple. The first category, "biomarker," deserves a more detailed subclassification. Biomarkers are still distant from surrogates and endpoints, which are studied only at the late or even final stages of the drug development process. Biomarkers are the key instruments of translational medicine. They are ideally established first in animal disease models and then used in the equivalent human diseases to support early confidence in the efficacy and safety of a new drug or intervention. It is obvious that their predictive value can range from almost useless (e.g., if one has to obtain serial brain slices, which will never become a human biomarker) to that of a surrogate (e.g., LDL-cholesterol, one of the few surrogates accepted by FDA), with most cases being somewhere in between.

Thus, a subclassification of the predictive value of biomarkers seems desirable and should be considered a major task in biomarker development (see Subchapters 3.3 and 3.4). Describing the mechanistic dimension of a biomarker, as done by Rolan and colleagues (2003), does not solve the problem of defining predictive biomarkers versus biomarkers without predictive potential, as there may be potent and weak biomarkers in all subcategories.

Another important set of attributes of biomarkers relates to their prospective use: the major divide is between efficacy and safety, and, thus, biomarkers are commonly attributed to one area or the other. A third operational category has to do with the increasing importance of personalized medicine issues: the right drug for the right patient. In this context, biomarkers have been specifically developed to predict responsiveness in a given patient; the most prominent example is HER2/neu expression as prerequisite for successful trastuzumab therapy in breast cancer treatment. The FDA has posted a rapidly growing list of valid genomic biomarkers in the context of approved drug labels (U.S. Department of Health and Human Services, 2014; see Table 3.4) demonstrating the rapidly increasing importance of these patient-oriented,

	CDR	HC	FDA	EMA	NICE	PBS	SMC
Saxagliptin (HbA1c)	no (e2)	N/S	yes (e)	yes (e1)	N/A	N/S	no (ref)
Sitagliptin (HbA1c)	no (e2)	N/S	yes (e)	N/S	N/A	N/S	no (e2)
Sitagliptin/Metformin (HbA1c)	N/S	N/A	N/S	yes (e1)	N/A	N/A	N/S
Ambrisentan (6MWD)	N/S	N/S	yes (used)	yes (e)	N/A	no (ref)	no (e2)
Sildenafil (6MWD)	no (e2)	N/A	N/S	yes (used)	N/A	N/S	no (e2)
Sitaxsentan (6MWD)	no (e2)	yes (e1)	N/A	yes (e)	N/A	N/S	N/A
Tadalafil (6MWD)	no (ref)	N/A	yes (used)	no (e1)	N/A	N/S	N/A
Treprostinil (6MWD)	no (e2)	N/A	no (e2)	N/A	N/A	N/S	N/A
Adefovir (composite)	no (e1)	N/A	yes (e)	yes (e)	yes (used)	no (e1)	N/S
Entecavir (composite)	N/S	N/S	yes (e2)	yes (guid)	yes (e1)	yes (e)	N/S
peg-IFN RB (SVR)	no (e1+e2)	N/A	N/S	yes (used)	yes (used)	N/A	N/S
Telbivudine (composite)	no (e2)	no (e1)	yes (e)	yes (guid)	yes (e1)	yes (e)	N/S
Tenofovir (composite)	no (e1)	N/A	N/A	yes (guid)	N/S	N/S	N/S

FIGURE 3.4 Comparison of international agencies: concerns with surrogate outcomes. *Y* axis: Drug submission. *X* axis: Agency. *No: no (e2) = implicit no "evidence 2"; no (ref) = implicit no "reference"; no (e) = explicit no "evidence 1"; no (e1 + e2) = explicit no "evidence 1" and implicit no "evidence 2"; Yes: yes (e1) = implicit yes "evidence 1"; yes (e2) = implicit yes "evidence 2"; yes (used) = implicit yes "used before"; yes (ref) = implicit yes "reference"; yes (e) = explicit yes; Not identified: N/S = no statement; N/A = not applicable; Red shade = negative statements of surrogate acceptability; Green shade = positive statements of surrogate acceptability; HbA1c = hemoglobin A1c; 6MWD= 6-minute walk distance; composite = histology, virology, serology; SVR = sustained virological response; CDR = Common Drug Review; HC = Health Canada; FDA=Food and Drug Administration; EMA = European Medicines Agency; NICE = National Institute for Health and Clinical Excellence; PBS = Pharmaceutical Benefit Scheme; SMC = Scottish Medicines Consortium. *(From Rocchi et al., 2013, Open Access.)*

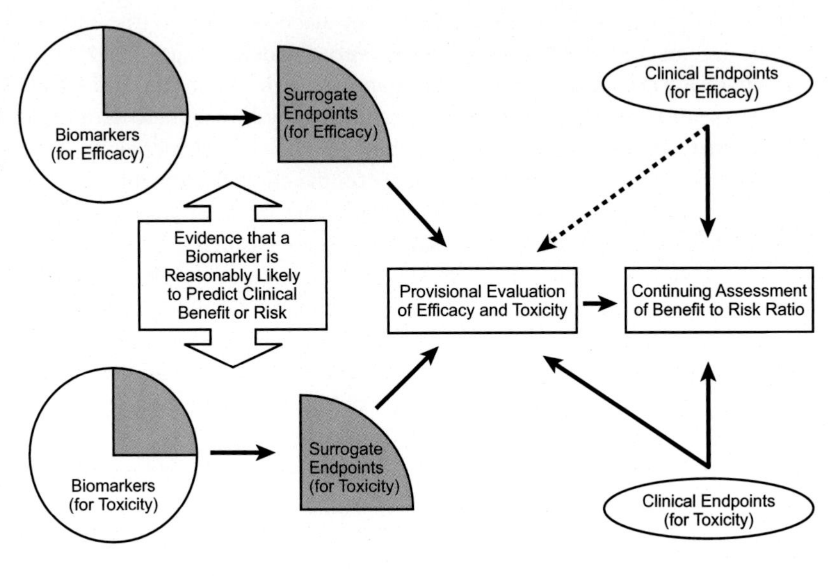

FIGURE 3.5 Conceptual model of the relationship of biomarkers, surrogate endpoints, and the process of evaluating therapeutic interventions according to the NIH biomarker definition group. *(From Biomarkers Definitions Working Group, 2001, reprinted by permission from Macmillan Publishers Ltd:* Clinical Pharmacology & Therapeutics, *69: 89–95, © 2001.)*

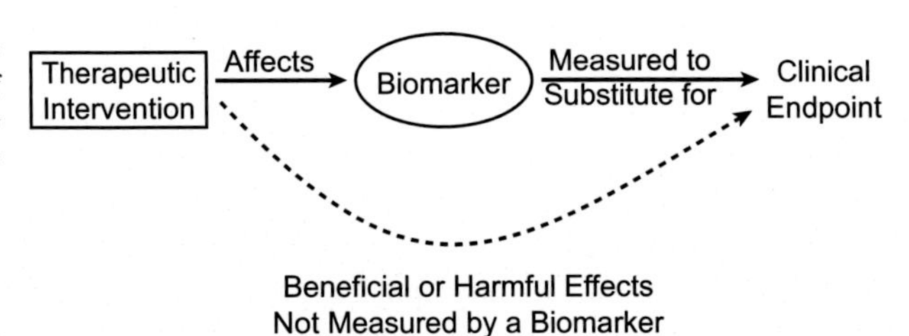

FIGURE 3.6 Conceptual model of the use of biomarkers in drug development: important efficacy and especially safety aspects may be missed if biomarker selection is not optimal. *(From Biomarkers Definitions Working Group, 2001, reprinted by permission from Macmillan Publishers Ltd:* Clinical Pharmacology & Therapeutics, *69: 89–95, © 2001.)*

TABLE 3.4 The FDA's list of valid genomic biomarkers in the context of approved drug labeling*

Drug	Therapeutic Area	HUGO Symbol	Referenced Subgroup	Labeling Sections
Abacavir	Infectious Diseases	HLA-B	HLA-B*5701 allele carriers	Boxed Warning, Contraindications, Warnings and Precautions, Patient Counseling Information
Ado-Trastuzumab Emtansine	Oncology	ERBB2	HER2 protein overexpression or gene amplification positive	Indications and Usage, Warnings and Precautions, Adverse Reactions, Clinical Pharmacology, Clinical Studies
Afatinib	Oncology	EGFR	EGFR exon 19 deletion or exon 21 substitution (L858R) mutation positive	Indications and Usage, Dosage and Administration, Adverse Reactions, Clinical Pharmacology, Clinical Studies, Patient Counseling Information
Amitriptyline	Psychiatry	CYP2D6	CYP2D6 poor metabolizers	Precautions
Anastrozole	Oncology	ESR1, PGR	Hormone receptor positive	Indications and Usage, Clinical Pharmacology, Clinical Studies
Aripiprazole	Psychiatry	CYP2D6	CYP2D6 poor metabolizers	Clinical Pharmacology, Dosage and Administration
Arsenic Trioxide	Oncology	PML/RARA	PML/RARα (t(15;17)) gene expression positive	Boxed Warning, Clinical Pharmacology, Indications and Usage, Warnings
Atomoxetine	Psychiatry	CYP2D6	CYP2D6 poor metabolizers	Dosage and Administration, Warnings and Precautions, Drug Interactions, Clinical Pharmacology
Atorvastatin	Endocrinology	LDLR	Homozygous familial hypercholesterolemia	Indications and Usage, Dosage and Administration, Warnings and Precautions, Clinical Pharmacology, Clinical Studies
Azathioprine	Rheumatology	TPMT	TPMT intermediate or poor metabolizers	Dosage and Administration, Warnings and Precautions, Drug Interactions, Adverse Reactions, Clinical Pharmacology
Belimumab	Autoimmune Diseases	BAFF/TNFSF13B	CD257 positive	Clinical Pharmacology, Clinical Studies
Boceprevir	Infectious Diseases	IFNL3	IL28B rs12979860 T allele carriers	Clinical Pharmacology
Bosutinib	Oncology	BCR/ABL1	Philadelphia chromosome (t(9;22)) positive	Indications and Usage, Adverse Reactions, Clinical Studies
Brentuximab Vedotin	Oncology	TNFRSF8	CD30 positive	Indications and Usage, Description, Clinical Pharmacology
Busulfan	Oncology	Ph Chromosome	Ph Chromosome negative	Clinical Studies
Capecitabine	Oncology	DPYD	DPD deficient	Contraindications, Warnings and Precautions, Patient Information
Carbamazepine (1)	Neurology	HLA-B	HLA-B*1502 allele carriers	Boxed Warning, Warnings and Precautions
Carbamazepine (2)	Neurology	HLA-A	HLA-A*3101 allele carriers	Boxed Warning, Warnings and Precautions
Carglumic Acid	Metabolic Disorders	NAGS	N-acetylglutamate synthase deficiency	Indications and Usage, Warnings and Precautions, Special Populations, Clinical Pharmacology, Clinical Studies
Carisoprodol	Rheumatology	CYP2C19	CYP2C19 poor metabolizers	Clinical Pharmacology, Special Populations

Continued

TABLE 3.4 The FDA's list of valid genomic biomarkers in the context of approved drug labeling*—cont'd

Drug	Therapeutic Area	HUGO Symbol	Referenced Subgroup	Labeling Sections
Carvedilol	Cardiology	CYP2D6	CYP2D6 poor metabolizers	Drug Interactions, Clinical Pharmacology
Celecoxib	Rheumatology	CYP2C9	CYP2C9 poor metabolizers	Dosage and Administration, Drug Interactions, Use in Specific Populations, Clinical Pharmacology
Cetuximab (1)	Oncology	EGFR	EGFR protein expression positive	Indications and Usage, Warnings and Precautions, Description, Clinical Pharmacology, Clinical Studies
Cetuximab (2)	Oncology	KRAS	KRAS codon 12 and 13 mutation negative	Indications and Usage, Dosage and Administration, Warnings and Precautions, Adverse Reactions, Clinical Pharmacology, Clinical Studies
Cevimeline	Dermatology	CYP2D6	CYP2D6 poor metabolizers	Drug Interactions
Chloroquine	Infectious Diseases	G6PD	G6PD deficient	Precautions
Chlorpropamide	Endocrinology	G6PD	G6PD deficient	Precautions
Cisplatin	Oncology	TPMT	TPMT intermediate or poor metabolizers	Clinical Pharmacology, Warnings, Precautions
Citalopram (1)	Psychiatry	CYP2C19	CYP2C19 poor metabolizers	Drug Interactions, Warnings
Citalopram (2)	Psychiatry	CYP2D6	CYP2D6 poor metabolizers	Drug Interactions
Clobazam	Neurology	CYP2C19	CYP2C19 poor metabolizers	Clinical Pharmacology, Dosage and Administration, Use in Specific Populations
Clomipramine	Psychiatry	CYP2D6	CYP2D6 poor metabolizers	Drug Interactions
Clopidogrel	Cardiology	CYP2C19	CYP2C19 intermediate or poor metabolizers	Boxed Warning, Dosage and Administration, Warnings and Precautions, Drug Interactions, Clinical Pharmacology
Clozapine	Psychiatry	CYP2D6	CYP2D6 poor metabolizers	Drug Interactions, Clinical Pharmacology
Codeine	Anesthesiology	CYP2D6	CYP2D6 poor metabolizers	Warnings and Precautions, Use in Specific Populations, Clinical Pharmacology
Crizotinib	Oncology	ALK	ALK gene rearrangement positive	Indications and Usage, Dosage and Administration, Drug Interactions, Warnings and Precautions, Adverse Reactions, Clinical Pharmacology, Clinical Studies
Dabrafenib (1)	Oncology	BRAF	BRAF V600E mutation positive	Indications and Usage, Dosage and Administration, Warnings and Precautions, Clinical Pharmacology, Clinical Studies, Patient Counseling Information
Dabrafenib (2)	Oncology	G6PD	G6PD deficient	Warnings and Precautions, Adverse Reactions, Patient Counseling Information
Dapsone (1)	Dermatology	G6PD	G6PD deficient	Indications and Usage, Precautions, Adverse Reactions, Patient Counseling Information

Dapsone (2)	Infectious Diseases	G6PD	G6PD deficient	Precautions, Adverse Reactions, Overdosage
Dasatinib	Oncology	BCR/ABL1	Philadelphia chromosome (t(9;22)) positive; T315I mutation-positive	Indications and Usage, Clinical Studies, Patient Counseling Information
Denileukin Diftitox	Oncology	IL2RA	CD25 antigen positive	Indications and Usage, Warnings and Precautions, Clinical Studies
Desipramine	Psychiatry	CYP2D6	CYP2D6 poor metabolizers	Drug Interactions
Dexlansoprazole (1)	Gastroenterology	CYP2C19	CYP2C19 poor metabolizers	Clinical Pharmacology, Drug Interactions
Dexlansoprazole (2)	Gastroenterology	CYP1A2	CYP1A2 genotypes	Clinical Pharmacology
Dextromethorphan and Quinidine	Neurology	CYP2D6	CYP2D6 poor metabolizers	Clinical Pharmacology, Warnings and Precautions, Drug Interactions
Diazepam	Psychiatry	CYP2C19	CYP2C19 poor metabolizers	Drug Interactions, Clinical Pharmacology
Doxepin	Psychiatry	CYP2D6	CYP2D6 poor metabolizers	Precautions
Drospirenone and Ethinyl Estradiol	Neurology	CYP2D6	CYP2D6 poor metabolizers	Clinical Pharmacology, Warnings and Precautions, Drug Interactions
Eltrombopag (1)	Hematology	F5	Factor V Leiden carriers	Warnings and Precautions
Eltrombopag (2)	Hematology	SERPINC1	Antithrombin III deficient	Warnings and Precautions
Erlotinib (1)	Oncology	EGFR	EGFR protein expression positive	Clinical Pharmacology
Erlotinib (2)	Oncology	EGFR	EGFR exon 19 deletion or exon 21 substitution (L858R) positive	Indications and Usage, Dosage and Administration, Clinical Pharmacology, Clinical Studies
Esomeprazole	Gastroenterology	CYP2C19	CYP2C19 poor metabolizers	Drug Interactions, Clinical Pharmacology
Everolimus (1)	Oncology	ERBB2	HER2 protein overexpression negative	Indications and Usage, Boxed Warning, Adverse Reactions, Use in Specific Populations, Clinical Pharmacology, Clinical Studies
Everolimus (2)	Oncology	ESR1	Estrogen receptor positive	Clinical Pharmacology, Clinical Studies
Exemestane	Oncology	ESR1	Estrogen receptor positive	Indications and Usage, Dosage and Administration, Clinical Studies, Clinical Pharmacology
Fluorouracil (1)	Dermatology	DPYD	DPD deficient	Contraindications, Warnings, Patient Information
Fluorouracil (2)	Oncology	DPYD	DPD deficient	Warnings
Fluoxetine	Psychiatry	CYP2D6	CYP2D6 poor metabolizers	Warnings, Precautions, Clinical Pharmacology
Flurbiprofen	Rheumatology	CYP2C9	CYP2C9 poor metabolizers	Clinical Pharmacology, Special Populations
Fluvoxamine	Psychiatry	CYP2D6	CYP2D6 poor metabolizers	Drug Interactions
Fulvestrant	Oncology	ESR1	Estrogen receptor positive	Indications and Usage, Clinical Pharmacology, Clinical Studies, Patient Counseling Information

Continued

TABLE 3.4 The FDA's list of valid genomic biomarkers in the context of approved drug labeling*—cont'd

Drug	Therapeutic Area	HUGO Symbol	Referenced Subgroup	Labeling Sections
Galantamine	Neurology	CYP2D6	CYP2D6 poor metabolizers	Special Populations
Glimepiride	Endocrinology	G6PD	G6PD deficient	Warning and Precautions
Glipizide	Endocrinology	G6PD	G6PD deficient	Precautions
Glyburide	Endocrinology	G6PD	G6PD deficient	Precautions
Ibritumomab Tiuxetan	Oncology	MS4A1	CD20 positive	Indications and Usage, Clinical Pharmacology, Description
Iloperidone	Psychiatry	CYP2D6	CYP2D6 poor metabolizers	Clinical Pharmacology, Dosage and Administration, Drug Interactions, Specific Populations, Warnings and Precautions
Imatinib (1)	Oncology	KIT	c-KIT D816V mutation negative	Indications and Usage, Dosage and Administration Clinical Pharmacology, Clinical Studies
Imatinib (2)	Oncology	BCR/ABL1	Philadelphia chromosome (t(9;22)) positive	Indications and Usage, Dosage and Administration, Clinical Pharmacology, Clinical Studies
Imatinib (3)	Oncology	PDGFRB	PDGFR gene rearrangement positive	Indications and Usage, Dosage and Administration, Clinical Studies
Imatinib (4)	Oncology	FIP1L1/PDGFRA	FIP1L1/PDGFRα fusion kinase (or CHIC2 deletion) positive	Indications and Usage, Dosage and Administration, Clinical Studies
Imipramine	Psychiatry	CYP2D6	CYP2D6 poor metabolizers	Drug Interactions
Indacaterol	Pulmonary	UGT1A1	UGT1A1 *28 allele homozygotes	Clinical Pharmacology
Irinotecan	Oncology	UGT1A1	UGT1A1*28 allele carriers	Dosage and Administration, Warnings, Clinical Pharmacology
Isosorbide and Hydralazine	Cardiology	NAT1-2	Slow acetylators	Clinical Pharmacology
Ivacaftor	Pulmonary	CFTR	CFTR G551D carriers	Indications and Usage, Adverse Reactions, Use in Specific Populations, Clinical Pharmacology, Clinical Studies
Lansoprazole	Gastroenterology	CYP2C19	CYP2C19 poor metabolizer	Drug Interactions, Clinical Pharmacology
Lapatinib	Oncology	ERBB2	HER2 protein overexpression positive	Indications and Usage, Clinical Pharmacology, Patient Counseling Information
Lenalidomide	Hematology	del (5q)	Chromosome 5q deletion	Boxed Warning, Indications and Usage, Clinical Studies, Patient Counseling
Letrozole	Oncology	ESR1, PGR	Hormone receptor positive	Indications and Usage, Adverse Reactions, Clinical Studies, Clinical Pharmacology
Lomitapide	Endocrinology	LDLR	Homozygous familial hypercholesterolemia and LDL receptor mutation deficient	Indication and Usage, Adverse Reactions, Clinical Studies

Mafenide	Infectious Diseases	G6PD	G6PD deficient	Warnings, Adverse Reactions
Maraviroc	Infectious Diseases	CCR5	CCR5 positive	Indications and Usage, Warnings and Precautions, Clinical Pharmacology, Clinical Studies, Patient Counseling Information
Mercaptopurine	Oncology	TPMT	TPMT intermediate or poor metabolizers	Dosage and Administration, Contraindications, Precautions, Adverse Reactions, Clinical Pharmacology
Methylene Blue	Hematology	G6PD	G6PD deficient	Precautions
Metoclopramide	Gastroenterology	CYB5R1-4	NADH cytochrome b5 reductase deficient	Precautions
Metoprolol	Cardiology	CYP2D6	CYP2D6 poor metabolizers	Precautions, Clinical Pharmacology
Mipomersen	Endocrinology	LDLR	Homozygous familial hypercholesterolemia and LDL receptor mutation deficient	Indication and Usage, Clinical Studies, Use in Specific Populations
Modafinil	Psychiatry	CYP2D6	CYP2D6 poor metabolizers	Drug Interactions
Mycophenolic Acid	Transplantation	HPRT1	HGPRT deficient	Precautions
Nalidixic Acid	Infectious Diseases	G6PD	G6PD deficient	Precautions, Adverse Reactions
Nefazodone	Psychiatry	CYP2D6	CYP2D6 poor metabolizers	Drug Interactions
Nilotinib (1)	Oncology	BCR/ABL1	Philadelphia chromosome (t(9 :22)) positive	Indications and Usage, Patient Counseling Information
Nilotinib (2)	Oncology	UGT1A1	UGT1A1*28 allele homozygotes	Warnings and Precautions, Clinical Pharmacology
Nitrofurantoin	Infectious Diseases	G6PD	G6PD deficient	Warnings, Adverse Reactions
Nortriptyline	Psychiatry	CYP2D6	CYP2D6 poor metabolizers	Drug Interactions
Ofatumumab	Oncology	MS4A1	CD20 positive	Indications and Usage, Clinical Pharmacology
Omacetaxine	Oncology	BCR/ABL1	BCR-ABL T315I	Clinical Pharmacology
Omeprazole	Gastroenterology	CYP2C19	CYP2C19 poor metabolizers	Dosage and Administration, Warnings and Precautions, Drug Interactions
Panitumumab (1)	Oncology	EGFR	EGFR protein expression positive	Indications and Usage, Warnings and Precautions, Clinical Pharmacology, Clinical Studies
Panitumumab (2)	Oncology	KRAS	KRAS codon 12 and 13 mutation negative	Indications and Usage, Clinical Pharmacology, Clinical Studies
Pantoprazole	Gastroenterology	CYP2C19	CYP2C19 poor metabolizers	Clinical Pharmacology, Drug Interactions, Special Populations
Paroxetine	Psychiatry	CYP2D6	CYP2D6 poor metabolizers	Clinical Pharmacology, Drug Interactions
Pazopanib	Oncology	UGT1A1	(TA)7/(TA)7 genotype (UGT1A1*28/*28)	Clinical Pharmacology, Warnings and Precautions

Continued

TABLE 3.4 The FDA's list of valid genomic biomarkers in the context of approved drug labeling*—cont'd

Drug	Therapeutic Area	HUGO Symbol	Referenced Subgroup	Labeling Sections
PEG-3350, Sodium Sulfate, Sodium Chloride, Potassium Chloride, Sodium Ascorbate, and Ascorbic Acid	Gastroenterology	G6PD	G6PD deficient	Warnings and Precautions
Peginterferon alfa-2b	Infectious Diseases	IFNL3	IL28B rs12979860 T allele carriers	Clinical Pharmacology
Pegloticase	Rheumatology	G6PD	G6PD deficient	Contraindications, Patient Counseling Information
Perphenazine	Psychiatry	CYP2D6	CYP2D6 poor metabolizers	Clinical Pharmacology, Drug Interactions
Pertuzumab	Oncology	ERBB2	HER2 protein overexpression positive	Indications and Usage, Warnings and Precautions, Adverse Reactions, Clinical Studies, Clinical Pharmacology
Phenytoin	Neurology	HLA-B	HLA-B*1502 allele carriers	Warnings
Pimozide	Psychiatry	CYP2D6	CYP2D6 poor metabolizers	Warnings, Precautions, Contraindications, Dosage and Administration
Ponatinib	Oncology	BCR/ABL1	Philadelphia chromosome (t(9;22)) positive, BCR –ABL T315I mutation	Indications and Usage, Warnings and Precautions, Adverse Reactions, Use in Specific Populations, Clinical Pharmacology, Clinical Studies
Prasugrel	Cardiology	CYP2C19	CYP2C19 poor metabolizers	Use in Specific Populations, Clinical Pharmacology, Clinical Studies
Pravastatin	Endocrinology	LDLR	Homozygous familial hypercholesterolemia and LDL receptor deficient	Clinical Studies, Use in Specific Populations
Primaquine	Infectious Diseases	G6PD	G6PD deficient	Warnings and Precautions, Adverse Reactions
Propafenone	Cardiology	CYP2D6	CYP2D6 poor metabolizers	Clinical Pharmacology
Propranolol	Cardiology	CYP2D6	CYP2D6 poor metabolizers	Precautions, Drug Interactions, Clinical Pharmacology
Protriptyline	Psychiatry	CYP2D6	CYP2D6 poor metabolizers	Precautions
Quinidine	Cardiology	CYP2D6	CYP2D6 poor metabolizers	Precautions
Quinine Sulfate	Infectious Diseases	G6PD	G6PD deficient	Contraindications, Patient Counseling Information
Rabeprazole	Gastroenterology	CYP2C19	CYP2C19 poor metabolizers	Drug Interactions, Clinical Pharmacology
Rasburicase	Oncology	G6PD	G6PD deficient	Boxed Warning, Contraindications

Rifampin, Isoniazid, and Pyrazinamide	Infectious Diseases	NAT1-2	Slow inactivators	Adverse Reactions, Clinical Pharmacology
Risperidone	Psychiatry	CYP2D6	CYP2D6 poor metabolizers	Drug Interactions, Clinical Pharmacology
Rituximab	Oncology	MS4A1	CD20 positive	Indication and Usage, Clinical Pharmacology, Description, Precautions
Rosuvastatin	Endocrinology	LDLR	Homozygous and Heterozygous familial hypercholesterolemia	Indications and Usage, Dosage and Administration, Clinical Pharmacology, Clinical Studies
Sodium Nitrite	Antidotal Therapy	G6PD	G6PD deficient	Warnings and Precautions
Succimer	Hematology	G6PD	G6PD deficient	Clinical Pharmacology
Sulfamethoxazole and Trimethoprim	Infectious Diseases	G6PD	G6PD deficient	Precautions
Tamoxifen (1)	Oncology	ESR1, PGR	Hormone receptor positive	Indications and Usage, Precautions, Medication Guide
Tamoxifen (2)	Oncology	F5	Factor V Leiden carriers	Warnings
Tamoxifen (3)	Oncology	F2	Prothrombin mutation G20210A	Warnings
Telaprevir	Infectious Diseases	IFNL3	IL28B rs12979860 T allele carriers	Clinical Pharmacology
Terbinafine	Infectious Diseases	CYP2D6	CYP2D6 poor metabolizers	Drug Interactions
Tetrabenazine	Neurology	CYP2D6	CYP2D6 poor metabolizers	Dosage and Administration, Warnings, Clinical Pharmacology
Thioguanine	Oncology	TPMT	TPMT poor metabolizer	Dosage and Administration, Precautions, Warnings
Thioridazine	Psychiatry	CYP2D6	CYP2D6 poor metabolizers	Precautions, Warnings, Contraindications
Ticagrelor	Cardiology	CYP2C19	CYP2C19 poor metabolizers	Clinical Studies
Tolterodine	Urology	CYP2D6	CYP2D6 poor metabolizers	Clinical Pharmacology, Drug Interactions, Warnings and Precautions
Tositumomab	Oncology	MS4A1	CD20 antigen positive	Indications and Usage, Clinical Pharmacology
Tramadol	Analgesic	CYP2D6	CYP2D6 poor metabolizers	Clinical Pharmacology
Trametinib	Oncology	BRAF	BRAF V600E/K mutation positive	Indications and Usage, Dosage and Administration, Adverse Reactions, Clinical Pharmacology, Clinical Studies, Patient Counseling Information
Trastuzumab	Oncology	ERBB2	HER2 protein overexpression positive	Indications and Usage, Warnings and Precautions, Clinical Pharmacology, Clinical Studies
Tretinoin	Oncology	PML/RARA	PML/RARα (t(15;17)) gene expression positive	Clinical Studies, Indications and Usage, Warnings

Continued

TABLE 3.4 The FDA's list of valid genomic biomarkers in the context of approved drug labeling*—cont'd

Drug	Therapeutic Area	HUGO Symbol	Referenced Subgroup	Labeling Sections
Trimipramine	Psychiatry	CYP2D6	CYP2D6 poor metabolizers	Drug Interactions
Valproic Acid (1)	Neurology	POLG	POLG mutation positive	Boxed Warning, Contraindications, Warnings and Precautions
Valproic Acid (2)	Neurology	NAGS, CPS1, ASS1, OTC, ASL, ABL2	Urea cycle enzyme deficient	Contraindications, Warnings and Precautions, Adverse Reactions, Medication Guide
Velaglucerase Alfa	Metabolic Disorders	GBA	Lysosomal glucocerebrosidase enzyme	Indication and Usage, Description, Clinical Pharmacology, Clinical Studies
Vemurafenib	Oncology	BRAF	BRAF V600E mutation positive	Indications and Usage, Warning and Precautions, Clinical Pharmacology, Clinical Studies, Patient Counseling Information
Venlafaxine	Psychiatry	CYP2D6	CYP2D6 poor metabolizers	Drug Interactions
Voriconazole	Infectious Diseases	CYP2C19	CYP219 intermediate or poor metabolizers	Clinical Pharmacology, Drug Interactions
Warfarin (1)	Cardiology or Hematology	CYP2C9	CYP2C9 intermediate or poor metabolizers	Dosage and Administration, Drug Interactions, Clinical Pharmacology
Warfarin (2)	Cardiology or Hematology	VKORC1	VKORC1 rs9923231 A allele carriers	Dosage and Administration, Clinical Pharmacology

**HUGO: Human Genome Organization; for abbreviations, see http://www.genenames.org/*
(From U.S. Department of Health and Human Services, 2014.)

personalized-medicine-type biomarkers. Of course, they aim at increasing efficacy and safety and, thus, could be easily categorized into either of the two areas of prospective use.

In addition, the FDA has issued a detailed documentation regarding the submission procedure for biomarkers (U.S. Department of Health and Human Services Food and Drug Administration, 2011) guiding the format of regulatory information in drug or device approval processes. It thus generalizes the requirements for valuable biomarkers that could now be any measure with proper documentation and evaluation.

The last classification of biomarkers is driven by technology: it is simply a description of what kind of measurement has to be performed to obtain the biomarker. The main groups worth mentioning are summarized in Table 3.5, which compiles major categories and prominent examples.

TABLE 3.5 Categorization of important biomarkers by technical condition and type of measure*

"Omics"
Proteomics or peptidomics
Genomics or pharmacogenomics
Metabolonomics
Histomics
Genetic markers
Serum markers, defined serum constituents (e.g., enzymes, proteins, electrolytes)
Histology
Microbiology
Functional tests, e.g.
Cardiovascular parameters (e.g., blood pressure)
Endothelial function
ECG
EEG
Neurological tests
Psychometric tests
Joint or muscle function tests
General: weight, fever, well-being
Imaging
X-ray, conventional, + contrast or staining
Tomography, magnetic imaging, X-ray, + contrast, staining, or receptor-staining
Functional imaging (e.g., fMRI)
PET
Ultrasound or Doppler, + contrast-enhancement (microbubbles)
Endoscopy
Clinical endpoints, e.g.
Death
MACE (major cardiovascular events)
Any disease-related measure by which treatment effect is thought to indicate disease modification according to normal clinical standards (what is normally used in daily clinical practice to find out whether a patient has improved).

**Listings are incomplete.*

As "omics" approaches did not really live up to expectations, imaging seems to have become the most valuable translational tool because—in many instances—it can be performed across species and has a rapidly evolving power in terms of resolution and signals detected.

This chapter shows that biomarkers can be classified by various approaches. Under the auspices of translational medicine, it is most important to classify them according to their predictive power. If this dimension is added to the simple NIH definition and if endpoints are understood as biomarkers as well, a useful classification could emerge. The technical classification is important in terms of methodological planning and investment. After the surge of investment into "omics" during the past 15–20 years (Bilello, 2005), imaging seems to be the focus of the next wave of investment. Because imaging parameters show considerable translational strength in general, this new awareness and investment flow are very promising. The imaging experts in the preclinical and clinical areas may become the most influential components of future drug development processes.

In the context of biomarkers, the related terms *risk factor* and *risk marker* should be wisely and specifically used. In general, they describe classified biomarkers that have a proven value on top of their basic function as biomarkers.

A *risk factor* is linked to a disease because of a causal relationship to major elements of the disease. In turn, this implies that therapeutic intervention affecting this biomarker has been shown to improve clinical outcomes. Typical examples are arterial hypertension or LDL-cholesterol. Not only is there a correlation between those parameters and morbidity or mortality, but intervention (by drugs) has been proven to change the parameter and clinical outcome in parallel.

A *risk marker* is associated with the disease (epidemiologically) but need not be causally linked; it may be a measure of the disease process, and, as such, it may represent only an epiphenomenon. Intervention aiming at this biomarker does not lead to improved clinical outcomes because there is no causal relationship between disease and marker. A prominent example unmasked by a failed drug is HDL-cholesterol (see Chapter 11), which in epidemiological analyses is clearly linked to cardiovascular risk but does not yield therapeutic success in all interventional settings (torcetrapib is the dramatic exception). There is no doubt that HDL-cholesterol is a risk marker, but its role as a risk factor needs further clarification and is not established at present.

REFERENCES

Bilello, J.A., 2005. The agony and ecstasy of "OMIC" technologies in drug development. Curr. Mol. Med. 5, 39–52.

Biomarkers Definitions Working Group, 2001. Biomarkers and surrogate endpoints: preferred definitions and conceptual framework. Clin. Pharmacol. Ther. 69, 89–95.

Echt, D.S., Liebson, P.R., Mitchell, L.B., Peters, R.W., Obias-Manno, D., Barker, A.H., et al. for the CAST-Investigators, 1991. Mortality and morbidity in patients receiving encainide, flecainide, or placebo. N. Engl. J. Med. 324, 781–788.

Food and Drug Modernization Act of 1997, 1997. Title 21 Code of Federal Regulations Part 314 Subpart H Section 314. 500.

Ledergerber, B., Flepp, M., Böni, J., Tomasik, Z., Cone, R.W., Lüthy, R., et al., 2000. Human immunodeficiency virus type 1 p24 concentration measured by boosted ELISA of heat-denatured plasma correlates with decline in CD4 cells, progression to AIDS, and survival: comparison with viral RNA measurement. J. Infect. Dis. 181, 1280–1288.

Rocchi, A., Khoudigian, S., Hopkins, R., Goeree, R., 2013. Surrogate outcomes: experiences at the Common Drug Review. Cost Eff. Resour. Alloc. 11, 31.

Rolan, P., Atkinson Jr., A.J., Lesko, L.J., Scientific Organizing Committee, Conference Report Committee, 2003. Use of biomarkers from drug discovery through clinical practice: report of the Ninth European Federation of Pharmaceutical Sciences Conference on Optimizing Drug Development. Clin. Pharmacol. Ther. 73, 284–291.

U.S. Department of Health and Human Services Food and Drug Administration, 2011. Guidance for Industry. E16 Biomarkers Related to Drug or Biotechnology Product Development: Context, Structure, and Format of Qualification Submissions. www.fda.gov/downloads/drugs/guidancecomplianceregulatoryinformation/guidances/ucm267449.pdf (accessed 24.02.14).

U.S. Department of Health and Human Services Food and Drug Administration, 2014. Table of Pharmacogenomic Biomarkers in Drug Labeling. http://www.fda.gov/drugs/scienceresearch/researchareas/pharmacogenetics/ucm083378.htm (accessed 24.02.14).

Chapter 3.3

Development of Biomarkers

Martin Wehling

The preceding chapters demonstrate that biomarkers may not be readily available when it comes to translational processes and that important decisions need to be based on biomarkers in drug or medical device development. One task is the assessment of existing biomarkers (see Subchapter 3.4), but if this assessment shows that current biomarkers are insufficient for the given task, this frequently leads to the need for the development of new ones. The sources of new biomarkers are mainly represented in the Tools subchapter (3.1). If biological material (blood, urine, tissue) is available—as in almost all cases—"omics" approaches are the ones tested most frequently. In general, "omics" tend to produce multiple readouts such as hundreds of up-or-down-regulated proteins (proteomics are the most prominent example), and it is very challenging to test such biomarkers and establish their utility.

Which ingredients are necessary for basing go-or-no-go decisions on biomarkers that have just been found to correlate with disease parameters and/or therapeutic interventions? Or in other words, which properties have to be established and proven for a biomarker to, for example, become a Food and Drug Administration (FDA)-approved genomic biomarker, as shown in Table 3.4?

There have been several different proposals that are partially specific to the targets to be measured. Pepe and colleagues (2001) proposed a five-step scheme for the development of an oncologic biomarker (Figure 3.7). In principle, the process resembles a drug development process, from preclinical stages through early- and late-stage clinical testing. Translation is a key element in this process, as one has to prove that biomarker differences are specific and reliable.

Vasan (2006) aligned key elements of biomarker development to these stages of biomarker development. After a biomarker is detected, it will be tested in small groups of patients (similar to proof-of-principle testing) and in retrospective studies; finally, prospective studies of its prognostic value, including its prognostic value for interventions, will be established. Phases 4 and 5 in this model are analogous to proof-of-concept testing in drug development. In the result segment of this table, basic requirements for valid and finally effective biomarkers are described: assay precision, reliability and sensitivity, reference limits, intra- or interindividual variation, and finally receiver operating characteristic (ROC) analyses and number-needed-to-screen analyses.

As specified further in the statistics chapter (Chapter 6), there are defined tests to attach key figures to a given biomarker and, thereby, describe some of its important features. Key elements of biomarker profiling aim to answer the following questions: What is the normal distribution of values of this biomarker? How do they differ between healthy individuals and patients (how well do they discriminate between health and disease, and is there an essential, obscuring overlap)? How useful is this biomarker for disease progression prediction? The latter question—in most cases—will be answered not by identifying a particular cutoff line (although this would be desirable) but by gradual increase of risk with a slurred border zone of transition. Again, LDL-cholesterol may serve as an example in that we know that risk reduction can be achieved even into ranges that have been termed normal up to the present; thus, there is no threshold at which the risk becomes zero. Accordingly, for treatment guidelines, cutoff lines (treat if above, do not treat if below) will be derived from consensus processes that comprise scientific evidence plus cost-effectiveness estimates and social aspects such as compliance. Figure 3.8 illustrates the ROC curves for prostate-specific antigen (PSA) concentrations and prostatic cancer detection at different points in time before diagnosis.

As one can see, sensitivity and specificity are highest if determined at the time of diagnosis, whereas increasing time intervals between sampling and cancer diagnosis lead to decreases in both parameters. This biomarker development sequence, in general, is reminiscent of the development of any established, clinically used laboratory test, for example, the test for blood glucose measurement. Development thus aims at the following key aspects:

- Assay standardization, meaning that methods are established and described in detail
- Assay reproducibility, accuracy, and availability (e.g., is the antibody required commercially available?)
- Knowledge of the distribution of biomarker values in the general population and in select subpopulations
- Abnormal levels, both in general and in subpopulations (e.g., women, children, elderly people)
- Correlation between biomarker levels and known disease risk factors

Principles of Translational Science in Medicine. http://dx.doi.org/10.1016/B978-0-12-800687-0.00014-1

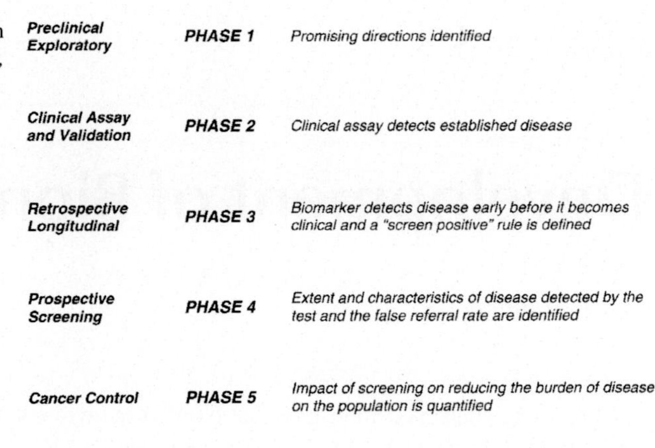

FIGURE 3.7 Proposed stages of biomarker development for the detection of early cancer. *(From Pepe et al., 2001,* J Natl Cancer Inst. *93, 1054–1061, reprinted by permission from Oxford University Press.)*

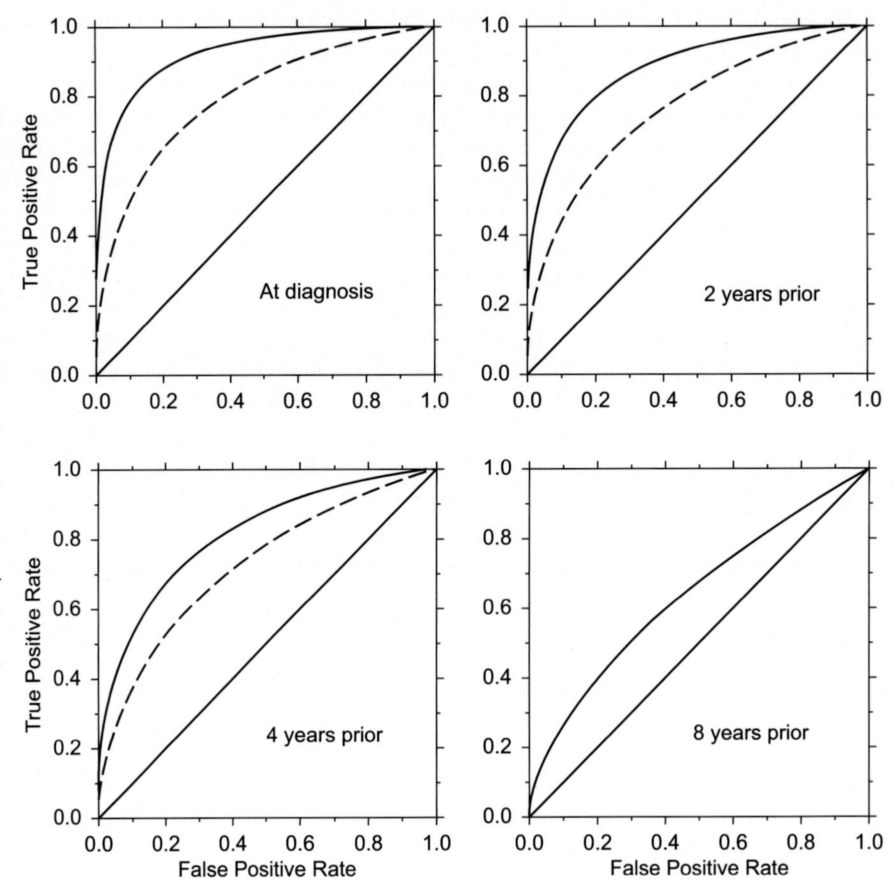

FIGURE 3.8 Receiver operating characteristic (ROC) curves for total prostate-specific antigen (PSA) and ratio of free to total PSA at various times before diagnosis. The thick solid lines represent total PSA, the dashed lines represent the PSA ratio, and the thin solid lines represent the line of identity. The figure shows that at time of diagnosis, PSA determinations (in particular, ratio determinations) have a high sensitivity and specificity. This is indicated by an almost rectangular ROC approaching 1 on the ordinate at small abscissa values. That means that, despite low false positive tests (high specificity), high sensitivity (true positive tests) is achieved. Or in other words, if a test shows positive results, prostate cancer is very likely to exist; if the test shows negative results, prostate cancer is very unlikely to exist. Thus, cancers that exist are not likely missed nor are cancers suspected if they do not exist. This strength of the test is increasingly lower if done at increasing time intervals before diagnosis. *(From Pepe et al., 2001,* J. Natl. Cancer Inst. *93, 1054–1061, reprinted by permission from Oxford University Press.)*

- Increased utility (whether great or incremental) of the new biomarker over currently established biomarkers
- Knowledge of the relation between the new biomarker and mechanisms of disease initiation or progression
- Knowledge of the relation between the biomarker and outcome prediction
- A superior multiple readout strategy using the new biomarker in combination with known biomarkers for overall testing for the disease
- Ability to predict the therapeutic course of action or the response to an agent by the new biomarker
- Cost-effectiveness

These questions will recur with modifications and additions when biomarker assessment approaches are examined in Subchapter 3.4. The Pepe model is focused not on drug or medical device development-related biomarkers but rather on oncologic disease markers (which can also be used to describe drug effects).

Another model or scheme that is more drug development-oriented is a proposal for pharmacogenomic biomarker development by the FDA (U.S. Food and Drug Administration, 2005; Figure 3.9). The key elements of this proposal

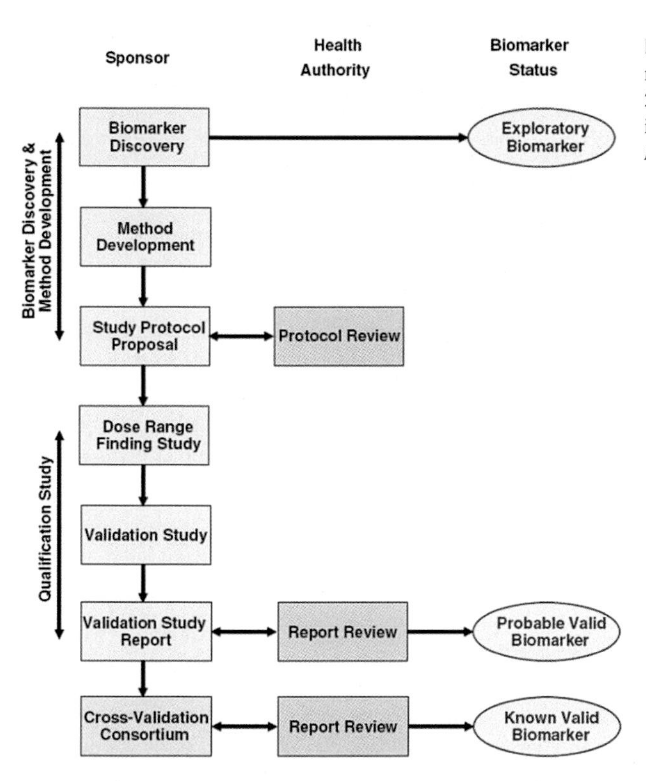

FIGURE 3.9 Schematic summary of the FDA recommendation for pharmacogenomic biomarker development (U.S. Food and Drug Administration 2005). Though principally independent from the drug development progress, it aims at facilitating it by developing novel biomarkers. *(From Marrer and Dieterle, 2007, reprinted by permission from Blackwell Publishing Ltd.)*

are essentially similar to the ones described previously, but this scheme aims at supporting drug development by novel markers under the premises of personalized medicine. It is the model still underlying the regulatory approval process for biomarkers as discussed previously (U.S. Department of Health and Human Services Food and Drug Administration, 2011, results shown in Table 3.4). It introduces staging terms describing the developmental stage of a novel biomarker:

- ***Pharmacogenetic test:*** *An assay intended to study interindividual variations in DNA sequence related to drug absorption and disposition (pharmacokinetics) or drug action (pharmacodynamics), including polymorphic variation in the genes that encode the functions of transporters, metabolizing enzymes, receptors, and other proteins.*
- ***Pharmacogenomic test:*** *An assay intended to study interindividual variations in whole-genome or candidate gene, single-nucleotide polymorphism (SNP) maps, haplotype markers, or alterations in gene expression or inactivation that may be correlated with pharmacological function and therapeutic response. In some cases, the* pattern or profile of change *is the relevant biomarker, rather than changes in individual markers.*
- ***Valid biomarker:*** *A biomarker that is measured in an analytical test system with well-established performance characteristics and for which there is an established scientific framework or body of evidence that elucidates the physiologic, toxicologic, pharmacologic, or clinical significance of the test results. The classification of biomarkers is context-specific. Likewise, validation of a biomarker is context-specific and the criteria for validation will vary with the intended use of the biomarker. The clinical utility (e.g., predict toxicity, effectiveness or dosing) and use of epidemiology/population data (e.g., strength of genotype–phenotype associations) are examples of approaches that can be used to determine the specific context and the necessary criteria for validation.*
- ***Known valid biomarker:*** *A biomarker that is measured in an analytical test system with well-established performance characteristics and for which there is widespread agreement in the medical or scientific community about the physiologic, toxicologic, pharmacologic, or clinical significance of the results.*
- ***Probable valid biomarker:*** *A biomarker that is measured in an analytical test system with well-established performance characteristics and for which there is a scientific framework or body of evidence that appears to elucidate the physiologic, toxicologic, pharmacologic, or clinical significance of the test results. A probable valid biomarker may not have reached the status of a known valid marker because, for example, of any one of the following reasons:*
  - *The data elucidating its significance may have been generated within a single company and may not be available for public scientific scrutiny.*
  - *The data elucidating its significance, although highly suggestive, may not be conclusive.*
  - *Independent verification of the results may not have occurred (U.S. Food and Drug Administration, 2005).*

Though probably somewhat biased for the use of genetic markers, as shown by the first two terms, this proposal conveys generalizability and may serve as a template for biomarkers from other sources as well. At least the categories "known valid biomarker" and "probable valid biomarker" seem useful, although their definition borders are not sharp and easily measurable in all cases. As noted in the definition for known valid biomarkers, this assessment is based on consensual processes that are accepted in the scientific community. But is this acceptance merely by common usage, or is it substantiated by data? Could widespread agreement be a result of company sponsoring?

These critical questions should point to the fact that an objective (or as objective as possible) assessment is considered superior to interindividual consensus agreements (a gut feeling consensus); this is the topic of Subchapter 3.4.

Figure 3.10 exemplifies the process of biomarker development for chronic graft rejection biomarkers. It summarizes the principal stages of drug development and codevelopment of a novel biomarker. After identification in discovery processes, this biomarker will be profiled and optimized as a useful, predictive biomarker; further tested in clinical trials to concentrate the patients (stratification) who may benefit most; used for adaptive trial designs (trial designs that will be changed while the trial is running, for example, the number of patients included may be changed on the basis of biomarker tests); and finally co-approved with the drug. That means that the approval of the drug includes the necessity to test the novel biomarker before its use, for example, HER2-testing with the use of trastuzumab (see Subchapter 3.2.).The evaluation of the novel biomarker will follow the FDA process for pharmacogenomic biomarker submissions (U.S. Food and Drug Administration, 2005, 2011). In the case of chronic graft rejection biomarkers as depicted in Figure 3.10, the NIH-led Clinical Trials in Organ Transplantation (CTOT) program (which is similar to the NCI Cancer Biomarker Project) facilitates the process. This program recently announced that CXCL9 may be a valuable stratification biomarker for kidney transplant rejection (Hricik et al., 2013), which could be useful in guiding clinical decision processes.

The Biomarkers Consortium, a private–public partnership founded in 2006 that includes the FDA, supports and organizes biomarker development (Buckman et al., 2007; Wilson et al., 2007). The Biomarkers Consortium plays a central role in this plot, and constantly announces new data on biomarker qualification studies, for example, on adiponectin as a biomarker to predict glycemic efficacy (Wagner et al., 2009). The NCI Cancer Biomarker Project has identified the following scientific components for detecting and developing biomarkers in the cancer area through the Early Detection Research Network (2008):

- Biomarker development laboratories
- Biomarker reference laboratories
- Clinical epidemiology and validation centers
- Data management and coordinating centers

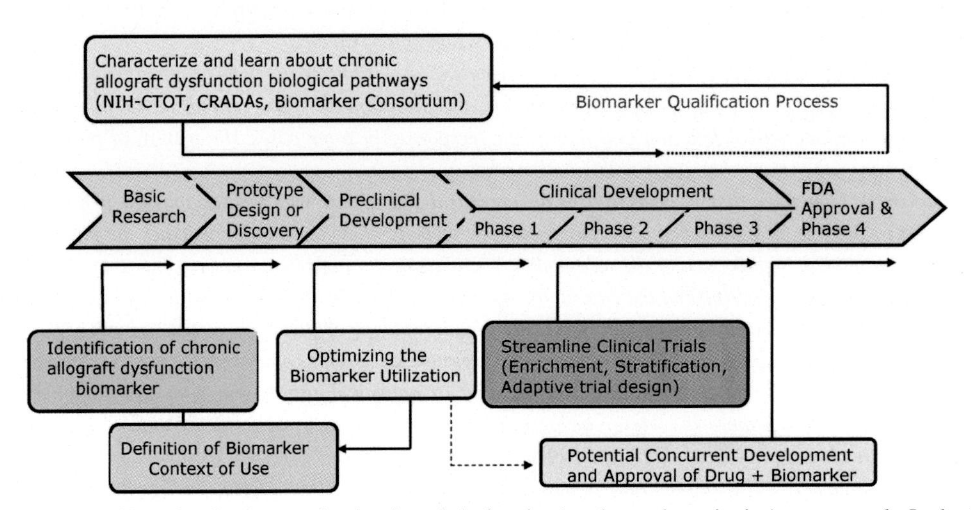

FIGURE 3.10 Scheme of the biomarker development for chronic graft dysfunction (renal transplant rejection) as an example. Its development parallels drug development, which is depicted in the central arrow. The graph shows the potential impact of novel biomarkers on drug development (streamlining clinical trials by patient concentration and stratification) and points to the possibility of codevelopment of drugs and biomarkers, resulting in concurring approval (i.e., the drug is given only when the novel biomarker reaches certain values). The qualification process involves the Biomarker Qualification Pilot Process of the FDA (see text) and, specific to this area, the NIH-CTOT (the Clinical Trials in Organ Transplantation study sponsored by the National Institutes of Health) and CRADAs (cooperative research and development agreements). *(From Burckart et al., 2008, reprinted by permission from Blackwell Publishing Ltd.)*

In a similar attempt, the Predictive Safety Testing Consortium (PSTC) was launched by the FDA and the C-Path Institute in March 2006 to assess new preclinical and clinical biomarkers for nephrotoxicity, hepatotoxicity, vascular injury, and genotoxic and nongenotoxic carcinogenicity. The PSTC presented 23 biomarkers for nephrotoxicity assessment to the FDA for qualification consideration in June 2007 (Burckart et al., 2008). Subsequently, the FDA accepted seven of them as valuable biomarkers for renal safety issues (Dieterle et al., 2010).

These activities demonstrate that biomarker development for efficacy and—maybe even more important—safety assessment has become a vital and essential part of drug development and underpins the translational efforts of industry and public institutions in the biomedical arena, especially the FDA, with biomarkers being considered the single most important vehicle of translational success.

REFERENCES

Buckman, S., Huang, S.-M., Murphy, S., 2007. Medical product development and regulatory science for the 21st century: the Critical Path vision and its impact on health care. Clin. Pharmacol. Ther. 81, 141–144.

Burckart, G.J., Amur, S., Goodsaid, F.M., Lesko, L.J., Frueh, F.W., Huang, S.M., et al., 2008. Qualification of biomarkers for drug development in organ transplantation. Am. J. Transplant. 8, 267–270.

Dieterle, F., Sistare, F., Goodsaid, F., Papaluca, M., Ozer, J.S., Webb, C.P., et al., 2010. Renal biomarker qualification submission: a dialog between the FDA-EMEA and Predictive Safety Testing Consortium. Nat. Biotechnol. 28, 455–462.

Early Detection Research Network, 2008. Science components. http://edrn.nci.nih.gov/about-edrn/scicomponents (accessed 16.01.08.).

Hricik, D.E., Nickerson, P., Formica, R.N., Poggio, E.D., Rush, D., Newell, K.A., et al., 2013. Multicenter validation of urinary CXCL9 as a risk-stratifying biomarker for kidney transplant injury. Am. J. Transplant. 13, 2634–2644.

Marrer, E., Dieterle, F., 2007. Promises of biomarkers in drug development—a reality check. Chem. Biol. Drug Des. 69, 381–394.

Pepe, M.S., Etzioni, R., Feng, Z., Potter, J.D., Thompson, M.L., Thornquist, M., et al., 2001. Phases of biomarker development for early detection of cancer. J. Natl. Cancer Inst. 93, 1054–1061.

U.S. Food and Drug Administration, 2005. Guidance for industry—pharmacogenomic data submissions. http://www.fda.gov/downloads/regulatoryinformation/guidances/ucm126957.pdf (accessed 24.02.14.).

U.S. Department of Health and Human Services Food and Drug Administration, 2011. Guidance for Industry. E16 biomarkers related to drug or biotechnology product development: context, structure, and format of qualification submissions. www.fda.gov/downloads/drugs/guidancecomplianceregulatoryinformation/guidances/ucm267449.pdf (accessed 24.02.14.).

Vasan, R.S., 2006. Biomarkers of cardiovascular disease-molecular basis and practical considerations. Circulation 113, 2335–2362.

Wagner, J.A., Wright, E.C., Ennis, M.M., Prince, M., Kochan, J., Nunez, D.J., et al., 2009. Utility of adiponectin as a biomarker predictive of glycemic efficacy is demonstrated by collaborative pooling of data from clinical trials conducted by multiple sponsors. Clin. Pharmacol. Ther. 86, 619–625.

Wilson, C., Schulz, S., Waldman, S.A., 2007. Biomarker development, commercialization, and regulation: individualization of medicine lost in translation. Clin. Pharmacol. Ther. 81, 153–155.

Chapter 3.4

Predictivity Classification of Biomarkers and Scores

Martin Wehling

Subchapters 3.1 to 3.3 clearly indicated that biomarkers and their development are the backbone of successful translation. A major challenge in this context is the assessment of biomarker quality. It is obvious that a biomarker may just reflect nonspecific alterations of broad physiological responses and its alterations could be induced by a manifold of influences. Thus, this marker is quite nonspecific and is prone to disturbances that may or may not have anything to do with the disease and the related intervention. Such a marker might be, for example, fever, C-reactive protein (CRP), or headache: those markers reflect inflammation at large (fever or CRP) or the state of feeling unwell (headache). One can easily understand that such markers at least as stand-alone biomarkers are not very helpful for designing and developing new drugs. These changes could have been induced by infections or psychological stress that might not have any relation to the intervention in question.

On the other hand, there are biomarkers that are ideally tightly correlated with outcome and highly specific for a given condition, and they may still be very simple and straightforward: tumor size serves as an example. If diagnosis has been proven by histology, any change in tumor size is a relevant biomarker for regression or progression, and, thus, the natural or interventional course of the disease. If size is determined by imaging techniques, as usual, the translational power is high, and drug effects can easily be followed across species borders. Failures of translation would have to do with tumor biology and animal models, but tumor size as such reliably measures disease progression or regression and therapeutic effects.

Although those two examples are located at the extremes of the spectrum of biomarker validity, there are more complex and unclear biomarker situations in which the assessment of the predictive value of a given biomarker is much more difficult. A prominent example is the problem of early biomarkers for beneficial drug effects on atherosclerotic plaque progression. To begin with, there are no good animal models for human atherosclerosis, which takes 30–50 years to develop, and conditions of arterial damage in animal models are very different from those in humans. Animal models of atherosclerosis (for a review, see Russell and Proctor, 2006) tend to be heterogeneous and, thus, regression of atherosclerosis in an animal model may not indicate that the same happens in human pathology. More troublesome is the fact that some species, namely rodents, seem to be protected from ischemic endpoints relating to atherosclerosis. Thus, even if atherosclerotic lesions develop in rodents, their importance for endpoints may be different from that of humans. Thus, the important goal is to find biomarkers that indicate antiatherosclerotic effects across species borders. Markers of inflammation, which is inherent to atherosclerotic processes, were seen as favorable biomarkers. They include high-sensitive CRP, which had been shown to correlate to cardiovascular events in humans but which has seen recent drawbacks and is still under critical discussion (Ben-Yehuda, 2007). Current thinking may even support the hypothesis that mild inflammation from independent sources could accelerate atherosclerosis (Grufman et al., 2014). Because this marker is still controversial in humans, related effects in insufficient animal models may be even more misleading and translationally irrelevant. Another route of assessment of atherosclerotic disease utilizes imaging techniques that are easily transferable and directly show the atherosclerotic plaque. As detailed in the subchapter on typical cardiovascular biomarkers (3.7.1), the main problem with imaging biomarkers is the assessment of plaque stability, because the leading problem is not its size, but its potential rupture. Thus, plaque size, location, and maybe composition (fat content) can be visualized by imaging (including in vivo staining), but it is still hard to predict which plaque will rupture (Briley-Saebo et al., 2007). Modern techniques are further developed in the hope to improve prediction, such as [18-F] fluorodeoxyglucose positron emission tomography/computed tomography (Mehta et al., 2012).

This example should draw our attention to a relevant area of biomarker development in which clearly predictive biomarkers are still lacking and major efforts are being undertaken to develop them. The Biomarkers Consortium (see Subchapter 3.3; Biomarkers Consortium, 2008; Briley-Saebo et al., 2007) has included related concepts in its portfolio. These examples should also demonstrate that biomarkers can range from very strong and predictive (not requiring any developmental investment), through unsatisfactory, to dismal or useless. Although this judgment seems straightforward for

Principles of Translational Science in Medicine. http://dx.doi.org/10.1016/B978-0-12-800687-0.00015-3

the examples given previously, an objective, interindividually reliable, reproducible, standardized, and valid process should be developed by which the value of a biomarker is assessed. If we accept that there should be a standardized assessment procedure in place, we need to define its remits.

In fact, the only true remit of this assessment is an answer to the question, How much does this biomarker or a biomarker panel (if they are concomitantly applied) contribute to the translation of preclinical findings into human use, both in terms of efficacy and safety? A simpler terminology would be *predictive value,* implying that a predictivity score should be aligned to a given biomarker. This predictive value is reflective of the pivotal role a biomarker plays in translational medicine. Thus, the *standardized assessment of the predictive biomarker value* would be a major remit of the science of translation in medicine to which this book is devoted.

In the subchapter on biomarker classification (3.2), some attempts to structure the universe of the large spectrum of potential biomarkers have been presented, including the technology-oriented mechanistic approach (Rolan et al., 2003) and the three-level approach by the NIH Biomarkers Definitions Working Group (2001). Although the first approach does not relate to the predictive value, the latter at least separates surrogates and clinical endpoints from the bulk of unclassified markers. Certainly, the predictive value of surrogates is thought to be close to 100% (almost all predictions from this type of marker into clinical endpoints had been correct), and the endpoints do not require prediction because they represent the top end of development.

The proposal for pharmacogenomic biomarker development by the FDA [U.S. Food and Drug Administration, 2005; new guidance U.S. Department of Health and Human Services Food and Drug Administration (2011)] goes one step further and introduces "valid biomarkers," subdivided into "known" and "probable" biomarkers, and provides simple definitions that, however, lack quantification. They rely on consensus (see Table 3.3) and, thus, require a good deal of gut feeling assessment.

However, there are only a handful of surrogates, and situations in drug or device development in which they can be used early on are rare. Thus, the vast majority of biomarkers require a true predictivity assessment, which is outlined in the following discussion. Vasan (2006) presented a list of questions to be answered for a given biomarker, which is recapitulated in Subchapter 3.3. Although those items represent major features of laboratory test development such as accuracy, reproducibility, stability, correlation to disease, and many others, the list lacks the major translational aspects (e.g., in how many species other than humans is the biomarker present and validated?). Also, it does not provide a score by which the overall value can be determined quantitatively. Although quantification of some variables such as stability or accuracy will be arbitrary at least in part, some attempt of overall utility scoring should be undertaken to be able to weigh biomarkers against each other. Notably, quantification has its limits, and measures such as "knock-outs" or "knock-ins"—that is, overruling of the other parameters—need to be installed in such an approach.

In this subchapter, the first attempt to provide a quantitative predictivity scoring system for biomarkers is described (Wehling, 2006, Table 3.6). The 10 items in the scoring system are related to animal data, human data, the proximity of the biomarker to the disease (e.g., causal relation, the disease constituent involved in pathophysiology), test parameters (sensitivity, specificity as integrated into predictivity, which has nothing to do with the overall predictive value of the biomarker but is a statistical measure of the test quality), and feasibility aspects (accessibility). The grading scale ranges from (0)1 to 5(6), with 0 being a knock-out feature and 6 being a knock-in feature, which indicates that the marker may be close to being a surrogate. The extreme scores overrun the other item scores. Following is an explanation of the 10 points of the scoring systems and how each point is scored (Modified from Wehling, 2006, reprinted by kind permission from Springer Science and Business Media):

1. Animal or in vitro testing: Are there data at this level? Yes = 1; animal plus human data = 5; only human data = 3. This item should underscore the fact that human data on this particular biomarker are highly relevant and reflected by a high score. The lowest score of 1 is given if at the time of assessment all considerations rely on animal or in vitro data only. It is obvious that in this situation, predictivity is merely speculative. A score of 0 would be given if no animal or in vitro data are available. This is rare but could reflect observational, nonsystematic material that is not sufficient to qualify a measure as a biomarker.
2. How many species have been tested positively? $\leq 1 = 1$; $2 = 3$; $\geq 3 = 5$. Obviously, if cross-species variation is low and several species show the same change with regards to the biomarker, scoring should be higher than in cases in which only one species or model shows expected effects.
3. Are the animal models sufficient to reflect human disease? If there is no human validation in terms of disease or mechanistic similarity, the score is 1; grading goes up to 5 if homology is perfect, for example, certain bacterial infections in which the in vitro testing is very similar to human testing because the infectious agent is the same. This decision should reflect the amount of scientific evidence as judged by the number of different groups involved, the consistency of the data, and the level of publication (whether the journal is peer-reviewed).
4. Is there corresponding clinical data—including genetics or histology—to judge the validity of the animal or in vitro data? This is not asking whether there are clinical studies with the biomarker in interventional studies, but whether there is evidence that the animal model or in vitro testing in general correlates with similar diseases or related condition.

TABLE 3.6 Predictive profiling of biomarkers

		Biomarker Profile	
1	Animal or in vitro		
2		How many species?	
3		Model sufficient to describe human disease	
4		Corresponding clinical data classification	
5	Human		
6		Data classification	
7	Pivotal disease constituent		
8	Predictability		
9	Accuracy or reproducibility		
10	Accessibility		

(Modified from Wehling, 2006, reprinted by kind permission from Springer Science and Business Media.)

Answering this question affirmatively does not mean that the biomarker has already been tested in the same disease; it means that there is evidence that the model is relevant. The score = 1 for a singular study, 2 for multiple studies; if the biomarker is correlated with intervention from other projects on similar targets, the score = 2 for a singular study, 3 for multiple studies; if the biomarker is correlated with interventional endpoint change from other projects on similar targets, the score = 4 for a singular study, 5 for multiple studies.

5. Are human data on this biomarker available that directly relate to the project or disease to be treated? Yes = 5; no (only animal data are available) = 1. This again underpins the importance of human data in this context. This is a yes-or-no item; no intermediate scores are allowed.
6. Human data classification: Does the available clinical material show correlation? If the biomarker is correlated with the disease, the score = 1 for a singular study, 2 for multiple studies; if the biomarker is correlated with intervention, the score = 2 for a singular study, 3 for multiple studies; if the biomarker is correlated with interventional endpoint change, the score = 4 for a singular study, 5 for multiple studies, *6 if it is a surrogate*. This score should be doubled to underscore that this pivotal feature has been derived from human studies.
7. Does the biomarker represent a pivotal disease constituent? In other words, is it essential to the pathophysiology of the disease to be assessed? No (it is likely to represent an epiphenomenon) = 1. A score up to 5 is given if it is a solitary, stand-alone contributor to the disease (e.g., absence of singular enzymes in monogenetic diseases, supplementation of gene product or its derivatives fully restores health).
8. What is the statistical predictability, which in a test describes its sensitivity and specificity (false positive and negative results)? > 60% = 1; > 75% = 2; > 85% = 3; > 90% = 4; > 95% = 5. If a precise percentage is not available, grade according to an estimate.
9. What is the accuracy or reproducibility of the biomarker assay? Poor (e.g., SD > 70%) = 1; a score of up to 5 is given if the accuracy or reproducibility is high (SD < 15%).
10. How accessible is the specimen? This is an important issue as well, as a specimen that is hard to obtain limits the use of a biomarker. Easy access (e.g., a by-product such as peripheral blood that requires no legal procedures or additional animal use) = 5; expensive animal work (e.g., monkey work), legal concerns, parameter measurement that has a relatively high impact on humans (e.g., muscle or liver biopsies) = 1. Inaccessibility in humans = 0, which is a knock-out (e.g., serial brain slices).

The final score can be interpreted as follows, regarding the predictive value of the biomarker:

- < 30 = a low-value biomarker; it may serve as a proof-of-mechanism marker.
- 31–40 = a medium-value biomarker; it may serve as a proof-of-principle marker.
- 41–50 = a high-value biomarker; this score is desirable for proof-of-principle markers, and a score of > 50 (maximum 55) means the biomarker is close to a proof-of-concept marker.
- A score of 60 or higher means that the biomarker is a proof-of-concept or surrogate marker; a score of 0 should prevent the biomarker from being considered to be able to promote the translational aspects of drug or device development.

Any single component described in the scoring system can overrun the scoring if it is considered to be overwhelmingly important. Thus, important parts of the scoring remain consensus decisions, although this should be limited to the reading of the score items. Of course, for example, the causal relationship of a biomarker to the disease may be more or less important, and scoring may thus be partially subjective. Subjectivity cannot be entirely excluded, and experts need to be involved in this process regardless.

This scoring system reflects two main assumptions: biomarkers that have no human validation are very weak, at least at that stage, because the human system is the ultimate test system for successful translation. The scoring system conveys the operational guidance to rapidly progress into human material, which could be genetics, retrospective analyses from tissue banks, results from other studies, and much more. The second assumption is that it is very important that the biomarker be tied into the disease process to become highly predictive. These two key features are supplemented by typical test parameters such as accuracy and (statistical) predictivity.

It is obvious that this proposal is only an initial attempt to give labels to biomarkers and to spur developmental aspects. Clearly, the scoring of a biomarker can be dramatically increased if, for example, a clinical trial shows its value, or if the third and fourth species seem to be in line with the first and second. Thus, this grading system should also be inspiring in the sense that a bad biomarker needs the hypothetical studies A, B, and C to become a better one if results are positive. On the other hand, in some cases one is surprised how limited the justification of the use of frequently used biomarkers (e.g., CRP in atherosclerosis) is when challenged by a structured approach like the one described previously.

REFERENCES

Ben-Yehuda, O., 2007. High-sensitivity C-reactive protein in every chart? The use of biomarkers in individual patients. J. Am. Coll. Cardiol. 49, 2139–2141.

Biomarkers Consortium, 2008. Project concepts in development. http://www.thebiomarkersconsortium.org/index.php?option=com_content&task=view&id=93&Itemid=143 (accessed 26.02.14.).

Biomarkers Definitions Working Group, 2001. Biomarkers and surrogate endpoints: preferred definitions and conceptual framework. Clin. Pharmacol. Ther. 69, 89–95.

Briley-Saebo, K.C., Mulder, W.J., Mani, V., Hyafil, F., Amirbekian, V., Aguinaldo, J.G., et al., 2007. Magnetic resonance imaging of vulnerable atherosclerotic plaques: current imaging strategies and molecular imaging probes. J. Magn. Reson. Imaging 26, 460–479.

Grufman, H., Gonçalves, I., Edsfeldt, A., Nitulescu, M., Persson, A., Nilsson, M., et al., 2014. Plasma levels of high-sensitive C-reactive protein do not correlate with inflammatory activity in carotid atherosclerotic plaques. J. Intern. Med. 275, 127–133.

Mehta, N.N., Torigian, D.A., Gelfand, J.M., Saboury, B., Alavi, A., 2012. Quantification of atherosclerotic plaque activity and vascular inflammation using [18-F] fluorodeoxyglucose positron emission tomography/computed tomography (FDG-PET/CT). J. Vis. Exp. 63, e3777.

Rolan, P., Atkinson Jr., A.J., Lesko, L.J., Scientific Organizing Committee, Conference Report Committee, 2003. Use of biomarkers from drug discovery through clinical practice: report of the Ninth European Federation of Pharmaceutical Sciences Conference on Optimizing Drug Development. Clin. Pharmacol. Ther. 73, 284–291.

Russell, J.C., Proctor, S.D., 2006. Small animal models of cardiovascular disease: tools for the study of the roles of metabolic syndrome, dyslipidemia, and atherosclerosis. Cardiovasc. Pathol. 15, 318–330.

U.S. Food and Drug Administration, 2005. Guidance for industry—pharmacogenomic data submissions. http://www.fda.gov/downloads/regulatoryinformation/guidances/ucm126957.pdf (accessed 24.02.14.).

U.S. Department of Health and Human Services Food and Drug Administration, 2011. Guidance for Industry. E16 biomarkers related to drug or biotechnology product development: context, structure, and format of qualification submissions. www.fda.gov/downloads/drugs/guidancecomplianceregulatoryinformation/guidances/ucm267449.pdf (accessed 24.02.14.).

Vasan, R.S., 2006. Biomarkers of cardiovascular disease-molecular basis and practical considerations. Circulation 113, 2335–2362.

Wehling, M., 2006. Translational medicine: can it really facilitate the transition of research "from bench to bedside"? Eur. J. Clin. Pharmacol. 62, 91–95.

Chapter 3.5

Case Studies

Martin Wehling

As a first step its ongoing validation process, this scoring system has been applied to biomarkers in the cardiovascular area. Table 3.7 shows the results of applying the proposed grading scheme for biomarkers to frequently used imaging procedures in the assessment of atherosclerotic lesions. They include magnetic resonance imaging (MRI) techniques, computer tomography, ultrasound (US) techniques, intima-media thickness determination, X-ray techniques, positron emission tomography (PET) techniques, single photon emission computed tomography (SPECT), and intravascular ultrasound (IVUS) techniques. All these methods are assessed in Table 3.7 according to their capacity to predict the beneficial effects of anti-atherosclerotic drugs in translational approaches. As can be seen, the scores range from low (18) to high (47). Intima-media thickness received the highest score because it has been shown in several human studies to predict cardiovascular outcome and the beneficial effects of interventions (for a review, see Hurst et al., 2007). It was termed *surrogate*, although this is not generally accepted because its determination by US is subjective and reproducibility is limited. Thus, it did not surmount the limit of 50 to be considered a surrogate or to at least come close to it.

At the other end, determination of intra-plaque lipid core, mainly by MRI in carotid arteries, scored lowest in this set of biomarkers. It has been postulated that this parameter can predict plaque instability and, thereby, identify the so-called vulnerable patient. The vulnerable patient is threatened by his or her unstable plaques, which may suddenly rupture, causing coronary thrombosis and myocardial infarction and death. At the time of initial assessment (early 2005), this hypothesis was almost exclusively driven by animal evidence. There were no human data clearly showing that in humans the extent of the lipid core of a plaque was predictive. This, however, dramatically changed when the ORION (*o*utcome of *r*osuvastatin treatment on carotid artery atheroma: a magnetic resonance *i*maging *o*bservatio*n*) trial was published (Underhill et al., 2008). This trial for the first time showed that intervention with rosuvastatin beneficially affected the lipid core; this represents the first human data on this type of intervention. If the assessment was done including ORION data (column "post-ORION" next to column "pre-ORION" in Table 3.7), the scores relating to human data (items 1, 2, 4, 5, 6, and 9 in Table 3.7) dramatically increase, thereby elevating the total score to 46. This is the second best score in this table, only surmounted by intima-media thickness.

This example demonstrates that assessment of biomarkers can unveil major gaps in the strength of a biomarker. The generation of data that fill the gap, as in this case, measurably changes the grading, which thus reflects successful biomarker development. The given example of scoring various imaging biomarkers in atherosclerosis does seem to identify stronger and weaker biomarkers, which could be a valuable help for decisions based on such biomarkers. If the grading process is further validated by its practical use, it may become a constant reminder to push biomarker development, because successful drug or device development relies on predictive biomarkers. Biomarker picking and development in certain areas could become more expensive than drug development, but this could represent a way to optimize costs because lower rates of late attrition could compensate or even overcompensate for increased biomarker development costs.

The proposed grading is an initial step in raising translational medicine to the rank of a science; it requires further refinement and development. Validation can only be derived from experiences with its practical impact. Researchers in this field will have to decide whether they want to continue with the merely subjective biomarker assessment approaches highlighted here—such as the gut feeling approach—or whether a structured approach such as the 10-step scoring system is superior. The author strongly believes that there is no alternative to structured, scientific approaches if translational medicine is to develop past the level of wishful thinking to become an essential, validated tool.

Principles of Translational Science in Medicine. http://dx.doi.org/10.1016/B978-0-12-800687-0.00016-5

TABLE 3.7 Predictive profiling of atherosclerosis imaging biomarkers according to the grading scheme described earlier*

	Biomarker profile	Total plaque burden (MRI)	Carotid plaque volume (MRI, US)	Coronary plaque volume (IVUS, MRI)	Stenosis (X-ray angiography, CT, MRI, IVUS, SPECT, PET)	Intra-plaque lipid core (MRI, US strain analysis), pre-ORION	Intra-plaque lipid core (MRI, US strain analysis), post-ORION	Intra-plaque thrombus (MRI, US)	Calcium scoring (X-ray, CT, US [for animals])	Intima-media thickness (US)
1	Are animal or in vitro data available?	5	5	5	5	1	5	1	5	5
2	How many species have been tested positively?	3	3	3	5	1	5	1	5	5
3	Are the animal models sufficient to reflect human disease?	2	2	2	2	2	2	2	3	3
4	Is there corresponding clinical data?	2	5	5	5	2	5	2	2	5
5	Are human data available?	5	5	5	5	1	5	1	5	5
6	Human data classification	4	10	10	10	2	10	2	4	10
7	Does the biomarker represent a pivotal disease constituent?	2	3	3	3	2	3	3	2	3
8	What is the statistical predictability?	1	2	3	3	1	3	2	4	3
9	What is the accuracy or reproducibility of the assay?	3	4	3	4	2	4	2	4	3
10	How accessible is the specimen?	3	4	1	1	4	4	4	3	5
	Score	30	43	40	43	18	46	20	37	47

**Wehling (2006a). The table depicts the individual scores and sums for frequently used biomarkers in the field. IVUS = intravascular ultrasound; MRI = magnetic resonance imaging; PET = positron emission tomography; SPECT = single photon emission computed tomography; US = ultrasound.*

(Modified from Wehling, 2006b, reprinted by permission from Adis International, Wolters Kluwer Health.)

REFERENCES

Hurst, R.T., Ng, D.W., Kendall, C., Khandheria, B., 2007. Clinical use of carotid intima-media thickness: review of the literature. J. Am. Soc. Echocardiogr. 20, 907–914.

Underhill, H.R., Yuan, C., Zhao, X.Q., Kraiss, L.W., Parker, D.L., Saam, T., et al., 2008. Effect of rosuvastatin therapy on carotid plaque morphology and composition in moderately hypercholesterolemic patients: a high-resolution magnetic resonance imaging trial. Am. Heart. J. 155 (584), e1–8.

Wehling, M., 2006a. Translational medicine: can it really facilitate the transition of research "from bench to bedside"? Eur. J. Clin. Pharmacol. 62, 91–95.

Wehling, M., 2006b. Translational science in medicine—implications for the pharmaceutical industry. Int. J. Pharm. Med. 20, 303–310.

Chapter 3.6

Biomarker Panels and Multiple Readouts

Andrea Padoan, Daniela Bernardi and Mario Plebani

INTRODUCTION

Powerful new technologies in diagnostic laboratory testing exist to offer unprecedented prospects for translating early and accurate diagnoses of human disorders into improvements in clinical care (Plebani and Laposata, 2006). High-throughput technologies, such as mass spectrometry and microarrays, allow measurements of proteins and RNA/DNA to be carried out in large numbers. The enormous amount of data produced during genetics and proteomics analysis, coupled with the inherent variation in the instrumentation used as well as the biological variation of the subjects, requires the utilization of sound experimental design, proper calibration of instruments, and appropriate biostatistics/bioinformatics methods in order to generate good-quality data by which valid conclusions could be drawn.

In particular, some strategies have to be used and applied to assure the appropriate interpretation and utilization of such laboratory information, given the complexity of large volumes of data and numbers, which potentially could create confusion in practicing physicians.

While the field is complex, presenting several areas of research and development, we would like to focus on some relevant issues:

- Sources of errors in multimarkers based proteomics biomarkers discovery studies.
- Use of biostatistics/computational methods to match the different spectral patterns of proteins and peptides identified in a particular group of patients (e.g., cancer) in comparison to other known patterns that characterize control groups.
- Most informative data reporting for results obtained using a multiparameter approach, both when the technique allows a simultaneous measurement on the same platform of different analytes, and when the results of various and related parameters come from different analytical platforms. The development of diagnostic algorithms, in fact, adds value to any *single* analytical result. The diagnostic performance of *some* biochemical markers might be greatly improved by combining them with clinical parameters in algorithms or scores. Moreover, diagnostic algorithms may also be developed by combining *many* markers using multimarker panels. Diagnostic test panels are typically composed of many individual laboratory tests, combined with or without patient medical conditions, and they are intended to help physicians to achieve more accurate and reliable diagnoses. For a more effective clinical utilization of the data, it is debatable if it is more valuable to report any single result (e.g., CA125 or HE4 alone), to provide a score that summarizes the obtained analytical result (e.g., ROMA; see below) or to define the disease status (e.g., presence/absence of prostate and pancreatic cancer, liver fibrosis), by using a bioinformatic method that combine many biochemical markers.

SOURCE OF ERRORS IN PROTEOMICS STUDIES

Clinical application of proteomics is based on the concept that a multimarker approach should be more accurate and reliable in identifying a disease state. However, the proteomic pattern diagnostics, sometimes called *serum proteomic profiling*, presents several challenges.

As underlined by Ransohoff (2005), the introduction of "discovery-based" research, in which high-throughput technologies allow the measurements of multiple proteins, mutations, and gene expression, *without a specific hypothesis*, threatens the validity of obtained data. The study *design* is the essential feature of small studies, because it minimizes problems from chance (random errors) and systematic biases. Small studies done well can effectively answer important questions and demonstrate a "proof-of-principle" about a molecular marker. Systematic biases are types of errors that occur nonrandomly during the analysis, while random errors can be caused by unknown and unpredictable changes. Problems caused by *systematic biases and chance* could arise in any point of the pre-analytical, analytical and post-analytical phases and are

Principles of Translational Science in Medicine. http://dx.doi.org/10.1016/B978-0-12-800687-0.00017-7

shared across them. A large sample size does not directly address biases, although it can reduce statistical uncertainty by providing a smaller confidence interval around a result and thereby reduce random error. Both sources of errors could be carefully minimized by (1) choosing an appropriate study design and (2) selecting adequate standard operating procedures (SOPs). SOPs are defined by international scientific organizations such as the Human Proteome Organization (HUPO). HUPO fosters the development of new technologies and techniques, and defines protocols to (a) standardize proteomics analyses and (b) achieve uniformity of results, such that different laboratories can use the data obtained interchangeably.

Pre-analytical conditions, such as collection practices, sample handling, and storage conditions, may differ from institution to institution, thus influencing the protein present in biological fluids. The use of specimens from multiple institutions is required to reliably demonstrate the real efficiency of protein profiling.

Analytical variables, such as inconsistency in conditions of instruments, could result in poor reproducibility and high measurement imprecision. The use of control material after calibration and throughout a run ensures good instrument performance and reliable data. However, relatively few studies (Colantonio and Chan, 2005) have focused on the way in which quality control, an essential and quality-related feature, should be incorporated into proteomic experimental protocols. Reproducibility studies performed with adequate control materials are, in fact, prerequisites for the safe introduction of the new techniques in clinical laboratory practice.

Finally, *post-analytical variables,* such as biological variability and reference ranges as well as decision limits, represent other key issues for a correct interpretation of the laboratory data.

Current limitations and open questions relate to *within-class biological variability*. It may concern possible diurnal variation in protein expression, making time standardization of sample collection mandatory. For any individual, many analytes fluctuate based on time of day, fasting state, or age. Although these fluctuations may be not clinically relevant, they do add an additional level of complexity in elucidating disease-induced protein changes from changes due to biorhythmic fluctuations. An evaluation of the effects of gender, age, pathophysiological conditions, and benign diseases is also important in understanding other possible effects on protein profiling expression.

As a result of the copious amount of data produced from proteomic analysis, use of bioinformatics tools is required to sift through the data in order to identify proteins of interest from mass spectrometric data. However, a further step required before translating research insight of proteomics into clinical practice is to identify and characterize any single peak detected in the proteomic profile. Again, bioinformatics represents a fundamental tool in order to identify proteins and protein fragments through databases available for specific proteins and post-translational modifications, such as Pfam (Finn et al., 2010), UCSC Proteome Browser (Hsu et al., 2005), Swiss-Prot, and UniProt (Wu et al., 2006).

STATISTICAL AND COMPUTATIONAL METHODS IN CLINICAL PROTEOMICS

Currently, there is a big debate in the statistical methods for the analysis of "-omics" data, such as proteomics or genomics data. This discussion is mostly due to the very high dimensionality of the obtained dataset, which usually contains a large number of variables (features) as compared to the low number of subjects. This type of dataset motivates the need for computational techniques in data analysis and for suitable methods to assess reproducibility and overfitting (Barla et al., 2008; Ransohoff, 2005).

Supervised and Unsupervised Computational Methods

Two useful broad categorizations of the techniques used are *supervised* learning techniques and *unsupervised* learning techniques (Barla et al., 2008). The two techniques are easily distinguished by the presence of external subjects' labels (patients' classification):

- The supervised learning techniques find the features that are correlated with the specific patient's group, thus creating predictive algorithms for these labels.
- In contrast, the unsupervised learning techniques find those features that are correlated across all the samples and operate independently of any external labels (e.g., pathological and healthy subjects).

These two machine-learning methods are generally used to answer different types of questions. In supervised learning, a typical goal is to obtain a set of variables (a process also known as *features selection*) that can be used reliably to make a diagnosis, predict future outcome, predict future response to pharmacologic intervention, or categorize that patient as part of a class of interest. In unsupervised learning, the typical application is to find either a completely novel cluster of peptide/proteins with putative common (but previously unknown) expression or, more commonly, to obtain clusters or *groups of features* that appear to have patterns *of similar expression*.

The major techniques used for unsupervised learning are principal component analysis and clustering determination. In contrast, multiple logistic regression, classification and regression trees (CARTs), support vector machines, and neural networks (NNWs) are widely used as supervised learning techniques.

Currently, superiority or inferiority of multiple logistic regression, support vector machines, and NNWs to each other is not demonstrated yet, although some important differences can be elucidated for logistic regression. Multiple logistic regressions present the advantage of providing a mathematical algorithm (equation) containing predictors (e.g., age, biomarkers) and their corresponding coefficients. Thus, through use of this equation, it is possible to directly derive the patient's probability of developing the disease. Differently, the results produced by CARTs, support vector machines, and NNWs define the patient outcome, so solely the presence or absence of the disease. For this reason, many researchers considered decision trees, support vector machines, and NNWs as "black boxes" rather than multiparameter approaches.

Receiver Operating Characteristic Analysis and Net Reclassification Improvement

In order to evaluate the clinical utility for both diagnostic and prognostic models of multiple logistic regression and of CARTs, one typically uses the receiver operating characteristic (ROC) curve. However, a strong risk predictor may have limited impact on the area under the curve, even if it alters predicted values. Moreover, ROC curve analyses are not applicable to all algorithms. Therefore, reliable measurements of the performance's improvement, achieved by including new predictors in algorithms, are not easily obtainable.

One of the methods recently suggested by Pencina, D'Agostino, and Vasan (2010) is the net reclassification improvement (NRI). It is based on the reclassification tables, obtained from the base algorithm (which included the best predictors) and the extended algorithm, the latter obtained after including additional new predictor variables, which would be tested for their real contributor to classification accuracy. The improvement in reclassification can be quantified as a sum of differences in proportions of individuals moving upward (from low to high risk) minus the proportion moving downward (from high to low risk) for people who had events, and the proportion of individuals moving down minus the proportion moving up for people who did not have events; this sum is called NRI. A simple statistical test can be used to determine the significance of the improvement. To complete the analysis, one can use the integrated discrimination improvement to test any potential increase in sensitivity, with respect to one minus specificity (Pencina et al., 2008).

Overfitting Issues

Despite the employing of these "computational methods" in data analysis, the overfitting problem remains unsolved and afflicts any approaches listed in the preceding sections. With the term *overfitting*, researchers generally mean the probability of finding a discriminatory pattern of features completely by chance. This probability becomes high when large numbers of variables are assessed for a small number of outcomes. The problem can be partially overcome by splitting the database in a *training set* (a randomly chosen patient subgroup is used) and in a *validation set* (the remaining patients are used to assess the classification accuracy) (Ransohoff, 2004). However, when the dataset contains a limited numbers of subjects, dividing the database in training and validation sets is not a convenient choice.

For this reason, *cross-validation* is applied to assess algorithm performance, without dividing the database. The two most used cross-validation methods are the k-fold cross-validation and the leave-one-out cross-validation.

MULTIPARAMETER APPROACH

A widely held viewpoint of research on biomarkers of disease is that no single marker can provide high enough discrimination between cases and noncases for clinical applications and that the use of multiple markers, combined in some type of algorithm, will be necessary to produce the required level of predictive power (see Figure 3.11).

The following chapters describe some examples of successful application of multiparametric approaches for biomarkers discovery.

Ovarian Cancer

Score

It is now recognized that ovarian cancer is a clinical heterogeneous disease, with many histotypes and different genetic mutations among these histotypes (Huang et al., 2012). Therefore, it is unlikely that any single marker will be sufficiently

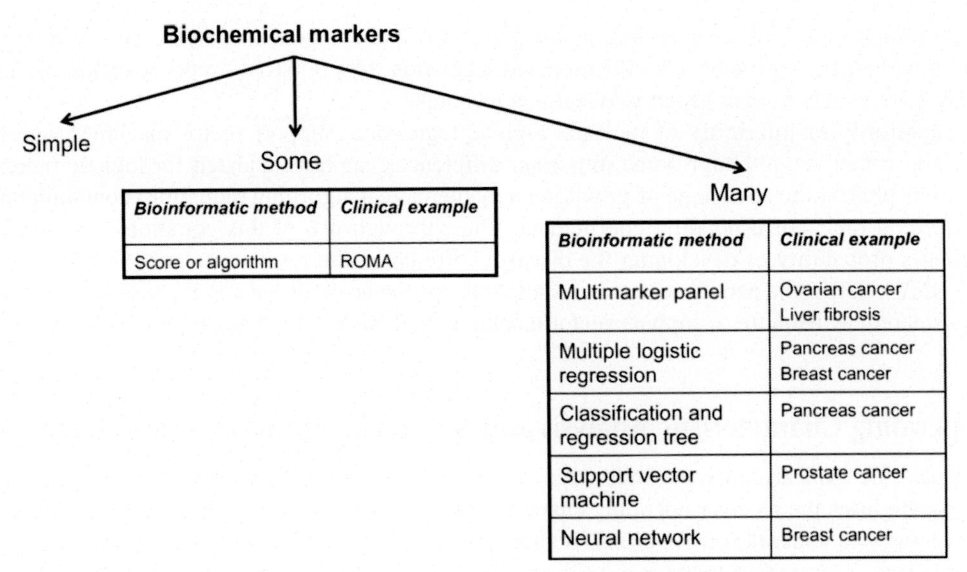

FIGURE 3.11 Bioinformatics methods and related clinical examples.

sensitive to provide an optimal initial screen (Badgwell and Bast, 2007). The use of multiple markers for the differential diagnosis of pelvic masses represents a clinical setting within the management of ovarian cancer where significant improvements could be achieved in the near future.

Currently, CA 125 is the commonly used serum tumor marker to predict the presence of a malignancy in women with a pelvic mass. However, Kirchhoff et al. (1990), upon differential cDNA screening of human epididymal tissues, observed that the human epididymal secretory protein E4 gene expression was upregulated in epithelial ovarian carcinomas. From that evidence, the derived protein HE4 has been extensively studied as biomarker for ovarian cancer. Up to now, several publications have demonstrated the superiority of HE4 over CA 125, especially in patients with early stage disease, in which HE4 has been shown to have greater sensitivity than CA 125 alone (Plebani and Melichar, 2011).

Recently, some multimarker strategy-based risk indexes have been developed. An example is the Risk of Ovarian Malignancy Algorithm (ROMA) that was introduced as a predictive index and basically takes into account the serum concentrations of both biomarkers (CA 125 and HE4) and the pre- or postmenopausal status (Moore et al., 2008). ROMA performances were initially demonstrated to be superior to CA 125 or HE4 alone in detecting ovarian cancer in patients with pelvic mass by a large prospective study (Table 3.8).

However, a recent meta-analysis of Wang et al. (2014) showed that CA 125, HE4, and ROMA performed equally in diagnosing ovarian cancer (Table 3.9).

Multimarker Panel

An example of a proteomic-derived diagnostic test panel is OVA1, which was cleared by the FDA in September 2009 (Zhang and Chan, 2010). The OVA1 is an in vitro diagnostic multivariate index assay (IVDMIA) for assessing ovarian cancer risk in women diagnosed with ovarian tumor prior to a planned surgery. OVA1 analyzes five serum biomarkers (CA 125, transthyretin, ApoA1, β-2 microglobulin, transferrin) by immunometric assays, and the results are combined through an algorithm to yield a single-valued index within the range of 0–10. A menopausal status-dependent cutoff is used to classify a patient into high- or low-risk group. In comparison to CA 125, the OVA1 panel provides an increase in sensitivity (94%), while specificity widely decreased to 35%. Thus, OVA1 does not seem to improve the diagnostic performance of CA 125 alone.

Breast Cancer

Multiple Logistic Regression

To look for fingerprints of cancer, Li and Colleagues (2002) analyzed protein profiles of serum samples of patients with or without breast cancer. Each mass peak was evaluated according to its contribution toward the maximal separation of the cancer patients from the noncancer controls. This evaluation led to the identification of three discriminatory markers

TABLE 3.8 Comparison of diagnostic performance between CA 125, HE4, and ROMA-derived composite score at sensitivity of 94%

	Specificity	AUC
CA 125		
Early	51.9%	0.85
Late	83.3%	0.96
All	62.2%	0.93
HE4		
Early	52.9%	0.86
Late	82.8%	0.96
All	63.2%	0.94
ROMA		
Early	61.8%	0.90
Late	88.8%	0.97
All	76.5%	0.95

Early = early stages, Late = late stages, All = all stages of ovarian cancer, based on FIGO classification; AUC = Area under the ROC curve.
(Modified from Karlsen et al., 2012.)

TABLE 3.9 AUC pooled results from the meta-analysis

	AUC (95% CI)
CA 125	0.87 (0.84–0.90)
HE4	0.89 (0.86–0.92)
ROMA	0.91 (0.88–0.93)

(BC1, BC2, BC3). Multivariate logistic regression was used to combine the three selected biomarkers to form a single-value composite index, achieving both high sensitivity (93%) and high specificity (91%) in detecting breast cancer from the non-cancer controls. ROC curve analysis of the composite index gave a much-improved area under the ROC curve (AUC) (0.972) compared with the AUCs from individual biomarkers, which were 0.85, 0.79, and 0.93 for BC1, BC2, and BC3, respectively.

Neural Network

An example of multiple marker combination, for the identification of a new panel of biomarkers from plasma protein profiling, is the application of the NNW, as reported by Zhang et al. (2013). They studied three batches of plasma samples, collected by using the same SOPs, which included healthy age-matched control samples and samples of women with breast cancer. From these batches, they found 1,423, 1,389, and 1,249 total proteins, while 246 proteins were found to be common between the three datasets. By applying the NNW method, they initially built a model of 60 markers, while the best five markers were used to determine the classification accuracy. By splitting the databases in training and testing sets to overcome the overfitting problems, they achieved a classification accuracy of 85%, with sensitivity and specificity of 82.5% and 87.5%.

Prostate Cancer

Support Vector Machines

Prostate cancer, like ovarian and breast cancer, has been demonstrated to be a heterogeneous disease at genomic levels by Barbieri and Tomlins (2014). Therefore, multimarker strategies appear particularly promising for this type of tumor.

Through proteomic profiling, Neuhaus et al. (2013) studied the seminal plasma of 125 patients: 70 with prostate cancer, 21 with benign prostate hyperplasia (BPH), 25 with chronic prostatitis, and 9 healthy controls. They firstly identified three panels of peptides: (a) 21 peptides differently expressed in prostate cancer/BPH versus inflammatory and healthy prostate; (b) 5 peptides, differently expressed in prostate cancer and BPH; and (c) 11 peptides, differently expressed in organ-confined or advanced disease. They applied the support vector machine to construct algorithms based on the panels of peptides found and split the database in training and testing to avoid overfitting problems. By combining the 21 and 5 peptide panels, they obtained 83% sensitivity and 67% specificity for diagnosing prostate cancer versus other conditions, although these performances were as high as that obtained by PSA (87% sensitivity and 59% specificity). They obtained better results (AUC = 0.99) for the identification of advanced versus organ-confined diseases.

Pancreatic Cancer

Classification and Regression Tree

Diabetes mellitus (DM) has been demonstrated to be associated with pancreatic cancer (PDAC), especially when of new onset. More interestingly, a strong temporal association underlined that diabetes may be a risk factor for PDAC when long standing, whereas it is considered a consequence of the tumor when of recent onset. Therefore, some authors suggested that patients presenting new-onset DM should undergo screening for PDAC, being "high-risk" patients. Padoan et al. (2013) studied serum peptidome profiling of PDAC patients, compared to that of DM or healthy subjects and chronic pancreatitis (ChrPa) and gastric cancer patients. They used both single- and multimarker strategies, by applying univariate statistical analyses as well as CART analyses. Results were compared with the established biomarker for PDAC that is Ca 19-9. The best classification accuracy was obtained for distinguishing diabetic patients from PDAC and ChrPa patients, achieving a cross-validated (leave-one-out cross-validation) percentage of classification rate of 65.4%.

Multiple Logistic Regression

Univariate analyses, performed in the same study, showed that serum Apo-A1 and complement C3 were altered in PDAC patients. These two candidate biomarkers were determined by a nephelometric assay, and the results were validated by using multiple logistic regressions. A logistic model containing Apo-A1 and C3 underlined that they performed better than CA 19-9 alone in discriminating PDAC patients with respect to diabetes mellitus patients.

Liver Fibrosis

Multimarker Panel

There is a need for noninvasive markers of liver fibrosis both for the huge number of patients with chronic liver diseases and for the limitations of the liver biopsy, which is still considered the gold standard for the diagnosis. In fact, liver biopsy not only shows a high incidence of clinical complications but is also affected by significant sampling and interpretative errors (Afdhal and Nunes, 2004).

The search of an individual biochemical marker with high diagnostic specificity and sensitivity (the holy grail) did not lead to definitive results. On the contrary, the combination of different and common biochemical tests such as the Fibrotest, APRI, and Forns (Forns et al., 2002; Imbert-Bismut et al., 2001; Wai et al., 2003) was found to assure satisfactory clinical performances. However, any single biochemical panel, per se, presents well-defined limitations. After the diagnostic accuracy of each individual method was first assessed using liver biopsy as the gold standard, they were combined with the aim of defining stepwise algorithms of higher diagnostic accuracy; these were to be used to identify cases with significant fibrosis and those with cirrhosis among patients presenting with chronic HCV infection.

The noninvasive marker panels evaluated include the AST-to-platelet ratio index (APRI) proposed by Wai et al. (2003), the Forns' index based on age, platelets, γGT, and cholesterol and more sophisticated models like Fibrotest. Fibrotest is based on five serological parameters including bilirubin, γGT, apolipoprotein A1 (ApoA1), alfa-2-macroglobulin (A2M), and haptoglobin. These different methods usually have been applied individually in the different validation studies. All of them have limitations. APRI and the Forns' index leave many patients unclassified, while Fibrotest is more expensive and uses two uncommon parameters. Furthermore, the diagnostic accuracy of most methods does not exceed 80%–85%. Therefore, APRI, the Forns' index, and Fibrotest have been combined in a consecutive series of patients with chronic hepatitis C in an attempt to improve diagnostic accuracy (Sebastiani et al., 2006). Algorithms have been developed that optimized diagnostic performances in the clinical management of patients with chronic HCV infection. In fact, the algorithm for detecting significant fibrosis in patients with elevated alanine transaminase (ALT) reached 94% accuracy, with no underestimation of fibrosis and limited overestimation (5%–6%). Similar performances were calculated for the algorithm developed for identifying significant fibrosis in patients with persistently normal ALT. The algorithm developed to identify cirrhosis also had 93%–95% accuracy and required liver biopsy in only around one-third of cases.

CONCLUSIONS

The increasingly advocated use of multimarker strategies for improving the early detection, diagnosis, prognosis, and monitoring of human diseases represents a major challenge, including the identification of valuable reporting strategies. Proteomics, like other "-omics" sciences, generates massive quantities of data and requires appropriate data-mining strategies to discover the most meaningful parts of these data. This process, which is like "looking for a needle in a haystack," often requires that biology, biochemistry, statistics, and bioinformatics share knowledge and responsibilities.

Moreover, the increased number of provided results may generate more confusion than added value. In fact, most of the results obtained from many studies failed to demonstrate a marked superiority with respect to the already established biomarkers. However, the explanation of these unsatisfactory results is manifold. At least in some cases, "-omics" data could not contain any valuable information helpful to predict patients' disease status; otherwise, data may contain a large quantity of noise, which overlaps with true signals, for example, for high biological variability. Other explanations are possible. "Heterogeneity" may really foster the inability of proteomics/genetics profiling to untangle the key elements that would, if known, help in predicting which patients will be diagnosed with a given disease.

The future of biomarker panels involves cooperation and collaboration between different professionals, such as basic researchers, clinicians, laboratory workers, and bioinformaticians, to translate potentially powerful technologies into clinical practice.

REFERENCES

Afdhal, N.H., Nunes, D., 2004. Evaluation of liver fibrosis: a concise review. Am. J. Gastroenterol. 99, 1160–1174.

Badgwell, D., Bast, R.C., 2007. Early detection of ovarian cancer. Dis. Markers 23, 397–410.

Barbieri, C.E., Tomlins, S.A., 2014. The prostate cancer genome: perspectives and potential. Urol. Oncol. 32, 53 e15–22.

Barla, A., Jurman, G., Riccadonna, S., Merler, S., Chierici, M., Furlanello, C., 2008. Machine learning methods for predictive proteomics. Brief Bioinform. 9, 119–128.

Colantonio, D.A., Chan, D.W., 2005. The clinical application of proteomics. Clin. Chim. Acta 357, 151–158.

Finn, R.D., Mistry, J., Tate, J., Coggill, P., Heger, A., Pollington, J.E., et al., 2010. The Pfam protein families database. Nucleic Acids Res. 38, D211–D222.

Forns, X., Ampurdanès, S., Llovet, J.M., Aponte, J., Quintó, L., Martínez-Bauer, E., et al., 2002. Identification of chronic hepatitis C patients without hepatic fibrosis by a simple predictive model. Hepatology 36, 986–992.

Hsu, F., Pringle, T.H., Kuhn, R.M., Karolchik, D., Diekhans, M., Haussler, D., et al., 2005. The UCSC Proteome Browser. Nucleic Acids Res. 33, D454–458.

Huang, R.Y., Chen, G.B., Matsumura, N., Lai, H.-C., Mori, S., Li, J., et al., 2012. Histotype-specific copy-number alterations in ovarian cancer. BMC Med. Genomics 5, 47.

Imbert-Bismut, F., Ratziu, V., Pieroni, L., Charlotte, F., Benhamou, Y., Poynard, T., et al., 2001. Biochemical markers of liver fibrosis in patients with hepatitis C virus infection: a prospective study. The Lancet 357, 1069–1075.

Karlsen, M.A., Sandhu, N., Høgdall, C., Christensen, I.J., Nedergaard, L., Lundvall, L., et al., 2012. Evaluation of HE4, CA125, risk of ovarian malignancy algorithm (ROMA) and risk of malignancy index (RMI) as diagnostic tools of epithelial ovarian cancer in patients with a pelvic mass. Gynecol. Oncol. 127, 379–383.

Kirchhoff, C., Osterhoff, C., Habben, I., Ivell, R., Kirchloff, C., 1990. Cloning and analysis of mRNAs expressed specifically in the human epididymis. Int. J. Androl. 13, 155–167.

Li, J., Zhang, Z., Rosenzweig, J., Wang, Y.Y., Chan, D.W., 2002. Proteomics and bioinformatics approaches for identification of serum biomarkers to detect breast cancer. Clin. Chem. 48, 1296–1304.

Moore, R.G., Brown, A.K., Miller, M.C., Skates, S., Allard, W.J., Verch, T., et al., 2008. The use of multiple novel tumor biomarkers for the detection of ovarian carcinoma in patients with a pelvic mass. Gynecol. Oncol. 108, 402–408.

Neuhaus, J., Schiffer, E., von Wilcke, P., Bauer, H.W., Leung, H., Siwy, J., et al., 2013. Seminal plasma as a source of prostate cancer peptide biomarker candidates for detection of indolent and advanced disease. PLoS ONE 8, e67514.

Padoan, A., Seraglia, R., Basso, D., Fogar, P., Sperti, C., Moz, S., et al., 2013. Usefulness of MALDI-TOF/MS identification of low-MW fragments in sera for the differential diagnosis of pancreatic cancer. Pancreas 42, 622–632.

Pencina, M.J., D'Agostino, R.B., D'Agostino, R.B., Vasan, R.S., 2008. Evaluating the added predictive ability of a new marker: from area under the ROC curve to reclassification and beyond. Stat. Med. 27, 157–172.

Pencina, M.J., D'Agostino, R.B., Vasan, R.S., 2010. Statistical methods for assessment of added usefulness of new biomarkers. Clin. Chem. Lab. Med. 48, 1703–1711.

Plebani, M., Laposata, M., 2006. Translational research involving new biomarkers of disease: a leading role for pathologists. Am. J. Clin. Pathol. 126, 169–171.

Plebani, M., Melichar, B., 2011. ROMA or death: advances in epithelial ovarian cancer diagnosis. Clin. Chem. Lab. Med. 49, 443–445.

Ransohoff, D.F., 2004. Rules of evidence for cancer molecular-marker discovery and validation. Nat. Rev. Cancer 4, 309–314.

Ransohoff, D.F., 2005. Bias as a threat to the validity of cancer molecular-marker research. Nat. Rev. Cancer 5, 142–149.

Sebastiani, G., Vario, A., Guido, M., Noventa, F., Plebani, M., Pistis, R., et al., 2006. Stepwise combination algorithms of non-invasive markers to diagnose significant fibrosis in chronic hepatitis C. J. Hepatol. 44, 686–693.

Wai, C.-T., Greenson, J.K., Fontana, R.J., Kalbfleisch, J.D., Marrero, J.A., Conjeevaram, H.S., et al., 2003. A simple noninvasive index can predict both significant fibrosis and cirrhosis in patients with chronic hepatitis C. Hepatology 38, 518–526.

Wang, J., Gao, J., Yao, H., Wu, Z., Wang, M., Qi, J., 2014. Diagnostic accuracy of serum HE4, CA125 and ROMA in patients with ovarian cancer: a meta-analysis. Tumour Biol. 35, 6127–6138.

Wu, C.H., Apweiler, R., Bairoch, A., Natale, D.A., Barker, W.C., Boeckmann, B., et al., 2006. The Universal Protein Resource (UniProt): an expanding universe of protein information. Nucleic Acids Res. 34, D187–191.

Zhang, Z., Chan, D.W., 2010. The road from discovery to clinical diagnostics: lessons learned from the first FDA-cleared in vitro diagnostic multivariate index assay of proteomic biomarkers. Cancer Epidemiol. Biomarkers Prev. 19, 2995–2999.

Zhang, F., Chen, J., Wang, M., Drabier, R., 2013. A neural network approach to multi-biomarker panel discovery by high-throughput plasma proteomics profiling of breast cancer. BMC Proc. 7, S10.

Chapter 3.7.1

Cardiovascular Biomarkers: Translational Aspects of Hypertension, Atherosclerosis, and Heart Failure in Drug Development

A. Rogier van der Velde, Wouter C. Meijers and Rudolf A. de Boer

HYPERTENSION

Introduction

Hypertension is one of the most common cardiovascular disorders; over the age of 60 years, around 60%–70% of the population is affected (Egan, Zhao, and Axon, 2010). Hypertension is defined as systolic blood pressure ≥ 140 mmHg and diastolic blood pressure ≥ 90 mmHg (Chobanian et al., 2003). Antihypertensive drugs are still among the most prescribed drugs by clinicians (Egan et al., 2010). Despite all the available treatment options, disease management is still not adequate. Only 50% of patients with hypertension have their blood pressure regulated properly (below 140/90 mmHg) (James et al., 2014), and often compliance to therapy for this (mostly) asymptomatic condition is low (Wang and Vasan, 2005). This low compliance can partially be explained by side effects of therapy that interfere with quality of life. In the end, hypertension remains one of the most important risk factors for stroke and heart disease (Greenland et al., 2003).

Hypertension is often regarded as a single disorder, but this is not the case. Instead, it is a complex and multifactorial disease resulting from the interaction of multiple environmental and genetically determined homeostatic control mechanisms (Takahashi and Smithies, 2004), and its etiology can be primary or secondary. The exact cause of primary (or essential) hypertension is unknown, but includes an increased activity of the renin–angiotensin–aldosterone system and increased activity of the sympathetic neural system (Staessen et al., 2003). The central role of the kidneys in blood pressure control is generally accepted, and high levels of sodium intake are associated with high blood pressure (Chien et al., 2008). Secondary hypertension can have several causes, including primary renal disease, endocrine disorders, or drug-induced causes.

Animal Models of Hypertension

The use of animal models that mimic the human hypertensive situation may offer valuable information in understanding the different routes of pathophysiology and studying new therapeutic interventions. In experimental hypertension research, several animal models exist, and rats are the most commonly used animals (Dornas and Silva, 2011). Animal models can be classified in two large groups: animals with primary hypertension and animals with secondary hypertension. Among animals with primary hypertension, genetically modified animals are most popular. One of the most used models, a phenotype-driven model derived from the Wistar strain, is the spontaneous hypertensive rat (SHR). In this model rats develop hypertension at 4–6 weeks after birth. The Dahl salt-sensitive rat is another example of a phenotype-driven model. These rats are derived from Sprague–Dawley rats and become hypertensive with a normal salt diet. Transgenic rat models are also available: TGR(mREN2)27 (REN2) transgenic rats show an overexpression of renin and therefore develop hypertension (Mullins, Peters, and Ganten, 1990).

Other nongenetic animal models of primary hypertension include the "cold model" and the "stress model". Different kinds of sensory stimulation, such as flashing lights or loud noises, can stress the animal, thus causing hypertension (Zimmerman and Frohlich, 1990). Also cold temperature (around 5°C) leads to hypertension in animals within 3 weeks (Sun, Cade, and Morales, 2002).

Principles of Translational Science in Medicine. http://dx.doi.org/10.1016/B978-0-12-800687-0.00018-9

TABLE 3.10 Animal models of hypertension and their characteristics

Rat model	Characteristic features
Primary hypertension	
REN2 rat	Overexpression of the renin gene
Spontaneously hypertensive rat (SHR)	Increased sympathetic tone
Dahl salt-sensitive rat	Fluid retention by salt administration
Cold/stress model	Stressor-induced hypertension
Secondary hypertension	
DOCA-salt hypertension rat	Salt retention and impaired renal function give rise to fluid retention
1-clip 2-kidney rat	Increased renin levels
Aortic banding	High aortic pressures

Secondary hypertension may be induced via surgery, for instance, in the 2K1C model (2 kidneys, 1 clip). In this model, unilateral constriction of the renal artery is accomplished, which will lead to a chronic increase of blood pressure within 2 weeks. This procedure can be performed in different kinds of animals (Wiesel et al., 1997). But there are also pharmacological options to create secondary hypertension. The DOCA-salt method is widely used. In this method a mineralocorticoid (deoxycorticosterone acetate) is administered to the animal (Terris et al., 1976), which in turn leads to increased RAAS activity and subsequent hypertension. (See Table 3.10).

Biomarkers of Hypertension: Clinical Markers of Hypertension

Hypertension may induce end organ damage in several organs like the heart (left ventricular hypertrophy), the kidneys (microalbuminuria), the brain (stroke), the eyes (retinal damage), and the vascular system (increased vascular wall thickness and arterial stiffness). A reliable method for measuring blood pressure is a 24-hour ambulatory blood pressure measurement, especially when there is a suspicion for white-coat hypertension or episodic hypertension (Weber et al., 2014). It is also possible to execute supportive studies on hypertension, for instance, to obtain myocardial wall thickness with MRI or echocardiography to assess left ventricular hypertrophy. Other options include measurements of pulse wave velocity in the aorta by means of carotid-femoral pulse wave analysis to estimate vascular stiffness. Ultrasound examination of carotid intima-media thickness can be used to measure vascular wall thickness. Furthermore, urine samples can be investigated for microalbuminuria, and retinal photographs can be made to evaluate retinal changes (Touyz and Burger, 2012).

Techniques for blood pressure measurements have improved in the past few years. A couple of techniques can be used, which also include nonstop measurements. The most used technique is an indirect method, the cuff technique, using a tail or limb to measure cuff pressure (Kurtz et al., 2005). This method has the disadvantage that animals should be restrained, which in theory could affect the blood pressure. Radiotelemetry techniques (catheters) are also available to perform direct measurements. Connected to external transducers, this technique can be used for nonstop blood pressure measurements in nonstressed, unrestrained animals (Pickering et al., 2005), and it resembles the human 24-hour ABPM.

When one is selecting a method for an animal study, indirect methods for measuring blood pressure are best suited for the screening of large numbers of animals to measure systolic hypertension and to show substantial group differences. This is both suitable for single measurements as well as measurements over time. As a direct method, radiotelemetry is recommended for quantification of the magnitude of hypertension, in quantifying relationships between blood pressure and other variables, or in measuring blood pressure variability (Kurtz et al., 2005). This method is also recommended when studying pharmacokinetics because nonstop measurements are available. It should be noted that during blood pressure measurements, all kinds of anesthesia should be avoided, since this clearly affects cardiovascular function (Vatner, 1978).

Biomarkers of Hypertension: (Blood-Borne) Biomarkers

RAAS Activity

Renin is a component of the renin–angiotensin–aldosterone system (RAAS) and is released during renal hypoperfusion or increased sympathetic activity (Nushiro, Ito, and Carretero, 1990). However, in hypertension, patients either are in a "low-renin state" or a "high-renin state," as proposed by Laragh and Sealey (2011). Low renin hypertension is a state in which sodium increases to a certain level, so the RAAS system becomes "switched off," and hypertension develops by volume retention. These patients will mostly benefit from diuretics or calcium antagonists (Alderman et al., 2010). Whereas in a "high renin state," renin is secreted in too high amounts relative to sodium levels, and patients will benefit more from ACE inhibitors. This implies that renin (or plasma renin activity) is not of use to assess severity of hypertension, but can help to distinguish what kind of therapy will be useful for the patient (Turner et al., 2010).

Endothelial Dysfunction

Endothelial dysfunction is one of the detrimental effects of hypertension (Quyyumi and Patel, 2010). Healthy endothelial vessel walls become stiffer, and vasorelaxation is diminished. As such, measures of endothelial dysfunction can be used as a marker for hypertension. One of the main causes for endothelial dysfunction is the reduced availability of nitric oxide (NO), resulting in increased oxidative stress in the endothelium (Feng and Hedner, 1990). Endothelial dysfunction can also precede hypertension, and less compliant vessels will establish higher blood pressures (Park and Schiffrin, 2001). Eventually, endothelial dysfunction leads to a proinflammatory, prothrombotic, vasoconstrictive state with increased cell adhesion and oxidative stress (Flammer and Luscher, 2010).

Markers of Vascular Resistance

As mentioned, NO is a marker of vascular resistance. Increased levels of NO inhibit oxidative stress and promote vasodilation and vascular health (Kawashima, 2004) and therefore can be used as an estimate for hypertension. Degradation products of NO, such as nitrite or nitrate, can also be used as surrogate measures (Shiekh et al., 2011). Prostacyclin is another marker of vascular resistance and is a member of the prostaglandin family. Prostacyclin is also produced by endothelial cells, determines vasorelaxation, and inhibits coagulation (Frolich, 1990). Its metabolite, 6-ketoprostaglandin F1α, is used mostly as an estimate for prostacyclin bioavailability and correlates with endothelial function (Doroszko, Andrzejak, and Szuba, 2011; Touyz and Burger, 2012).

Markers of Inflammation

Hypertension promotes the release of cytokines and is accompanied with low-grade inflammation (Savoia and Schiffrin, 2007). Therefore, markers of inflammation have been studied for their use as biomarkers for hypertension. It is thought that individuals with inflammatory activation represent a more severe hypertensive state, implicating that inflammation might play a role in the pathophysiology of hypertension (Harrison et al., 2011). TNF-α is an important cytokine released during inflammation. It is secreted by monocytes, macrophages, and vascular cells (Locksley, Killeen, and Lenardo, 2001). Its levels are increased in hypertension (Chrysohoou et al., 2004), and reduction of blood pressure decreases TNF-α levels (Koh et al., 2003). IL-6, another proinflammatory cytokine produced by macrophages and vascular wall cells, is also elevated in patients with hypertension and normalized after blood pressure reduction (Chae et al., 2001; Vazquez-Oliva et al., 2005). Cell adhesion also plays an important part in inflammation, providing attachment of inflammatory cells to the tissue. Levels of these cell adhesive molecules, for example, vascular cell adhesion molecule (VCAM) and intercellular adhesion molecule (ICAM), are also increased in patients with hypertension, and levels of these molecules have shown to correlate with severity of hypertension (Preston et al., 2002).

Markers of Coagulation

Hypertension results in endothelial dysfunction, which is reflected by a prothrombotic state (Endemann and Schiffrin, 2004). In this regard, factors of coagulation reflect endothelial (dys)function or hypertension. For instance, von Willebrand factor plasma levels are elevated in patients with hypertension and correlate with severity of endothelial dysfunction (Spencer et al., 2002). Also other factors of coagulation like PAI-1, which inhibits fibrinolysis, or P-selectin, involved in platelet activation, are increased in hypertension and reduced by antihypertensive treatment (Binder et al., 2002; Spencer et al., 2002).

MicroRNAs

MicroRNAs are small, noncoding nucleotides, which regulate gene expression at the post-transcriptional level and can regulate multiple downstream genes. Deregulation of microRNA results in diminished function (Jamaluddin et al., 2011). Out of the endothelial wall, multiple microRNAs are expressed. Among them, the most valuable ones in hypertension are miR-126, miR-217, miR-15, miR-21, and miR-122 (Batkai and Thum, 2012; Kriegel, Mladinov, and Liang, 2012). Other micro RNAs, like miR-155, are targeting the RAAS system, whereas miR-29b is involved in the fibrotic pathway. MicroRNAs are a source of new potential targets; however, these targets still need to be validated in clinical studies (Touyz and Burger, 2012). (See Table 3.11).

TABLE 3.11 Biomarkers of hypertension

Clinical biomarkers	Blood pressure measurements
	Arterial stiffness
	LV mass
Blood-borne biomarkers	
RAAS activity	Renin
Vascular resistance	Nitric oxide
	Prostacyclin
Inflammation	TNF-α
	IL-6
	ICAM/VCAM
Coagulation	Von Willebrand factor
	PAI-1
	P-selectin
MicroRNAs	miR-126, miR-217, miR-15, miR-21, miR-155, miR-29b

Conclusion

To study hypertension in animal models, appropriate, well-studied models are available. They are an accepted measure for registration. However, the research area of hypertension is not very dynamic. The reason may be that a good clinical marker is already available: blood pressure measurements. It is easy to obtain and is noninvasive, and 24-hour measurements are a reliable measure of hypertension. Therefore, the need for other biomarkers may be limited in the field of hypertension. A good blood-borne biomarker for hypertension is missing, but one could question if this is really necessary. The potential value of blood-borne biomarkers could lie in the fact that they offer insights in the pathogenesis of hypertension. Renin or aldosterone could be of additional interest for choosing therapy; however, it does not reflect severity of hypertension. Prothrombotic or proinflammatory markers could be of use, but their specificity is questionable. MicroRNAs could be potential biomarkers of hypertension or may act as a therapeutic target. Clinical studies are needed to unravel the potential of these relatively new markers in hypertension.

ATHEROSCLEROSIS

Introduction

Atherosclerosis is a chronic inflammatory disease, characterized by lipid-containing inflammatory lesions of large- and medium-sized arteries, mostly occurring at sites with disturbed laminar flow such as branch points (Moore and Tabas, 2011). Via this process of slow progressive lesion formation, atherosclerosis gives rise to cerebrovascular and coronary artery disease and luminal narrowing of arteries. Infiltrated immune cells, like macrophages, take up cholesterol and are transformed into foam cells. They form a necrotic core in the growing atherosclerotic plaque. In the development of these atherosclerotic plaques, three different stages can be distinguished. First, major lipid accumulation occurs, which is accompanied by reduced amounts of smooth muscle cells; this is the vulnerable plaque stage. Fibroblasts will produce collagen to stabilize the plaque. This stage is characterized by a thin fibrous cap, which separates the blood stream from the lipid core. At this time the fibrous cap is prone to rupture if subjected to strain. Second, this plaque grows more extensive and is now obstructing the vascular lumen—the stable plaque stage. Because this plaque contains collagen-producing cells, it is stable, and plaque rupture or thrombosis does not occur. This kind of plaque may lead to signs or symptoms of angina pectoris. The third stage is the erosive plaque stage. This stage is characterized by formation of thrombosis, and acute myocardial infarction may occur at this stage.

TABLE 3.12 Differences in the phenotype and response of LDLR-/- and apoE-/- mice

	LDLR$^{-/-}$	apoE$^{-/-}$
Develops atherosclerosis on chow	>1 year	Yes
Prominent lipoproteins	VLDL, LDL	Remnants
Effect of hepatic lipase deficiency	Increased atherosclerosis	Decreased atherosclerosis
Correlation VLDL levels with aortic root atherosclerosis	Yes	No

Animal Models

In the study of atherosclerosis, the mouse is the most frequently used species. Nevertheless, heart rate, total plasma cholesterol, sites of atherosclerosis, and the development time differ between mice and human. The most commonly used mice models are those in which genetic ablation of apoE or the LDL receptor is used. Disruption of the apoE gene causes increased levels of circulating cholesterol without any changes in diet and rapid progression of atherosclerosis. An atherogenic diet would further exacerbate plaque growth (Zadelaar et al., 2007). The LDL-receptor knockout requires an atherogenic diet to develop atherosclerotic lesions. Also a “double knock” exists, which is a crossing between the ApoE and the LDL-receptor knockouts. Some variation exists between these two models, which are summarized in Table 3.12.

Limitations of these different animal models are also present: high cholesterol levels are needed to induce atherosclerosis (small lesions); plaque rupture does not occur; and cholesteryl ester transfer protein (CETP), a potential target in humans, is not expressed in mice. Except for these limitations, it should be stated that new treatment options tested in these models are administered in early stages of lesion formation, whereas in real life, treatment would be started after a cardiovascular event and not in early development of atherosclerotic lesions.

Biomarkers for Atherosclerosis

A very important feature in atherosclerosis is the switch to plaque destabilization or rupture, because this will ultimately lead to an atherosclerotic event. Many biomarkers have been studied to adequately predict this phenomenon.

CRP

C-reactive protein (CRP) is a member of the pentraxin family and one of the most studied proinflammatory molecules. It is synthesized by hepatocytes, and the production is primarily stimulated by interleukin-6 (IL-6). Recently, local CRP production in the atherosclerotic plaques also was suggested (Burke et al., 2002). Epidemiological studies have provided evidence for CRP to predict future cardiovascular risk in different patient cohorts (healthy, stable angina, acute coronary syndrome, post-myocardial infarction, and metabolic syndrome). In the EPIC-Norfolk study, CRP was among the strongest variables predicting risk of coronary heart disease (CHD) (Boekholdt et al., 2006).

IL-6

IL-6 is a single chain glycoprotein, produced by monocytes, endothelial cells, and adipose tissue (Rattazzi et al., 2003). Injection of IL-6 in mice led to enhanced fatty lesion development (Huber et al., 1999). Levels of IL-6 are higher at sites of coronary plaque rupture compared to IL-6 in systemic circulation (Maier et al., 2005). The clinical importance of IL-6 is based on the predictive power and the fact that patients with elevated levels might benefit the most from an early invasive strategy (Lindmark et al., 2001).

IL-18

IL-18 is a proinflammatory cytokine, expressed in many cell types (Gracie, Robertson, and McInnes, 2003). It enhances expression of MMPs (Ishida et al., 2004) and induces the immune response, both important in atherosclerosis formation. In apoE knockout mice, inhibition of IL-18 led to less atherosclerosis. Increased levels of IL-18 are associated with plaque destabilization and vulnerability (Mallat et al., 2001). Until now, no prognostic value has been described regarding elevated IL-18 levels (Blankenberg, Barbaux, and Tiret, 2003).

Oxidized LDL (Ox-LDL)

Ox-LDL plays a central role in the development and progression of atherosclerosis. Oxidative modification is seen as the most significant event in early lesion formation. It is hypothesized that ox-LDL is crucial in the transition from stable to vulnerable plaque. This function could be due to the endothelial receptor LOX-1 (Li et al., 2003), which can induce MMP-1 an MMP-9. In large surveys, ox-LDL was the strongest predictor of events in CHD.

Glutathion Peroxidase

Glutathion peroxidase (GPx) is a selenium-containing enzyme with antioxidant properties. In GPx knockout mice, oxLDL levels were increased (Guo et al., 2001). Furthermore, activity of GPx is inversely associated with the risk of future fatal and nonfatal CV events in patients with CHD (Blankenberg et al., 2003a).

Lipoprotein-Associated Phospholipase A2

Lipoprotein-associated phospholipase (Lp-PLA2) is a new and emerging biomarker for atherosclerosis. It is secreted by monocytes, macrophages, T-lymphocytes, and mast cells. It is upregulated in coronary lesions that are prone to rupture (Zalewski and Macphee, 2005). Lp-PLA2 is mainly bound to LDL and can enhance LDL oxidation. It is regarded as an important factor that is able to turn LDL into ox-LDL and induce a local inflammatory response. Clinically, Lp-PLA2 was associated with coronary and cerebrovascular events, independent of different potential confounders (Oei et al., 2005). More importantly, this enzyme could serve as a therapeutic target. Azetidiones and pyromidones seem to have clinical benefit via inhibition of Lp-PLA2.

Matrix Metalloproteinases

Matrix metalloproteinases (MMPs) are crucial in providing a balance in extracellular matrix remodeling. MMPs are widely expressed in monocytes/macrophages, endothelial cells, smooth muscle cells, fibroblasts, and neoplastic cells (Jones, Sane, and Herrington, 2003). MMPs mostly emerge at the shoulder region of the fibrous cap (Galis et al., 1994), which hypothetically leads to destabilization of the atherosclerotic plaque. MMP-9 at baseline is also associated with CV death (Blankenberg et al., 2003b).

Monocyte Chemoattractant Protein-1

Monocyte chemoattractant protein-1 (MCP-1) is the most important chemokine that regulates migration and infiltration of monocytes/macrophages. MCP-1 is highly expressed in atherosclerotic plaques in response to cytokines, growth factors, oxLDL, and CD40L. In experimental studies, MCP-1 was significantly correlated with the amount of atherosclerosis and macrophage infiltration (Namiki et al., 2002). MCP-1 gene therapy has shown to inhibit both atherosclerosis as well as macrophage infiltration in the apoE-knockout mice (Inoue et al., 2002). Although in a large dataset MCP-1 elevation preceded CHD events, unfortunately no independent risk prediction was observed (Herder et al., 2006).

Imaging

Measurement of atherosclerotic progression is an ideal surrogate marker because it is predictive of future cardiovascular events. There are many vascular imaging technologies, both established and emerging, that permit investigators to collect information on vascular structure and development of atherosclerosis. Four vascular imaging technologies can be used to show the atherosclerotic lesion: ultrasound, magnetic resonance imaging, computed tomography, and positron emission tomography. Importantly, atherosclerosis is a systemic arterial disorder, as documented in numerous postmortem studies. Patients who develop atherosclerosis in one part of the vascular bed (e.g., heart) will also develop this in other vascular beds (e.g., limbs).

Carotid Intima-Media Thickness

Carotid intima-media thickness (CIMT) was one of the first available methods to display the atherosclerotic lesion and is commonly used in clinical trials as an endpoint (Thoenes et al., 2007). Intimal thickening is one of the first signs of atherosclerosis and can be measured with CIMT. This B-mode ultrasound can identify the boundaries of the tunica intima and the adventitia media. It is a noninvasive and relatively cheap technique to use. CIMT predicts vascular events in several patient cohorts (Simons et al., 1999), and reduction of CIMT leads to a reduction in cardiovascular events (Crouse et al., 2007). However, hypertension or aging also could be the main reason for intima thickening and therefore does not represent atherosclerotic progression, which is regarded as a major limitation of CIMT. CIMT can also not be used to predict coronary events (O'Leary et al., 1999).

Contrast-Enhanced Ultrasound

Contrast-enhanced ultrasound (CEUS) gives additional information about the plaque composition and inflammation state compared to CIMT (Tsutsui et al., 2004). CEUS is able to show the microvasculature within the plaque, which is associated with rupture. This technique could be used in a clinical setting to monitor therapy or select patients eligible for intervention.

Intravascular Ultrasound

The catheter-based intravascular ultrasound (IVUS) technique is able to detect the presence of atherosclerosis more sensitively and measures the extent of plaque formation more precisely compared to CIMT and CEUS. Nowadays, IVUS is used in clinical trials to monitor disease progression and treatment. The CETP inhibitor, torcetrapib, showed no regression with IVUS but did have LDL-lowering effects and HDL elevation. This enhances the need for imaging techniques in clinical trials. A newly developed technique, the virtual-histology IVUS (VH-IVUS), is capable of detecting fibrous, fibrolipidic, calcified, and calcified-necrotic regions from plaques (Nasu et al., 2006). However, a huge limitation of this technique is the restricted invasiveness.

Magnetic Resonance Imaging

Magnetic resonance imaging (MRI) can provide information about different atherosclerotic lesions at the same time (Sanz and Fayad, 2008), primarily focusing on the carotid artery and the aorta. Using contrast weightings (T1, T2, and proton-density weightings), a lot of different information can be obtained about the plaque formation. After addition of gadolinium, more information can be gathered about the vascularity and inflammatory status of the plaque.

Computed Tomography

Calcium accumulation in the plaque is most easily detected with computed tomography (CT), and addition of coronary artery calcium scoring to current clinical risk-scoring systems has been shown to provide additional predictive information (Detrano et al., 2008). The amount of coronary calcium detected with CT was significantly correlated with the amount of calcium that was histologically present (Sangiorgi et al., 1998). CT is able to differentiate between lipid-rich and fibrous plaques, and combined with CT/SPECT it is likely that this can further enhance plaque imaging. The drawback of CT imaging is radiation exposure, which would limit the use (for serial measurements).

Optical Coherence Tomography

Optical coherence tomography (OCT) is a catheter-based technique using infrared light for tissue characterization. Although limited in depth of penetration, this technique has a very high resolution of the plaque. Recently, this technique was shown to be able to discriminate between plaque rupture and erosion in humans in vivo.

Conclusion

Atherosclerosis is mostly studied in mice. However, drug development in these models of atherosclerosis is limited due to the large biological differences that exist between mice and humans in the development of atherosclerosis. When one is choosing an animal model to study atherosclerosis, genetic ablation models of apoE or the LDL receptor are the models of choice. Human trials showed that besides blood-borne biomarkers, imaging techniques could help to identify effective treatment options. Blood-borne biomarkers do have their drawbacks; however, imaging techniques provide noninvasive serial measurements that can be performed easily, are reliable, and can be executed at low cost.

HEART FAILURE

Introduction

Heart failure is a clinical syndrome characterized by an impaired ability of the heart to eject or fill with blood. Due to an aging population and better treatment options, there is an increasing prevalence of heart failure, and this represents a new epidemic in cardiovascular disease: it affects nearly 23 million people worldwide nowadays (Liu and Eisen, 2014) or > 10% of people above 70 years old (Mosterd and Hoes, 2007). The main causes of heart failure include ischemic heart disease, dilated cardiomyopathy, valvular heart disease, and hypertensive heart disease (Baldasseroni et al., 2002). Because of increasing prevalence, heart failure remains an interesting topic for pharmaceutical companies, and therefore, many clinical trials are conducted in heart failure patients. Despite this extensive research, mortality and morbidity remain very high and have not improved much during recent years, with a one-year mortality around 30% after hospitalization for heart failure (Kosiborod et al., 2006; Liu and Eisen, 2014).

Animal Models in Heart Failure

Several animal models can be used to study different kinds of heart failure. Each model represents a different etiology of heart failure, which tries to mimic the human conditions of heart failure. However, identical replications of human heart failure are scarce.

Hypertensive Heart Disease

Hypertensive heart disease is characterized by persistent pressure overload, concentric hypertrophy, with a normal ejection fraction and normal or decreased end-diastolic volume (Levy et al., 1990). These patients are more often female, are older, have a higher BMI, and more often have diabetes. With persistent, prolonged hypertension, this form of compensated hypertrophy makes the transition to heart failure with an increase in LV end-systolic and end-diastolic volumes, decreased LVEF, increased LV chamber stiffness, and impaired LV relaxation (Frohlich et al., 1992). Hypertensive heart disease can induce heart failure without a significant decrease in LVEF, and diastolic dysfunction is an important contributor in decompensated heart failure.

Animal models of hypertensive heart disease include a canine nephrectomy model (Morioka and Simon, 1982), and other variants of this model are also currently used (Shapiro et al., 2008). Models using smaller animals include the SHR (Pfeffer et al., 1982), the TGR(mREN2)27 rat (Pinto et al., 2000), aortic banding (Litwin et al., 1995), or genetically modified mice (Molkentin et al., 1998).

Valvular Lesions

Valvular lesions can result in heart failure. Aortic stenosis causes LV pressure overload and mitral regurgitation causes LV volume overload. Both conditions will eventually result in fluid retention and fatigue, caused by diastolic dysfunction, left ventricular hypertrophy, and eventually decompensated systolic failure.

Progressive aortic constriction has been performed in large animals such as pigs, dogs, and sheep (Houser et al., 2012) and highly resembles the features of human aortic constriction. Transverse aortic constriction (TAC) in the mouse is the most commonly used model of aortic constriction in small animals. In this model, fixed aortic constriction causes an acute increase in LV pressure. This may cause significant postoperative mortality due to acute hemodynamic instability (Teekakirikul et al., 2010). The model is suitable for research with genetically modified mice. However, TAC in mice is not a progressive disease like in humans. For studying progressive modifications in heart failure, large animal models are more suitable.

Few animal models are available to study mitral regurgitation, but there are canine models available to study this. This model results in severe contractile dysfunction, LV dilation, and eccentric LVH with myocyte lengthening. In smaller animals, like rodents, it was not possible to induce mitral regurgitation. Therefore, to study LV volume overload, induction of aortic insufficiency or aortocaval fistula has been described as an alternative (Truter et al., 2009; Wei et al., 2003).

Dilated Cardiomyopathies

Dilated cardiomyopathy (DCM) is characterized by dilatation of the ventricles, elevated filling pressures, eccentric myocyte hypertrophy, impaired diastolic filling, and systolic dysfunction (Beltrami et al., 1995). DCM is often caused by ischemic heart disease or long-term hypertension (Wexler et al., 2009).

Several models are available to study DCM in large animals. DCM can be induced by coronary ligation in pigs, dogs, and sheep (Blom et al., 2005; Jugdutt and Menon, 2004; van der Velden et al., 2004). Left coronary artery microembolization can be used in dogs and sheep (Monreal et al., 2004; Saavedra et al., 2002). Furthermore, it is also possible to induce DCM using pacing-induced tachycardia (Kaye et al., 2007) or after serial administration of intravenous doxorubicin (Bartoli et al., 2011).

In rodents, DCM can be induced using coronary ligation or myocardial infarction (Pfeffer et al., 1979). Furthermore, several transgenic overexpression and knockout models are available that cause DCM (Chu, Haghighi, and Kranias, 2002; Wang, Bohlooly-Y, and Sjoquist, 2004). Also, infusion of doxorubicin (Weinstein, Mihm, and Bauer, 2000) or isoproterenol (Oudit et al., 2003) will result in a DCM phenotype. The SHR is also useful to study DCM (Itter et al., 2004).

Restrictive Cardiomyopathies

In restrictive cardiomyopathy (RCM), a restrictive filling pattern exists, and this is accompanied by a normal ventricular ejection fraction. Relatively small increases in ventricular filling volumes are associated with exorbitantly increased diastolic pressures (Kushwaha, Fallon, and Fuster, 1997). Ventricular chamber size and wall thickness usually are normal. A common feature is excessive remodeling of the myocardial extracellular matrix, either by pathological protein deposition or replacement fibrosis caused by cardiomyocyte death (Kushwaha et al., 1997). Causes of RCM often include systemic diseases like sarcoidosis, scleroderma, amyloidosis, and hemochromatosis, but could also be radiation-induced or genetically induced (Kushwaha et al., 1997).

Rodents are the most popular animals to study RCM. In rodents, RCM can be induced by iron overload, eosinophilic myocarditis, scleroderma, radiation fibrosis, or amyloidosis (Bartfay et al., 1999; Bashey et al., 1993; Boerma and Hauer-Jensen, 2010; Hirasawa et al., 2007; Teng et al., 2001). Furthermore, hemojuvelin (JUV-/-) knockout mice exist that mimic the clinical features of hereditary hemochromatosis and therefore RCM (Huang et al., 2005). Also two different cTnI mutations have produced mouse strains that resemble familial RCM. (See Table 3.13).

TABLE 3.13 Animal models of heart failure

Etiology	Intervention	Species
Hypertensive heart disease	Nephrectomy	Dog
	Spontaneous hypertensive rat (SHR)	Rat
	REN2 rat	Rat
	Aortic banding	Rat
	Genetically modified	Mouse
Valvular lesions	Progressive aortic constriction	Dog, pig, sheep, cat
	Transverse aortic constriction	Mouse
	Mitral regurgitation	Dog
	Aortic insufficiency	Rabbit
	Aortocaval fistula	Rat
Dilated cardiomyopathy	Coronary ligation	Dog, pig, sheep, rat, mouse
	Coronary microembolization	Dog, sheep
	Toxic infusion (doxorubicin, isoproterenol)	Mouse, rat, dog, sheep, cow
	Pacing-induced tachycardia	Dog, pig, sheep
	Transgenic overexpression	Mouse
	Knockout models	Mouse
	Spontaneous hypertensive rat (SHR)	Rat
Restrictive cardiomyopathy	Spontaneously occurring RCM	Cat
	Amyloidosis	Monkey, cow, cat, mouse, rat
	Iron overload	Mouse, gerbil, guinea pig
	Radiation fibrosis	Rat
	Scleroderma	Mouse
	Eosinophilic myocarditis	Mouse
	Hereditary hemochromatosis	Mouse
	Sarcomeric protein mutations	Mouse

Clinical Markers for Heart Failure

Echocardiography

Left ventricular hypertrophy (LVH) is associated with development of HF with either HFpEF or HFrEF (Milani et al., 2011). Although the exact mechanism by which LVH impacts LV function is not totally known, LVH and myocardial fibrosis may adversely affect diastolic dysfunction. In a post hoc analysis of the TOPCAT study (Shah et al., 2014), including 3,445 HF patients with LVEF >45%, LV hypertrophy, elevated LV filling pressure, and higher pulmonary artery pressure were predictive for HF rehospitalizations and mortality. E/A ratio was the most strongly associated (diastolic) parameter with the composite outcome, especially an E/A ratio > 1.6. In this study neither LV volumes nor LVEF was predictive for worse outcome.

Six-Minute Walking Test

Six-minute walking distance (6MWD) measurements are often performed to assess exercise capacity. It has repeatedly been shown that a reduced (<300 meters) 6MWD is associated with increased mortality and morbidity in chronic HF patients (Davies et al., 2010; Shah et al., 2001). In a study evaluating hospital readmission due to disease progression, the readmission rate was reported to be significantly higher in chronic HF patients with a 6MWD < 200 meters compared to patients with a 6MWD > 200 meters (Alahdab et al., 2009).

(Blood-Borne) Biomarkers of Heart Failure

Heart failure is a multifactorial disease; therefore, several biomarkers do exist that reflect different pathophysiological mechanisms of heart failure. These biomarkers can be used to better predict diagnosis or prognosis of patients, or to better treat patients in a more personalized matter. Because heart failure is a complex disease, it is difficult to identify the perfect biomarker because multiple pathways are involved. Therefore, a combination of biomarkers from different biological processes may help to give insights into the subtype of heart failure.

Markers for Cardiac Remodeling

After myocardial injury, cardiac remodeling may occur. Cardiac remodeling is composed of various biological processes, including apoptosis, hypertrophy of cardiomyocytes, fibrosis, and inflammation. Established biomarkers of heart failure include natriuretic peptides (NT-proBNP), GDF-15, ST2, galectin-3, MR-proADM, and NGAL, and each biomarker represents a different spectrum of cardiac remodeling (Figure 3.12).

ST2 and galectin-3 are both markers that are involved in fibrosis. ST2 is released from cardiomyocytes and fibroblasts upon volume overload and consists of two isoforms: ST2L (a transmembrane receptor) and soluble ST2 (sST2). ST2L has antifibrotic and antihypertrophic effects and acts via IL-33 (Sanada et al., 2007). sST2 is a circulating form of ST2, which acts as a decoy receptor, and can bind IL-33 to block its effect. Via this pathway, sST2 is associated with poor prognosis in

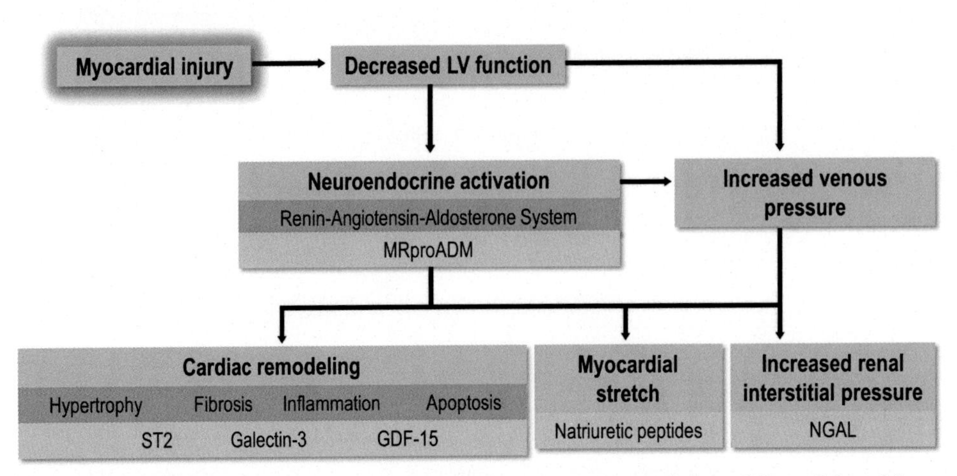

FIGURE 3.12 Biomarkers reflect the different pathophysiological mechanisms that are present in heart failure. *(Adapted from* van der Velde et al., 2014*).*

patients with HF (Shimpo et al., 2004); higher levels of sST2 identify patients with more progressive cardiac remodeling (Shah and Januzzi, 2010).

Galectin-3 is released during cardiac stress by activated macrophages in the heart and stimulates fibrosis. It is produced not solely in the heart, but in the entire body, and plays an important role in cell–cell adhesion. Besides fibrosis, galectin-3 is also involved in tumor growth, metastasis, and inflammation. As a fibrotic factor, galectin-3 has shown to predict outcome in patients with acute and chronic heart failure (Meijers et al., 2014) and has more prognostic value in patients with HFPEF (de Boer et al., 2011). After myocardial infarction (MI), galectin-3 can predict infarct size and LVEF (Mayr et al., 2013). In the general population, elevated galectin-3 levels can also identify patients who are at risk for developing cardiovascular disease (de Boer et al., 2012; Ho et al., 2012).

GDF-15 is upregulated in cardiomyocytes upon various cardiovascular events, such as ischemia or pressure overload. It is colocalized with macrophages, but also circulates through the bloodstream and is not cardio-specific (Kempf et al., 2006). GDF-15 can be regarded as a "good player," since it has shown to bear anti-inflammatory and antihypertrophic effects. GDF-15 has also shown to be elevated in heart failure and is an independent predictor in chronic HF patients, as well as in patients with HFPEF (Kempf et al., 2007). More prospective data are needed to confirm these results in patients with acute heart failure.

Markers for Myocardial Stretch

Myocardial stress is increased in heart failure, due to elevated filling pressures. As a result of elevated intra-cardiac pressures, natriuretic peptides are released from cardiomyocytes. N-terminal pro B-type natriuretic peptide (NT-proBNP) is now widely used as a diagnostic and prognostic marker in heart failure, and is regarded as the "gold standard." NT-proBNP plays a very important role in the emergency department to establish the diagnosis of heart failure, with a sensitivity of 92% and specificity of 84% (Maisel et al., 2008). Normal levels of NT-proBNP (< 100 pg/mL) basically exclude significant heart disease. NT-proBNP can also be measured to estimate risk in heart failure: a 50% reduction of NT-proBNP levels led to a 50% reduction in events (Januzzi et al., 2011). Moreover, NT-proBNP levels at discharge are better prognostic indicators, compared to plasma levels during hospital admission (Waldo et al., 2008).

Mid-regional pro-atrial natriuretic peptide (MR-proANP) is another type of natriuretic peptide that is mainly released from the atrial cardiomyocytes. MR-proANP is also a strong prognostic marker in heart failure, and can be compared to NT-proBNP (Gegenhuber et al., 2007). For long-term prognosis, MR-proANP showed to be an even better predictor than NT-proBNP in a head-to-head comparison (Seronde et al., 2013). Drawbacks of natriuretic peptide measurements are that they are affected by renal function and obesity. Renal failure will lead to increased plasma levels of natriuretic peptides (Mueller et al., 2005), and obese patients tend to show lower levels of natriuretic peptides via a mechanism that is not completely understood (Das et al., 2005).

Neurohormonal Activation

In heart failure, all kinds of neurohormonal processes are activated to compensate for the reduced cardiac output. Following increased cardiac wall stress, midregional pro-adrenomedullin (MR-proADM) is released by myocardial and endothelial tissue. MR-proADM induces natriuresis and vasodilatation, via cAMP and NO, to decrease afterload. This peptide hormone was found to be an important biomarker of heart failure; in a comparison with other biomarkers for heart failure, MR-proADM was the best predictor for 30-day mortality (Maisel et al., 2011a). Certainly more studies are needed to replace natriuretic peptides as the "gold standard," but MR-proADM absolutely is an important biomarker that reflects the neurohormonal activation in heart failure.

Markers for Inflammation

Inflammation is also regarded as an important contributor to heart failure. It is known that inflammation after myocardial infarction contributes to increased cardiac remodeling via activation of profibrotic factors (Frangogiannis, 2014). C-reactive protein (CRP) is one of the best markers of inflammation and is released after stimulation with IL-6 (Du Clos and Mold, 2004). CRP levels are often elevated in patients with heart failure, and its levels are associated with higher mortality or rehospitalization for heart failure (Kalogeropoulos et al., 2014). A huge drawback of CRP is the (a)specificity for heart failure. In the presence of infections, the value for patients with heart failure becomes unreliable (Lourenco et al., 2010).

Renal Injury

In patients with heart failure, renal insufficiency is very common, especially during hospitalization for acute decompensated heart failure. Furthermore, worsening renal function is associated with poor outcome in heart failure (Damman et al., 2007).

TABLE 3.14 Biomarkers of heart failure

Clinical markers	
Blood-borne biomarkers	Echocardiography 6-minute walking test
Cardiac remodeling	GDF-15
	Galectin-3
	ST2
Cardiomyocyte stretch	NT-proBNP
	MR-proANP
Neuroendocrine activation	MR-proADM
Renal injury	NGAL

Neutrophil gelatinase-associated lipocalin (NGAL) is released from the renal tubular cells in response to injury. An interesting feature of NGAL is the ability to detect renal damage in an earlier stage compared to creatinine (Maisel et al., 2011b). Patients with high NGAL levels have a higher risk to die, compared to patients with low NGAL levels (Aghel et al., 2010). Therefore, assessing tubular damage can be helpful for risk prediction of heart failure patients. (See Table 3.14).

Conclusion

Many different models are available to study heart failure. Heart failure has different kinds of etiology, and for each of them, animal models are available. Clinical biomarkers such as the 6MWD test and echocardiographic parameters are useful to provide prognostic insights. To establish diagnosis of heart failure, NT-proBNP (or other natriuretic peptides) is the biomarker of choice. For risk prediction, other biomarkers also do have their value. Timing seems to be crucial; whereas galectin-3 is a strong prognostic marker for short-term prognosis, natriuretic peptides are especially useful to predict long-term prognosis (>1 year), as was shown in two large cohorts (Gegenhuber et al., 2007; Seronde et al., 2013). Overall, natriuretic peptides remain important predictors for outcome in heart failure. Besides natriuretic peptides, MR-proADM is an interesting candidate for risk prediction in heart failure. Ultimately, heart failure biomarkers (clinical and blood-borne) can help us to diagnose heart failure, but also to identify individuals with distinct pathophysiologic pathways of heart failure and to estimate prognosis.

REFERENCES

Aghel, A., Shrestha, K., Mullens, W., Borowski, A., Tang, W.H., 2010. Serum neutrophil gelatinase-associated lipocalin (NGAL) in predicting worsening renal function in acute decompensated heart failure. J. Card. Fail. 16, 49–54.

Alahdab, M.T., Mansour, I.N., Napan, S., Stamos, T.D., 2009. Six minute walk test predicts long-term all-cause mortality and heart failure rehospitalization in African-American patients hospitalized with acute decompensated heart failure. J. Card. Fail. 15, 130–135.

Alderman, M.H., Cohen, H.W., Sealey, J.E., Laragh, J.H., 2010. Pressor responses to antihypertensive drug types. Am. J. Hypertens. 23, 1031–1037.

Baldasseroni, S., Opasich, C., Gorini, M., Lucci, D., Marchionni, N., Marini, M., et al., 2002. Left bundle-branch block is associated with increased 1-year sudden and total mortality rate in 5517 outpatients with congestive heart failure: a report from the Italian network on congestive heart failure. Am. Heart J. 143, 398–405.

Bartfay, W.J., Dawood, F., Wen, W.H., Lehotay, D.C., Hou, D., Bartfay, E., et al., 1999. Cardiac function and cytotoxic aldehyde production in a murine model of chronic iron-overload. Cardiovasc. Res. 43, 892–900.

Bartoli, C.R., Brittian, K.R., Giridharan, G.A., Koenig, S.C., Hamid, T., Prabhu, S.D., 2011. Bovine model of doxorubicin-induced cardiomyopathy. J. Biomed. Biotechnol. 2011. 758736.

Bashey, R.I., Philips, N., Insinga, F., Jimenez, S.A., 1993. Increased collagen synthesis and increased content of type VI collagen in myocardium of tight skin mice. Cardiovasc. Res. 27, 1061–1065.

Batkai, S., Thum, T., 2012. MicroRNAs in hypertension: mechanisms and therapeutic targets. Curr. Hypertens. Rep. 14, 79–87.

Beltrami, C.A., Finato, N., Rocco, M., Feruglio, G.A., Puricelli, C., Cigola, E., et al., 1995. The cellular basis of dilated cardiomyopathy in humans. J. Mol. Cell Cardiol. 27, 291–305.

Binder, B.R., Christ, G., Gruber, F., Grubic, N., Hufnagl, P., Krebs, M., et al., 2002. Plasminogen activator inhibitor 1: physiological and pathophysiological roles. News Physiol. Sci. 17, 56–61.

Blankenberg, S., Barbaux, S., Tiret, L., 2003. Adhesion molecules and atherosclerosis. Atherosclerosis 170, 191–203.

Blankenberg, S., Rupprecht, H.J., Bickel, C., Torzewski, M., Hafner, G., Tiret, L., et al., 2003a. Glutathione peroxidase 1 activity and cardiovascular events in patients with coronary artery disease. N. Engl. J. Med. 349, 1605–1613.

Blankenberg, S., Rupprecht, H.J., Poirier, O., Bickel, C., Smieja, M., Hafner, G., et al., 2003b. Plasma concentrations and genetic variation of matrix metalloproteinase 9 and prognosis of patients with cardiovascular disease. Circulation 107, 1579–1585.

Blom, A.S., Mukherjee, R., Pilla, J.J., Lowry, A.S., Yarbrough, W.M., Mingoia, J.T., et al., 2005. Cardiac support device modifies left ventricular geometry and myocardial structure after myocardial infarction. Circulation 112, 1274–1283.

Boekholdt, S.M., Hack, C.E., Sandhu, M.S., Luben, R., Bingham, S.A., Wareham, N.J., et al., 2006. C-reactive protein levels and coronary artery disease incidence and mortality in apparently healthy men and women: the EPIC-Norfolk Prospective Population Study 1993–2003. Atherosclerosis 187, 415–422.

Boerma, M., Hauer-Jensen, M., 2010. Preclinical research into basic mechanisms of radiation-induced heart disease. Cardiol. Res. Pract. http://dx.doi.org/10.4061/2011/858262.

Burke, A.P., Tracy, R.P., Kolodgie, F., Malcom, G.T., Zieske, A., Kutys, R., et al., 2002. Elevated C-reactive protein values and atherosclerosis in sudden coronary death: association with different pathologies. Circulation 105, 2019–2023.

Chae, C.U., Lee, R.T., Rifai, N., Ridker, P.M., 2001. Blood pressure and inflammation in apparently healthy men. Hypertension 38, 399–403.

Chien, K.L., Hsu, H.C., Chen, P.C., Su, T.C., Chang, W.T., Chen, M.F., et al., 2008. Urinary sodium and potassium excretion and risk of hypertension in Chinese: report from a community-based cohort study in Taiwan. J. Hypertens. 26, 1750–1756.

Chobanian, A.V., Bakris, G.L., Black, H.R., Cushman, W.C., Green, L.A., Izzo Jr., J.L., et al., 2003. The seventh report of the Joint National Committee on Prevention, Detection, Evaluation, and Treatment of High Blood Pressure: The JNC 7 report. JAMA 289, 2560–2572.

Chrysohoou, C., Pitsavos, C., Panagiotakos, D.B., Skoumas, J., Stefanadis, C., 2004. Association between prehypertension status and inflammatory markers related to atherosclerotic disease: the ATTICA study. Am. J. Hypertens. 17, 568–573.

Chu, G., Haghighi, K., Kranias, E.G., 2002. From mouse to man: understanding heart failure through genetically altered mouse models. J. Card. Fail. 8, S432–S449.

Crouse 3rd, J.R., Raichlen, J.S., Riley, W.A., Evans, G.W., Palmer, M.K., O'Leary, D.H., et al., 2007. Effect of rosuvastatin on progression of carotid intima-media thickness in low-risk individuals with subclinical atherosclerosis: the METEOR trial. JAMA 297, 1344–1353.

Damman, K., Navis, G., Voors, A.A., Asselbergs, F.W., Smilde, T.D., Cleland, J.G., et al., 2007. Worsening renal function and prognosis in heart failure: systematic review and meta-analysis. J. Card. Fail. 13, 599–608.

Das, S.R., Drazner, M.H., Dries, D.L., Vega, G.L., Stanek, H.G., Abdullah, S.M., et al., 2005. Impact of body mass and body composition on circulating levels of natriuretic peptides: results from the Dallas Heart Study. Circulation 112, 2163–2168.

Davies, E.J., Moxham, T., Rees, K., Singh, S., Coats, A.J., Ebrahim, S., et al., 2010. Exercise training for systolic heart failure: Cochrane systematic review and meta-analysis. Eur. J. Heart Fail. 12, 706–715.

de Boer, R.A., Lok, D.J., Jaarsma, T., van der Meer, P., Voors, A.A., Hillege, H.L., et al., 2011. Predictive value of plasma galectin-3 levels in heart failure with reduced and preserved ejection fraction. Ann. Med. 43, 60–68.

de Boer, R.A., van Veldhuisen, D.J., Gansevoort, R.T., Muller Kobold, A.C., van Gilst, W.H., Hillege, H.L., et al., 2012. The fibrosis marker galectin-3 and outcome in the general population. J. Intern. Med. 272, 55–64.

Detrano, R., Guerci, A.D., Carr, J.J., Bild, D.E., Burke, G., Folsom, A.R., et al., 2008. Coronary calcium as a predictor of coronary events in four racial or ethnic groups. N. Engl. J. Med. 358, 1336–1345.

Dornas, W.C., Silva, M.E., 2011. Animal models for the study of arterial hypertension. J. Biosci. 36, 731–737.

Doroszko, A., Andrzejak, R., Szuba, A., 2011. Role of the nitric oxide metabolic pathway and prostanoids in the pathogenesis of endothelial dysfunction and essential hypertension in young men. Hypertens. Res. 34, 79–86.

Du Clos, T.W., Mold, C., 2004. C-reactive protein: an activator of innate immunity and a modulator of adaptive immunity. Immunol. Res. 30, 261–277.

Egan, B.M., Zhao, Y., Axon, R.N., 2010. US trends in prevalence, awareness, treatment, and control of hypertension, 1988–2008. JAMA 303, 2043–2050.

Endemann, D.H., Schiffrin, E.L., 2004. Endothelial dysfunction. J. Am. Soc. Nephrol. 15, 1983–1992.

Feng, Q., Hedner, T., 1990. Endothelium-derived relaxing factor (EDRF) and nitric oxide (NO). I. physiology, pharmacology and pathophysiological implications. Clin Physiol. 10, 407–426.

Flammer, A.J., Luscher, T.F., 2010. Three decades of endothelium research: from the detection of nitric oxide to the everyday implementation of endothelial function measurements in cardiovascular diseases. Swiss Med. Wkly. 140, w13122.

Frangogiannis, N.G., 2014. The inflammatory response in myocardial injury, repair, and remodelling. Nat. Rev. Cardiol. 11, 255–265.

Frohlich, E.D., Apstein, C., Chobanian, A.V., Devereux, R.B., Dustan, H.P., Dzau, V., et al., 1992. The heart in hypertension. N. Engl. J. Med. 327, 998–1008.

Frolich, J.C., 1990. Prostacyclin in hypertension. J. Hypertens. Suppl. 8 (4), S73–S78.

Galis, Z.S., Sukhova, G.K., Lark, M.W., Libby, P., 1994. Increased expression of matrix metalloproteinases and matrix degrading activity in vulnerable regions of human atherosclerotic plaques. J. Clin. Invest. 94, 2493–2503.

Gegenhuber, A., Struck, J., Dieplinger, B., Poelz, W., Pacher, R., Morgenthaler, N.G., et al., 2007. Comparative evaluation of B-type natriuretic peptide, mid-regional pro-A-type natriuretic peptide, mid-regional pro-adrenomedullin, and copeptin to predict 1-year mortality in patients with acute destabilized heart failure. J. Card. Fail. 13, 42–49.

Gracie, J.A., Robertson, S.E., McInnes, I.B., 2003. Interleukin-18. J. Leukoc. Biol. 73, 213–224.

Greenland, P., Knoll, M.D., Stamler, J., Neaton, J.D., Dyer, A.R., Garside, D.B., et al., 2003. Major risk factors as antecedents of fatal and nonfatal coronary heart disease events. JAMA 290, 891–897.

Guo, Z., Van Remmen, H., Yang, H., Chen, X., Mele, J., Vijg, J., et al., 2001. Changes in expression of antioxidant enzymes affect cell-mediated LDL oxidation and oxidized LDL-induced apoptosis in mouse aortic cells. Arterioscler. Thromb. Vasc. Biol. 21, 1131–1138.
Harrison, D.G., Guzik, T.J., Lob, H.E., Madhur, M.S., Marvar, P.J., Thabet, S.R., et al., 2011. Inflammation, immunity, and hypertension. Hypertension 57, 132–140.
Herder, C., Baumert, J., Thorand, B., Martin, S., Lowel, H., Kolb, H., et al., 2006. Chemokines and incident coronary heart disease: results from the MONICA/KORA Augsburg Case-Cohort Study, 1984–2002. Arterioscler. Thromb. Vasc. Biol. 26, 2147–2152.
Hirasawa, M., Ito, Y., Shibata, M.A., Otsuki, Y., 2007. Mechanism of inflammation in murine eosinophilic myocarditis produced by adoptive transfer with ovalbumin challenge. Int. Arch. Allergy Immunol. 142, 28–39.
Ho, J.E., Liu, C., Lyass, A., Courchesne, P., Pencina, M.J., Vasan, R.S., et al., 2012. Galectin-3, a marker of cardiac fibrosis, predicts incident heart failure in the community. J. Am. Coll. Cardiol. 60, 1249–1256.
Houser, S.R., Margulies, K.B., Murphy, A.M., Spinale, F.G., Francis, G.S., Prabhu, S.D., et al., 2012. Animal models of heart failure: a scientific statement from the American Heart Association. Circ. Res. 111, 131–150.
Huang, F.W., Pinkus, J.L., Pinkus, G.S., Fleming, M.D., Andrews, N.C., 2005. A mouse model of juvenile hemochromatosis. J. Clin. Invest. 115, 2187–2191.
Huber, S.A., Sakkinen, P., Conze, D., Hardin, N., Tracy, R., 1999. Interleukin-6 exacerbates early atherosclerosis in mice. Arterioscler. Thromb. Vasc. Biol. 19, 2364–2367.
Inoue, S., Egashira, K., Ni, W., Kitamoto, S., Usui, M., Otani, K., et al., 2002. Anti-monocyte chemoattractant protein-1 gene therapy limits progression and destabilization of established atherosclerosis in apolipoprotein E-knockout mice. Circulation 106, 2700–2706.
Ishida, Y., Migita, K., Izumi, Y., Nakao, K., Ida, H., Kawakami, A., et al., 2004. The role of IL-18 in the modulation of matrix metalloproteinases and migration of human natural killer (NK) cells. FEBS Lett. 569, 156–160.
Itter, G., Jung, W., Juretschke, P., Schoelkens, B.A., Linz, W., 2004. A model of chronic heart failure in spontaneous hypertensive rats (SHR). Lab. Anim. 38, 138–148.
Jamaluddin, M.S., Weakley, S.M., Zhang, L., Kougias, P., Lin, P.H., Yao, Q., et al., 2011. miRNAs: roles and clinical applications in vascular disease. Expert Rev. Mol. Diagn. 11, 79–89.
James, P.A., Oparil, S., Carter, B.L., Cushman, W.C., Dennison-Himmelfarb, C., Handler, J., et al., 2014. 2014 evidence-based guideline for the management of high blood pressure in adults: report from the panel members appointed to the Eighth Joint National Committee (JNC 8). JAMA 311, 507–520.
Januzzi Jr., J.L., Rehman, S.U., Mohammed, A.A., Bhardwaj, A., Barajas, L., Barajas, J., et al., 2011. Use of amino-terminal pro-B-type natriuretic peptide to guide outpatient therapy of patients with chronic left ventricular systolic dysfunction. J. Am. Coll. Cardiol. 58, 1881–1889.
Jones, C.B., Sane, D.C., Herrington, D.M., 2003. Matrix metalloproteinases: a review of their structure and role in acute coronary syndrome. Cardiovasc. Res. 59, 812–823.
Jugdutt, B.I., Menon, V., 2004. Valsartan-induced cardioprotection involves angiotensin II type 2 receptor upregulation in dog and rat models of in vivo reperfused myocardial infarction. J. Card. Fail. 10, 74–82.
Kalogeropoulos, A.P., Tang, W.H., Hsu, A., Felker, G.M., Hernandez, A.F., Troughton, R.W., et al., 2014. High-sensitivity C-reactive protein in acute heart failure: insights from the ASCEND-HF trial. J. Card. Fail. 20, 319–326.
Kawashima, S., 2004. The two faces of endothelial nitric oxide synthase in the pathophysiology of atherosclerosis. Endothelium 11, 99–107.
Kaye, D.M., Preovolos, A., Marshall, T., Byrne, M., Hoshijima, M., Hajjar, R., et al., 2007. Percutaneous cardiac recirculation-mediated gene transfer of an inhibitory phospholamban peptide reverses advanced heart failure in large animals. J. Am. Coll. Cardiol. 50, 253–260.
Kempf, T., Eden, M., Strelau, J., Naguib, M., Willenbockel, C., Tongers, J., et al., 2006. The transforming growth factor-beta superfamily member growth-differentiation factor-15 protects the heart from ischemia/reperfusion injury. Circ. Res. 98, 351–360.
Kempf, T., von Haehling, S., Peter, T., Allhoff, T., Cicoira, M., Doehner, W., et al., 2007. Prognostic utility of growth differentiation factor-15 in patients with chronic heart failure. J. Am. Coll. Cardiol. 50, 1054–1060.
Koh, K.K., Ahn, J.Y., Han, S.H., Kim, D.S., Jin, D.K., Kim, H.S., et al., 2003. Pleiotropic effects of angiotensin II receptor blocker in hypertensive patients. J. Am. Coll. Cardiol. 42, 905–910.
Kosiborod, M., Lichtman, J.H., Heidenreich, P.A., Normand, S.L., Wang, Y., Brass, L.M., et al., 2006. National trends in outcomes among elderly patients with heart failure. Am. J. Med. 119, 616.e1–616.e7.
Kriegel, A.J., Mladinov, D., Liang, M., 2012. Translational study of microRNAs and its application in kidney disease and hypertension research. Clin. Sci. (Lond) 122, 439–447.
Kurtz, T.W., Griffin, K.A., Bidani, A.K., Davisson, R.L., Hall, J.E., AHA Council on High Blood Pressure Research, Professional and Public Education Subcommittee, 2005. Recommendations for blood pressure measurement in animals: summary of an AHA scientific statement from the Council on High Blood Pressure Research, Professional and Public Education Subcommittee. Arterioscler. Thromb. Vasc. Biol. 25, 478–479.
Kushwaha, S.S., Fallon, J.T., Fuster, V., 1997. Restrictive cardiomyopathy. N. Engl. J. Med. 336, 267–276.
Laragh, J.H., Sealey, J.E., 2011. The plasma renin test reveals the contribution of body sodium-volume content (V) and renin-angiotensin (R) vasoconstriction to long-term blood pressure. Am. J. Hypertens. 24, 1164–1180.
Levy, D., Garrison, R.J., Savage, D.D., Kannel, W.B., Castelli, W.P., 1990. Prognostic implications of echocardiographically determined left ventricular mass in the Framingham Heart Study. N. Engl. J. Med. 322, 1561–1566.
Li, D., Liu, L., Chen, H., Sawamura, T., Ranganathan, S., Mehta, J.L., 2003. LOX-1 mediates oxidized low-density lipoprotein-induced expression of matrix metalloproteinases in human coronary artery endothelial cells. Circulation 107, 612–617.
Lindmark, E., Diderholm, E., Wallentin, L., Siegbahn, A., 2001. Relationship between interleukin 6 and mortality in patients with unstable coronary artery disease: effects of an early invasive or noninvasive strategy. JAMA 286, 2107–2113.

Litwin, S.E., Katz, S.E., Weinberg, E.O., Lorell, B.H., Aurigemma, G.P., Douglas, P.S., 1995. Serial echocardiographic-Doppler assessment of left ventricular geometry and function in rats with pressure-overload hypertrophy. Chronic angiotensin-converting enzyme inhibition attenuates the transition to heart failure. Circulation 91, 2642–2654.

Liu, L., Eisen, H.J., 2014. Epidemiology of heart failure and scope of the problem. Cardiol. Clin. 32, 1–8, vii.

Locksley, R.M., Killeen, N., Lenardo, M.J., 2001. The TNF and TNF receptor superfamilies: integrating mammalian biology. Cell 104, 487–501.

Lourenco, P., Paulo Araujo, J., Paulo, C., Mascarenhas, J., Frioes, F., Azevedo, A., et al., 2010. Higher C-reactive protein predicts worse prognosis in acute heart failure only in noninfected patients. Clin. Cardiol. 33, 708–714.

Maier, W., Altwegg, L.A., Corti, R., Gay, S., Hersberger, M., Maly, F.E., et al., 2005. Inflammatory markers at the site of ruptured plaque in acute myocardial infarction: locally increased interleukin-6 and serum amyloid A but decreased C-reactive protein. Circulation 111, 1355–1361.

Maisel, A., Mueller, C., Adams Jr., K., Anker, S.D., Aspromonte, N., Cleland, J.G., et al., 2008. State of the art: using natriuretic peptide levels in clinical practice. Eur. J. Heart Fail. 10, 824–839.

Maisel, A., Mueller, C., Nowak, R.M., Peacock, W.F., Ponikowski, P., Mockel, M., et al., 2011a. Midregion prohormone adrenomedullin and prognosis in patients presenting with acute dyspnea: results from the BACH (Biomarkers in Acute Heart Failure) trial. J. Am. Coll. Cardiol. 58, 1057–1067.

Maisel, A.S., Mueller, C., Fitzgerald, R., Brikhan, R., Hiestand, B.C., Iqbal, N., et al., 2011b. Prognostic utility of plasma neutrophil gelatinase-associated lipocalin in patients with acute heart failure: the NGAL EvaLuation Along with B-type NaTriuretic Peptide in acutely decompensated heart failure (GALLANT) trial. Eur. J. Heart Fail. 13, 846–851.

Mallat, Z., Corbaz, A., Scoazec, A., Besnard, S., Leseche, G., Chvatchko, Y., et al., 2001. Expression of interleukin-18 in human atherosclerotic plaques and relation to plaque instability. Circulation 104, 1598–1603.

Mayr, A., Klug, G., Mair, J., Streil, K., Harrasser, B., Feistritzer, H.J., et al., 2013. Galectin-3: relation to infarct scar and left ventricular function after myocardial infarction. Int. J. Cardiol. 163, 335–337.

Meijers, W.C., Januzzi, J.L., deFilippi, C., Adourian, A.S., Shah, S.J., van Veldhuisen, D.J., et al., 2014. Elevated plasma galectin-3 is associated with near-term rehospitalization in heart failure: a pooled analysis of 3 clinical trials. Am. Heart J. 167, 853–860.

Milani, R.V., Drazner, M.H., Lavie, C.J., Morin, D.P., Ventura, H.O., 2011. Progression from concentric left ventricular hypertrophy and normal ejection fraction to left ventricular dysfunction. Am. J. Cardiol. 108, 992–996.

Molkentin, J.D., Lu, J.R., Antos, C.L., Markham, B., Richardson, J., Robbins, J., et al., 1998. A calcineurin-dependent transcriptional pathway for cardiac hypertrophy. Cell 93, 215–228.

Monreal, G., Gerhardt, M.A., Kambara, A., Abrishamchian, A.R., Bauer, J.A., Goldstein, A.H., 2004. Selective microembolization of the circumflex coronary artery in an ovine model: dilated, ischemic cardiomyopathy and left ventricular dysfunction. J. Card. Fail. 10, 174–183.

Moore, K.J., Tabas, I., 2011. Macrophages in the pathogenesis of atherosclerosis. Cell. 145, 341–355.

Morioka, S., Simon, G., 1982. Echocardiographic evidence for early left ventricular hypertrophy in dogs with renal hypertension. Am. J. Cardiol. 49, 1890–1895.

Mosterd, A., Hoes, A.W., 2007. Clinical epidemiology of heart failure. Heart 93, 1137–1146.

Mueller, C., Laule-Kilian, K., Scholer, A., Nusbaumer, C., Zeller, T., Staub, D., et al., 2005. B-type natriuretic peptide for acute dyspnea in patients with kidney disease: insights from a randomized comparison. Kidney Int. 67, 278–284.

Mullins, J.J., Peters, J., Ganten, D., 1990. Fulminant hypertension in transgenic rats harbouring the mouse Ren-2 gene. Nature 344, 541–544.

Namiki, M., Kawashima, S., Yamashita, T., Ozaki, M., Hirase, T., Ishida, T., et al., 2002. Local overexpression of monocyte chemoattractant protein-1 at vessel wall induces infiltration of macrophages and formation of atherosclerotic lesion: synergism with hypercholesterolemia. Arterioscler. Thromb. Vasc. Biol. 22, 115–120.

Nasu, K., Tsuchikane, E., Katoh, O., Vince, D.G., Virmani, R., Surmely, J.F., et al., 2006. Accuracy of in vivo coronary plaque morphology assessment: a validation study of in vivo virtual histology compared with in vitro histopathology. J. Am. Coll. Cardiol. 47, 2405–2412.

Nushiro, N., Ito, S., Carretero, O.A., 1990. Renin release from microdissected superficial, midcortical, and juxtamedullary afferent arterioles in rabbits. Kidney Int. 38, 426–431.

Oei, H.H., van der Meer, I.M., Hofman, A., Koudstaal, P.J., Stijnen, T., Breteler, M.M., et al., 2005. Lipoprotein-associated phospholipase A2 activity is associated with risk of coronary heart disease and ischemic stroke: the Rotterdam Study. Circulation 111, 570–575.

O'Leary, D.H., Polak, J.F., Kronmal, R.A., Manolio, T.A., Burke, G.L., Wolfson Jr., S.K., 1999. Carotid-artery intima and media thickness as a risk factor for myocardial infarction and stroke in older adults. Cardiovascular Health Study Collaborative Research Group. N. Engl. J. Med. 340, 14–22.

Oudit, G.Y., Crackower, M.A., Eriksson, U., Sarao, R., Kozieradzki, I., Sasaki, T., et al., 2003. Phosphoinositide 3-kinase gamma-deficient mice are protected from isoproterenol-induced heart failure. Circulation 108, 2147–2152.

Park, J.B., Schiffrin, E.L., 2001. Small artery remodeling is the most prevalent (earliest?) form of target organ damage in mild essential hypertension. J. Hypertens. 19, 921–930.

Pfeffer, J.M., Pfeffer, M.A., Mirsky, I., Braunwald, E., 1982. Regression of left ventricular hypertrophy and prevention of left ventricular dysfunction by captopril in the spontaneously hypertensive rat. Proc. Natl. Acad. Sci. U.S.A. 79, 3310–3314.

Pfeffer, M.A., Pfeffer, J.M., Fishbein, M.C., Fletcher, P.J., Spadaro, J., Kloner, R.A., et al., 1979. Myocardial infarct size and ventricular function in rats. Circ. Res. 44, 503–512.

Pickering, T.G., Hall, J.E., Appel, L.J., Falkner, B.E., Graves, J., Hill, M.N., et al., 2005. Recommendations for blood pressure measurement in humans and experimental animals: Part 1: Blood pressure measurement in humans: a statement for professionals from the Subcommittee of Professional and Public Education of the American Heart Association Council on High Blood Pressure Research. Circulation 111, 697–716.

Pinto, Y.M., Pinto-Sietsma, S.J., Philipp, T., Engler, S., Kossamehl, P., Hocher, B., et al., 2000. Reduction in left ventricular messenger RNA for transforming growth factor beta(1) attenuates left ventricular fibrosis and improves survival without lowering blood pressure in the hypertensive TGR(mRen2)27 rat. Hypertension 36, 747–754.

Preston, R.A., Ledford, M., Materson, B.J., Baltodano, N.M., Memon, A., Alonso, A., 2002. Effects of severe, uncontrolled hypertension on endothelial activation: soluble vascular cell adhesion molecule-1, soluble intercellular adhesion molecule-1 and von Willebrand factor. J. Hypertens. 20, 871–877.

Quyyumi, A.A., Patel, R.S., 2010. Endothelial dysfunction and hypertension: cause or effect? Hypertension 55, 1092–1094.

Rattazzi, M., Puato, M., Faggin, E., Bertipaglia, B., Zambon, A., Pauletto, P., 2003. C-reactive protein and interleukin-6 in vascular disease: culprits or passive bystanders? J. Hypertens. 21, 1787–1803.

Saavedra, W.F., Tunin, R.S., Paolocci, N., Mishima, T., Suzuki, G., Emala, C.W., et al., 2002. Reverse remodeling and enhanced adrenergic reserve from passive external support in experimental dilated heart failure. J. Am. Coll. Cardiol. 39, 2069–2076.

Sanada, S., Hakuno, D., Higgins, L.J., Schreiter, E.R., McKenzie, A.N., Lee, R.T., 2007. IL-33 and ST2 comprise a critical biomechanically induced and cardioprotective signaling system. J. Clin. Invest. 117, 1538–1549.

Sangiorgi, G., Rumberger, J.A., Severson, A., Edwards, W.D., Gregoire, J., Fitzpatrick, L.A., et al., 1998. Arterial calcification and not lumen stenosis is highly correlated with atherosclerotic plaque burden in humans: a histologic study of 723 coronary artery segments using nondecalcifying methodology. J. Am. Coll. Cardiol. 31, 126–133.

Sanz, J., Fayad, Z.A., 2008. Imaging of atherosclerotic cardiovascular disease. Nature 451, 953–957.

Savoia, C., Schiffrin, E.L., 2007. Vascular inflammation in hypertension and diabetes: molecular mechanisms and therapeutic interventions. Clin. Sci. (Lond) 112, 375–384.

Seronde, M.F., Gayat, E., Logeart, D., Lassus, J., Laribi, S., Boukef, R., et al., 2013. Comparison of the diagnostic and prognostic values of B-type and atrial-type natriuretic peptides in acute heart failure. Int. J. Cardiol. 168, 3404–3411.

Shah, A.M., Claggett, B., Sweitzer, N.K., Shah, S.J., Anand, I.S., O'Meara, E., et al., 2014. Cardiac structure and function and prognosis in heart failure with preserved ejection fraction: findings from the echocardiographic study of the treatment of preserved cardiac function heart failure with an aldosterone antagonist (TOPCAT) trial. Circ. Heart Fail. 7, 740–751.

Shah, M.R., Hasselblad, V., Gheorghiade, M., Adams Jr., K.F., Swedberg, K., Califf, R.M., et al., 2001. Prognostic usefulness of the six-minute walk in patients with advanced congestive heart failure secondary to ischemic or nonischemic cardiomyopathy. Am. J. Cardiol. 88, 987–993.

Shah, R.V., Januzzi Jr., J.L., 2010. ST2: a novel remodeling biomarker in acute and chronic heart failure. Curr. Heart Fail. Rep. 7, 9–14.

Shapiro, B.P., Owan, T.E., Mohammed, S., Kruger, M., Linke, W.A., Burnett Jr., J.C., et al., 2008. Mineralocorticoid signaling in transition to heart failure with normal ejection fraction. Hypertension 51, 289–295.

Shiekh, G.A., Ayub, T., Khan, S.N., Dar, R., Andrabi, K.I., 2011. Reduced nitrate level in individuals with hypertension and diabetes. J. Cardiovasc. Dis. Res. 2, 172–176.

Shimpo, M., Morrow, D.A., Weinberg, E.O., Sabatine, M.S., Murphy, S.A., Antman, E.M., et al., 2004. Serum levels of the interleukin-1 receptor family member ST2 predict mortality and clinical outcome in acute myocardial infarction. Circulation 109, 2186–2190.

Simons, P.C., Algra, A., Bots, M.L., Grobbee, D.E., van der Graaf, Y., 1999. Common carotid intima-media thickness and arterial stiffness: indicators of cardiovascular risk in high-risk patients. The SMART study (Second Manifestations of ARTerial disease). Circulation 100, 951–957.

Spencer, C.G., Gurney, D., Blann, A.D., Beevers, D.G., Lip, G.Y., ASCOT Steering Committee, Anglo-Scandinavian Cardiac Outcomes Trial, 2002. Von Willebrand factor, soluble P-selectin, and target organ damage in hypertension: a substudy of the Anglo-Scandinavian Cardiac Outcomes Trial (ASCOT). Hypertension 40, 61–66.

Staessen, J.A., Wang, J., Bianchi, G., Birkenhager, W.H., 2003. Essential hypertension. Lancet 361, 1629–1641.

Sun, Z., Cade, R., Morales, C., 2002. Role of central angiotensin II receptors in cold-induced hypertension. Am. J. Hypertens. 15, 85–92.

Takahashi, N., Smithies, O., 2004. Human genetics, animal models and computer simulations for studying hypertension. Trends Genet. 20, 136–145.

Teekakirikul, P., Eminaga, S., Toka, O., Alcalai, R., Wang, L., Wakimoto, H., et al., 2010. Cardiac fibrosis in mice with hypertrophic cardiomyopathy is mediated by non-myocyte proliferation and requires Tgf-beta. J. Clin. Invest. 120, 3520–3529.

Teng, M.H., Yin, J.Y., Vidal, R., Ghiso, J., Kumar, A., Rabenou, R., et al., 2001. Amyloid and nonfibrillar deposits in mice transgenic for wild-type human transthyretin: a possible model for senile systemic amyloidosis. Lab. Invest. 81, 385–396.

Terris, J.M., Berecek, K.H., Cohen, E.L., Stanley, J.C., Whitehouse Jr., W.M., Bohr, D.F., 1976. Deoxycorticosterone hypertension in the pig. Clin. Sci. Mol. Med. Suppl. 3, 303s–305s.

Thoenes, M., Oguchi, A., Nagamia, S., Vaccari, C.S., Hammoud, R., Umpierrez, G.E., et al., 2007. The effects of extended-release niacin on carotid intimal media thickness, endothelial function and inflammatory markers in patients with the metabolic syndrome. Int. J. Clin. Pract. 61, 1942–1948.

Touyz, R.M., Burger, D., 2012. Biomarkers in hypertension. In: Berbari, A.J., Mancia, G. (Eds.), Special issues in hypertension. Springer, New York, pp. 237–246.

Truter, S.L., Catanzaro, D.F., Supino, P.G., Gupta, A., Carter, J., Herrold, E.M., et al., 2009. Differential expression of matrix metalloproteinases and tissue inhibitors and extracellular matrix remodeling in aortic regurgitant hearts. Cardiology 113, 161–168.

Tsutsui, J.M., Xie, F., Cano, M., Chomas, J., Phillips, P., Radio, S.J., et al., 2004. Detection of retained microbubbles in carotid arteries with real-time low mechanical index imaging in the setting of endothelial dysfunction. J. Am. Coll. Cardiol. 44, 1036–1046.

Turner, S.T., Schwartz, G.L., Chapman, A.B., Beitelshees, A.L., Gums, J.G., Cooper-DeHoff, R.M., et al., 2010. Plasma renin activity predicts blood pressure responses to beta-blocker and thiazide diuretic as monotherapy and add-on therapy for hypertension. Am. J. Hypertens. 23, 1014–1022.

van der Velde, A.R., Meijers, W.C., de Boer, R.A., 2014. Biomarkers for risk prediction in acute decompensated heart failure. Curr Heart Fail Rep. 11, 246–259.

van der Velden, J., Merkus, D., Klarenbeek, B.R., James, A.T., Boontje, N.M., Dekkers, D.H., et al., 2004. Alterations in myofilament function contribute to left ventricular dysfunction in pigs early after myocardial infarction. Circ. Res. 95. e85–e95.

Vatner, S.F., 1978. Effects of anesthesia on cardiovascular control mechanisms. Environ. Health Perspect. 26, 193–206.

Vazquez-Oliva, G., Fernandez-Real, J.M., Zamora, A., Vilaseca, M., Badimon, L., 2005. Lowering of blood pressure leads to decreased circulating interleukin-6 in hypertensive subjects. J. Hum. Hypertens. 19, 457–462.

Waldo, S.W., Beede, J., Isakson, S., Villard-Saussine, S., Fareh, J., Clopton, P., et al., 2008. Pro-B-type natriuretic peptide levels in acute decompensated heart failure. J. Am. Coll. Cardiol. 51, 1874–1882.

Wang, Q.D., Bohlooly, Y.M., Sjoquist, P.O., 2004. Murine models for the study of congestive heart failure: implications for understanding molecular mechanisms and for drug discovery. J. Pharmacol. Toxicol. Methods 50, 163–174.

Wang, T.J., Vasan, R.S., 2005. Epidemiology of uncontrolled hypertension in the United States. Circulation 112, 1651–1662.

Weber, M.A., Schiffrin, E.L., White, W.B., Mann, S., Lindholm, L.H., Kenerson, J.G., et al., 2014. Clinical practice guidelines for the management of hypertension in the community a statement by the American Society of Hypertension and the International Society of Hypertension. J. Hypertens. 32, 3–15.

Wei, C.C., Lucchesi, P.A., Tallaj, J., Bradley, W.E., Powell, P.C., Dell'Italia, L.J., 2003. Cardiac interstitial bradykinin and mast cells modulate pattern of LV remodeling in volume overload in rats. Am. J. Physiol. Heart Circ. Physiol. 285, H784–H792.

Weinstein, D.M., Mihm, M.J., Bauer, J.A., 2000. Cardiac peroxynitrite formation and left ventricular dysfunction following doxorubicin treatment in mice. J. Pharmacol. Exp. Ther. 294, 396–401.

Wexler, R.K., Elton, T., Pleister, A., Feldman, D., 2009. Cardiomyopathy: an overview. Am. Fam. Physician 79, 778–784.

Wiesel, P., Mazzolai, L., Nussberger, J., Pedrazzini, T., 1997. Two-kidney, one clip and one-kidney, one clip hypertension in mice. Hypertension 29, 1025–1030.

Zadelaar, S., Kleemann, R., Verschuren, L., de Vries-Van der Weij, J., van der Hoorn, J., Princen, H.M., et al., 2007. Mouse models for atherosclerosis and pharmaceutical modifiers. Arterioscler. Thromb. Vasc. Biol. 27, 1706–1721.

Zalewski, A., Macphee, C., 2005. Role of lipoprotein-associated phospholipase A2 in atherosclerosis: biology, epidemiology, and possible therapeutic target. Arterioscler. Thromb. Vasc. Biol. 25, 923–931.

Zimmerman, R.S., Frohlich, E.D., 1990. Stress and hypertension. J. Hypertens. Suppl. 8, S103–S107. http://dx.doi.org/10.1016/B978-0-12-800687-0.00018-9.

Chapter 3.7.2

Biomarkers in Oncology

Jon Cleland, Faisal Azam, Rachel Midgley and David J. Kerr

Modern developments in molecular genetics have led oncologists not only to define tumors according to their anatomical site, stage, and histological type, but also to appreciate the heterogeneity between tumors according to their specific molecular, genetic, or immunologic subtype (La Thangue and Kerr, 2011). This therefore empowers oncologists to ideally provide more individually tailored treatment to particular patients, and their specific tumor subtype, by using highly targeted therapies, as opposed to more broad-spectrum and nonspecific cytotoxic agents (Kerr and Shi, 2013).

When we consider some of the first targeted therapies to be clinically approved, such as imatinib in the treatment of chronic myeloid leukemia and the greatly improved 5-year survival (Deininger and Drucker, 2003), it would appear that an exciting dawn of scientifically eloquent targeted treatments for cancer is upon us. However, the process of developing and, indeed, using such targeted agents is fraught with difficulty. The large development costs, potential for serious toxicity, and the fact that inherently only particular subgroups of patients are likely to benefit at all mean that such treatments are highly scrutinized by regulatory authorities ahead of approval (Raftery, 2009).

Effective biomarkers therefore must be identified and then utilized appropriately to aid clinical decision making and ensure targeted agents are given to patients in whom they will be suitably efficacious. For instance, in patients with resected colorectal cancer, the QUASAR study clarified mismatch repair deficiency (MMR-D) as a prognostic biomarker, resulting in changes to guidelines regarding use of adjuvant chemotherapy (Kerr and Midgley, 2010).

In addition to aiding clinical decision making in the future, development of biomarkers also leads us to retrospectively question the validity of the data derived from previous clinical trials and the subsequent conclusions drawn. For example, retrospective analysis of the PETACC 3 trial (Van Cutsem et al., 2009) demonstrates a statistically significant difference in (previously unidentified) poor prognosis T4 tumors in just one of the two arms of this trial. Hence, by identifying new markers for patient stratification, we may find flaws in the design of previous clinical trials, leading us to question the plausibility of the findings.

It is clear that it is imperative to develop and validate partner biomarkers in association with new agents, from the beginning of the drug development pathway, that is, right from phase I trial onwards. Such utilization has made a supportive contribution to dose selection and primary endpoint measurement in early-phase cancer trials. It helps identify the most appropriate patients, provides proof-of-concept for target modulation, helps test the underlying hypothesis, helps in dose selection and schedule, and may predict clinical outcomes. Only 5% of oncology drugs that enter the clinic make it to marketing approval. Use of biomarkers should reduce this high level of attrition and bring forward key decisions (e.g., fail fast), thereby reducing the spiraling cost of drug development and increasing the likelihood of getting innovative and active drugs to cancer patients (Sarker and Workman, 2007). Biomarkers may yield information earlier than current imaging techniques and thus provide a potentially more efficient way of following up on patients receiving therapy. Furthermore, they may predict a more biologically appropriate target dosage of a particular agent in order to adjust doses and reduce toxicity without compromising efficacy.

The characteristics of a good biomarker (Bhatt et al., 2007) are as follows:

- It should be cost-effective.
- It should be of low expression or circulation in normal individuals.
- It should be accessible by noninvasive means such as blood, urine, and so on.
- It should be accurately reproducible in multiple clinical centers.

Although noninvasive means of following up on patients would be most appropriate, there is still much interest in obtaining tissue biopsies from patients receiving targeted therapies so that the molecular features of the tumor after exposure can be studied.

Principles of Translational Science in Medicine. http://dx.doi.org/10.1016/B978-0-12-800687-0.00019-0

In the absence of biopsy material, however, a lot of information can still be gathered. For example, biomarkers of anti-angiogenic therapy may be cytokines and circulating endothelial cells in peripheral blood. Many surface markers are used to define circulating endothelial cells by multiparameter flow cytometry. Beerepoot and colleagues (2004) found that patients with progressing cancer have higher levels of mature circulating endothelial cells than healthy people or patients with a stable disease. Shaked and colleagues (2006) demonstrated an increase in circulating endothelial progenitors early in the course of the therapy, so sampling at multiple time points is required for interpretation of data. Circulating proteins are vascular endothelial growth factors (VEGFs), basic fibroblast growth factors (bFGFs), endostatin, and thrombospondin 1 (TSP1). In non-small cell lung cancer (NSCLC), VEGF levels are high, whereas TSP1 levels are low, which may trigger and sustain tumor angiogenesis. A high level of serum VEGF at the time of presentation in NSCLC predicts worse survival (Arkadiusz et al., 2005).

Measurement of target inhibition should ideally be undertaken in a tumor before and after administering a drug, and sometimes more than two biopsies are needed. The tumor needs to be accessible, and the patient must consent to repeated biopsies. This limits the number of patients available to such studies. Normal tissues like buccal mucosa, peripheral mononuclear cells, blood, and skin can be used to measure the changes in molecular effects; for example, in a trial of the aromatase inhibitor anastrazole, peripheral blood oestradiol levels were the endpoint (Ploudre et al., 1995). Similarly, in a trial of metastatic gastrointestinal stromal tumor patients, monocytes were assessed to measure the biologic effects of the targeted agent (sunitinib malate), inhibiting the VEGF and vascular endothelial growth factor receptor (VEGFR), and were found to be modulated by treatment, suggesting that these measurements may be effective pharmacodynamic (PD) markers for sunitinib (Norden-Zfoni et al., 2007).

OTHER PATHWAYS THAT MAY BE SUITABLE FOR BIOMARKER-ASSISTED DEVELOPMENT

Receptor tyrosine kinases (RTKs) play an important role in the regulation of cell proliferation, cell differentiation, and intracellular signaling processes. High RTK activity is associated with a range of human cancers. Therefore, inhibition of these receptors represents a potent approach to the treatment of various cancers. The epidermal growth factor receptor (EGFR) is one of the RTKs that regulate cell proliferation. The EGFR has extracellular ligand-binding domains to which the epidermal growth factor (EGF) binds. Subsequently, their intracellular autophosphorylation domains in the C-terminal region are trans-autophosphorylated, and the catalytic domains of EGFR are activated. Active EGFR recruits and phosphorylates signaling proteins, such as Grb2, which links cell surface receptors to the Ras/mitogen-activated protein kinase (MAPK) signaling cascade. Ultimately, this leads to intranuclear mitogenesis and cell proliferation caused by activation of the Ras/MAPK signaling pathway. A clinical scenario illustrating the importance of the EGFR pathway is high-grade gliomas in which the pathway is commonly amplified. Abnormal signaling from the receptor tyrosine kinase is believed to contribute to the malignant phenotypes seen in these tumors. Targeted therapy with small molecule inhibitors of the receptor tyrosine kinase may improve the treatment of these highly aggressive brain tumors. Treatment with the EGFR kinase inhibitors ZD1839 (Gefitinib) with high specificity for EGFR resulted in significant suppression of EGFR autophosphorylation, even with very low levels of the drug (Bin et al., 2003). Patients with tumors with the wild-type KRAS gene achieve disruption of cellular growth signaling through the EGFR pathway when EGFR-targeted therapies are administered. Conversely, tumors with mutated KRAS genes have the ability to continue signaling through the proliferation cascade within the EGFR axis, even when EGFR inhibitor therapy is applied. Thus, patients with mutated KRAS genes will not respond to EGFR inhibitors. Examples include cetuximab and panitumumab in colorectal cancer and erlotinib and gefitinib in non-small cell lung cancer. Such biomarker-guided decision making is now directly implemented in international guidelines and practice for colorectal cancer treatment (Mack, 2009). The prescribing label for such drugs is evolving as more mutations in Ras (Kras and NRas) are identified. This has the effect of decreasing the potential population for treatment, decreasing population risk and toxicity, increasing effectiveness in the eventually treated population, and decreasing cost. These are the ultimate goals of any effective segregating biomarker.

It has taken a long time to discover this differential effect of the EGFR inhibitors, and this discovery necessitated many expensive trials and much retrospective analysis before the biomarker was truly appreciated and refined. We hope that the cultural shift will continue and, in the future, testing and validation of suspected biomarkers will be undertaken prospectively and earlier in the preclinical development stages of drug testing, before any patient is even exposed to the drugs. Further validation can then take place during phase I and phase II trials. When such biomarkers are tested in clinical trials, the endpoint (e.g., the level of expression of protein); the methods of measurement (e.g., mRNA or protein); and the standard operating procedures governing processing, fixing, transport, and storage of the tissue should be clearly defined before the trial starts. The amount of tissue required should also be clearly specified, as this makes a significant difference in terms of ethical and practical applicability. Preclinical studies should guide the magnitude of target inhibition and its impact on tumor growth and the proportion of patients having the effect with a given dose to conclude the recommended dose.

An excellent example of more promptly conducting early phase trials involving biomarkers is the BRAF V600E inhibitors in melanoma. BRAF is a serine-threonine kinase that activates MAP/ERK kinase signaling to alter gene activation. Following identification of BRAF mutations in 60% of melanoma patients (Miller and Mihm, 2006), and development of BRAF V600E inhibitors, patients were then recruited to phase I trials based on the presence of this mutation. Initial results appear promising, with a reported 81% response rate in the selected patients in the phase I trial (Flaherty et al., 2010).

However, this is just one example, and there are many different ways that PD markers can be used within a phase I trial of novel agents or targeted therapies. PD measurement can be included as the primary endpoint in trial protocols to assess the magnitude of a biological effect and its relationship to the dose of the drug administered. If there is considerable variation in PD effects among patients, then more patients need to be treated on the same dose level to assess the differences. A detailed PD assessment at all dose levels is necessary for calculating the recommended dose. The phase I trial may be an opportunity to learn about targets if targets are not fully defined before the trial starts. Markers that look interesting can be further developed in the study at later stages or in subsequent studies. Where the optimal biological dose has not been fully defined, PD effect may be studied at two different dose levels, that is, the maximum tolerated dose and a lower well-tolerated dose. It is important that preclinical studies define the appropriate target, PD marker, and desired biological outcome, demonstrating, for example, the dose-dependent changes in a signaling pathway in tumors that have responded in the preclinical setting. Sometimes these agents have more than one target or biological effect. Antitumor effects seen in the preclinical studies may not be associated with the same target, so phase I trials are effective in studying those agents; for example, sorafenib was thought to be a specific inhibitor of Raf-1 kinase (Lyons et al., 2001) but was later found to act on several other kinases such as the VEGFR, the platelet-derived growth factor receptor (PDGFR) (Wilhelm et al., 2003), and so on, and thus was found to be effective against renal cell cancer.

A greater number of patients may be required up front in these elegant trials, compared to a "standard design" with the primary endpoint being toxicity. For example, for a trial to show a dose response in target inhibition, rising from 40% at one dose level to 90% at a higher dose level, 17 patients will need to be treated at each dose level (Korn et al., 2001). The biopsy or assay procedures should be clearly explained in the patient information sheet and informed consent document. Maintaining a close link between the clinic and the PD laboratory is clearly very important. Laboratory scientists, radiologists, and molecular pharmacology teams should be involved in protocol writing and development.

TISSUE BIOPSIES

Tumor biopsy will give the most direct and relevant information about the target inhibition before and after a treatment but may be quite challenging from a pragmatic or ethical viewpoint. Some tumors such as superficial melanomas, head and neck cancers, and cervical cancers are more easily accessible than others. Therefore, alternatives to biopsying the tumor itself have been explored. In a gefitinib phase I trial, skin was biopsied as a marker to assess EGFR tyrosine kinase inhibitor, and inhibition of EGFR activation was seen at all doses in all patients (Baselga et al., 2002).

Preclinical studies should provide some guidance about the tissue that is to be biopsied, the amount needed, and the condition and processing of the tissue. The primary endpoint should be defined up front in cases in which several assays are taken. Assays used to measure PD endpoint in plasma, normal tissue, and tumor samples include ELISAs, immunohistochemistry, Northern and Western blotting, real-time polymerase chain reactions, and gene expression arrays. Most of these are refined and developed for a specific target. Sometimes the pharmaceutical company that is developing the novel agent also develops the assays of PD development. However, often good collaboration between an academic scientific group and the pharmaceutical company will bring about the speediest development of the appropriate assays. If the PD measurement is the primary endpoint of the trial, then the assay needs to be validated to good clinical laboratory practice standards.

The need for such close collaboration and validation of assays has been highlighted by a debate regarding the measurement of HER2 levels. It is well established that HER2 gene expression determines efficacy of the targeted trastuzumab as adjuvant therapy (Slamon et al., 2001). However, assays involving fluorescence in situ hybridization (FISH) to measure HER2 indicate that conventional assays are suboptimal, with up to 20% giving an incorrect result (Perez et al., 2002). Thus, it is plausible that an inaccurate assay could have been used to influence clinical decision making in patients with breast cancer, which shows how imperative it is that assays for such biomarkers are sufficiently validated.

Biomarkers are thus essential to allow identification of novel therapeutic targets, stratify patients according to likely efficacy, and monitor response in order to deliver the new-age philosophy of individually tailored and targeted cancer treatment. However, it is imperative to also validate such markers sufficiently, before utilizing them to influence clinical practice, and to fully understand their potential applications.

In particular, the prospect of retrospectively applying new biomarker assays to existing trial data to reveal previously unidentified statistical biases and subsequently masked outcomes is fascinating. This is demonstrated eloquently when retrospectively identifying T4 tumors in the PETACC3 trial and uncovering a statistically significant bias between the patients recruited to the two arms of the trial. Hence, we are led to question the validity of the outcomes of this trial, and it is interesting to contemplate in how many other trials retrospective analysis with biomarkers could invalidate and indeed alter the outcome.

REFERENCES

Arkadiusz, Z., Dudek, A.Z., Mahaseth, H., 2005. Circulating angiogenic cytokines in patients with advanced non small cell lung cancer: correlation with treatment response and survival. Cancer Invest. 23, 193–200.

Baselga, J., Rischin, D., Ranson, M., Calvert, H., Raymond, E., Kieback, D.G., et al., 2002. Phase I safety, pharmacokinetic, and pharmacodynamic trial of ZD1839, a selective oral epidermal growth factor receptor tyrosine kinase inhibitor, in patients with five selected solid tumor types. J. Clin. Oncol. 20, 4292–4302.

Beerepoot, L.V., Mehra, N., Vermaat, J.S., Zonnenberg, B.A., Gebbink, M.F., Voest, E.E., 2004. Increased levels of viable circulating endothelial cells are an indicator of progressive disease in cancer patients. Ann. Oncol. 15, 139–145.

Bhatt, R.S., Seth, P., Sukhatme, V.P., 2007. Biomarkers for monitoring antiangiogenic therapy. Clin. Cancer Res. 13, 777s–780s.

Bin, L., Chang, C.M., Min, Y., McKenna, W.G., Shu, H.K., 2003. Resistance to small molecules inhibitors of epidermal growth factor receptors in malignant gliomas. Can. Res. 63, 7443–7450.

Deininger, M.W., Drucker, B.J., 2003. Specific targeted therapy of chronic myelogenous leukaemia with imatinib. Pharmacol. Rev. 55, 401–423.

Flaherty, K.T., Puzanov, I., Kim, K.B., Ribas, A., McArthur, G.A., Sosman, J.A., et al., 2010. Inhibition of mutated, activated BRAF in metastatic melanoma. N. Engl. J. Med. 363, 809–819.

Kerr, D.J., Midgley, R., 2010. Defective mismatch repair in colon cancer; a prognostic or predictive biomarker? J. Clin. Oncol. 28, 3210–3212.

Kerr, D.J., Shi, Y., 2013. Tailoring treatment and trials to prognosis. Nat. Rev. Clin. Oncol. 10, 429–430.

Korn, E.L., Arbuck, S.G., Pluda, J.M., Simon, R., Kaplan, R.S., Christian, M.C., 2001. Clinical trial design for cytostatic agents: are new approaches needed? J. Clin. Oncol. 19, 265–272.

La Thangue, N.B., Kerr, D.J., 2011. Predictive biomarkers: a paradigm shift towards personalized cancer medicine. Nat. Rev. Clin. Oncol. 8, 587–596.

Lyons, J.F., Wilhelm, S., Hibner, B., Bollag, G., 2001. Discovery of a novel Raf kinase inhibitor. Endor. Relat. Cancer 8, 219–225.

Mack, G.S., 2009. FDA holds court on post hoc data linking KRAS status to drug response. Nat. Biotechnol. 27, 110–112.

Miller, A.J., Mihm Jr, M.C., 2006. Melanoma. N. Engl. J. Med. 355, 51–65.

Norden-Zfoni, A., Desai, J., Manola, J., Beudy, P., Force, J., Maki, R., et al., 2007. Blood based biomarkers of SU11248 activity and clinical outcome in patients with metastatic imatinib resistant gastrointestinal stromal tumour. Clin. Cancer Res. 13, 2643–2650.

Perez, E.A., Roche, P.C., Jenkins, R.B., Reynolds, C.A., Halling, K.C., Ingle, J.N., et al., 2002. HER2 testing in patients with breast cancer; poor correlation between weak positivity by immunohistochemistry and gene amplification by fluorescence in situ hybridization. Mayo Clin. Proc. 77, 148–154.

Ploudre, P.V., Dyroff, M., Dowsett, M., Demers, L., Yates, R., Webster, A., 1995. Arimidex, a new oral, once a day aromatase inhibitor. J. Steroid. Biochem. Mol. Biol. 53, 175–179.

Raftery, J., 2009. NICE and the challenge of cancer drugs. BMJ 388, b67.

Sarker, D., Workman, P., 2007. Pharmacodynamic biomarker for molecular cancer therapeutics. Adv. Cancer Res. 13, 6545–6548.

Shaked, Y., Ciarrocchi, A., Franco, M., Lee, C.R., Man, S., Cheung, A.M., et al., 2006. Therapy induced acute recruitment of circulating endothelial progenitor cells to tumours. Science 313, 1785–1787.

Slamon, D.J., Leyland-Jones, B., Shak, S., Fuchs, H., Paton, V., Bajamonde, A., et al., 2001. Use of chemotherapy plus a monoclonal antibody against HER2 for metastatic breast cancer that overexpresses HER2. N. Engl. J. Med. 344, 783–792.

Van Cutsem, E., Labianca, R., Bodoky, G., Barone, C., Aranda, E., Nordlinger, B., et al., 2009. Randomised phase III trial comparing biweekly infusional fluorouracil/leucovorin alone or with irinotecan in the adjuvant treatment of stage III colon cancer: PETACC-3. J. Clin. Oncol. 27, 3117–3125.

Wilhelm, S., Carter, C., Tang, L.Y., 2003. Bay 43-9006 exhibits broad spectrum anti tumour activity and targets Raf/MEK/ERK pathway and receptor tyrosine kinases involved in tumour progression and angiogenesis. Clin. Cancer Res. 9, A78.

Chapter 3.7.3

Translational Imaging Research

Lars Johansson

WHY IMAGING?

Imaging has gained widespread acceptance in both academia and industry as a research tool in both basic research and the clinical setting.[1] The main features of imaging that attract researchers are its noninvasive capabilities and its ability to monitor interventions in a longitudinal setting. The possibility of allowing each animal or human subject to be its own control in longitudinal studies dramatically reduces the sample size required for any study compared to a histological approach, in which only group comparisons can be made. This is well in line with the three *R*s in preclinical research—refine, reduce, and replace—as imaging will reduce the number of animals required for a conclusion to be drawn. In the clinical setting, several imaging modalities offer noninvasive quantification of the disease of interest. There are multiple examples throughout disease areas such as neuroscience, cardiovascular diseases, oncology, respiratory diseases, and many others.

DIFFERENCES BETWEEN TRANSLATIONAL IMAGING AND CONVENTIONAL IMAGING

Using the definition of *translational medicine* found in Wikipedia, as described in the introduction to this book, the equivalent definition of *translational imaging* would be as follows: translational imaging typically refers to the translation of imaging in basic research into the imaging of therapeutic effects in real patients. This definition clearly distinguishes the use of imaging in translational research from the conventional use of imaging for diagnostic purposes. In the conventional use of imaging for diagnostic purposes, the sensitivity and specificity of the test are most important, whereas, as the preceding definition states, in translational imagining, the aim is to measure the therapeutic effect. If we want to quantify a therapeutic effect—described, for example, as area, volume, permeability, uptake, velocity, and so on—the test must be precise and powerful in order to design a reliable experiment with a fair chance of success. In order to determine the power of the test, we need to know its reproducibility and hence understand the underlying properties of the test. This also enables us to separate the variations of the technology platforms from the biological variations. The hardware manufacturers are typically not interested in these features, and this information is therefore often missing. So if we want to use existing imaging techniques for translational research, we are often forced to start the process by determining the reproducibility of the test before we can apply it to any meaningful experiments. Sometimes also published data on the reproducibility of tests are not enough because of the constant drive to improve the sensitivity of tests and their ability to detect low levels and concentrations of the analytes of interest. An example of this is shown in a publication by Szczepaniak and colleagues (2005) in which they investigated the reproducibility of hepatic lipid measurements by magnetic resonance spectroscopy (MRS). They concluded a coefficient of variance of 8% over the entire interval of lipid levels ranging from about 0.5% to 20%. This is fine if we want to assess baseline levels or changes in lipid levels in obese or diabetic subjects. However, it is not sufficient if we want to assess changes in hepatic lipid contents in normal lean subjects as safety biomarkers, as these subjects have a much lower amount of hepatic lipid content and hence a larger coefficient of variance because we are approaching the detection limit of the test. This example clearly shows that, before we can apply imaging in any model or human setting, we need to be aware of the characteristics of the animals or subjects we want to image.

VALIDATION OF IMAGING AND BACK-TRANSLATION

Several of the new imaging techniques currently in use have been developed directly in humans. However, in order for any imaging modality to be used in research, proper validation is required. In many cases this can be performed by simple

1. This chapter reprinted and slightly modified from the first edition.

Principles of Translational Science in Medicine. http://dx.doi.org/10.1016/B978-0-12-800687-0.00020-7

comparison of the imaging data with histological sections from excised specimens. However, for some techniques—for example, vessel wall imaging in the coronary arteries or myocardial infarct size determination—this may not be possible in the clinical setting. In these cases histological validation is impossible simply because of the difficulties of obtaining such specimens. In such cases the validation may have to take place in a preclinical setting; thus, the back-translation of the techniques becomes just as important as the forward translation. An example of this kind of validation study is the work by Kim and colleagues in which the magnetic resonance imaging (MRI)-based technique for infarct size determination in humans was published and later validated in animals (Kim et al., 1999).

IMAGING MODALITIES

In 1901 Wilhelm Conrad Röntgen was awarded the first Nobel Prize in physics in recognition of the extraordinary services he rendered by the discovery of X-rays, which—at least in Europe—were later named after him. Now, more than a hundred years later, great developments in imaging technology, especially during the last three to four decades, have resulted in additional Nobel Prizes awarded for the discovery of new imaging concepts such as computed tomography (CT) and MRI. Also, nuclear medicine imaging techniques including single photon emission computed tomography (SPECT) and positron emission tomography (PET) as well as ultrasound-based techniques have emerged as very useful tools in daily clinical practice. The physical concepts of the various imaging techniques differ dramatically, as do the characteristics of the imaging techniques. This section focuses on the imaging modalities that have been used for both animal and clinical investigations.

X-Ray

X-ray imaging is based on the fact that tissue will absorb photons from an X-ray beam in relation to the electron density of the tissue. This means that bones absorb more photons than lean tissue does. The number of photons passing through the body of interest will then be detected either by film or now by image detectors that convert the body's direct attenuation of the photons into digital images. The resulting images are a two-dimensional projection of a three-dimensional structure. A disadvantage of X-ray-based techniques is the ionizing radiation, which limits the usefulness of these techniques for longitudinal studies.

Computed Tomography

The term *tomography* originates from the word *tomos* (slice) in Greek and *graphein* (to write). It is now used for all modalities in which slices of the body are generated. CT is based on the traditional X-ray technology, but, to generate slices through the body, the X-ray source and the detector are rotated around the body and, through use of a mathematical operation called back-projection, the content in the slice can be calculated.

Ultrasound

Ultrasound imaging is obtained by measuring the reflection of high-frequency or ultrasonic sound waves. By knowing the position of the ultrasound generator, the computer can calculate the distance to the reflection or absorbing structure, and by that, it can generate two- or three-dimensional images of the body. Ultrasound is considered to be safe due to the lack of ionizing radiation.

Magnetic Resonance Imaging

MRI can be fully explained only by the quantum mechanic properties of a spin. For this short introduction, however, classical physics will suffice. For normal imaging, the physical properties of protons are utilized to create contrast between different tissues. The proton density and the water content of the tissue, in particular, will generate the contrast in the images. MRI is especially useful in the differentiation between different soft tissues such as gray versus white matter or tumors versus normal tissue. It requires a strong magnetic field to polarize the protons and a radio frequency transmitter to excite the protons. Following the excitation, the protons in the tissue will transmit a radio frequency signal that can be detected by an RF-antenna (coil), and the computer will calculate the images by two- or three-dimensional Fourier transforms. MRI has the advantage that there is no ionizing radiation, and it is therefore well suited to longitudinal studies in which repeated measurements are required. The sensitivity of MRI is fair—in the mille- to micro-molar range.

Magnetic Resonance Spectroscopy

Whereas MRI can image the position of a lesion with high accuracy, MRS has the ability to tell us something about the tissue's biochemical properties. This is achieved by taking advantage of the fact that protons and other nuclei have somewhat different resonance frequencies, depending on what molecule they are situated in and how they are positioned in that molecule. By collecting all frequencies in the tissue response following an excitation, we are able to determine the relative concentrations of metabolites in a specific tissue.

Gamma Camera and SPECT

Gamma cameras are based on the detection of gamma rays emitted from radionuclides. Radionuclides can be ingested or injected into the body. The camera accumulates counts of gamma photons, which are detected by crystals in the camera. Just like an X-ray, the gamma camera will yield a two-dimensional projection of a three-dimensional object. A tomographic version of the gamma camera is called SPECT, which yields slices through the body. Because the detection techniques of gamma cameras and SPECT are based on the same concept, the same radioisotopes can be used for both techniques. Commonly used isotopes include technetium-99m, iodine-123, and indium-111.

Positron Emission Tomography

PET is based on the concept that isotopes undergo positron emission decay. When the positron hits an electron, two gamma photons are emitted in opposite directions. The system then detects the arrival of the photons, and if they coincide within a few nanoseconds, they are considered to originate from a positron emission and are included in the creation of the image. Because radiolabeled isotopes do not exist naturally in the body, they are administered into a biologically active molecule, and its distribution will hence depend on the metabolism of that molecule. The most common molecule used for PET imaging is fluorodeoxyglucose, a glucose analogue labeled with ^{18}F. It will be taken up by cells in relation to their glucose uptake. Radiolabeling of molecules can also be done by other isotopes such as ^{11}C, ^{13}N, or ^{15}O. The sensitivity of PET is very high—in the nanomolar range. Due to this high sensitivity, very low doses are required for detection. This has led to an increased interest in using this technique for the labeling of new drugs to study biodistribution, receptor occupancy, and other important pharmacological variables. If the amount injected is low enough, the requirements of the existing guidelines dictating that toxicology be performed before injection into humans are reduced. This is called the microdosing concept (Bergström, Grahnén, and Långström, 2003).

CHARACTERISTICS OF VARIOUS IMAGING MODALITIES

Because the imaging modalities described here are based on completely different physical principles, the characteristics of the various imaging modalities differ a lot. Although X-ray-based techniques and MRI have the highest resolution (< 1 mm), their sensitivity is relatively small (in the millimolar range). On the other hand, nuclear medicine techniques such as PET and SPECT have a very high sensitivity (in the nanomolar range) but a relatively low resolution (3–10 mm). This means that the applications of the various techniques differ substantially, as shown in Figure 3.13.

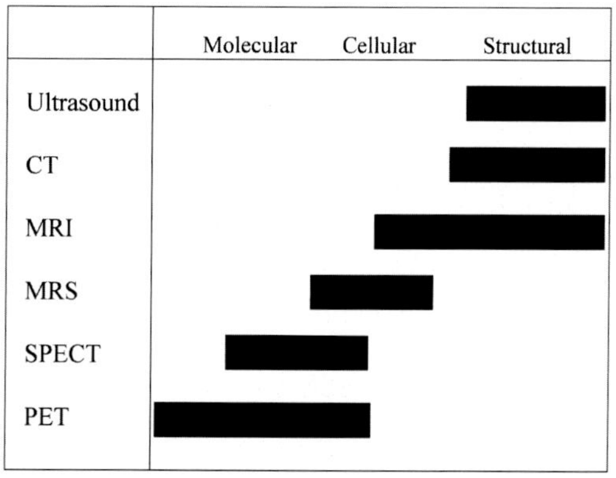

Molecular Cellular Structural
Ultrasound
CT
MRI
MRS
SPECT
PET

FIGURE 3.13 Graph showing the characteristics of the various imaging modalities.

As Figure 3.13 describes, there is a trade-off between resolution and sensitivity among the different methods. To overcome these trade-offs, the hardware manufacturers have recently developed combinations of these technologies, such as SPECT-CT, PET-CT, and, most recently, PET-MRI. These so-called multimodality systems combine the excellent structural information from CT and MRI with the metabolic and molecular information from the nuclear medicine techniques.

EXAMPLES OF TRANSLATIONAL IMAGING IN VARIOUS DISEASE AREAS

In the following sections, examples of translational imaging approaches are given from various disease areas. They are not meant to be complete overviews of imaging in each disease area; rather they are just examples. These examples focus on the methods that can be applied in both animals and humans. There are multiple examples of dedicated imaging probes for CT, MRI, and nuclear medicine techniques designed for specific diseases tested in various animal models. However, most of them require full clinical development before they can be used in humans and are therefore not readily available. The exception to this are probes for PET, which fall under the microdosing concept, as described previously.

Cardiovascular Medicine

In the cardiovascular field, MRI, CT, and ultrasound are frequently used for structural imaging of the heart and vessels. MRI and ultrasound, in particular, have been extensively used in longitudinal studies due to their lack of ionizing radiation. A special version of ultrasound called intravascular ultrasound, which utilizes intravascular probes, has been extensively used in drug trials in humans (Nicholls et al., 2006). However, the soft tissue contrast is limited, and the structural components of the plaques are better visualized by MRI. Plaque composition is of interest because it has been shown that various features of the plaque, such as lipid-rich necrotic cores, fibrous cap thickness, and inflammation, are associated with cardiovascular events (Takaya et al., 2006). MRI of plaques has been used in several intervention studies with statins in which both plaque size (Lima et al., 2004) and plaque composition (Underhill et al., 2008) have been studied with positive results. In addition, it has also been shown that 18F-FDG is taken up in atherosclerotic plaques (Tawakol et al., 2006) and that statins may be able to reduce the inflammation as detected by PET-CT (Tahara et al., 2006). Both of these techniques can be used in both preclinical and clinical settings.

MRI is nowadays considered the gold standard for assessment of left ventricular mass and function. This method has been extensively used in trials in subjects with heart failure and has also recently been used in subjects with myocardial protection (Piot et al., 2008). As described earlier in this chapter, these methods often have to be validated in a preclinical setting.

Metabolic Medicine

With the pandemic increase in cases of obesity, imaging has attracted a lot of interest with regard to imaging metabolic disorders. Imaging offers the potential to study adipose tissue distribution, and CT was developed in the 1980s for this purpose (Sjöström et al., 1986). Since then, a rapid development has taken place, and MRI is now frequently used to assess adipose tissue distribution in a highly automated way (Kullberg et al., 2007), as well as organ lipid content using MRS. All of these techniques can be performed in both animals and humans. Several studies have been published deploying these techniques in interventional drug studies (Eriksson et al., 2008; Juurinen et al., 2008) as well as surgical intervention (Johansson et al., 2008). An example of the latter is shown in Figure 3.14.

Currently, several attempts are being made to develop specific PET-ligands to measure Betacell mass. In these cases, the translational approach is essential for these methods to be successful because, as described earlier, the validation of these studies is extremely complicated in a clinical setting but potentially easier in a preclinical setting. The potential problem that arises is that, in order for this technique to succeed, the same targets must be expressed in animals and humans, and they must be regulated by glucose in a similar fashion.

Oncology

In oncology studies, imaging has been used for several years, and many experiences have been collected. Traditionally, imaging has been used to measure tumor size in interventional trials using the National Cancer Institute's response evaluation criteria in solid tumors (RECIST) (2008). Lately, the use of dynamic contrast enhancement with MRI also has proven to be a useful tool in drug development. The method is based on the fact that tumors typically have a high blood volume and increased vessel permeability. The contrast agents will then give a high response due to an increased tissue distribution and a faster leakage into the interstitial space. A good translational example of this approach is given in two papers investigating the use of a VEGF signal inhibitor in animals (Bradley et al., 2008) and humans (Drevs et al., 2007).

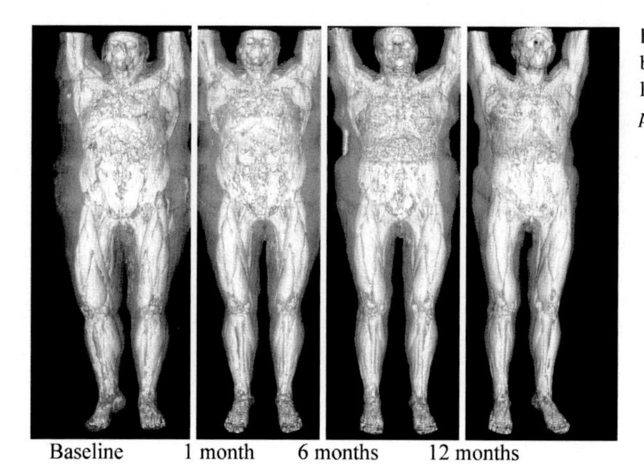

FIGURE 3.14 A subject at baseline and at 1, 6, and 12 months after gastric bypass surgery. The subcutaneous adipose tissue is shown in dark gray, and the lean tissue is shown in light gray. *(Johansson et al., 2008, reprinted by kind permission from Springer Science and Business Media.)*

With the introduction of integrated PET-CT scanners around the year 2000, technology has gained acceptance in clinical oncology. There are multiple studies using FDG-PET in clinical trials. However, there is still a need for quantitative approaches; examples of these are discussed in the paper by Castell and Cook (2008).

Neuroscience

Structural imaging has a long tradition in neuroscience. It has been applied in multiple interventional trials in the central nervous system field. In particular, researchers in the field of neurodegenerative disorders have established structural imaging using MRI as a tool to study hippocampal and whole-brain atrophy. These methods are currently in use in large multicenter trials.

PET has the ability to trace a radiolabeled drug in vivo. Recent advancement in radiochemistry now enables many drugs to be labeled with commonly used radionuclides such as ^{11}C and ^{18}F. After they are injected, brain imaging can be used as confirmation of efficacy, and whole-body imaging can be used for assessing biodistribution and hence safety. There are, however, several obstacles to successfully labeling a drug, as described in the paper by Lee and Farde (2006). These obstacles include radiochemistry, biochemistry, and exposure properties. However, if successfully developed, drug-labeled ligands offer the possibility to study receptor binding (occupancy) in preclinical models and hence guide the dosing of the drug in a clinical trial. Results from clinical studies can also be back-translated into animal models. This iterative process may improve the predictive value of the animal models.

CONCLUSION

Imaging has developed at a fast pace throughout the last few decades and has become accepted as an important biomarker in basic research and drug development. This chapter described imaging methods and examples of how they have been deployed in various research regimes. It is important to remember that the primary use of imaging biomarkers is for decision making in research, although a few imaging biomarkers are approved for regulatory purposes. For a promising imaging biomarker to become a regulatory endpoint, very large investments may be required. These investments typically cannot be taken on by a single party but require collaborative effort among industry, academia, and regulatory bodies.

REFERENCES

Bergström, M., Grahnén, A., Långström, B., 2003. Positron emission tomography microdosing: a new concept with application in tracer and early clinical drug development. Eur. J. Clin. Pharmacol. 59, 357–366.

Bradley, D.P., Tessier, J.J., Lacey, T., Scott, M., Jürgensmeier, J.M., Odedra, R., et al., 2008. Examining the acute effects of cediranib (RECENTIN, AZD2171) treatment in tumor models: a dynamic contrast-enhanced MRI study using gadopentate. Magn. Reson. Imaging 27, 377–384.

Castell, F., Cook, G.J., 2008. Quantitative techniques in 18FDG PET scanning in oncology. Br. J. Cancer 98, 1597–1601.

Drevs, J., Siegert, P., Medinger, M., Mross, K., Strecker, R., Zirrgiebel, U., et al., 2007. Phase I clinical study of AZD2171, an oral vascular endothelial growth factor signaling inhibitor, in patients with advanced solid tumors. J. Clin. Oncol. 25, 3045–3054.

Eriksson, J.W., Jansson, P.A., Carlberg, B., Hägg, A., Kurland, L., Svensson, M.K., et al., 2008. Hydrochlorothiazide, but not candesartan, aggravates insulin resistance and causes visceral and hepatic fat accumulation: the mechanisms for the diabetes preventing effect of candesartan (MEDICA) study. Hypertension 52, 1030–1037.

Johansson, L., Roos, M., Kullberg, J., Weis, J., Ahlström, H., Sundbom, M., et al., 2008. Lipid mobilization following Roux-en-Y gastric bypass examined by magnetic resonance imaging and spectroscopy. Obes. Surg. 18, 1297–1304.

Juurinen, L., Kotronen, A., Granér, M., Yki-Järvinen, H., 2008. Rosiglitazone reduces liver fat and insulin requirements and improves hepatic insulin sensitivity and glycemic control in patients with type 2 diabetes requiring high insulin doses. J. Clin. Endocrinol. Metab. 93, 118–124.

Kim, R.J., Fieno, D.S., Parrish, T.B., Harris, K., Chen, E.L., Simonetti, O., et al., 1999. Relationship of MRI delayed contrast enhancement to irreversible injury, infarct age, and contractile function. Circulation 100, 1992–2002.

Kullberg, J., Ahlström, H., Johansson, L., Frimmel, H., 2007. Automated and reproducible segmentation of visceral and subcutaneous adipose tissue from abdominal MRI. Int. J. Obes. (Lond). 31, 1806–1817.

Lee, C.M., Farde, L., 2006. Using positron emission tomography to facilitate CNS drug development. Trends Pharmacol. Sci. 27, 310–316.

Lima, J.A., Desai, M.Y., Steen, H., Warren, W.P., Gautam, S., Lai, S., 2004. Statin-induced cholesterol lowering and plaque regression after 6 months of magnetic resonance imaging-monitored therapy. Circulation 110, 2336–2341.

National Cancer Institute, 2008. Cancer imaging program: imaging response criteria. http://imaging.cancer.gov/clinicaltrials/imaging (accessed 2008).

Nicholls, S.J., Sipahi, I., Schoenhagen, P., Crowe, T., Tuzcu, E.M., Nissen, S.E., 2006. Application of intravascular ultrasound in anti-atherosclerotic drug development. Nat. Rev. Drug. Discov. 5, 485–492.

Piot, C., Croisille, P., Staat, P., Thibault, H., Rioufol, G., Mewton, N., et al., 2008. Effect of cyclosporine on reperfusion injury in acute myocardial infarction. N. Engl. J. Med. 359, 473–481.

Sjöström, L., Kvist, H., Cederblad, A., Tylén, U., 1986. Determination of total adipose tissue and body fat in women by computed tomography, 40K, and tritium. Am. J. Physiol. 250, E736–745.

Szczepaniak, L.S., Nurenberg, P., Leonard, D., Browning, J.D., Reingold, J.S., Grundy, S., et al., 2005. Magnetic resonance spectroscopy to measure hepatic triglyceride content: prevalence of hepatic steatosis in the general population. Am. J. Physiol. Endocrinol. Metab. 288, E462–468.

Tahara, N., Kai, H., Ishibashi, M., Nakaura, H., Kaida, H., Baba, K., et al., 2006. Simvastatin attenuates plaque inflammation: evaluation by fluorodeoxyglucose positron emission tomography. J. Am. Coll. Cardiol. 48, 1825–1831.

Takaya, N., Yuan, C., Chu, B., Saam, T., Underhill, H., Cai, J., et al., 2006. Association between carotid plaque characteristics and subsequent ischemic cerebrovascular events: a prospective assessment with MRI—initial results. Stroke 37, 818–823.

Tawakol, A., Migrino, R.Q., Bashian, G.G., Bedri, S., Vermylen, D., Cury, R.C., et al., 2006. In vivo 18F-fluorodeoxyglucose positron emission tomography imaging provides a noninvasive measure of carotid plaque inflammation in patients. J. Am. Coll. Cardiol. 48, 1818–1824.

Underhill, H.R., Yuan, C., Zhao, X.Q., Kraiss, L.W., Parker, D.L., Saam, T., et al., 2008. Effect of rosuvastatin therapy on carotid plaque morphology and composition in moderately hypercholesterolemic patients: a high-resolution magnetic resonance imaging trial. Am. Heart. J. 155, 584. e1–8.

Chapter 3.7.4

Translational Medicine in Psychiatry: Challenges and Imaging Biomarkers

Andreas Meyer-Lindenberg, Heike Tost and Emanuel Schwarz

BIOLOGICAL TREATMENT OF PSYCHIATRIC DISORDERS

Even a cursory view at global morbidity and mortality statistics makes it clear that psychiatric disorders are among the most common and most disabling conditions in medicine. In the WHO statistics measuring the number of years of disability due to illness, 4 of the top 10 disease categories (addiction, depression, schizophrenia, and anxiety disorders) fall into the domain of psychiatry, and, combined, mental disorders cause more disability than any other class of medical illness in Americans between ages 15 and 44 (Insel and Scolnick, 2006). Up to 40% of the population must expect to have a psychiatric condition necessitating treatment at least once in their life, and 20% are being treated at any given time. The direct and especially indirect costs (including absenteeism and presenteeism at work) of mental illness to society are staggering, most prominently so for disorders like schizophrenia that often have a chronic course and afflict patients in adolescence or early adulthood, interfering with education and the ability to compete for employment on the open market.

It is widely accepted that mental disorders, although they manifest themselves in the domain of psychology and social behavior, nevertheless have a strong and often predominant biological basis: many diseases in the purview of psychiatry—especially the most severe ones, such as schizophrenia, bipolar disorder, and depression—are highly heritable, indicating that the major contributor of disease risk is, in fact, genetic (Meyer-Lindenberg and Weinberger, 2006; Sullivan et al., 2012).

Even for disorders in which the impact of (usually adverse) life events is clear, such as anxiety disorders, depression, or borderline personality disorder, heritability typically still accounts for around 30%–50% of disease risk. This is, in principle, good news in the sense that it enables a biological approach to these illnesses that has proved powerful in other areas of medicine and that adds to the armamentarium of psychiatry. To address this spectrum of biological, environmental, and individual risk factors, psychiatrists are called upon to choose wisely from available psychotherapeutical, social or rehabilitative, and biological interventions that need to be tailored to the individual patients if the burden of mental illness is to be combated effectively. In this chapter, we focus on biological interventions.

Given their frequency and severity, it should not come as a surprise that psychiatric or psychotropic medications account for a large proportion of the budget spent on drugs in industrialized nations, and, as a group, they are usually among the top two or three spending categories. Antipsychotics, antidepressants, mood stabilizers such as anticonvulsants, and anti-anxiety agents account for the majority of this expense, reflecting the importance and prevalence of schizophrenia, depression, bipolar disorder, and anxiety disorders. A large body of evidence in the literature shows that these drugs are effective in treating these illnesses and useful in managing anxiety, impulsivity, and mood in many others. Does this indicate that patients, psychiatrists, and the general public get a satisfactory return on their investment? Most researchers in the field would agree that this is not the case.

For some disorders such as autism and anorexia nervosa, biomedical treatments for the core symptoms are absent. For others, such as schizophrenia, current treatments are efficacious, but only for circumscribed symptom clusters and to a degree that often falls short of recovery. It is clear that conventional and atypical antipsychotics reduce the so-called positive psychotic symptoms of schizophrenia, an indispensable component of the current therapy for this disease, but it is considerably less clear whether they are helpful for cognitive deficits or negative symptoms (Lieberman and Stroup, 2011) Two large-scale trials, the Clinical Antipsychotic Treatment of Intervention Effectiveness (CATIE) trial (Lieberman et al., 2005a) and the Cost Utility of the Latest Antipsychotic Drugs in Schizophrenia Study (CUTLASS) trial (Jones et al., 2006),

Principles of Translational Science in Medicine. http://dx.doi.org/10.1016/B978-0-12-800687-0.00021-9

found that only about one in four patients with chronic schizophrenia continued either a first- or second-generation antipsychotic throughout an 18-month trial, and the data did not show a difference in efficacy between older and newer (so-called second-generation antipsychotic) drugs. A large European trial, the European First Episode Schizophrenia Trial (EUFEST), also found no differences in efficacy, although it did find better adherence to second-generation drugs. However, this trial was open, and therefore, patient and therapist bias might have contributed to this finding (Kahn et al., 2008). In a 5-year follow-up of 118 schizophrenic patients after their first episode of psychosis, only 13.7% met criteria for full remission lasting 2 years or more. This rate of remission is disappointingly low. In essence, schizophrenia remains a chronic, disabling disorder, just as it was a century ago. The most dramatic confirmation of this contention is the fact that the mortality of this illness has not declined in the area of pharmacotherapy and may, in fact, have increased (Murray-Thomas et al., 2013).

Things may not be much better in the clinical practice of the treatment of depression. Although antidepressants reduce the symptoms of major depressive disorder, these drugs are significantly better than placebo only after 6 weeks of administration. More importantly, many studies show that patients on medication continue to have substantial depressive symptoms at the end of the trial, and a large proportion of patients require a trial of several drugs, dose adjustment, or comedication.

Insel and Scolnick (2006) reviewed the portfolio of available treatments for schizophrenia and depression and concluded that the number of mechanistically distinct drug classes (as opposed to drugs that have the same mechanism of action but different side effect profiles and chemical formulation) available to treat these disorders had remained essentially stagnant since the 1950s. This appeared to be a problem of translational success specific to psychiatry, because the number of drug classes used to treat hypertension, which was used as a comparison complex frequent disorder from internal medicine, was similar in the 1950s but rose by a factor of 10 in the following years. As a consequence, the therapeutic arsenal for these psychiatric disorders lagged behind that for hypertension by almost an order of magnitude. Despite the fact that investments into R&D have increased considerably throughout the industry, and that funding for life sciences research in academia has also grown precipitously in many countries, the output of novel substances (the "cures") has not kept up. It seems therefore reasonable to ask whether there are obstacles to translation and innovation in psychiatry. If the answer is yes, what can be done to overcome them?

SPECIFIC CHALLENGES OF TRANSLATION IN PSYCHIATRY

Unknown Pathophysiology

A fundamental problem specific to psychiatry is immediately visible: psychiatry lags behind almost any other branch of medicine in the understanding of the pathophysiology of the illnesses it is called upon to treat. One reason for this is, of course, that the "organ" in which mental illness plays out is the human brain, whose complexity far exceeds that of most other biological phenomena. However, enormous strides have been made in understanding brain function on the molecular, cellular, and systems level, which again calls into question why these advances in neuroscience have been so slow in translating into treatment (Insel, 2014; Insel and Gogtay, 2014). One considerable problem here may be the nature of psychiatric disease entities themselves. An illness such as depression or schizophrenia is currently defined by an account of symptoms such as depressed mood, sleep disturbance, or hallucinations, which are elicited from the patient in clinical interview or observed, combined with certain criteria regarding the course of the illness (e.g., specifying a minimum duration of symptoms), impairment of function, and exclusion criteria (e.g., requiring the exclusion of certain somatic illnesses that may present similarly). These diagnostic criteria are manualized and applied internationally, leading to a high degree of reliability that is necessary to, for example, compare incidence rates or conduct a clinical study multinationally. Although they are reliable, psychiatric disease entities such as depression or schizophrenia nevertheless have no biological parameters that establish these diagnoses: there is no laboratory test for schizophrenia or bipolar disorder (Kapur, Phillips, and Insel, 2012). This means that the biological validity of these phenotypes is likely to be low, in the sense that it is very likely that several as yet incompletely characterized biological processes could underlie the clinical entity we now call depression. Seen this way, psychiatric disorders are closer on a categorical level to dyspnea or blindness, in the sense that these are also direct clinical observables that are biologically strongly heterogeneous, and not on the level of, for example, left ventricular failure or macular degeneration—disorders that can cause these clinical conditions but can be biologically verified or excluded. This should not be construed to say that psychiatric disease entities are arbitrary or even somehow not "real"—they not only cause enormous suffering, as detailed previously, but also are "good enough" to exhibit very high heritability, showing a clear biological foundation that, however, we still have to successfully parse.

This state of affairs creates considerable problems for translation: modern translational medicine starts from an account of the molecular pathophysiology of the illness and uses it, through the techniques discussed in this text, to discover and validate new treatments. This necessitates the identification of drug targets and the screening of candidate substances in vivo and in vitro, including in animal models, before they can proceed to clinical studies. To the considerable degree that the information about molecular pathophysiology is lacking in psychiatry, this approach will be difficult.

A second problem—one that is (almost) specific to and definitely characteristic of psychiatry—is that many psychiatric disorders display features that are specific to humans: for example, schizophrenia and bipolar disorder often have an impact on aspects of formal thought and language. Although we believe that we can identify and create states of fear and drug dependence in animal models (Stewart and Kalueff, 2014), this becomes less obvious when we turn to depression and seems intractable for a disease such as schizophrenia: what would be the mouse model for hallucinations?

Stigma and the Second Translation

Even if a novel treatment modality becomes available, it needs to be accepted by patients and physicians and implemented into routine care. This second translation also faces specific problems in psychiatry, as most psychiatric disorders carry a degree of stigma and public discrimination unlike that of most somatic disorders. Schizophrenia is a notable example of such a stigmatized condition (Mestdagh and Hansen, 2014). This reduces not only the quality of life but the likelihood that a patient given this diagnosis will seek and maintain treatment, leading to worse outcomes, especially in disorders that are chronic and may require maintenance therapy. This carries relevant side effects that necessitate a considerable degree of patient adherence.

Recent data clearly show that service delivery in psychiatry is deficient. The *Schizophrenia* Patient Outcomes Research Team (PORT) study demonstrated that evidence-based treatments are severely underused (Lehman et al., 2004). A meta-analysis of psychoeducation showed that although medium effect sizes in preventing relapse can be obtained, at least in the first 12 months (Lincoln, Wilhelm, and Nestoriuc, 2007), only a minority of patients receive this treatment. An even starker picture is painted by a 10-year follow-up study of depression in Japan (Furukawa et al., 2008), in which the majority of patients received inadequate medical treatment and not a single participant received psychotherapy, despite the fact that clearly cost-effective treatment modalities exist.

NEW BIOMARKERS FOR TRANSLATION IN PSYCHIATRY

It has already been stated that the search for new treatments must start from an understanding of the molecular pathophysiology of the disorder. That is less obvious than it seems. For the past five decades, the biological study of mental illness has been dominated by the study of the action of the available drugs. This research has uncovered many fundamental and important aspects of these disorders, such as the relevance of the dopamine system, a discovery that was honored by Nobel prizes awarded to Greengard and Carlsson. However, two main problems with this approach exist for translation. First, understanding how drugs work is helpful only insofar as these drugs target the core pathophysiology of the disorders. For example, understanding how insulin works is certainly useful in understanding aspects of diabetes, but it is unlikely to lead to an understanding of the autoimmune basis of type I diabetes. Second, and more important to the current context, using animal and human models based on currently available drugs to screen candidate substances is essentially a recipe for developing "me-too compounds." For example, behavioral tests based on a dopaminergic blockade, such as catalepsy, are currently used in the screening process for antipsychotic substances, inducing a bias toward drugs with this mechanism, which they then share with currently available antipsychotics. To move beyond this process, translational research in psychiatry needs a push toward target identification through pathophysiology. What biomarkers can be derived here?

A first point of departure is in genetics. The prior waves of genomewide association studies (GWAS) have identified numerous frequent genetic variants associated with both schizophrenia and bipolar disorder, and other diseases such as autism are about to follow (Sullivan et al., 2012). Understanding the mechanism of action of these genetic variants will have a high potential to identify new treatment targets. Second, a novel GWAS result has been the identification of rare copy number variants under negative selection that increase the risk of psychiatric disorders by an order of magnitude and affect, as for example in the case of chromosome 15q11.2 deletions, brain structure in a pattern consistent with that observed in first-episode psychosis (Stefansson et al., 2014). Here again, dissection of the contribution of individual genes in these microdeleted regions is likely to considerably advance our understanding of the molecular architecture of psychiatric disorders, especially if combined with "omics" approaches. Because even if the gene product of a given risk gene variant is not by itself druggable, proteomics can be used to identify interaction partners that are. This gene-driven proteomic mining approach is likely to gain considerable traction when the interactomes of several risk genes are considered jointly, because interaction partners shared between risk genes can be expected to have a higher likelihood of being associated with the disorder. First examples of this approach are appearing in the literature. Identification of druggable targets, in turn, can be a point of departure to set in motion the machinery of high-throughput drug discovery that is described elsewhere in this volume and that has been instrumental to mechanistically developing new drugs in the somatic medical disciplines.

Finally, neuroimaging enables an approach specific to psychiatry. The ability to measure details of the structure and function of the living human brain has been instrumental in advancing our knowledge of the systems that are involved in the pathophysiology of these diseases (Meyer-Lindenberg, 2010). Understanding these neural systems, which, as we will see, are distributed networks processing information connecting areas of prefrontal cortex with regions of the limbic system and striatum and midbrain, can aid drug discovery in three ways: first, by understanding the neural systems implicated in psychiatric disorders, a new generation of veridical animal models can be created. For example, if we find that interactions between the hippocampal formation and the prefrontal cortex are important for schizophrenia, mimicking this situation in a rodent is a more tractable problem than taking the point of departure from human-specific symptoms such as thought disorder. Second, the explosion of information available about the molecular architecture of the rodent and human brain makes the construction of priors or even the selection of specific molecular targets for drug discovery possible in some instances. Third, an understanding of the neural systems involved in a given disorder may enable a new generation of go-or-no-go decisions in early drug development by allowing an early statement about whether a novel compound impacts a relevant neural system by using neuroimaging in healthy probands or in patients (Insel and Gogtay, 2014). All of these approaches gain, as we will see, considerable additional power from the ability of neuroimaging to identify neural systems that are impacted by genetic variation linked to psychiatric disorders. This so-called imaging genetics approach joins genetic and neural systems information to a powerful tool whose capabilities for drug discovery, response prediction, and individualized medicine are starting to become tested. Given the pivotal and specific status of imaging biomarkers in psychiatry, the following discussion provides an overview of the developing field of imaging biomarkers, focusing on the paradigmatic disorder, schizophrenia.

IMAGING BIOMARKERS IN SCHIZOPHRENIA

In the past 15 years, numerous cognitive, functional, morphological, and metabolic anomalies of the brain have been described in schizophrenia patients, suggesting an overall heterogeneous disease entity rather than a circumscribed pathology. At the same time, major progress has been made with respect to the delineation of a number of core schizophrenia phenomena, especially disturbances of dopaminergic neurotransmission, frontal lobe efficiency, and neural plasticity. This section reviews the current scientific knowledge on magnetic resonance imaging (MRI) biomarkers in schizophrenia. In doing so, it attempts to give special consideration to the neurocognitive domains most critically affected by the disorder, while examining advances in the visualization of antipsychotic medication effects and the impact of schizophrenia susceptibility genes on the neural system level. Due to the sheer volume of published results, this overview does not claim to be all-inclusive but rather focuses on selected areas of this vital and still-expanding research field.

Structural Brain Biomarkers

The original pathogenetic theory of Kraepelin postulates that schizophrenia is a progressive neurodegenerative disease state with an atypically early age of onset (Kraepelin, 1919). Over the past several decades, this historic assumption and the repeated finding of ventricular enlargement in affected patient populations has stimulated a large number of postmortem studies attempting to shed light on the question of whether a primarily degenerative process is involved in the development of the disease (Henn and Braus, 1999; Nasrallah, 1982; Stevens, 1982). In the past several decades, these studies evidenced a variety of subtle histopathological changes in the brains of schizophrenia patients, especially aberrantly located neurons in the entorhinal cortex (Arnold, Ruscheinsky, and Han, 1997; Jakob and Beckmann, 1986), smaller perikarya of cortical pyramidal neurons (Benes et al., 1991; Benes, Sorensen, and Bird, 1991; Rajkowska, Selemon, and Goldman-Rakic, 1998), and a reduction in dendritic arborization (Garey et al., 1998; Glantz and Lewis, 2000). These data were complemented by evidence suggesting alterations in volume, density, or total number of neurons in several subcortical structures, especially the basal ganglia, thalamus, hippocampus, and amygdala (Kreczmanski et al., 2007; Walker et al., 2002; Zaidel, Esiri, and Harrison, 1997). Major neuropathological hallmarks of a neuronal degenerative process were, however, lacking, especially the proliferation of astrocytes and microglia typically seen in the context of a neural degenerative process (e.g., reactive gliosis) (Benes, 1999; Longson, Deakin, and Benes, 1996). This observation led to the formulation of alternative pathogenetic disease models that have postulated a (nonprogressive) prenatal disturbance in neurodevelopment as the main pathogenetic mechanism. One influential hypothesis assumes that schizophrenia emerges from intrauterine disturbances of temporolimbic-prefrontal network formation, resulting in the manifestation of overt clinical symptoms in early adulthood (Benes, 1993; Lewis and Levitt, 2002; Weinberger, 1987). The concept is supported by data showing that candidate susceptibility genes of schizophrenia (Brandon et al., 2004; Sei et al., 2007) and known epigenetic risk factors (Meyer, Yee, and Feldon, 2007) interfere with neuronal migration processes during central nervous system (CNS) development.

Since the middle of the 1980s, the availability of MRI has allowed for the noninvasive examination of structural brain biomarkers in schizophrenia patients (see Wright et al., 2000 and Honea et al., 2005 for a comprehensive review). Earlier studies in the field demonstrated global abnormalities in terms of increased ventricular size (Johnstone et al., 1976) and smaller mean cerebral volume (Gur et al., 1999; Ward et al., 1996). Subsequent regional analyses were performed by manually delineating a priori defined regions of interest (ROI) on MRI scans, an approach that yielded one of the most consistent findings in schizophrenia research, a bilateral decrease in hippocampal gray matter volume (Henn and Braus, 1999). Available meta-analyses suggest that the mean hippocampal gray matter (GM) volume in schizophrenia patients is decreased to 95% of the volume of normal controls (Wright et al., 2000). In contrast, other subcortical structures like the basal ganglia have repeatedly been reported to show a volumetric increase in patient populations. This particular observation, however, is most likely explained by the effects of long-term exposure to neuroleptic agents, as this finding has not been observed in antipsychotic-naïve patients (Chua et al., 2007; Keshavan et al., 1998) and has proven to be reversed when the medication regimen is switched from conventional neuroleptics to atypical antipsychotic substances (e.g., olanzapine) (Lang et al., 2004; Scheepers et al., 2001). A recent meta-analysis of 317 studies investigating brain volume alterations in schizophrenia showed decreased intracranial and total brain volume as well as gray matter structures in medicated patients, whereas reductions in the caudate nucleus and thalamus were more pronounced in antipsychotic-naïve patients (Haijma et al., 2013). This study has shown that longer duration of illness and higher dose of antipsychotic medication at time of scanning were associated with stronger gray matter volume reductions. This evidence is consistent with longitudinal brain-volumetric change in schizophrenia, which is supported by meta-analyses of region-of-interest studies that have indicated that the disorder is associated with progressive structural brain abnormalities affecting both gray and white matter and ranging from 0.07% to 0.59% in annualized percentage change compared to control subjects (Olabi et al., 2011; Vita et al., 2012). Meta-analysis has also shown that such progressive gray matter changes are regionally specific affecting especially the left hemisphere and the superior temporal structures and are particularly active in the first stages of schizophrenia (Vita et al., 2012). It is interesting to note that relapse duration (but not number) has been found to be associated with brain volume decreases in schizophrenia, highlighting the importance of relapse prevention and early intervention if relapse occurs, as well as the potential utility of MRI-derived biomarkers for such clinical applications (Andreasen et al., 2013). In the past few decades, the implementation of new automated processing techniques has allowed for the large-scale analysis of structural MRI datasets. Voxel-based morphometry (VBM) is a fully automated method that allows for the unbiased analysis of the structural differences of the whole brain on a voxel-by-voxel basis (Ashburner and Friston, 2000). This method has been shown to be sensitive to subtle regional changes in tissue volume and concentration that are otherwise inaccessible to standard imaging techniques (Krams et al., 1999; Maguire et al., 2000; Valfre et al., 2008). This automated approach is superior to conventional ROI analyses in terms of regional sensitivity and reliability of structural findings and is especially helpful in situations in which the pathology does not reflect traditional neuroanatomical boundaries. Further methodological advances like the development of fully automated whole-brain segmentation methods have also facilitated the volumetric analysis of subcortical structures in large cohorts (Meda et al., 2008). One of our own large-scale, automated MRI segmentation studies (Goldman et al., 2008), which was conducted in 221 healthy controls, 169 patients with schizophrenia, and 183 unaffected siblings, demonstrated a bilateral decrease in hippocampal and cortical gray matter volume in schizophrenia patients compared to the healthy controls. Moreover, evidence of the heritability of cortical and hippocampal volume reductions was derived.

A thorough meta-analysis of VBM studies identified GM reductions in the bilateral superior temporal gyrus (STG) (left: 57% of studies, right: 50% of studies), left medial temporal lobe (69% of studies), and left medial and inferior frontal gyrus (50% of studies) as the most consistent regional findings in schizophrenia (Honea et al., 2005). Other studies have shown that some of these morphological abnormalities are already observable at disease onset and are thus unlikely to be solely attributable to the effects of illness chronicity or medication (Gur et al., 1998; Keshavan et al., 1998; Molina et al., 2006; Szeszko et al., 2003). Similarly, meta-analysis has shown that frontotemporal structural alterations determined through VBM are already present in presymptomatic individuals at high risk of developing schizophrenia (Chan et al., 2011). Longitudinal studies performed in first-episode (Cahn et al., 2002; DeLisi et al., 1997; Gur et al., 1998; Lieberman et al., 2001), chronic (Davis et al., 1998; Mathalon et al., 2001), and childhood-onset (Rapoport et al., 1999) patients suggest a progressive development of GM volumetric reduction over the course of the illness. Longitudinal voxel-based morphometry analysis of first-episode antipsychotic-naïve schizophrenia patients followed up for 4 years also found progressive gray matter changes, some of which were related to functional outcome (Mané et al., 2009). In view of the postmortem data reviewed previously, the neurobiological significance of the observed magnetic resonance (MR) signal changes remains to be elucidated (Weinberger and McClure, 2002). Although most likely of neurodevelopmental origin, the implications of the evidenced volumetric brain changes are still a matter of controversy, especially regarding the question of whether the evidenced alterations in the GM and cerebrospinal fluid (CSF) compartments reflect an actual loss of neurons or, instead, reductions in neuropil and neuronal size (Harrison, 1999). Importantly also, most patient samples do not only vary by disease status from healthy individuals, but also by a variety of other confounds linked to brain volume changes such as the exposure to alcohol (Taki et al., 2006), tobacco (Brody et al., 2004), social environmental risk factors (Gianaros et al., 2007), and medication. The last aspect is of particular importance, as both the acute (Tost et al., 2010) and chronic exposure to antipsychotics (Ho et al., 2011) appear to have profound effects on structural neuroimaging measures, which complicates the interpretation of existing data.

While many of these questions remain currently unresolved, it is interesting to note that readouts of different structural abnormalities have been combined into diagnostic classification rules using machine learning approaches, and initial evidence exists that such rules can not only discriminate accurately between schizophrenia patients and controls (Nieuwenhuis et al., 2012; Schnack et al., 2014; Takayanagi et al., 2011), but also when compared against relevant differential diagnoses, such as bipolar disorder (Schnack et al., 2014). This is especially interesting in the context of high-risk prediction, where it has been shown that frontal cortex volume reductions can predict conversion to full-blown psychosis in ultra-high-risk individuals (Dazzan et al., 2012; McIntosh et al., 2011).

Functional Imaging Markers in Schizophrenia

Since the early 1990s, functional brain anomalies in mental disease states have been predominantly examined by means of one neuroimaging technique: functional magnetic resonance imaging (fMRI). The popularity of the method can be explained by numerous favorable attributes, especially the broad availability of clinical scanners, the noninvasiveness of the technique, and the broad spectrum of complementary MRI applications that can offer additional insights into the structure and biochemistry of the living brain. Concurrent with the growing popularity of this technique, the development of fMRI experiments has been refined from simple paradigms with blockwise stimulation to rapid, event-related task designs. On the data analysis side, sophisticated methods for connectivity analyses of neural network interactions have been developed (Meyer-Lindenberg et al., 2005a; Stein et al., 2007) Advancements in processing technology and speed also paved the way for real-time fMRI applications that allow providing patients with neurofeedback and the possibility to modulate their own functional brain activity, a promising approach toward a new treatment modality for patients with mental illness (Weiskopf, 2012). In the meantime, the success of fMRI has given rise to an enormous amount of published fMRI studies in schizophrenia research. The following section reviews major findings in the neurocognitive domains most critically affected by the disorder.

Auditory and Language Processing

Abnormalities of the STG and associated language areas of the temporal and frontal lobe have emerged as some of the most prominent biomarkers in schizophrenia research. In the past two decades, structural and functional deficits of these regions have been extensively examined in the context of auditory hallucinations, a cardinal positive symptom of schizophrenia that has been conceptualized as dysfunctional processing of silent inner articulations (David and Cutting, 1994). Early milestone work in schizophrenia research demonstrated the spontaneous activity of speech areas during hallucinatory experiences, a fact that explains why these internal voices are accepted as real, despite arising in the absence of any external sensory stimulation (Dierks et al., 1999; Ffytche et al., 1998; McGuire, Shah, and Murray, 1993; Silbersweig et al., 1995). Subsequently, several fMRI studies have provided evidence of a functional deficit that affects multiple levels of auditory perception and language processing. On a lower level, a diminished response of the primary auditory cortex to external speech has been observed during hallucinatory experiences, a fact that is best explained by the competition of physiological and pathological processes for limited neural processing resources (David et al., 1996; Woodruff et al., 1997). It should be noted that the extent of STG dysfunction during speech processing seems to predict the severity of the patient's thought disorder, a clinical symptom that manifests as irregularities in speech (Weinstein et al., 2006). On the neural networks level, an impaired functional coupling of the STG and anterior cingulate gyrus (ACG) seems to be at the core of the misattribution of the patient's own voice versus alien voices that frequently characterizes patients with auditory hallucinations (Mechelli et al., 2007).

In terms of brain structure, scientific evidence points to an anatomical correlate of the acoustic hallucinations in schizophrenia. Early morphometric studies repeatedly reported a decrease in the physiological leftward asymmetry of the planum temporale, a higher-order auditory processing area that overlaps with Wernicke's area (Kwon et al., 1999; Shapleske et al.,

1999; Sumich et al., 2005). Voxel-based morphometric studies have confirmed a significant gray matter decrease in the STG that predicted the severity of experienced auditory hallucinations (Aguilar et al., 2008) as well as several other symptom dimensions quantified using the Positive and Negative Syndrome Scale (Lui et al., 2009). Also, gray matter reductions of the STG have been shown to be progressive during the transition period from an ultra-high-risk state to first-episode schizophrenia with acute psychosis (Takahashi et al., 2009), an effect that could not be observed in longitudinal assessment of chronic schizophrenia patients (Yoshida et al., 2009). Temporal volumetric alterations have also been identified by a meta-analysis of antipsychotic-naïve VBM studies to potentially underlie the clinical onset of psychotic symptoms (Fusar-Poli et al., 2011). Moreover, folding abnormalities of the STG and Broca's area, as well as microstructural changes of the main fibers connecting the white matter tract to the frontal lobe (arcuate fasciculus), seem to promote the emergence of hallucinatory symptoms (Cachia et al., 2008; Catani et al., 2011; Gaser et al., 2004; Hubl et al., 2004; Neckelmann et al., 2006; Steinmann, Leicht, and Mulert, 2014).

Motor Functioning

Disturbances of psychomotor functions are well-known features of schizophrenia and range from subtle deficits like neurological soft signs to extensive behavioral abnormalities like stereotypia and catatonia (Schroeder et al., 1991; Vrtunski et al., 1986). In the early years of schizophrenia imaging research, the examination of motor dysfunctions was very popular, usually involving a block-design approach in which simple repetitive motor activities alternated with rest conditions (e.g., finger tapping, pronation-supination). In addition to other findings, hypoactivations of the primary and higher-order supplementary motor and premotor cortices have been reported in schizophrenia patients (Braus et al., 1999; Buckley et al., 1997; Mattay et al., 1997; Schröder et al., 1995; Schröder et al., 1999; Wenz et al., 1994). Moreover, within the highly lateralized motor system, a physiologically abnormal symmetry (i.e., reduced laterality) of recruited neural resources has emerged as a cardinal fMRI biomarker of motor system dysfunctions in schizophrenia (Bertolino et al., 2004a; Mattay et al., 1997; Rogowska, Gruber, and Yurgelun-Todd, 2004). On the neural networks level, a functional deficiency of transcallosal glutamergic projections has been suggested (Mattay et al., 1997); this assumption is in line with current neurodevelopmental disease models and has been encouraged by corresponding electrophysiological findings (Hoppner et al., 2001).

In recent years, fMRI studies in the motor domain have been scarce. The pronounced impact of antipsychotic agents on the examined circuitry and the lack of studies in unmedicated patient samples offer two potential explanations for this fact (Tost et al., 2006). There is evidence suggesting that both schizophrenia patients and their unaffected relatives are characterized by a lack of striatal activation increase during movement anticipation (Vink et al., 2006). Although this finding is suggestive of an endophenotypic trait marker, it still awaits replication. On the structural level, the functional findings are complemented by reports of gray matter volume decreases and impaired structural integrity of the connecting white matter tracts of the higher-order processing areas of the motor system (Exner et al., 2006; Goghari et al., 2005; Hirjak et al., 2014; Matsui et al., 2002; Walther et al., 2011).

Working Memory

Convergent evidence from several neuroscience disciplines (genetic, molecular, cellular, physiological, and neuroimaging) suggests that schizophrenia is a genetically complex disorder of brain development that promotes downstream disturbances in dopaminergic neurotransmission and subsequent impairments in prefrontal cortical efficiency. As a result, schizophrenia patients suffer from a variety of higher-order cognitive functions known to be dependent on the integrity of the dorsolateral and medial prefrontal lobe, especially cognitive flexibility, selective attention, response inhibition, and working memory (WM). Previous experimental work in animals and humans has shown that mesocortical dopamine (DA), especially the stimulation of the dopamine D1 receptor subtype, plays a crucial role in the modulation of WM (dys)functions (Fuster 1990; Goldman-Rakic, 1995; Williams and Goldman-Rakic, 1995). According to this evidence, an inverted U-shaped relationship exists between the amount of D1 receptor stimulation, WM-related activation of prefrontal cortex (PFC) neurons, and prefrontal cognitive efficacy (Callicott et al., 1999; Williams and Goldman-Rakic, 1995). Although balanced D1 dopaminergic tone seems to augment the robustness of PFC network representations by making them less susceptible to background neural "noise," states of excessive or lacking dopaminergic drive seem to weaken the system's robustness to interfering stimuli and, as a result, promote the development of cognitive deficits and of psychotic symptoms (Durstewitz, Seamans, and Sejnowski, 2000; Mattay et al., 2003).

In the past decade, the neurobiological basis of WM dysfunctions in schizophrenia has been extensively examined with fMRI. On a general level, this cognitive asset can be divided into maintenance and manipulation subprocesses. Classical WM paradigms challenge the active storage and online manipulation of information, often assessed in the form of so-called *n*-back tasks, in which patients are asked to monitor an ongoing sequence of stimulus presentations and respond to items that match the one presented *n* stimuli before. The majority of studies reported disorder-related dysfunctions of the dorsolateral prefrontal cortex (DLPFC) (Brodmann areas 46 and 9), as well as anomalies in the functional coupling of this area to the medial temporal lobe (Meyer-Lindenberg et al., 2001; Meyer-Lindenberg et al., 2005a). The precise pathophysiological background of these findings is still, however, a matter of debate, as diverse anomalies in the form of hypofunctions (Andreasen et al., 1997; Barch et al., 2001; Paulman et al., 1990; Schneider et al., 2007; Weinberger, Berman, and Zec, 1986), increased activations (Callicott et al., 2000; Potkin et al., 2009a), and combined hyper- and hypoactive states (Callicott et al., 2003; Karlsgodt et al., 2007; Kim et al., 2009; Lee et al., 2008) have been reported. This inconsistency has raised questions about the traditional theory of pure functional "hypofrontality" in schizophrenia and has stimulated the formulation of more complex models of prefrontal cortex dysfunction (Callicott et al., 2003; Callicott et al., 2000; Manoach, 2003; Potkin et al., 2009a). In another study reporting combined hyper- and hypoactive states, increased and decreased DLPFC and parietal activity was associated with higher and lower performance during the fMRI task, respectively, and this has been suggested to support the theory that a combined hyper- and hypoactivation may reflect a continuum of behavioral performance in schizophrenia (Karlsgodt et al., 2007). FMRI work in first-episode schizophrenia patients confirmed a preferential impairment of the manipulation component of WM and reported a disproportional engagement of the ventrolateral prefrontal cortex (VLPFC, Brodmann areas 44, 45, and 47) during task performance (Tan et al., 2005). This finding has been interpreted as a deficiency in the functional specialization and hierarchical organization of the prefrontal cortex. The deficit is thought to manifest itself in terms of reduced efficiency in DLPFC functioning, which in turn triggers the compensatory recruitment of hierarchically inferior and less specialized areas in the VLPFC (Tan et al., 2006). Recent evidence suggests a relationship between working memory capacity at the beginning of the illness and the extent of the subsequent prefrontal cortical thinning during course of the illness (Gutiérrez-Galve et al., 2014). Furthermore, working-memory-related activation and connectivity features have been found in unaffected relatives, suggesting their usefulness as intermediate phenotypes for genetic investigations.

Selective Attention

On a general level, selective attention describes the mental capacity to maintain a behavioral or cognitive set in the face of distracting or competing stimuli. The term covers several cognitive subprocesses, especially the top-down sensitivity control, competitive selection, and automatic bottom-up filtering for salient stimuli (Knudsen, 2007). Since Bleuler's first clinical descriptions of schizophrenia (Bleuler, 1950), attentional deficits have been considered core symptoms of the condition that eventually leads to overt psychopathological symptoms like thought disorder, incoherence of speech, and disorganized behavior. Previous neuroimaging research in psychiatry has adopted different experimental techniques to examine the neurobiological correlates of these deficits, the most popular being the so-called continuous performance test (CPT). The term describes a heterogeneous array of nonstandardized paradigms in which participants are meant to selectively respond to target presentations, while inhibiting responses to nontargets. Besides these basic features, most CPTs additionally challenge other cognitive subprocesses like visual attention (e.g., degraded CPTs), cognitive interference monitoring, or working memory (e.g., contingent CPTs like CPT-AX, CPT-IP, and CPT-double-T). The heterogeneity of CPT (and other) paradigms must be taken into account when interpreting available fMRI data on attentional dysfunctions in schizophrenia.

Many studies in the field have used variations of a contingent CPT to examine the neural correlates of attentional dysfunctions in schizophrenia. Most of these studies reported functional anomalies of the DLPFC (Lesh et al., 2013; MacDonald and Carter, 2003; Perlstein et al., 2003; Volz et al., 1999) and VLPFC (Eyler et al., 2004) in schizophrenia patients and their unaffected first-grade relatives (Diwadkar et al., 2011), findings that are most likely related to the moderate working memory demands of this particular task type. Similar findings have been observed in first-episode neuroleptic-naïve patients during CPT task performance, suggesting that this functional biomarker is at the core of the disorder and not merely a medication-induced phenomenon (Barch et al., 2001). In line with this assumption, Delawalla et al. (2008) observed task-related hyperactivation of DLPFC in unaffected siblings of schizophrenia patients, suggesting that this deficit resembles an intermediate phenotype of the genetic susceptibility to the illness. In contrast, most of the studies that used paradigms challenging interference monitoring observed dysfunctions of the ACG. Reported anomalies in schizophrenia include regional ACG hypoperfusion (Carter et al., 1997), disturbed interregional connectivity to the prefrontal cortex (Eyler et al., 2004; Honey et al., 2005), and the absence or dislocation of activation foci within the ACG (Heckers et al., 2004). Evidence from structural MRI studies suggests that the observed functional and behavioral deficits relate to a morphological and microstructural impairment (Nazeri et al., 2013), in particular the ACG and its main white matter fiber tract, the cingulum bundle (Artiges et al., 2006; Kubicki et al., 2003; Salgado-Pineda et al., 2004; Spalletta et al., 2014; Sun et al., 2003; Yücel et al., 2002).

IMAGING OF GENETIC SUSCEPTIBILITY FACTORS

Schizophrenia is a highly heritable mental disorder with a complex genetic architecture. Current evidence suggests that multiple genetic risk variants, each accounting for only a small increment of the risk of the development of the disorder, interact both with one another and with the environment. On the neural systems level, this interaction interferes with the functional properties of multiple brain circuits that, in turn, shape a variety of different cognitive, emotional, and behavioral functions and dysfunctions (see Figure 3.15 for an illustration of these concepts) (Meyer-Lindenberg and Weinberger, 2006).

Multiple genetic risk variants, through interactions with one another and with the environment, affect multiple neural systems linked to several neuropsychological and behavioral domains that are impaired, in differing proportions, in psychiatric diseases. As examples, the following genetic variants are depicted (chromosomal variation in parentheses): GRM3 single nucleotide polymorphism 4 (7q21.1–q21.2) (Egan et al., 2004), dopamine receptor D2 (DRD2) Taq 1a (11q23) (Cohen et al., 2005), catechol-O-methyltransferase (COMT) Val66Met (22q11.2) (Egan et al., 2001; Meyer-Lindenberg et al., 2006b), serotonin transporter length polymorphism (5-HTTLPR/SLC6A4) (17q11.1–q12) (Hariri et al., 2002; Pezawas et al., 2005), and monoamine oxidase A variable number tandem repeat (MAOA VNTR) (Xp11.23) (Meyer-Lindenberg et al., 2006a). These variants are shown to affect a circuit that links the PFC with the midbrain (MB) and striatum (caudate and putamen); this circuit is (1) relevant to schizophrenia, and a circuit that connects the amygdala (AM) with regulatory cortical and limbic areas, and (2) implicated in depression and anxiety. These circuits, in turn, are shown to mediate risk for schizophrenia, depression, and various neuropsychological functions: Brodmann's area 25 (BA 25), hippocampal formation (HF), and orbitofrontal cortex (OFC).

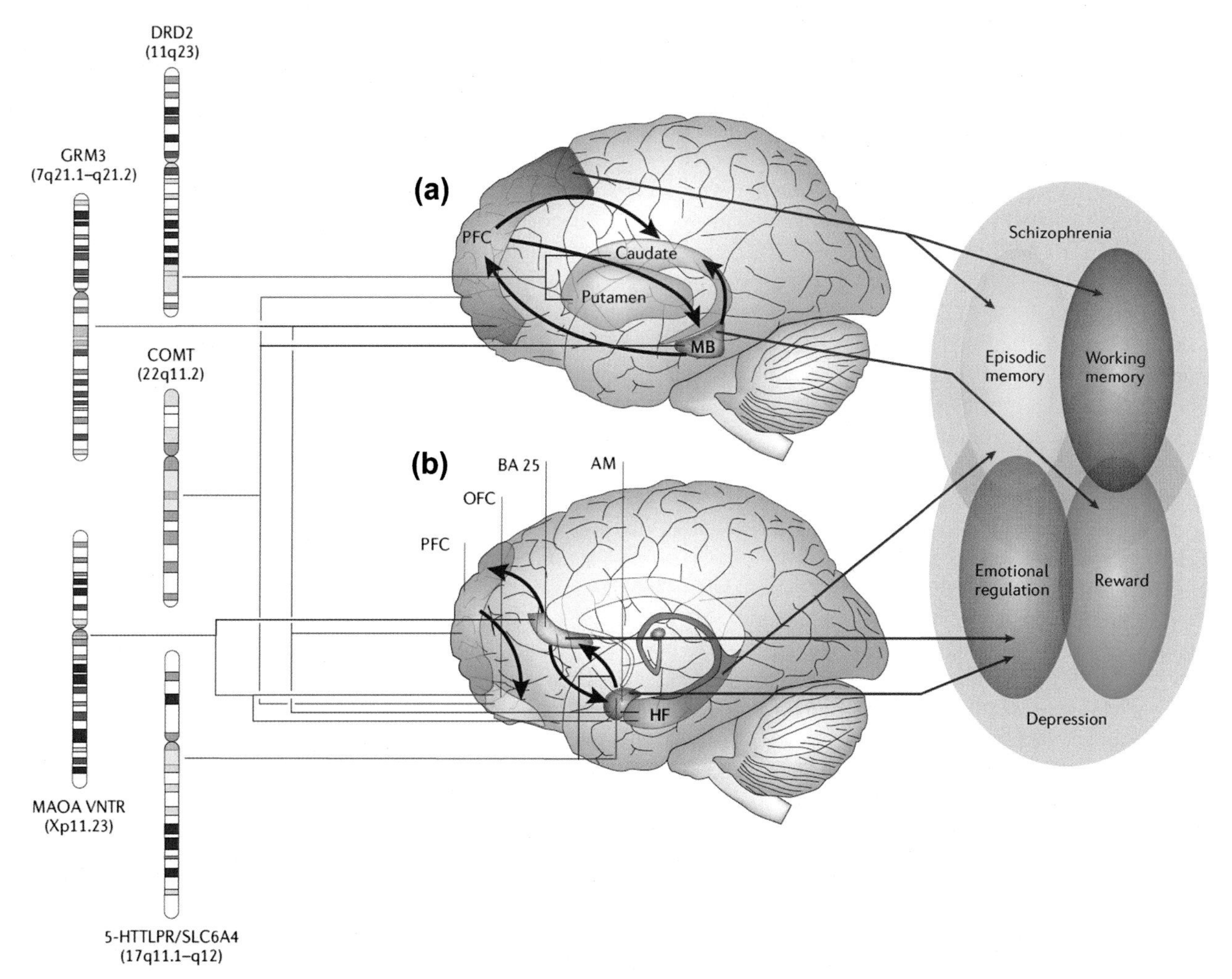

FIGURE 3.15 Depiction of the complex path from genes to behavioral and disease phenotype. *(From Meyer-Lindenberg and Weinberger, 2006, reprinted by permission from Macmillan Publishers Ltd:* Nat. Rev. NeuroSci. *7: 818–827, © 2006.)*

Clearly, there is no one-to-one mapping between risk gene variants and neural system mechanisms, or between neural mechanisms and psychopathology, a fact that renders the identification of valid biomarkers on the basis of genetics alone difficult. Imaging genetics, the characterization of susceptibility gene mechanisms on the neural systems level using multimodal neuroimaging, has proven a successful research strategy for overcoming this obstacle. Many gene variants associated with an enhanced risk of the development of schizophrenia are frequent in healthy individuals. The imaging genetics approach assumes that the penetrance of gene effects is greater at the neurobiological level than at the level of complex behavior, and that these gene effects are traceable at the neural systems level in carriers of risk alleles, even when no overt clinical signs of the disease phenotype are expressed. In recent years, the attempt to genetically dissect schizophrenia with neuroimaging methods has led to the characterization of several intermediate phenotypes (i.e., core pathophysiologic characteristics observable at the level of the neural substrate that bridge the gap between genetic variation and psychiatric symptoms).

The strongest evidence for the efficacy of this approach arises from intermediate phenotype studies on catechol-O-methyltransferase (COMT), a major enzyme for the degradation of catecholamines in the central nervous system. The COMT gene is located in a chromosomal region that has been implicated in schizophrenia linkage studies (Owen, Williams, and O'Donovan, 2004; Talkowski et al., 2008); in the 22q11 deletion syndrome (22q11DS) (Karayiorgou and Gogos, 2004), which is a hemideletion syndrome with a 30-fold increased risk of developing a schizophrenia-like illness (Murphy, 2002); and, more recently, in copy number variation in sporadic schizophrenia (Stark et al., 2008). Due to the lack of dopamine transporters in the PFC, the regulation of extracellular dopamine levels in the PFC is critically dependent on COMT functioning (Lewis et al., 2001). As has been previously shown, a common val$^{(108/158)}$met substitution in the COMT gene interferes with the thermostability of the transcribed protein, leading to significantly decreased enzyme efficacy (Chen et al., 2004). Studies from our laboratory have demonstrated that the val$^{(108/158)}$met coding variant in the COMT gene impacts PFC functional measures during working memory performance (Egan et al., 2001), modulates subjects' performance in neuropsychological tests challenging executive functioning (Goldberg et al., 2003), and influences the cortical response to amphetamine in healthy subjects (Mattay et al., 2003). These data extend basic evidence for an inverted-U functional response curve to increasing dopamine signaling in the PFC, and they validate the concept of prefrontal cortical inefficiency as the key endomechanism promoting the risk of schizophrenia. According to this model, the COMT genotype places individuals at predictable points along the inverted U-shaped curve that links prefrontal dopaminergic stimulation, neuronal activity, and PFC efficiency. Homozygotes for the val-encoding allele are thought to be positioned to the left of met allele carriers at a point of decreased PFC efficiency, whereas the met allele carriers seem to be optimally located near the peak of that curve (val/val: COMT efficacy ↑, synaptic DA ↓; met/met: COMT efficacy ↓, synaptic DA ↑). The genetic risk associated with the COMT val$^{(108/158)}$met coding variant is thought to be mediated by a reduced signal-to-noise ratio in the PFC, an idea that is supported by the finding that WM-related and WM-unrelated PFC activity are inversely related to neuroimaging markers of midbrain DA synthesis, which in turn is directionally dependent on the COMT val$^{(108/158)}$met genotype (Meyer-Lindenberg et al., 2005b).

Subsequent work from our laboratory (Meyer-Lindenberg et al., 2007) indicates that a frequent haplotype of the gene encoding of the dopamine- and cAMP-regulated phosphoprotein DARPP-32 (PPP1R1B) is associated with the risk of schizophrenia, and that it impacts measures of frontostriatal structure, function, and cognition. DARPP-32 is a major target for dopamine-activated adenylyl cyclase and serves as an important functional switch integrating the multiple information streams that converge onto dopaminoceptive neurons in the striatum (i.e., striatal neurotransmitters, neuropeptides, and neurosteroids) (Svenningsson et al., 2004). It has been shown that DARPP-32 is a key node in the final common pathway of psychotomimetic action in the prefrontal cortex and striatum (Svenningsson et al., 2003), making it an attractive candidate gene for schizophrenia. Our imaging genetics study showed a pronounced and convergent effect of genetic variation in PPP1R1B on the function and volume of the striatum, and related measures of frontal-striatal connectivity. Moreover, a pronounced impact of PPP1R1B variation on a wide range of prefrontal cognitive measures was observed. These findings might suggest that PPP1R1B contributes to the risk of schizophrenia by disturbed gating (Swerdlow, Geyer, and Braff, 2001) of fronto-striatal information flow. Interestingly, AKT1, another key molecule in a noncanonical signal transduction pathway for dopaminergic neurons, showed an impact on fronto-striatal circuitry as well (Tan et al., 2008), establishing a convergent neural signature for postsynaptic dopaminergic neurotransmission.

In addition to associations between candidate genetic schizophrenia risk variants and brain activity, we have demonstrated that a common genetic variant in *ZNF804A* that has been linked to both schizophrenia as well as bipolar disorder through GWAS shows gene-dose dependent associations with functional coupling (correlated activity) of the DLPFC with hippocampus, mirroring findings in patients (Meyer-Lindenberg et al., 2001), their unaffected first-grade relatives (Rasetti et al., 2011), and genetic mouse models of schizophrenia (Sigurdsson et al., 2010). These results indicated that even in the absence of changes in regional activity, imaging genetics approaches have utility in uncovering disturbed connectivity patterns as neurogenetic risk mechanisms for psychiatric illnesses (Esslinger et al., 2009). The hypothesis that *ZNF804A* modulates schizophrenia risk through impaired brain connectivity but not structure is consistent with recent studies showing that variation in this gene does not affect total or regional brain volumes in healthy young adults (Cousijn et al., 2012) as well as schizophrenia (Donohoe et al., 2011).

Finally, an interesting avenue for discovering genetic risk variants for mental illnesses within an imaging genetics framework is the direct use of neuroimaging readouts as quantitative traits for GWAS. Based on this strategy, Potkin et al., showed that using the mean activation in the DLPFC in response to the Sternberg working memory task as a quantitative trait for GWAS analysis, genetic variants with significant genotype-diagnosis interactions could be identified (Potkin et al., 2009b).

CHARACTERIZATION OF ANTIPSYCHOTIC DRUG EFFECTS

The standard medical treatment for schizophrenia is antipsychotic medication. First-generation antipsychotics like haloperidol or fluphenazine (also referred to as conventional antipsychotics or major tranquilizers) belong to a class of antipsychotic drugs developed in the 1950s. Due primarily to their D_2-receptor blocking properties, this substance class is very efficient at relieving positive symptoms but also bears the potential for unwanted side effects such as extrapyramidal side effects (EPS) or tardive dyskinesia. In the past 15 years, available treatment options for schizophrenia patients were extended substantially by the development of second-generation, or "atypical," antipsychotic drugs (e.g., clozapine, olanzapine). The term *atypical antipsychotics* refers to a biochemically heterogeneous group of drugs characterized by the absence of EPS and increased efficacy compared to conventional neuroleptics for the treatment of negative symptoms and cognitive deficits (see Miyamoto et al., 2005 for a comprehensive review). Previous evidence has suggested that the degree and timing of the clinical response to atypical antipsychotics is modulated by the patient's genetic profile for dopamine catabolic enzymes. As discussed earlier, the met variant of the val158met COMT gene polymorphism inactivates prefrontal dopamine at a slower rate (Lachman et al., 1996), a mechanism that is associated with a greater improvement of negative symptoms (Bertolino et al., 2004b). Bertolino and colleagues (Bertolino et al., 2007) replicated this finding and provided additional evidence of a faster response to atypical antipsychotics in the met allele carriers. Although the precise mechanism of these effects is not yet fully understood, this finding suggests that met allele–carrying patients may have greater benefit from olanzapine treatment because of their relative excess of prefrontal cortical dopamine. On the neural systems level, several fMRI studies have provided evidence suggesting a favorable impact of atypical antipsychotics on previously "disturbed" functional brain patterns. Bertolino and Colleagues (2004a) examined the impact of olanzapine on motor loop functioning and observed an alleviation of the sensorimotor hypoactivations in the unmedicated state. Similarly, Stephan and Colleagues (2001) reported a restoration of functional cerebellar connectivity during motor task performance under olanzapine. A more recent fMRI study on the effects of antipsychotic treatment on motor-task-related activation differences in schizophrenia reported a drug-dose dependent diminishing effect of antipsychotics on activity in cortical and subcortical motor and default mode networks (Abbott et al., 2011). Other relevant performance domains that have been associated with favorable functional effects under atypical antipsychotics include working memory (Meisenzahl et al., 2006), verbal fluency (Jones et al., 2004), and conflict detection (Snitz et al., 2005). Another interesting avenue of investigation has been the assessment of antipsychotic effects on resting state neural activity using fMRI. In this context it has been shown that 6-weeks second-generation antipsychotic treatment resulted in widespread increased regional synchronous neural activity, especially in the bilateral prefrontal and parietal cortex, left superior temporal cortex, and right caudate nucleus and that this was correlated with a reduction of clinical symptoms, and a widespread attenuation in functional connectivity (Lui et al., 2009). Similar restorative effects of antipsychotics on brain function in the visual attention network have been reported in a recent longitudinal study in first-episode schizophrenia patients (Keedy et al., 2014).

The differential effect of atypical versus typical antipsychotic drugs has been examined mainly in cross-sectional designs. Several studies examining fMRI task performance related to motor functioning (Braus et al., 1999), working memory (Honey et al., 2005), and prepulse inhibition (Kumari et al., 2007) have suggested a superior effect of second-generation antipsychotics over classical neuroleptics on functional brain measures in schizophrenia. On the structural brain level, evidence derived with different MRI techniques supports this notion. In a multicenter, longitudinal study in 161 patients with first-episode psychosis, Lieberman and Colleagues (2005b) observed a reduction of global brain GM volume over time in patients treated with haloperidol, but not in patients treated with olanzapine or healthy control subjects. Similarly, previous MRI spectroscopy studies focusing on N-acetylaspartate and phospholipid metabolism have suggested a favorable impact of atypical antipsychotics in the PFC that is superior to the effect of traditional neuroleptics (Bertolino et al., 2001; Braus et al., 2001; Ende et al., 2000; Smesny et al., 2012; Szulc et al., 2005). The neurobiological basis for this differential drug effect, however, remains unknown. Possible explanations include a greater therapeutic effect of atypical antipsychotics on disease-inherent brain morphological changes, as well as the possibility of a neurotoxic effect associated with the application of conventional neuroleptics.

Only a few neuroimaging studies have examined the impact of antipsychotic drugs on functional brain measures in healthy volunteers. In a double-blind cross-over drug challenge, Abler, Erk, and Walter (2007) examined the influence of olanzapine (5 mg) on reward-related brain activations and observed a drug-related activation decrease in the ventral striatum, ACG, and inferior frontal cortex. Our own prior work in the visual (Brassen et al., 2003) and motor system (Tost et al., 2006) on the effect of a single dose of haloperidol (5 mg/kg) confirmed that antipsychotic drugs have a primarily dampening effect on functional brain measures. During motor task performance, for example, a significant drug-related activation decrease in the dorsal striatum and an increased lateralization of motor cortex activations were observed. In line with previous work in the field, a preferential interference of haloperidol with cognitive measures challenging mental and behavioral flexibility (related to dopamine D_2 receptor functioning) was evidenced, while working memory performance (related to dopamine D_1 receptor functioning) remained unaffected (Tost et al., 2006). These results emphasize the need to control for medication effects in fMRI studies examining psychiatric populations.

CONCLUSIONS AND FUTURE DIRECTIONS

Among research disciplines, neuroscience has introduced the most fundamental changes in the way in which mental disease states are conceptualized and pursued today. In terms of funding, modern-day psychiatric research increasingly faces the challenge of bridging the gap between theoretical concepts and practical solutions, while focusing available resources on questions that will likely lead to future therapeutic applications. Nowhere is this more obvious than in translational medicine: the arduous, slow, empirical process of finding new treatments must be properly incentivized for this work to be done by the current and next generation of psychiatric neuroscientists. This will also require thinking outside the box regarding the interactions of clinical and industrial research; neither of the two can effectively go the entire way to a mechanistically new drug alone, and instead of focusing on ethical questions regarding payments to physician-researchers in late or promotional phases of psychiatric drugs, it is urgently necessary to create an ethically secure environment in which academia and industry can interact more closely. An exemplary initiative in this regard is the Innovative Medicines Initiative (IMI) of the European Union, which brings together EFPIA pharmaceutical firms with consortia formed from academia and small and medium enterprises to tackle drug discovery, or the ECNP Medicines chest project, which provides access by academic researchers to compounds that are currently not or no longer pursued by the companies that originally developed them. The practical need for this development is founded not only on the tremendous emotional burden that psychiatric illness causes in the affected individuals and their families, but also on the enormous treatment expenses that mental disorders impose on our society in general, an amount that exceeds an estimated $70 billion per year in the United States alone (Office of the Surgeon General, 1999). As we have seen, in the past decades, neuroimaging methods have provided unique insights into the core neuropathophysiological processes associated with the development and treatment of schizophrenia. As a crucial part of most multimodal research approaches, these techniques have helped to characterize the mechanisms that translate disease vulnerability from the genetic level to the molecular, cellular, and neural systems level, as well as to the level of overt behavioral disturbances. Within this translational framework, the development and application of neuroimaging methods is expected to pioneer future improvements in disorder prevention, diagnosis, and treatment.

REFERENCES

Abbott, C., Juarez, M., White, T., Gollub, R.L., Pearlson, G.D., Bustillo, J., et al., 2011. Antipsychotic dose and diminished neural modulation: a multi-site fMRI study. Prog. Neuropsychopharmacol. Biol. Psychiatry 35, 473–482.

Abler, B., Erk, S., Walter, H., 2007. Human reward system activation is modulated by a single dose of olanzapine in healthy subjects in an event-related, double-blind, placebo-controlled fMRI study. Psychopharmacology (Berl) 191, 823–833.

Aguilar, E.J., Sanjuan, J., Garcia-Marti, G., Lull, J.J., Robles, M., 2008. MR and genetics in schizophrenia: focus on auditory hallucinations. Eur. J. Radiol. 67, 434–439.

Andreasen, N.C., Liu, D., Ziebell, S., Vora, A., Ho, B.C., 2013. Relapse duration, treatment intensity, and brain tissue loss in schizophrenia: a prospective longitudinal MRI study. Am. J. Psychiatry 170, 609–615.

Andreasen, N.C., O'Leary, D.S., Flaum, M., Nopoulos, P., Watkins, G.L., Boles Ponto, L.L., et al., 1997. Hypofrontality in schizophrenia: distributed dysfunctional circuits in neuroleptic-naïve patients. Lancet 349 (9067), 1730–1734.

Arnold, S.E., Ruscheinsky, D.D., Han, L.Y., 1997. Further evidence of abnormal cytoarchitecture of the entorhinal cortex in schizophrenia using spatial point pattern analyses. Biol. Psychiatry 42, 639–647.

Artiges, E., Martelli, C., Naccache, L., Bartres-Faz, D., Leprovost, J.B., Viard, A., et al., 2006. Paracingulate sulcus morphology and fMRI activation detection in schizophrenia patients. Schizophr. Res. 82, 143–151.

Ashburner, J., Friston, K.J., 2000. Voxel-based morphometry—the methods. Neuroimage 11, 805–821.

Barch, D.M., Carter, C.S., Braver, T.S., Sabb, F.W., MacDonald III, A., Noll, D.C., et al., 2001. Selective deficits in prefrontal cortex function in medication-naïve patients with schizophrenia. Arch. Gen. Psychiatry 58, 280–288.

Benes, F.M., 1993. The relationship between structural brain imaging and histopathologic findings in schizophrenia research. Harv. Rev. Psychiatry 1, 100–109.

Benes, F.M., 1999. Evidence for altered trisynaptic circuitry in schizophrenic hippocampus. Biol. Psychiatry 46, 589–599.

Benes, F.M., McSparren, J., Bird, E.D., SanGiovanni, J.P., Vincent, S.L., 1991. Deficits in small interneurons in prefrontal and cingulate cortices of schizophrenic and schizoaffective patients. Arch. Gen. Psychiatry 48, 996–1001.

Benes, F.M., Sorensen, I., Bird, E.D., 1991. Reduced neuronal size in posterior hippocampus of schizophrenic patients. Schizophr. Bull. 17, 597–608.

Bertolino, A., Blasi, G., Caforio, G., Latorre, V., De Candia, M., Rubino, V., et al., 2004a. Functional lateralization of the sensorimotor cortex in patients with schizophrenia: effects of treatment with olanzapine. Biol. Psychiatry 56, 190–197.

Bertolino, A., Caforio, G., Blasi, G., De Candia, M., Latorre, V., Petruzzella, V., et al., 2004b. Interaction of COMT (Val(108/158)Met) genotype and olanzapine treatment on prefrontal cortical function in patients with schizophrenia. Am. J. Psychiatry 161, 1798–1805.

Bertolino, A., Caforio, G., Blasi, G., Rampino, A., Nardini, M., Weinberger, D.R., et al., 2007. COMT Val158Met polymorphism predicts negative symptoms response to treatment with olanzapine in schizophrenia. Schizophr. Res. 95, 253–255.

Bertolino, A., Callicott, J.H., Mattay, V.S., Weidenhammer, K.M., Rakow, R., Egan, M.F., et al., 2001. The effect of treatment with antipsychotic drugs on brain N-acetylaspartate measures in patients with schizophrenia. Biol. Psychiatry 49, 39–46.

Bleuler, E., 1950. Dementia praecox or the group of schizophrenias (1911). International Universities Press, New York.

Brandon, N.J., Handford, E.J., Schurov, I., Rain, J.C., Pelling, M., Duran-Jimeniz, B., et al., 2004. Disrupted in schizophrenia 1 and Nudel form a neurodevelopmentally regulated protein complex: implications for schizophrenia and other major neurological disorders. Mol. Cell. Neurosci. 25, 42–55.

Brassen, S., Tost, H., Hoehn, F., Weber-Fahr, W., Klein, S., Braus, D.F., 2003. Haloperidol challenge in healthy male humans: a functional magnetic resonance imaging study. Neurosci. Lett. 340, 193–196.

Braus, D.F., Ende, G., Weber-Fahr, W., Demirakca, T., Henn, F.A., 2001. Favorable effect on neuronal viability in the anterior cingulate gyrus due to long-term treatment with atypical antipsychotics: an MRSI study. Pharmacopsychiatry 34, 251–253.

Braus, D.F., Ende, G., Weber-Fahr, W., Sartorius, A., Krier, A., Hubrich-Ungureanu, P., et al., 1999. Antipsychotic drug effects on motor activation measured by functional magnetic resonance imaging in schizophrenic patients. Schizophr. Res. 39, 19–29.

Brody, A.L., Olmstead, R.E., London, E.D., Farahi, J., Meyer, J.H., Grossman, P., et al., 2004. Smoking-induced ventral striatum dopamine release. Am. J. Psychiatry 161, 1211–1218.

Buckley, P.F., Friedman, L., Wu, D., Lai, S., Meltzer, H.Y., Haacke, E.M., et al., 1997. Functional magnetic resonance imaging in schizophrenia: initial methodology and evaluation of the motor cortex. Psychiatry. Res. Neuroimaging 74, 13–23.

Cachia, A., Paillere-Martinot, M.L., Galinowski, A., Januel, D., de Beaurepaire, R., Bellivier, F., et al., 2008. Cortical folding abnormalities in schizophrenia patients with resistant auditory hallucinations. Neuroimage 39, 927–935.

Cahn, W., Hulshoff Pol, H.E., Lems, E.B., van Haren, N.E., Schnack, H.G., van der Linden, J.A., et al., 2002. Brain volume changes in first-episode schizophrenia: a 1-year follow-up study. Arch. Gen. Psychiatry 59, 1002–1010.

Callicott, J., Mattay, V., Verchinski, B.A., Marenco, S., Egan, M.F., Weinberger, D.R., 2003. Complexity of prefrontal cortical dysfunction in schizophrenia: more than up or down. Am. J. Psychiatry 160, 2209–2215.

Callicott, J.H., Bertolino, A., Mattay, V.S., Langheim, F.J., Duyn, J., Coppola, R., et al., 2000. Physiological dysfunction of the dorsolateral prefrontal cortex in schizophrenia revisited. Cereb. Cortex 10, 1078–1092.

Callicott, J.H., Mattay, V.S., Bertolino, A., Finn, K., Coppola, R., Frank, J.A., et al., 1999. Physiological characteristics of capacity constraints in working memory as revealed by functional MRI. Cereb. Cortex 9, 20–26.

Carter, C.S., Mintun, M., Nichols, T., Cohen, J.D., 1997. Anterior cingulate gyrus dysfunction and selective attention deficits in schizophrenia: [15O]H2O PET study during single-trial Stroop task performance. Am. J. Psychiatry 154, 1670–1675.

Catani, M., Craig, M.C., Forkel, S.J., Kanaan, R., Picchioni, M., Toulopoulou, T., et al., 2011. Altered integrity of perisylvian language pathways in schizophrenia: relationship to auditory hallucinations. Biol. Psychiatry 70, 1143–1150.

Chan, R.C.K., Di, X., McAlonan, G.M., Gong, Q.Y., 2011. Brain anatomical abnormalities in high-risk individuals, first-episode, and chronic schizophrenia: An activation likelihood estimation meta-analysis of illness progression. Schizophr. Bull. 37, 177–188.

Chen, J., Lipska, B.K., Halim, N., Ma, Q.D., Matsumoto, M., Melhem, S., et al., 2004. Functional analysis of genetic variation in catechol-O-methy ltransferase (COMT): effects on mRNA, protein, and enzyme activity in postmortem human brain. Am. J. Hum. Genet. 75, 807–821.

Chua, S.E., Cheung, C., Cheung, V., Tsang, J.T., Chen, E.Y., Wong, J.C., et al., 2007. Cerebral grey, white matter and CSF in never-medicated, first-episode schizophrenia. Schizophr. Res. 89, 12–21.

Cohen, M.X., Young, J., Baek, J.M., Kessler, C., Ranganath, C., 2005. Individual differences in extraversion and dopamine genetics predict neural reward responses. Brain Res. Cogn. Brain Res. 25, 851–861.

Cousijn, H., Rijpkema, M., Harteveld, A., Harrison, P.J., Fernández, G., Franke, B., Arias-Vásquez, A., 2012. Schizophrenia risk gene ZNF804A does not influence macroscopic brain structure: an MRI study in 892 volunteers. Mol. Psychiatry 17, 1155–1157.

David, A., Cutting, J., 1994. The neuropsychological origin of auditory hallucinations. Lawrence Erlbaum Associates, Hove, Sussex.

David, A.S., Woodruf, P.W., Howard, R., Mellers, J.D., Brammer, M., Bullmore, E., et al., 1996. Auditory hallucinations inhibit exogenous activation of auditory association cortex. Neuroreport 7, 932–936.

Davis, K.L., Buchsbaum, M.S., Shihabuddin, L., Spiegel-Cohen, J., Metzger, M., Frecska, E., et al., 1998. Ventricular enlargement in poor-outcome schizophrenia. Biol. Psychiatry 43, 783–793.

Dazzan, P., Soulsby, B., Mechelli, A., Wood, S.J., Velakoulis, D., Phillips, L.J., et al., 2012. Volumetric abnormalities predating the onset of schizophrenia and affective psychoses: an MRI study in subjects at ultrahigh risk of psychosis. Schizophr. Bull. 38, 1083–1091.

Delawalla, Z., Csernansky, J.G., Barch, D.M., 2008. Prefrontal cortex function in nonpsychotic siblings of individuals with schizophrenia. Biol. Psychiatry 63 (5), 490–497.

DeLisi, L.E., Sakuma, M., Tew, W., Kushner, M., Hoff, A.L., Grimson, R., 1997 Schizophrenia as a chronic active brain process: a study of progressive brain structural change subsequent to the onset of schizophrenia. Psychiatry Res. 74, 129–140.

Dierks, T., Linden, D.E., Jandl, M., Formisano, E., Goebel, R., Lanfermann, H., et al., 1999. Activation of Heschl's gyrus during auditory hallucinations. Neuron 22, 615–621.

Diwadkar, V.A., Segel, J., Pruitt, P., Murphy, E.R., Keshavan, M.S., Radwan, J., et al., 2011. Hypo-activation in the executive core of the sustained attention network in adolescent offspring of schizophrenia patients mediated by premorbid functional deficits. Psychiatry Res 192, 91–99.

Donohoe, G., Rose, E., Frodl, T., Morris, D., Spoletini, I., Adriano, F., et al., 2011. ZNF804A risk allele is associated with relatively intact gray matter volume in patients with schizophrenia. Neuroimage 54, 2132–2137.

Durstewitz, D., Seamans, J.K., Sejnowski, T.J., 2000. Dopamine-mediated stabilization of delay-period activity in a network model of prefrontal cortex. J. Neurophysiol. 83, 1733–1750.

Egan, M.F., Goldberg, T.E., Kolachana, B.S., Callicott, J.H., Mazzanti, C.M., Straub, R.E., et al., 2001. Effect of COMT Val108/158 Met genotype on frontal lobe function and risk for schizophrenia. Proc. Natl. Acad. Sci. U. S. A. 98, 6917–6922.

Egan, M.F., Straub, R.E., Goldberg, T.E., Yakub, I., Callicott, J.H., Hariri, A.R., et al., 2004. Variation in GRM3 affects cognition, prefrontal glutamate, and risk for schizophrenia. Proc. Natl. Acad. Sci. U. S. A. 101, 12604–12609.

Ende, G., Braus, D.F., Walter, S., Weber-Fahr, W., Soher, B., Maudsley, A.A., et al., 2000. Effects of age, medication, and illness duration on the N-acetyl aspartate signal of the anterior cingulate region in schizophrenia. Schizophr. Res. 41, 389–395.

Esslinger, C., Walter, H., Kirsch, P., Erk, S., Schnell, K., Arnold, C., et al., 2009. Neural mechanisms of a genome-wide supported psychosis variant. Science 324, 605.

Exner, C., Weniger, G., Schmidt-Samoa, C., Irle, E., 2006. Reduced size of the pre-supplementary motor cortex and impaired motor sequence learning in first-episode schizophrenia. Schizophr. Res. 84, 386–396.

Eyler, L.T., Olsen, R.K., Jeste, D.V., Brown, G.G., 2004. Abnormal brain response of chronic schizophrenia patients despite normal performance during a visual vigilance task. Psychiatry Res 130, 245–257.

Ffytche, D.H., Howard, R.J., Brammer, M.J., David, A., Woodruff, P., Williams, S., 1998. The anatomy of conscious vision: an fMRI study of visual hallucinations. Nat. Neurosci. 1, 738–742.

Furukawa, T.A., Fujita, A., Harai, H., Yoshimura, R., Kitamura, T., Takahashi, K., 2008. Definitions of recovery and outcomes of major depression: results from a 10-year follow-up. Acta. Psychiatr. Scand. 117, 35–40.

Fusar-Poli, P., Borgwardt, S., Crescini, A., Deste, G., Kempton, M.J., Lawrie, S., Mc Guire, P., Sacchetti, E., 2011. Neuroanatomy of vulnerability to psychosis: a voxel-based meta-analysis. Neurosci. Biobehav. Rev. 35 (5), 1175–1185.

Fuster, J.M., 1990. Prefrontal cortex and the bridging of temporal gaps in the perception-action cycle. Ann. N. Y. Acad. Sci. 608, 318–329; discussion 330–316.

Garey, L.J., Ong, W.Y., Patel, T.S., Kanani, M., Davis, A., Mortimer, A.M., et al., 1998. Reduced dendritic spine density on cerebral cortical pyramidal neurons in schizophrenia. J. Neurol. Neurosurg. Psychiatry 65, 446–453.

Gaser, C., Nenadic, I., Volz, H.P., Buchel, C., Sauer, H., 2004. Neuroanatomy of "hearing voices": a frontotemporal brain structural abnormality associated with auditory hallucinations in schizophrenia. Cereb. Cortex 14, 91–96.

Gianaros, P.J., Horenstein, J.A., Cohen, S., Matthews, K.A., Brown, S.M., Flory, J.D., et al., 2007. Perigenual anterior cingulate morphology covaries with perceived social standing. Soc. Cogn. Affect. NeuroSci. 2, 161–173.

Glantz, L.A., Lewis, D.A., 2000. Decreased dendritic spine density on prefrontal cortical pyramidal neurons in schizophrenia. Arch. Gen. Psychiatry 57, 65–73.

Goghari, V.M., Lang, D.J., Flynn, S.W., Mackay, A.L., Honer, W.G., 2005. Smaller corpus callosum subregions containing motor fibers in schizophrenia. Schizophr. Res. 73, 59–68.

Goldberg, T.E., Egan, M.F., Gscheidle, T., Coppola, R., Weickert, T., Kolachana, B.S., et al., 2003. Executive subprocesses in working memory: relationship to catechol-O-methyltransferase Val158Met genotype and schizophrenia. Arch. Gen. Psychiatry 60, 889–896.

Goldman, A.L., Pezawas, L., Mattay, V.S., Fischl, B., Verchinski, B.A., Zoltick, B., et al., 2008. Heritability of brain morphology related to schizophrenia: a large-scale automated magnetic resonance imaging segmentation study. Biol. Psychiatry 63, 475–483.

Goldman-Rakic, P.S., 1995. Cellular basis of working memory. Neuron 14, 477–485.

Gur, R.E., Maany, V., Mozley, P.D., Swanson, C., Bilker, W., Gur, R.C., 1998. Subcortical MRI volumes in neuroleptic-naïve and treated patients with schizophrenia. Am. J. Psychiatry 155, 1711–1717.

Gur, R.E., Turetsky, B.I., Bilker, W.B., Gur, R.C., 1999. Reduced gray matter volume in schizophrenia. Arch. Gen. Psychiatry 56, 905–911.

Gutiérrez-Galve, L., Chu, E., Leeson, V., Price, G., Barnes, T., Joyce, E., et al., 2014. A longitudinal study of cortical changes and their cognitive correlates in patients followed up after first-episode psychosis. Psychol. Med. 3, 1–12.

Haijma, S.V., Van Haren, N., Cahn, W., Koolschijn, P.C., Hulshoff Pol, H.E., Kahn, R.S., 2013. Brain volumes in schizophrenia: a meta-analysis in over 18,000 subjects. Schizophr. Bull. 39, 1129–1138.

Hariri, A.R., Mattay, V.S., Tessitore, A., Kolachana, B., Fera, F., Goldman, D., et al., 2002. Serotonin transporter genetic variation and the response of the human amygdala. Science 297, 400–403.

Harrison, P.J., 1999. The neuropathology of schizophrenia: a critical review of the data and their interpretation. Brain 122, 593–624.

Heckers, S., Weiss, A.P., Deckersbach, T., Goff, D.C., Morecraft, R.J., Bush, G., 2004. Anterior cingulate cortex activation during cognitive interference in schizophrenia. Am. J. Psychiatry 161, 707–715.

Henn, F.A., Braus, D.F., 1999. Structural neuroimaging in schizophrenia: an integrative view of neuromorphology. Eur. Arch. Psychiatry Clin. NeuroSci. 249 (Suppl. 4), 48–56.

Hirjak, D., Wolf, R., Wilder-Smith, E., Kubera, K., Thomann, P., 2014. Motor abnormalities and basal ganglia in schizophrenia: evidence from structural magnetic resonance imaging. Brain Topogr. May 31 [Epub ahead of print].

Ho, B.C., Andreasen, N.C., Ziebell, S., Pierson, R., Magnotta, V., 2011. Long-term antipsychotic treatment and brain volumes: a longitudinal study of first-episode schizophrenia. Arch. Gen. Psychiatry 68, 128–137.

Honea, R., Crow, T.J., Passingham, D., Mackay, C.E., 2005. Regional deficits in brain volume in schizophrenia: a meta-analysis of voxel-based morphometry studies. Am. J. Psychiatry 162, 2233–2245.

Honey, G.D., Pomarol-Clotet, E., Corlett, P.R., Honey, R.A., McKenna, P.J., Bullmore, E.T., et al., 2005. Functional dysconnectivity in schizophrenia associated with attentional modulation of motor function. Brain 128, 2597–2611.

Hoppner, J., Kunesch, E., Grossmann, A., Tolzin, C.J., Schulz, M., Schlafke, D., et al., 2001. Dysfunction of transcallosally mediated motor inhibition and callosal morphology in patients with schizophrenia. Acta. Psychiatr. Scand. 104, 227–235.

Hubl, D., Koenig, T., Strik, W., Federspiel, A., Kreis, R., Boesch, C., et al., 2004. Pathways that make voices: white matter changes in auditory hallucinations. Arch. Gen. Psychiatry 61, 658–668.

Insel, T.R., 2014. The NIMH Research Domain Criteria (RDoC) Project: precision medicine for psychiatry. Am. J. Psychiatry 171, 395–397.

Insel, T.R., Gogtay, N., 2014. National institute of mental health clinical trials: new opportunities, new expectations. JAMA Psychiatry 71, 745–746.

Insel, T.R., Scolnick, E.M., 2006. Cure therapeutics and strategic prevention: raising the bar for mental health research. Mol. Psychiatry 11, 11–17.

Jakob, H., Beckmann, H., 1986. Prenatal developmental disturbances in the limbic allocortex in schizophrenics. J .Neural. Transm. 65, 303–326.

Johnstone, E.C., Crow, T.J., Frith, C.D., Husband, J., Kreel, L., 1976. Cerebral ventricular size and cognitive impairment in chronic schizophrenia. Lancet 2, 924–926.

Jones, H.M., Brammer, M.J., O'Toole, M., Taylor, T., Ohlsen, R.I., Brown, R.G., et al., 2004. Cortical effects of quetiapine in first-episode schizophrenia: a preliminary functional magnetic resonance imaging study. Biol. Psychiatry 56, 938–942.

Jones, P.B., Barnes, T.R., Davies, L., Dunn, G., Lloyd, H., Hayhurst, K.P., et al., 2006. Randomized controlled trial of the effect on quality of life of second- vs first-generation antipsychotic drugs in schizophrenia: Cost Utility of the Latest Antipsychotic Drugs in Schizophrenia Study (CUTLASS 1). Arch. Gen. Psychiatry 63, 1079–1087.

Kahn, R.S., Fleischhacker, W.W., Boter, H., Davidson, M., Vergouwe, Y., Keet, I.P., et al., 2008. Effectiveness of antipsychotic drugs in first-episode schizophrenia and schizophreniform disorder: an open randomised clinical trial. Lancet 371, 1085–1097.

Kapur, S., Phillips, A.G., Insel, T.R., 2012. Why has it taken so long for biological psychiatry to develop clinical tests and what to do about it? Mol. Psychiatry 17, 1174–1179.

Karayiorgou, M., Gogos, J.A., 2004. The molecular genetics of the 22q11-associated schizophrenia. Brain Res. Mol. Brain Res 132, 95–104.

Karlsgodt, K.H., Glahn, D.C., van Erp, T.G., Therman, S., Huttunen, M., Manninen, M., et al., 2007. The relationship between performance and fMRI signal during working memory in patients with schizophrenia, unaffected co-twins, and control subjects. Schizophr. Res. 89, 191–197.

Keedy, S.K., Reilly, J.L., Bishop, J.R., Weiden, P.J., Sweeney, J.A., 2014. Impact of antipsychotic treatment on attention and motor learning systems in first-episode schizophrenia. Schizophr. Bull. June 3 [Epub ahead of print].

Keshavan, M.S., Rosenberg, D., Sweeney, J.A., Pettegrew, J.W., 1998. Decreased caudate volume in neuroleptic-naïve psychotic patients. Am. J. Psychiatry 155, 774–778.

Kim, D.I., Manoach, D.S., Mathalon, D.H., Turner, J.A., Mannell, M., Brown, G.G., et al., 2009. Dysregulation of working memory and default-mode networks in schizophrenia using independent component analysis, an fBIRN and MCIC study. Hum. Brain. Mapp. 30, 3795–3811.

Knudsen, E.I., 2007. Fundamental components of attention. Annu. Rev. NeuroSci. 30, 57–78.

Kraepelin, E., 1919. Dementia praecox and paraphrenia. Livingstone, Edinburgh.

Krams, M., Quinton, R., Ashburner, J., Friston, K.J., Frackowiak, R.S., Boulou, P.M., et al., 1999. Kallmann's syndrome: mirror movements associated with bilateral corticospinal tract hypertrophy. Neurology 52, 816–822.

Kreczmanski, P., Heinsen, H., Mantua, V., Woltersdorf, F., Masson, T., Ulfig, N., et al., 2007. Volume, neuron density and total neuron number in five subcortical regions in schizophrenia. Brain 130, 678–692.

Kubicki, M., Westin, C.F., Nestor, P.G., Wible, C.G., Frumin, M., Maier, S.E., et al., 2003. Cingulate fasciculus integrity disruption in schizophrenia: a magnetic resonance diffusion tensor imaging study. Biol. Psychiatry 54, 1171–1180.

Kumari, V., Antonova, E., Geyer, M.A., Ffytche, D., Williams, S.C., Sharma, T., 2007. A fMRI investigation of startle gating deficits in schizophrenia patients treated with typical or atypical antipsychotics. Int. J. Neuropsychopharmacol. 10, 463–477.

Kwon, J.S., McCarley, R.W., Hirayasu, Y., Anderson, J.E., Fischer, I.A., Kikinis, R., et al., 1999. Left planum temporale volume reduction in schizophrenia. Arch. Gen. Psychiatry 56, 142–148.

Lachman, H.M., Papolos, D.F., Saito, T., Yu, Y.M., Szumlanski, C.L., Weinshilboum, R.M., 1996. Human catechol-O-methyltransferase pharmacogenetics: description of a functional polymorphism and its potential application to neuropsychiatric disorders. Pharmacogenetics 6, 243–250.

Lang, D.J., Kopala, L.C., Vandorpe, R.A., Rui, Q., Smith, G.N., Goghari, V.M., et al., 2004. Reduced basal ganglia volumes after switching to olanzapine in chronically treated patients with schizophrenia. Am. J. Psychiatry 161, 1829–1836.

Lee, J., Folley, B.S., Gore, J., Park, S., 2008. Origins of spatial working memory deficits in schizophrenia: an event-related fMRI and near-infrared spectroscopy study. PLoS One 3, e1760.

Lehman, A.F., Kreyenbuhl, J., Buchanan, R.W., Dickerson, F.B., Dixon, L.B., Goldberg, R., et al., 2004. The Schizophrenia Patient Outcomes Research Team (PORT): updated treatment recommendations 2003. Schizophr. Bull. 30, 193–217.

Lesh, T.A., Westphal, A.J., Niendam, T.A., Yoon, J.H., Minzenberg, M.J., Ragland, J.D., et al., 2013. Proactive and reactive cognitive control and dorsolateral prefrontal cortex dysfunction in first episode schizophrenia. Neuroimage. Clin. 2, 590–599.

Lewis, D.A., Levitt, P., 2002. Schizophrenia as a disorder of neurodevelopment. Annu. Rev. NeuroSci. 25, 409–432.

Lewis, D.A., Melchitzky, D.S., Sesack, S.R., Whitehead, R.E., Auh, S., Sampson, A., 2001. Dopamine transporter immunoreactivity in monkey cerebral cortex: regional, laminar, and ultrastructural localization. J. Comp. Neurol. 432, 119–136.

Lieberman, J., Chakos, M., Wu, H., Alvir, J., Hoffman, E., Robinson, D., et al., 2001. Longitudinal study of brain morphology in first episode schizophrenia. Biol. Psychiatry 49, 487–499.

Lieberman, J., Stroup, T., 2011. The NIMH-CATIE Schizophrenia Study: what did we learn? Am. J. Psychiatry 168, 770–775.

Lieberman, J.A., Stroup, T.S., McEvoy, J.P., Swartz, M.S., Rosenheck, R.A., Perkins, D.O., et al., 2005a. Effectiveness of antipsychotic drugs in patients with chronic schizophrenia. N. Engl. J. Med. 353, 1209–1223.

Lieberman, J.A., Tollefson, G.D., Charles, C., Zipursky, R., Sharma, T., Kahn, R.S., et al., 2005b. Antipsychotic drug effects on brain morphology in first-episode psychosis. Arch. Gen. Psychiatry 62, 361–370.

Lincoln, T.M., Wilhelm, K., Nestoriuc, Y., 2007. Effectiveness of psychoeducation for relapse, symptoms, knowledge, adherence and functioning in psychotic disorders: a meta-analysis. Schizophr. Res. 96, 232–245.

Longson, D., Deakin, J.F., Benes, F.M., 1996. Increased density of entorhinal glutamate-immunoreactive vertical fibers in schizophrenia. J. Neural. Transm. 103, 503–507.

Lui, S., Deng, W., Huang, X., Jiang, L., Ma, X., Chen, H., et al., 2009. Association of cerebral deficits with clinical symptoms in antipsychotic-naïve first-episode schizophrenia: an optimized voxel-based morphometry and resting state functional connectivity study. Am. J. Psychiatry 166, 196–205.

MacDonald III, A.W., Carter, C.S., 2003. Event-related FMRI study of context processing in dorsolateral prefrontal cortex of patients with schizophrenia. J. Abnorm. Psychol. 112, 689–697.

Maguire, E.A., Gadian, D.G., Johnsrude, I.S., Good, C.D., Ashburner, J., Frackowiak, R.S., et al., 2000. Navigation-related structural change in the hippocampi of taxi drivers. Proc. Natl. Acad. Sci. U. S. A. 97, 4398–4403.

Mane, A., Falcon, C., Mateos, J.J., Fernandez-Egea, E., Horga, G., Lomena, F., et al., 2009. Progressive gray matter changes in first episode schizophrenia: a 4-year longitudinal magnetic resonance study using VBM. Schizophr. Res. 114, 136–143.

Manoach, D.S., 2003. Prefrontal cortex dysfunction during working memory performance in schizophrenia: reconciling discrepant findings. Schizophr. Res. 60, 285–298.

Mathalon, D.H., Sullivan, E.V., Lim, K.O., Pfefferbaum, A., 2001. Progressive brain volume changes and the clinical course of schizophrenia in men: a longitudinal magnetic resonance imaging study. Arch. Gen. Psychiatry 58, 148–157.

Matsui, M., Yoneyama, E., Sumiyoshi, T., Noguchi, K., Nohara, S., Suzuki, M., et al., 2002. Lack of self-control as assessed by a personality inventory is related to reduced volume of supplementary motor area. Psychiatry Res. 116, 53–61.

Mattay, V.S., Callicott, J.H., Bertolino, A., Santha, A.K., Tallent, K.A., Goldberg, T.E., et al., 1997. Abnormal functional lateralization of the sensorimotor cortex in patients with schizophrenia. Neuroreport 8, 2977–2984.

Mattay, V.S., Goldberg, T.E., Fera, F., Hariri, A.R., Tessitore, A., Egan, M.F., et al., 2003. Catechol O-methyltransferase val158–met genotype and individual variation in the brain response to amphetamine. Proc. Natl. Acad. Sci. U. S. A. 100, 6186–6191.

McGuire, P.K., Shah, G.M., Murray, R.M., 1993. Increased blood flow in Broca's area during auditory hallucinations in schizophrenia. Lancet 342, 703–706.

McIntosh, A.M., Owens, D.C., Moorhead, W.J., Whalley, H.C., Stanfield, A.C., Hall, J., et al., 2011. Longitudinal volume reductions in people at high genetic risk of schizophrenia as they develop psychosis. Biol. Psychiatry 69, 953–958.

Mechelli, A., Allen, P., Amaro Jr., E., Fu, C.H., Williams, S.C., Brammer, M.J., et al., 2007. Misattribution of speech and impaired connectivity in patients with auditory verbal hallucinations. Hum. Brain. Mapp. 28, 1213–1222.

Meda, S.A., Giuliani, N.R., Calhoun, V.D., Jagannathan, K., Schretlen, D.J., Pulver, A., et al., 2008. A large scale (N = 400) investigation of gray matter differences in schizophrenia using optimized voxel-based morphometry. Schizophr. Res. 101, 95–105.

Meisenzahl, E.M., Scheuerecker, J., Zipse, M., Ufer, S., Wiesmann, M., Frodl, T., et al., 2006. Effects of treatment with the atypical neuroleptic quetiapine on working memory function: a functional MRI follow-up investigation. Eur. Arch. Psychiatry Clin. NeuroSci. 256, 522–531.

Mestdagh, A., Hansen, B., 2014. Stigma in patients with schizophrenia receiving community mental health care: a review of qualitative studies. Soc. Psychiatry. Psychiatr. Epidemiol. 49, 79–87.

Meyer, U., Yee, B.K., Feldon, J., 2007. The neurodevelopmental impact of prenatal infections at different times of pregnancy: the earlier the worse? Neuroscientist 13, 241–256.

Meyer-Lindenberg, A., 2010. From maps to mechanisms through neuroimaging of schizophrenia. Nature 468, 194–202.

Meyer-Lindenberg, A., Buckholtz, J.W., Kolachana, B., Hariri, A.R., Pezawas, L., Blasi, G., et al., 2006a. Neural mechanisms of genetic risk for impulsivity and violence in humans. Proc. Natl. Acad. Sci. U. S. A. 103, 6269–6274.

Meyer-Lindenberg, A., Hariri, A.R., Munoz, K.E., Mervis, C.B., Mattay, V.S., Morris, C.A., et al., 2005a. Neural correlates of genetically abnormal social cognition in Williams syndrome. Nat. NeuroSci. 8, 991–993.

Meyer-Lindenberg, A., Kohn, P.D., Kolachana, B., Kippenhan, S., McInerney-Leo, A., Nussbaum, R., et al., 2005b. Midbrain dopamine and prefrontal function in humans: interaction and modulation by COMT genotype. Nat. NeuroSci. 8, 594–596.

Meyer-Lindenberg, A., Nichols, T., Callicott, J.H., Ding, J., Kolachana, B., Buckholtz, J., et al., 2006b. Impact of complex genetic variation in COMT on human brain function. Mol. Psychiatry 11, 867–877.

Meyer-Lindenberg, A., Poline, J.B., Kohn, P.D., Holt, J.L., Egan, M.F., Weinberger, D.R., et al., 2001. Evidence for abnormal cortical functional connectivity during working memory in schizophrenia. Am. J. Psychiatry 158, 1809–1817.

Meyer-Lindenberg, A., Straub, R.E., Lipska, B.K., Verchinski, B.A., Goldberg, T., Callicott, J.H., et al., 2007. Genetic evidence implicating DARPP-32 in human frontostriatal structure, function, and cognition. J. Clin. Invest. 117, 672–682.

Meyer-Lindenberg, A., Weinberger, D.R., 2006. Intermediate phenotypes and genetic mechanisms of psychiatric disorders. Nat. Rev. NeuroSci. 7, 818–827.

Miyamoto, S., Duncan, G.E., Marx, C.E., Lieberman, J.A., 2005. Treatments for schizophrenia: a critical review of pharmacology and mechanisms of action of antipsychotic drugs. Mol. Psychiatry 10, 79–104.

Molina, V., Sanz, J., Sarramea, F., Luque, R., Benito, C., Palomo, T., 2006. Dorsolateral prefrontal and superior temporal volume deficits in first-episode psychoses that evolve into schizophrenia. Eur. Arch. Psychiatry Clin. NeuroSci. 256, 106–111.

Murphy, K.C., 2002. Schizophrenia and velo-cardio-facial syndrome. Lancet 359, 426–430.

Murray-Thomas, T., Jones, M.E., Patel, D., Brunner, E., Shatapathy, C.C., Motsko, S., et al., 2013. Risk of mortality (including sudden cardiac death) and major cardiovascular events in atypical and typical antipsychotic users: a study with the general practice research database. Cardiovasc. Psychiatry Neurol. 2013, 247486.

Nasrallah, H.A., 1982. Neuropathology of the corpus callosum in schizophrenia. Br. J. Psychiatry 141, 99–100.

Nazeri, A., Chakravarty, M.M., Felsky, D., Lobaugh, N.J., Rajji, T.K., Mulsant, B.H., et al., 2013. Alterations of superficial white matter in schizophrenia and relationship to cognitive performance. Neuropsychopharmacology 38, 1954–1962.

Neckelmann, G., Specht, K., Lund, A., Ersland, L., Smievoll, A.I., Neckelmann, D., et al., 2006. Mr morphometry analysis of grey matter volume reduction in schizophrenia: association with hallucinations. Int. J. NeuroSci. 116, 9–23.

Nieuwenhuis, M., van Haren, N.E., Hulshoff Pol, H.E., Cahn, W., Kahn, R.S., Schnack, H.G., 2012. Classification of schizophrenia patients and healthy controls from structural MRI scans in two large independent samples. Neuroimage 61, 606–612.

Office of the Surgeon General, 1999. Mental health: a report of the surgeon general. http://www.surgeongeneral.gov/library/mentalhealth/home.html (accessed 06.08.14.).

Olabi, B., Ellison-Wright, I., McIntosh, A.M., Wood, S.J., Bullmore, E., Lawrie, S.M., 2011. Are there progressive brain changes in schizophrenia? a meta-analysis of structural magnetic resonance imaging studies. Biol. Psychiatry 70, 88–96.

Owen, M.J., Williams, N.M., O'Donovan, M.C., 2004. The molecular genetics of schizophrenia: new findings promise new insights. Mol. Psychiatry 9, 14–27.

Paulman, R.G., Devous Sr., M.D., Gregory, R.R., Herman, J.H., Jennings, L., Bonte, F.J., et al., 1990. Hypofrontality and cognitive impairment in schizophrenia: dynamic single-photon tomography and neuropsychological assessment of schizophrenic brain function. Biol. Psychiatry 27, 377–399.

Perlstein, W.M., Dixit, N.K., Carter, C.S., Noll, D.C., Cohen, J.D., 2003. Prefrontal cortex dysfunction mediates deficits in working memory and prepotent responding in schizophrenia. Biol. Psychiatry 53, 25–38.

Pezawas, L., Meyer-Lindenberg, A., Drabant, E.M., Verchinski, B.A., Munoz, K.E., Kolachana, B.S., et al., 2005. 5-HTTLPR polymorphism impacts human cingulate-amygdala interactions: a genetic susceptibility mechanism for depression. Nat. NeuroSci. 8, 828–834.

Potkin, S.G., Turner, J.A., Brown, G.G., McCarthy, G., Greve, D.N., Glover, G.H., et al., 2009a. Working memory and DLPFC inefficiency in schizophrenia: the FBIRN study. Schizophr. Bull. 35, 19–31.

Potkin, S.G., Turner, J.A., Guffanti, G., Lakatos, A., Fallon, J.H., Nguyen, D.D., et al., 2009b. A genome-wide association study of schizophrenia using brain activation as a quantitative phenotype. Schizophr. Bull. 35, 96–108.

Rajkowska, G., Selemon, L.D., Goldman-Rakic, P.S., 1998. Neuronal and glial somal size in the prefrontal cortex: a postmortem morphometric study of schizophrenia and Huntington disease. Arch. Gen. Psychiatry 55, 215–224.

Rapoport, J.L., Giedd, J.N., Blumenthal, J., Hamburger, S., Jeffries, N., Fernandez, T., et al., 1999. Progressive cortical change during adolescence in childhood-onset schizophrenia: a longitudinal magnetic resonance imaging study. Arch. Gen. Psychiatry 56, 649–654.

Rasetti, R., Sambataro, F., Chen, Q., Callicott, J.H., Mattay, V.S., Weinberger, D.R., 2011. Altered cortical network dynamics: a potential intermediate phenotype for schizophrenia and association with ZNF804A. Arch. Gen. Psychiatry 68, 1207–1217.

Rogowska, J., Gruber, S.A., Yurgelun-Todd, D.A., 2004. Functional magnetic resonance imaging in schizophrenia: cortical response to motor stimulation. Psychiatry. Res. Neuroimaging 130, 227–243.

Salgado-Pineda, P., Junque, C., Vendrell, P., Baeza, I., Bargallo, N., Falcon, C., et al., 2004. Decreased cerebral activation during CPT performance: structural and functional deficits in schizophrenic patients. Neuroimage 21, 840–847.

Scheepers, F.E., de Wied, C.C., Hulshoff Pol, H.E., van de Flier, W., van der Linden, J.A., Kahn, R.S., 2001. The effect of clozapine on caudate nucleus volume in schizophrenic patients previously treated with typical antipsychotics. Neuropsychopharmacology 24, 47–54.

Schnack, H.G., Nieuwenhuis, M., van Haren, N.E., Abramovic, L., Scheewe, T.W., Brouwer, R.M., et al., 2014. Can structural MRI aid in clinical classification? A machine learning study in two independent samples of patients with schizophrenia, bipolar disorder and healthy subjects. Neuroimage 84, 299–306.

Schneider, F., Habel, U., Reske, M., Kellermann, T., Stöcker, T., Shah, N.J., et al., 2007. Neural correlates of working memory dysfunction in first-episode schizophrenia patients: an fMRI multi-center study. Schizophr. Res. 89, 198–210.

Schröder, J., Essig, M., Baudendistel, K., Jahn, T., Gerdsen, I., Stockert, A., et al., 1999. Motor dysfunction and sensorimotor cortex activation changes in schizophrenia: a study with functional magnetic resonance imaging. Neuroimage 9, 81–87.

Schröder, J., Wenz, F., Schad, L.R., Baudendistel, K., Knopp, M.V., 1995. Sensorimotor cortex and supplementary motor area changes in schizophrenia: a study with functional magnetic resonance imaging. Br. J .Psychiatry 167, 197–201.

Schroeder, J., Niethammer, R., Geider, F.J., Reitz, C., Binkert, M., Jauss, M., et al., 1991. Neurological soft signs in schizophrenia. Schizophr. Res. 6, 25–30.

Sei, Y., Ren-Patterson, R., Li, Z., Tunbridge, E.M., Egan, M.F., Kolachana, B.S., et al., 2007. Neuregulin1-induced cell migration is impaired in schizophrenia: association with neuregulin1 and catechol-O-methyltransferase gene polymorphisms. Mol. Psychiatry 12, 946–957.

Shapleske, J., Rossell, S.L., Woodruff, P.W., David, A.S., 1999. The planum temporale: a systematic, quantitative review of its structural, functional and clinical significance. Brain Res. Brain Res. Rev. 29, 26–49.

Sigurdsson, T., Stark, K.L., Karayiorgou, M., Gogos, J.A., Gordon, J.A., 2010. Impaired hippocampal-prefrontal synchrony in a genetic mouse model of schizophrenia. Nature 464, 763–767.

Silbersweig, D.A., Stern, E., Frith, C., Cahill, C., Holmes, A., Grootoonk, S., et al., 1995. A functional neuroanatomy of hallucinations in schizophrenia. Nature 378 (6553), 176–179.

Smesny, S., Langbein, K., Rzanny, R., Gussew, A., Burmeister, H.P., Reichenbach, J.R., et al., 2012. Antipsychotic drug effects on left prefrontal phospholipid metabolism: a follow-up 31P-2D-CSI study of haloperidol and risperidone in acutely ill chronic schizophrenia patients. Schizophr. Res. 138, 164–170.

Snitz, B.E., MacDonald III, A., Cohen, J.D., Cho, R.Y., Becker, T., Carter, C.S., 2005. Lateral and medial hypofrontality in first-episode schizophrenia: functional activity in a medication-naïve state and effects of short-term atypical antipsychotic treatment. Am. J. Psychiatry 162, 2322–2329.

Spalletta, G., Piras, F., Piras, F., Caltagirone, C., Orfei, M.D., 2014. The structural neuroanatomy of metacognitive insight in schizophrenia and its psychopathological and neuropsychological correlates. Human Brain Mapp. 35, 4729–4740.

Stark, K.L., Xu, B., Bagchi, A., Lai, W.S., Liu, H., Hsu, R., et al., 2008. Altered brain microRNA biogenesis contributes to phenotypic deficits in a 22q11-deletion mouse model. Nat. Genet. 40, 751–760.

Stefansson, H., Meyer-Lindenberg, A., Steinberg, S., Magnusdottir, B., Morgen, K., Arnarsdottir, S., et al., 2014. CNVs conferring risk of autism or schizophrenia affect cognition in controls. Nature 505, 361–366.

Stein, J.L., Wiedholz, L.M., Bassett, D.S., Weinberger, D.R., Zink, C.F., Mattay, V.S., et al., 2007. A validated network of effective amygdala connectivity. Neuroimage 36, 736–745.

Steinmann, S., Leicht, G., Mulert, C., 2014. Interhemispheric auditory connectivity: structure and function related to auditory verbal hallucinations. Front. Hum. NeuroSci. 8, 55.

Stephan, K.E., Magnotta, V.A., White, T., Arndt, S., Flaum, M., O'Leary, D.S., et al., 2001. Effects of olanzapine on cerebellar functional connectivity in schizophrenia measured by fMRI during a simple motor task. Psychol. Med. 31, 1065–1078.

Stevens, J.R., 1982. Neuropathology of schizophrenia. Arch. Gen. Psychiatry 39, 1131–1139.

Stewart, A.M., Kalueff, A.V., 2014. Developing better and more valid animal models of brain disorders. Behav. Brain. Res. December 30 [Epub ahead of print].

Sullivan, P.F., Daly, M.J., O'Donovan, M., 2012. Genetic architectures of psychiatric disorders: the emerging picture and its implications. Nat. Rev. Genet. 13, 537–551.

Sumich, A., Chitnis, X.A., Fannon, D.G., O'Ceallaigh, S., Doku, V.C., Faldrowicz, A., et al., 2005. Unreality symptoms and volumetric measures of Heschl's gyrus and planum temporal in first-episode psychosis. Biol. Psychiatry 57, 947–950.

Sun, Z., Wang, F., Cui, L., Breeze, J., Du, X., Wang, X., et al., 2003. Abnormal anterior cingulum in patients with schizophrenia: a diffusion tensor imaging study. Neuroreport 14, 1833–1836.

Svenningsson, P., Nishi, A., Fisone, G., Girault, J.A., Nairn, A.C., Greengard, P., 2004. DARPP-32: an integrator of neurotransmission. Annu. Rev. Pharmacol. Toxicol. 44, 269–296.

Svenningsson, P., Tzavara, E.T., Carruthers, R., Rachleff, I., Wattler, S., Nehls, M., et al., 2003. Diverse psychotomimetics act through a common signaling pathway. Science 302, 1412–1415.

Swerdlow, N.R., Geyer, M.A., Braff, D.L., 2001. Neural circuit regulation of prepulse inhibition of startle in the rat: current knowledge and future challenges. Psychopharmacology. (Berl) 156, 194–215.

Szeszko, P.R., Goldberg, E., Gunduz-Bruce, H., Ashtari, M., Robinson, D., Malhotra, A.K., et al., 2003. Smaller anterior hippocampal formation volume in antipsychotic-naïve patients with first-episode schizophrenia. Am. J. Psychiatry 160, 2190–2197.

Szulc, A., Galinska, B., Tarasow, E., Dzienis, W., Kubas, B., Konarzewska, B., et al., 2005. The effect of risperidone on metabolite measures in the frontal lobe, temporal lobe, and thalamus in schizophrenic patients: a proton magnetic resonance spectroscopy (1H MRS). Pharmacopsychiatry 38, 214–219.

Takahashi, T., Wood, S.J., Yung, A.R., Soulsby, B., McGorry, P.D., Suzuki, M., et al., 2009. Progressive gray matter reduction of the superior temporal gyrus during transition to psychosis. Arch. Gen. Psychiatry 66, 366–376.

Takayanagi, Y., Takahashi, T., Orikabe, L., Mozue, Y., Kawasaki, Y., Nakamura, K., et al., 2011. Classification of first-episode schizophrenia patients and healthy subjects by automated MRI measures of regional brain volume and cortical thickness. PLoS One 6, e21047.

Taki, Y., Kinomura, S., Sato, K., Goto, R., Inoue, K., Okada, K., et al., 2006. Both global gray matter volume and regional gray matter volume negatively correlate with lifetime alcohol intake in non-alcohol-dependent Japanese men: A volumetric analysis and a voxel-based morphometry. Alcohol. Clin. Exp. Res. 30, 1045–1050.

Talkowski, M.E., Kirov, G., Bamne, M., Georgieva, L., Torres, G., Mansour, H., et al., 2008. A network of dopaminergic gene variations implicated as risk factors for schizophrenia. Hum. Mol. Genet. 17, 747–758.

Tan, H.Y., Choo, W.C., Fones, C.S., Chee, M.W., 2005. FMRI study of maintenance and manipulation processes within working memory in first-episode schizophrenia. Am. J. Psychiatry 162, 1849–1858.

Tan, H.Y., Nicodemus, K.K., Chen, Q., Li, Z., Brooke, J.K., Honea, R., et al., 2008. Genetic variation in AKT1 is linked to dopamine-associated prefrontal cortical structure and function in humans. J. Clin. Invest. 118, 2200–2208.

Tan, H.Y., Sust, S., Buckholtz, J.W., Mattay, V.S., Meyer-Lindenberg, A., Egan, M.F., et al., 2006. Dysfunctional prefrontal regional specialization and compensation in schizophrenia. Am. J. Psychiatry 163, 1969–1977.

Tost, H., Braus, D.F., Hakimi, S., Ru, M., Vollmert, C., Hohn, F., et al., 2010. Acute D2 receptor blockade induces rapid, reversible remodeling in human cortical-striatal circuits. Nat. NeuroSci. 13, 920–922.

Tost, H., Meyer-Lindenberg, A., Klein, S., Schmitt, A., Hohn, F., Tenckhoff, A., et al., 2006. D2 antidopaminergic modulation of frontal lobe function in healthy human subjects. Biol. Psychiatry 60, 1196–1205.

Valfre, W., Rainero, I., Bergui, M., Pinessi, L., 2008. Voxel-based morphometry reveals gray matter abnormalities in migraine. Headache 48, 109–117.

Vink, M., Ramsey, N.F., Raemaekers, M., Kahn, R.S., 2006. Striatal dysfunction in schizophrenia and unaffected relatives. Biol. Psychiatry 60, 32–39.

Vita, A., De Peri, L., Deste, G., Sacchetti, E., 2012. Progressive loss of cortical gray matter in schizophrenia: a meta-analysis and meta-regression of longitudinal MRI studies. Transl. Psychiatry 2, e190.

Volz, H., Gaser, C., Hager, F., Rzanny, R., Ponisch, J., Mentzel, H., et al., 1999. Decreased frontal activation in schizophrenics during stimulation with the continuous performance test—a functional magnetic resonance imaging study. Eur. Psychiatry 14, 17–24.

Vrtunski, P.B., Simpson, D.M., Weiss, K.M., Davis, G.C., 1986. Abnormalities of fine motor control in schizophrenia. Psychiatry Res. 18, 275–284.

Walker, M.A., Highley, J.R., Esiri, M.M., McDonald, B., Roberts, H.C., Evans, S.P., et al., 2002. Estimated neuronal populations and volumes of the hippocampus and its subfields in schizophrenia. Am. J. Psychiatry 159, 821–828.

Walther, S., Federspiel, A., Horn, H., Razavi, N., Wiest, R., Dierks, T., et al., 2011. Alterations of white matter integrity related to motor activity in schizophrenia. Neurobiol. Dis. 42, 276–283.

Ward, K.E., Friedman, L., Wise, A., Schulz, S.C., 1996. Meta-analysis of brain and cranial size in schizophrenia. Schizophr. Res. 22, 197–213.

Weinberger, D.R., 1987. Implications of normal brain development for the pathogenesis of schizophrenia. Arch. Gen. Psychiatry 44, 660–669.

Weinberger, D.R., Berman, K.F., Zec, R.F., 1986. Physiologic dysfunction of dorsolateral prefrontal cortex in schizophrenia. I. regional cerebral blood flow evidence. Arch. Gen. Psychiatry 43, 114–124.

Weinberger, D.R., McClure, R.K., 2002. Neurotoxicity, neuroplasticity, and magnetic resonance imaging morphometry: what is happening in the schizophrenic brain? Arch. Gen. Psychiatry 59, 553–558.

Weinstein, S., Werker, J.F., Vouloumanos, A., Woodward, T.S., Ngan, E.T., 2006. Do you hear what I hear? Neural correlates of thought disorder during listening to speech in schizophrenia. Schizophr. Res. 86, 130–137.

Weiskopf, N., 2012. Real-time fMRI and its application to neurofeedback. Neuroimage 62, 682–692.

Wenz, F., Schad, L.R., Knopp, M.V., Baudendistel, K.T., Flomer, F., Schroder, J., et al., 1994. Functional magnetic resonance imaging at 1.5 T: activation pattern in schizophrenic patients receiving neuroleptic medication. Magn. Reson. Imaging 12, 975–982.

Williams, G.V., Goldman-Rakic, P.S., 1995. Modulation of memory fields by dopamine D1 receptors in prefrontal cortex. Nature 376, 572–575.

Woodruff, P.W., Wright, I.C., Bullmore, E.T., Brammer, M., Howard, R.J., Williams, S.C., et al., 1997. Auditory hallucinations and the temporal cortical response to speech in schizophrenia: a functional magnetic resonance imaging study. Am. J. Psychiatry 154, 1676–1682.

Wright, I.C., Rabe-Hesketh, S., Woodruff, P.W., David, A.S., Murray, R.M., Bullmore, E.T., 2000. Meta-analysis of regional brain volumes in schizophrenia. Am. J. Psychiatry 157, 16–25.

Yoshida, T., McCarley, R.W., Nakamura, M., Lee, K., Koo, M.S., Bouix, S., et al., 2009. A prospective longitudinal volumetric MRI study of superior temporal gyrus gray matter and amygdala-hippocampal complex in chronic schizophrenia. Schizophr. Res. 113, 84–94.

Yücel, M., Pantelis, C., Stuart, G.W., Wood, S.J., Maruff, P., Velakoulis, D., et al., 2002. Anterior cingulate activation during Stroop task performance: a PET to MRI coregistration study of individual patients with schizophrenia. Am. J. Psychiatry 159, 251–254.

Zaidel, D.W., Esiri, M.M., Harrison, P.J., 1997. Size, shape, and orientation of neurons in the left and right hippocampus: investigation of normal asymmetries and alterations in schizophrenia. Am. J. Psychiatry 154, 812–818.

Chapter 4

Early Clinical Trial Design

Chapter 4.1

Methodological Studies

Faisal Azam, Rhiana Newport, Rebecca Johnson, Rachel Midgley and David J. Kerr

CONVENTIONAL PHASE I TRIAL METHODOLOGY

Initially, new agents are given to human beings in a single dose to assess the tolerability, safety, pharmacokinetics (PK), and pharmacodynamics (PD) effects and to compare these results with those gleaned in the preclinical setting. The dose is escalated with concurrent observation of the aforementioned parameters in order to define a dose and schedule that can be recommended for phase II efficacy studies.

It is extremely important for phase I trials to be designed in such a way as to include the appropriate number and types of patients in order to minimize the risk of making inaccurate dose recommendations; otherwise, considerable additional time and effort will need to be invested when the drug is undergoing phase II investigations to further refine dosing.

Aims

The primary aim of a first-in-human phase I trial is to define the maximum tolerated dose (MTD) and to recommend the correct dose for further studies. Other aims are to describe the pattern of toxicity, PK profile, antitumor effect, and PD endpoints and to describe any relationship between dose and effects on toxicity.

Design

Because development of a new cancer drug is dependent on the phase I trial, it is important that the design, conduct, and analysis of phase I trials are rigorous. If phase I trials are not properly conducted—for example, if the trial includes patients who differ significantly in characteristics compared to those who will be treated in later studies, or if the trial schedule is created as per convenience rather than as per the data—the overall time taken for drug development may increase and/or the future of the new drug may be threatened. Gemcitabine is a case in point (Eisenhauer et al., 2006, example cited with permission from Oxford University Press). In a phase I trial of Gemcitabine, it was recognized early on that the schedule played an important role in the pattern of the side effects of the drug. The daily schedule, repeated five times, caused severe flu-like symptoms, and so only small doses could be administered (O'Rourke et al., 1994). When doses were given in a 30-minute infusion for 3 out of every 4 weeks, more of the drug could be administered per cycle, and thrombocytopenia was the dose-limiting toxicity (DLT). The latter schedule was thus selected for further development, and the recommended dose for phase II was 790 mg/m^2 (Abbruzzese et al., 1991). The population enrolled in the study was dominated by heavily pretreated patients, and dose escalation was very conservative, resulting in a large number of patients being treated at lower doses, which led to significant toxic effects. Only a few patients were treated with the recommended doses. When the drug was evaluated at the recommended dose in fitter patients, very few side effects were seen. Doses were first escalated to 1,000 mg/m^2 and then to 1,250 mg/m^2 (Anderson et al., 1994; Cormier et al., 1994; Mertens et al., 1993). Finally, the phase I trial was repeated in non-small-cell lung cancer patients at a recommended dose of 2,200 mg/m^2 (Fosella et al., 1997). At the same time, the dose was increased by administering the drug at a fixed-dose-rate infusion for increasingly longer periods of time (Brand et al., 1997). This approach was based on information gained earlier that indicated that there was a fixed rate at which gemcitabine could be converted into its active form. And so a phase II trial in pancreatic cancer compared the maximal dose of gemcitabine delivered as a 1,500 mg/m^2 dose at a 10 mg/m^2-per-minute infusion and 2,200 mg/m^2 at a 30-minute bolus (Tempero et al., 2003). In this trial, there was no observed impact on the primary endpoint of time to treatment failure, but there was a survival advantage (5 months vs 8 months) for a fixed-dose-rate arm, suggesting that the drug delivery approach needs further evaluation. This example illustrates that the starting dose of the phase I weekly

Principles of Translational Science in Medicine. http://dx.doi.org/10.1016/B978-0-12-800687-0.00022-0

schedule was too low because the patients enrolled were not representative of those enrolled in the phase II trial. This added time to the development process.

Let us now consider some important aspects of developing a phase I trial.

Patient Entry Criteria

Patient entry criteria should be carefully elucidated in the protocol. Some of the criteria used to select entry of patients into phase I trials are described in the following sections.

Performance Status

Patients who are selected for phase I trials should have a good to moderate performance status. This is critical; otherwise, it might not be possible to distinguish the drug effects from the disease-related symptoms if the patient is terminally ill. The tools developed by the World Health Organization (Miller et al., 1981), the Eastern Cooperative Oncology Group (ECOG) (Oken et al., 1982), and D. A. Karnofsky (Karnofsky and Burchenal, 1949) are some of the performance status assessment tools widely used to assess patients' performance status.

Cancer Type

Most phase I trials are conducted on a mixture of solid tumors. Some trials are based on specific antibodies. Some agents are antibodies or small molecular weight inhibitors against receptors that are present in specific cancers; trials of these agents should be restricted to patients with cancers likely to express these receptors. One example of this is cetuximab, which acts on epidermal growth factor receptors (EGFRs). Table 4.1 gives some other examples of targets.

Generally, agents being tested in phase I trials should initially be tried in patients with a case of recurrent, advanced, or metastatic cancer that has been treated previously with the standard treatments and for whom no further standard treatment is available. In addition, it is common practice to include only patients whose disease can be clinically or radiologically documented, although some would argue that this is less relevant in phase I trials in which the primary endpoint is dose finding based on toxicity.

Laboratory Investigations within Appropriate Limits as Entry Criteria

Phase I trials will often restrict entry of patients to those with adequate liver, kidney, and hematologic parameters. These parameters can vary from trial to trial and may vary according to whether the patient has a known liver metastatic disease. For example, the inclusion criteria for liver enzymes might be an increase up to twofold in the upper normal limit or an increase up to fivefold in the presence of hepatic metastatic disease. Although these limits are set to maintain patient safety, one could argue that studies should include some patients with at least moderate hepatic or renal impairment in order to find out whether these patients can be included in future development of the agent. It is often patients with the worst liver function tests who have the most to gain from novel drugs and agents, especially in the cancer setting.

Most commonly, pregnant and breast-feeding women are also excluded from phase I trials. This is done to protect the unborn or newborn child.

TABLE 4.1 Examples of novel agents and their targets

Target receptors	Drugs
EGFR	Cetuximab, gefitinib, erlotinib, sorafenib
HER 2	Trastuzumab
Angiogenesis: VEGF	Bevacizumab
Angiogenesis: VEGFR	Sunitinib, vatalanib
mTOR	Temsirolimus
ABL/c-kit	Imatinib mesylate

Special Drug Administration or Procedures

Some trials require measurement of the PD effect of the drug on a tumor, necessitating a fresh tumor biopsy. If this is the case, the patient clearly must have a tumor that is accessible to biopsy, the doctor must have the appropriate expertise to perform the biopsy safely, and the patient must consent to the procedure. Alternately, if a drug has to be administered intravenously, the patient will require intravenous access. This access can be obtained through a PICC line if repeat sampling is necessary and if the patient gives appropriate consent.

Patient Consent

Any patient entering a trial must be given a careful explanation of the aims of the trial, of the drugs involved, of any known or expected side effects, and of expected investigations and visits to the hospital that may be required. The patient should read the language-appropriate patient information sheet and should be given a period of time to digest this information and the opportunity to ask questions prior to signing the consent form.

Calculation of the Starting Dose

A wealth of preclinical information is utilized to inform the phase I trial starting dose and design. Statistical and other research has been undertaken to improve the safety, ethics, and efficiency of trial conduct and dose escalation methodologies. Usually, the murine LD10 is converted from mg/kg to mg/m^2 using standard conversion factors. The human starting dose is generally one-tenth the mouse LD10 equivalent, unless testing in a nonrodent species shows that this dose is excessively toxic. In this case, one-third to one-sixth the lowest toxic dose equivalent in the more sensitive species becomes the starting dose of human trials. Other factors such as clinical toxicity data from analogues of experimental agent, data from in vitro studies comparing the sensitivity of human and nonhuman cell lines, and data on drug protein binding in human and rodent plasma also help in determining the starting dose.

The Food and Drug Administration (FDA) suggests a method for calculating the safe starting dose in humans (U.S. Food and Drug Administration, 2005). According to this method, to convert the no observed adverse effect level (NOAEL) dose from the toxicology studies to the human equivalent dose (HED) as per body surface area, researchers must select the HED from the most appropriate species. Then a tenfold safety factor is applied to give a maximum recommended starting dose (MRSD), which is subsequently adjusted on the basis of the predicted pharmacological action of the investigational medicinal product (IMP).

Another method of calculating the safe starting dose of a new drug is based on the minimal anticipated biologic effect level (Early Stage Clinical Trial Taskforce, 2006), which takes into account the potency, mechanism of actions, dose–response data from animals and humans in vitro, concentration response data from animals in vivo, species specificity, PK, PD, calculated target occupancy versus concentration, and concentration of the target and target cells in humans in vivo. If those methods give different estimates, the lowest value is selected, and a margin of safety is built into the actual starting dose.

Dose Escalation

Initially, the dose is usually doubled for each increment if, after the data have been carefully assessed, the previous dose has been found to be safe. Alternately, serial measurements of the concentration of the new drug in the bloodstream during the trial can allow increases in dose to be guided by exposure to the new drug. It is the responsibility of the investigators, sponsor's physician, and sponsor's expert in PK to review all the data, including the preclinical data, before increasing the dose. If there is any concern about the tolerability, an intermediate dose can be administered, in order to avoid exceeding the NOAEL.

Dose escalation proceeds until a prespecified endpoint is reached. If toxicity is the primary endpoint, it continues until a minimum number experience a DLT. When DLT happens, the next lower dose is studied to get more information; this will be the recommended dose for further studies. If the lower dose still reveals excessive DLT events, then the dose is further de-escalated.

If the primary endpoint is a pharmacological measure, escalation will continue until a minimum level, as described in the protocol, is reached. Results of PK analyses and, if possible, PD analyses must be available after each dose level to make accurate decisions about further escalation.

Number of Patients Required for Dose Administration

The number of patients dosed on one occasion and the interval between dosing each patient or cohort of patients depends on the route of administration and the type of trial. Usually, a minimum of three patients per dose level is enrolled.

For high-risk agents, only one subject is administered the agent in the first instance. Often intravenous doses are given slowly over several hours. For low-risk agents or oral administration, cohorts of up to three patients are dosed over short intervals, rather than a single individual patient with a prolonged clinical review.

If the trial is starting at a dose at which no toxicity is observed, then it may be appropriate to enter just one patient per dose level for the first few dose levels (Eisenhauer et al., 2000). As escalation proceeds and the risk of toxicity increases, the number of patients per level is increased to three; this certainly applies once a toxicity of at least grade 2 has been observed in at least one patient. The number of patients per dose level increases to six if one of the three patients has DLT in a protocol-prescribed observation period.

However, if the drug is expected to have large inter-patient variability—for example, if there is likely to be large differences in oral bioavailability—then three patients per dose level is recommended from the first dose level onward. Furthermore, in the later expanded dose levels, patients who are as representative as possible of those likely to be tested in the phase II trial should be enrolled.

Stopping Rules

Phase I trials are stopped if significant DLT is observed. The following are a few examples of the stopping rules employed (Eisenhauer et al., 2006, cited with permission from Oxford University Press):

- *Rule of 3:* Stop if one toxicity is observed among the three patients. There is a high probability of stopping before a target dose is reached.
- *Rule of 3+3:* Stop if at least two out of six patients experience a DLT. The probability of stopping below the target dose is 34%, and the probability of stopping above the target dose is 29%. This can be improved by adding a third cohort of three patients if two out of six patients experience a DLT at a certain dose level. The expected sample size is 11.7 patients.
- *Rule of 3+3+3:* Stop as soon as at least three out of nine patients experience a DLT. This rule has a higher probability of stopping at a higher dose level than the target. The expected sample size is 16.1 patients.
- *Rule of 2+4:* If one DLT is observed in the first cohort of two patients, then an additional cohort of four patients is added. A high probability of stopping at a higher dose level is expected, but the sample size required is 12.6 patients.
- *Rule of 3+1+1:* If one DLT is observed among three patients, one more patient is added to the next cohort. This rule has a tendency of stopping at higher doses.

The Rule of 3+3 is the most widely used, but more aggressive rules may be more efficient if the drug is not expected to be very toxic.

MEASURING ENDPOINTS

The standard primary endpoint of a phase I trial is to find a recommended dose for future trials. Other endpoints can be employed—for example, those that assess dose-related effects of the drug on the body. These are known as pharmacodynamic endpoints and can include measures like molecular changes, antitumor effects, and toxicity.

Toxicity

Most anticancer drugs target the DNA, so doses that result in antitumor effects are also likely to be toxic for normal cells. The highest tolerable dose is usually the effective dose in clinical studies. In phase I trials, the dose of the IMP is usually escalated in cohorts of patients until predefined criteria are met. These are called dose-limiting toxic effects. DLT is a severe but reversible organ toxicity.

In 1981, the WHO published the first toxicity criteria assessment tool (Miller et al., 1981) to assess drug toxicities. Grades were used, ranging from grade 0 (normal with no adverse effect) to grade 5 (fatal). Similarly, in 1982 the U.S. National Cancer Institute (NCI) developed the common toxicity criteria (CTC), which are now commonly used to record adverse events in trials.

The phase I trial protocol must define the DLT with precise laboratory levels or grades for hematologic, renal, and hepatic events or should refer to a standard tool such as the NCI's CTC. If the patient misses more than a prespecified

TABLE 4.2 Dose-limiting toxicities used in phase I trials

Event	Dose-limiting toxicities
Neutrophil	Grade 4 (<0.5 × 10^9 / l)
Platelets	Grade 4 (<25 × 10^9 / l)
Liver function tests	AST or ALT grade 3
Myelosuppression	Febrile neutropenia or grade 3 infection with grade 3 or 4 neutropenia, thrombocytopenic bleeding Grade 3 or greater nausea and/or vomiting despite antiemetic therapy Other CTC: grade 3 or higher toxicities

number of doses due to toxicity, this toxicity is considered to be a DLT. Table 4.2 gives some of the DLTs used in phase I trials. The trial design should explain the number of patients who must experience DLT before further dose escalations or dose stops. Usually, two patients out of three or six must experience DLT for the escalation to stop.

The dose level at which DLT is seen in the number of patients required to prevent further dose escalations is called the MTD; it is also referred as the maximum administered dose (MAD). The recommended phase II dose (RD) is usually at a dose level below that at which escalation has stopped. Toxic effects at this dose may still be severe, but their frequency is expected to be lower than that seen at the MTD.

Pharmacodynamic Endpoints

Many novel anticancer agents target the intracellular and extracellular pathways and differ from the cytotoxics because of their preclinical dose–effect relationship (Eisenhauer, 1998). In these cases, once the target is saturated by the drug, the antitumor effect can plateau, but toxicity may not if the target level is higher in normal tissues. Clearly, then, maximal toxicity may not have been reached at a dose level necessary for maximum efficacy (Parulekar and Eisenhauer, 2002). So the endpoint in these cases would often be pharmacodynamic—for example, molecular biomarkers measured in repeated tissue biopsies, peripheral blood mononuclear cells, buccal mucosa, or skin.

MECHANISM-ORIENTED TRIAL DESIGN

The last decade has seen an extraordinary increase in the number of "druggable" targets discovered by molecular and cell biologists. The role of the enzymes and signal transduction pathways involved in the control of the cell cycle, proliferation, invasion, metastasis, and angiogenesis has been clarified. Not surprisingly, given their centrality to control of cellular homeostasis, many of these new targets have been kinases. This has yielded a new approach to drug development in which attempts are made to titrate drug dose to inhibition of the mechanistic target and to separate the biologically effective and MTD doses. This leads us to consider three related new principles of phase I clinical trial design: proof-of-mechanism (PoM), proof-of-principle (PoP), and proof-of-concept (PoC).

Proof-of-Mechanism

If the novel anticancer agent has been developed as a specific inhibitor of an individual kinase, then it is possible to construct an assay that reflects the degree of target inhibition, thus providing PoM. We have developed a potent, specific inhibitor of PIM-1 kinase, an enzyme widely expressed in a range of adenocarcinomas, the overexpression of which has been shown to be a poor prognostic factor. BAD, one of the proteins involved in induction of apoptosis, is a downstream target of PIM-1, and inhibition of the enzyme leads to a significant and dose-dependent reduction in phosphorylation of BAD. This therefore could be developed as a PD biomarker for the drug. Cell culture studies showed that reduction in phosphorylation of BAD to 90% of baseline levels was associated with induction of apoptotic cell death, suggesting a target that might be adopted as a mechanistic surrogate in a clinical trial. Although in an ideal world we would attempt to take tumor biopsies (at the baseline and at some period after drug dosing) to assess the biomarker, in practice this proves difficult, and in most cases samples are taken of peripheral tissues that express the target (e.g., circulating lymphocytes, skin biopsies, hair follicles, buccal mucosal scrapes). There are a few studies in which research groups have attempted to correlate parallel expression of biomarkers on tumors and surrogate tissues, but these are usually small, underpowered, and rather inconclusive. This has not, however, diminished the frequency with which surrogate tissues are used to provide PoM. PIM-1 kinase is expressed on lymphocytes, and it has been possible to develop a Western blot assay that detects phospho-BAD in the buffy coat of lymphocytes. This means that blood samples taken at intervals after drug dosing can be split into serum for drug concentration estimates and PK modeling and the buffy coat for measuring the decrease in the phosphorylation of BAD. Mathematical models can be constructed relating drug concentration to effect in the clinical setting, and reflecting back to the preclinical in vitro cell systems. PoM can now be taken to sophisticated levels; however, the wide interindividual variation in PK and PD variables can make interpretation difficult when only three patients are entered at each dose level.

Proof-of-Principle

PoP extends the mechanistic observations to include evidence of consequent phenotypic charge. For example, inhibition of a kinase that is essential for survival could lead to induction of apoptosis. There are several markers of apoptosis that could be used in the clinics to correlate the degree of inhibition of the drug target with the extent of cell death—for example, tumor biopsy and microscopic assessment of the fraction of apoptotic cells, disaggregation of the tumor biopsy and fluorescent cell sorting for the TUNEL assay, induction of caspases, and fragmentation of DNA characteristic of apoptosis. Other assays are being developed as serum markers of apoptosis and positron emission tomography (PET)-based imaging using novel tracers that rely on the increased cellular permeability associated with apoptosis, but they have not yet been validated.

Another example of PoP is the examination of the vascular axis following inhibition of the vascular endothelial growth factor (VEGF). Antibodies that bind to and biologically neutralize VEGF (such as Avastin™) and small molecular weight inhibitors that inhibit the VEGF receptor have been developed as inhibitors of angiogenesis, one of the crucial processes involved in tumor progression and metastases. The impact of VEGF inhibition on angiogenesis can be assessed by tumor biopsy and fluorescent microscopic staining of the microvasculature. The sophisticated imaging techniques depend on the capacity of VEGF inhibitors to reduce tumor vascular permeability. This is concluded from complex mathematical models derived from dynamic contrast-enhanced MRI scans that have been correlated with the degree of VEGF inhibition.

Proof-of-Concept

The ultimate PoC, of course, is whether the novel agent induces a worthwhile tumor response according to general agreed rules (e.g., RECIST criteria) in which tumor volume is assessed using conventional scanning methods (CT scan or MRI) or more developmental techniques such as PET scanning. By WHO criteria, tumor responses are categorized as complete (disappearance of all measurable disease), partial (a greater than 50% reduction in the bidimensional product of the tumor masses), or stable disease (in which the measurable disease remains the same, within the margins of ±25%). Radiological responses themselves are relatively poor surrogates of survival but would be accepted as a PoC.

CAN WE MAKE GO-OR-NO-GO DECISIONS AT THE END OF PHASE I?

It has been estimated that there are 1,000 novel chemical entities in oncology waiting to enter the clinic. This means that there is increasing pressure to make early go-or-no-go decisions on further drug development. Classically, this decision would be made at the end of phase II using the Gehan two-stage trial; for example, if no tumor responses were seen in the first 14 patients treated, the drug would be deemed inactive (assuming that an active drug would have a tumor response rate of 20% or better). The problem with this approach is that multiple phase II trials, each of a separate tumor type, would be required before a go-or-no-go decision could be made. An alternate model would be to use the PK and PD data gathered during phase I trials. For example, if the maximum plasma concentration of the drug was found to be in the nanomolar range in the clinic setting, and if all the preclinical data indicated that micromolar-range concentrations were required for sufficient induction of the appropriate phenotypic changes, then it would be reasonable to conclude that the experimental drug was unlikely ever to be effective. Similarly, if experimental evidence strongly suggested that 90% of enzyme activity had to be inhibited to induce a significant degree of apoptosis, and if clinical PoM data showed that only 5% target inhibition was achieved, then, again, it would be reasonable to halt drug development. So, it is distinctly possible that these elegant mechanistic endpoints can play an important role in preventing further development of "lame-duck" drugs. These algorithms are presented in Figure 4.1.

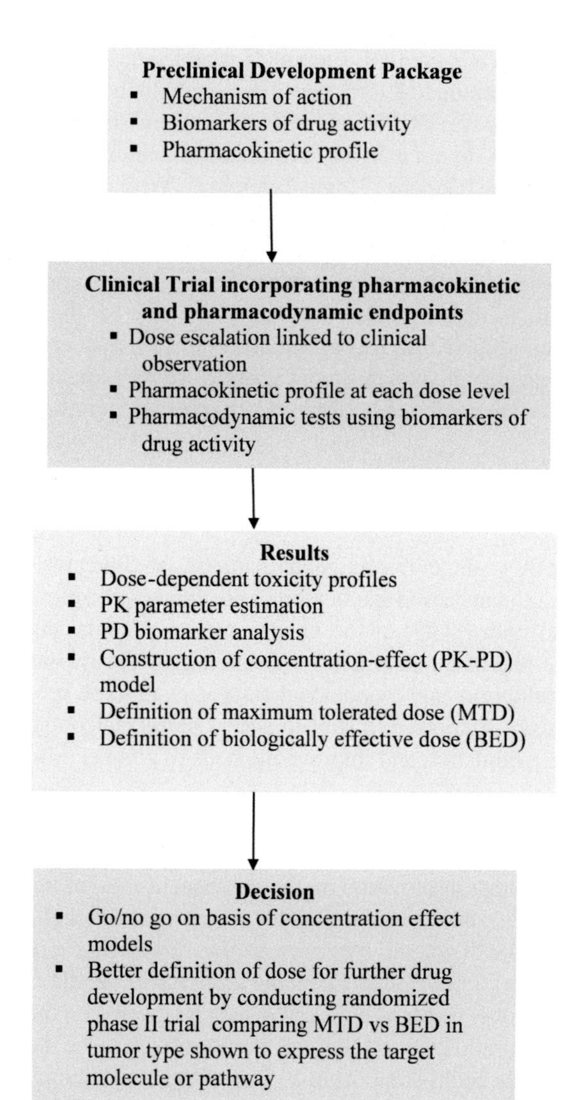

FIGURE 4.1 An early-phase drug development paradigm.

PHASE II TRIALS

The industry standard for phase II trial design in oncology has been predicated on demonstrating a tumor response rate of 20% (number of patients with complete remission [CR] plus partial remission [PR] as 20% of the total patient population treated) and selecting the minimum number of patients with a specified tumor type, stage, and prior number of treatments to be recruited to the trial. Gehan has shown that the probability of rejecting a drug with this level of activity is less than 0.05 if no responses are seen in the first 14 patients treated.

PERSONALIZED MEDICINE

Individualizing therapy through understanding which patients will respond to particular drugs will allow doctors to stop ineffective therapies in unresponsive patients during the crucial therapeutic window when alternative anticancer treatment should be initiated, while obviously eliminating associated side effects, expense, and waste of time for terminally ill patients. Omission of patients unlikely to benefit from specific therapies will enhance the costs–benefit analysis and make it more likely that the bodies governing health care finance will approve expensive drugs for use. The KRAS gene is one example of a situation in which individualized therapy is likely to improve clinical practice in the very near future.

The KRAS gene has a function in the intracellular signal transduction along the EGFR axis. It mediates proliferation signaling intracellularly downstream from the EGFR. Patients whose tumors harbor the wild-type KRAS gene demonstrate effective disruption of cellular growth signaling through the EGFR pathway when they receive EGFR-targeted therapies such as cetuximab. As a result, these patients achieve a greater response rate when the EGFR antibody is added to standard chemotherapy, compared to standard chemotherapy alone. On the other hand, mutated KRAS genes demonstrate the ability to continue signaling to the proliferation cascade within the EGFR axis, even when EGFR inhibitors such as cetuximab are administered. Consequently, patients with tumors harboring such mutant RAS do not gain any extra benefit from the additional administration of cetuximab, compared to receiving chemotherapy alone (Genetics Home Reference, 2008). Around 30% of colorectal cancer patients have tumors with the KRAS mutation, which makes them ineligible for EGFR-targeted therapies like cetuximab and panitumumab (Amado et al., 2008). Removing these patients from the trial reduces financial cost overall and reduces the number of patients exposed to the side effects of likely useless therapy. Recent data suggest that extension of RAS mutational screening to include additional variants further improves the therapeutic ratio for these EGFR-targeted therapies (Douillard et al., 2013). Although there has been a significant focus on markers to select patients who are more likely to respond to treatment, there is growing interest in identifying patients most at risk of serious toxicity and therefore improve risk/benefit ratios by reducing the incidence of life-threatening side effects. Consequently, germline markers of toxicity are appearing in the literature (Church et al., 2014).

OPEN ACCESS CLINICAL TRIALS

Drug discovery has been steadily declining in productivity over the past 30 years; currently, more than 80% of clinical trials fail for pioneer drug targets. There are many reasons for this, the dominant reason likely being poor target validation. A new public–private initiative led by the Structural Genomics Consortium (SGC) of the Universities of Toronto and Oxford aim to provide structural models of proteins of interest to the pharmaceutical and biotechnology industries as potential drug targets and make this information, along with chemical inhibitors and monoclonal antibodies to these proteins, freely available to the oncology community (academia and industry). Authors of this open access partnership have hypothesized that this approach will greatly improve the quality of target validation and improve the odds of a novel drug making it into the clinic (Edwards et al., 2009).

Is it possible to conceive of open access phase I clinical trials of these new agents, overcoming the prohibitive secrecy that most pharma employ in drug development? This would be the clinical equivalent of the experiment being performed by the SGC and would make clinical data publicly available in real time so that information on the risk/benefit ratio of the novel target could be shared with others in the field. Increasing access to clinical data brings obvious concerns about confidentiality and poses outstanding questions over how far the reporting of data will go; will it be necessary to document every blood test, every scan? If data are shared and to be used by others, this places a huge pressure on the accuracy of the data.

Performing a phase I trial with a clinical probe—akin to the chemical probes released by the SGC—is predicated on the likelihood that the probe, which will not be patent protected, allows clinical validation of the target but is unlikely to be the final, best-in-class agent that gets market approval. Financing for such trials could come from a consortium of academia, charities, and industry, which share a common goal of getting a greater number of effective medicines into wider use. Obviously, numerous issues still need to be addressed by the wider sharing of data, but this may be one means of modifying early phase trial design to better improve the efficiency of cancer drug development.

REFERENCES

Abbruzzese, J.L., Grunewald, R., Wees, E.A., Gravel, D., Adams, T., Nowak, B., et al., 1991. A phase I clinical, plasma, and cellular pharmacology study of gemcitabine. J. Clin. Oncol. 9, 491–498.

Amado, R., Wolf, M., Peeters, M., Van Cutsem, E., Siena, S., Freeman, D.J., et al., 2008. Wild-type KRAS is required for panitumumab efficacy in patients with metastatic colorectal cancer. J. Clin. Oncol. 10, 1626–1634.

Anderson, H., Lund, B., Bach, F., Thatcher, N., Walling, J., Hansen, H.H., 1994. Single agent activity of weekly gemcitabine in advanced non small lung cancer: a phase II study. J. Clin. Oncol. 12, 1821–1826.

Brand, R., Capadno, M., Tempero, M., 1997. A phase I trial of weekly gemcitabine administered as a prolonged infusion in patients with pancreatic cancer and other solid tumours. Invest. New Drugs 15, 331–341.

Church, D., Kerr, R., Domingo, E., Rosmarin, D., Palles, C., Maskell, K., Tomlinson, I., Kerr, D., 2014. 'Toxgnostics': an unmet need in cancer medicine. Nat. Rev. Cancer 14, 440–445.

Cormier, Y., Eisenhauer, E., Muldal, A., Gregg, R., Ayoub, J., Goss, G., et al., 1994. Gemcitabine is an active new agent in previously untreated small cell lung cancer (SCLC): a study of the National Cancer Institute of Canada Clinical Trials Group. Ann. Oncol. 5, 283–285.

Douillard, J.Y., Oliner, K.S., Siena, S., Tabernero, J., Burkes, R., Barugel, M., et al., 2013. Panitumumab-FOLFOX4 treatment and RAS mutations in colorectal cancer. N. Engl. J. Med. 369, 1023–1034.

Early Stage Clinical Trial Taskforce, 2006. Joint ABPI/BIA report (accessed 02.12.08.). Referred to on http://www.abpi.org.uk/media-centre/newsreleases/2006/Pages/240706b.aspx http://www.abpi.org.uk/information/pdfs/BIAABPI_taskforce2.pdf (accessed 22.08.14.).

Edwards, A.M., Bountra, C., Kerr, D.J., Willson, T.M., 2009. Open access chemical and clinical probes to support drug discovery. Nat. Chem. Biol. 5, 436–440.

Eisenhauer, E.A., 1998. Phase I and II trials of novel anti-cancer agents: endpoints, efficacy and existentialism. Ann. Oncol. 9, 1047–1052.

Eisenhauer, E.A., O'Dwyer, P.J., Christian, M., Humphrey, J.S., 2000. Phase I clinical trial design in cancer drug development. J. Clin. Oncol. 18, 684–692.

Eisenhauer, E.A., Twelves, C., Buyse, M., 2006. Phase I Cancer Clinical Trials: A Practical Guide. Oxford University Press, Oxford.

Fosella, F.V., Lippman, S.M., Shin, D.M., Tarassoff, P., Calayag-Jung, M., Perez-Soler, R., et al., 1997. Maximum tolerated dose defined single agent gemcitabine: a phase I dose escalation study in chemotherapy naïve patients with advanced non small cell lung cancer. J. Clin. Oncol. 15, 310–316.

Genetics Home Reference, 2008. KRAS. http://ghr.nlm.nih.gov/gene=kras (accessed 11.08.14.).

Karnofsky, D.A., Burchenal, J.H., 1949. The clinical evaluation of chemotherapeutic agents in cancer. In: Macleod, C.M. (Ed.), Evaluation of chemotherapeutic agents. Columbia University Press, New York, pp. 199–205.

Mertens, W.C., Eisenhauer, E.A., Moore, M., Venner, P., Stewart, D., Muldal, A., et al., 1993. Gemcitabine in advanced renal cell cancer: a phase II study of National Cancer Institute of Canada Clinical Trials Group. Ann. Oncol. 4, 331–332.

Miller, A.B., Hoogstraten, B., Staquet, M., Winkler, A., 1981. Reporting results of cancer treatment. Cancer 47, 207–214.

Oken, M.M., Creech, R.H., Tormey, D.C., Horton, J., Davis, T.E., McFadden, E.T., et al., 1982. Toxicity and response criteria of the Eastern Cooperative Oncology Group. Am. J. Clin. Oncol. 5, 649–655.

O'Rourke, T.J., Brown, T.D., Havlin, K., Kuhn, J.G., Craig, J.B., Burris, H.A., et al., 1994. Phase I clinical trial of gemcitabine given as intravenous bolus on 5 consecutive days. Eur. J. Cancer 30A, 417–418.

Parulekar, W.R., Eisenhauer, E.A., 2002. Novel endpoints and design of early clinical trials. Ann. Oncol. 13, 139–143.

Tempero, M., Plunkett, W., van Haperen, R., Hainsworth, J., Hochster, H., Lenzi, R., et al., 2003. Randomised phase II comparisons of dose intense gemcitabine: thirty minutes infusion and fixed dose rate infusion in patients with pancreatic adeno carcinoma. J. Clin. Oncol. 21, 3402–3408.

U.S. Food and Drug Administration, 2005. Guidance for industry. Estimating the maximum safe starting dose in initial clinical trials for therapeutics in adult healthy volunteers. http://www.fda.gov/downloads/Drugs/GuidanceComplianceRegulatoryInformation/Guidances/UCM078932.pdf (accessed 11.08.14).

Chapter 4.2

The Pharmaceutical R&D Productivity Crisis: Can Exploratory Clinical Studies Be of Any Help?

Cecilia Karlsson

The past decades have witnessed decreasing pharmaceutical R&D productivity. The costs of developing new drugs have increased, as have total R&D expenditures. The rate of introduction of new molecular entities (NMEs) has at best remained constant, and the attrition rates, especially in late-phase clinical trials, have increased (Pammolli, Magazzini, and Riccaboni, 2011). Today, the average cost to bring an NME to market is estimated to be $1.8 billion, and the cost is constantly increasing (Paul et al., 2010).

TRADITIONAL DRUG DEVELOPMENT

The traditional drug development process can be divided into two major parts: drug discovery and clinical development (Figure 4.2). In drug discovery, the preclinical part of the drug development process, large numbers of molecules are generated and characterized. The goal is to identify the most promising compounds for further development, using, for example, in vitro testing. If the results are promising, greater amounts of the compounds are prepared and tested in animal studies for efficacy and safety. Finally, one compound is commonly selected for an Investigational New Drug (IND) application. Thus, many years and resources are spent on compounds for which no human data are available.

The first human studies, the phase I studies, are normally conducted in healthy volunteers. Results from the target population will usually not be obtained prior to phases II and III. This means that it takes additional years after entering clinical development before human data from the target population is generated.

The clinical development accounts for approximately 63% of the costs for each NME launched, and the majority of the cost, 53%, is spent from phase II to launch (Paul et al., 2010). A vast majority of the investigational products that enter clinical trials fail (Kola and Landis, 2004; U.S. Food and Drug Administration, 2004). Recent publications indicate that phase II and phase III attrition rates are unacceptably high: 66% in phase II and 30% in phase III (Paul et al., 2010). In fact, phase II attrition rates have not improved at all since the previous analysis done over the period 1991–2000, demonstrating 62% attrition in phase II and 45% attrition in phase III (Kola and Landis, 2004). There are several reasons for failures: pharmacokinetic (PK) properties, efficacy, clinical safety, toxicology, cost of goods, formulation, commercial, etc. (Kola and Landis, 2004). A recent analysis of phase II and phase III attritions in 2011–2012 shows that the majority of failures between phase II and submission were due to lack of efficacy (56%) or safety issues (28%) (Arrowsmith and Miller, 2013). The proportion of efficacy failures was higher in phase II (59%) compared to phase III and beyond (52%). On the other hand, attritions due to safety were higher in phase III and beyond (35%) compared with phase II (22%). The late attrition due to safety is most likely due to the fact that some safety issues only become apparent in longer studies including a large number of patients (Arrowsmith and Miller, 2013).

Two key approaches have been suggested to reduce phase II and phase III attrition. The first approach is to improve target selection, with selection of more validated targets. The second approach is to pursue early clinical studies demonstrating target engagement or efficacy to drive an earlier clinical attrition (Kola and Landis, 2004; Paul et al., 2010). It has been claimed that reducing the attrition rate of drug candidates in clinical development would be the greatest challenge and opportunity for the viability of pharmaceutical R&D. Sensitivity analyses have shown that reducing phase II and phase III attrition would be the strongest levers for improved R&D efficiency and decreased cost per NME (Paul et al., 2010). A key question would be how to generate human data earlier, providing an opportunity to close projects early that do not hold promise and focus resources on promising compounds.

Principles of Translational Science in Medicine. http://dx.doi.org/10.1016/B978-0-12-800687-0.00023-2

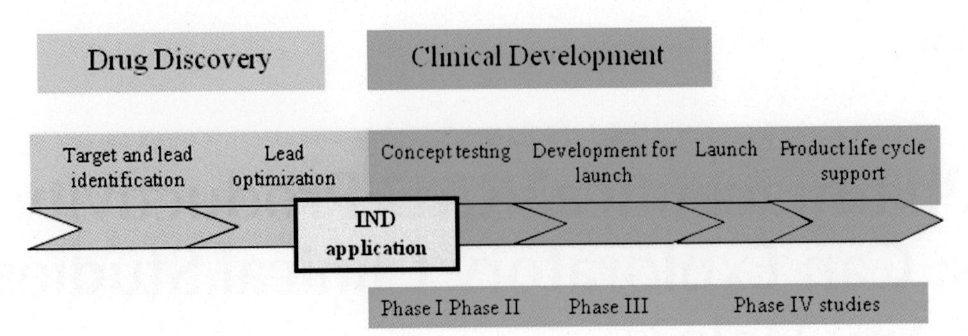

FIGURE 4.2 . Depiction of the two major parts of the traditional drug development process: drug discovery and clinical development.

IND APPLICATION

Before any human studies can be initiated, an IND application containing, among other things, information on any risks anticipated based on the results of pharmacological and toxicological data collected during studies of the drug in animals must be submitted to a regulatory authority. These early tests are normally resource intensive, and these resources may even be wasted on products whose profiles may be determined unacceptable when evaluated in humans. In a standard IND application used for regulatory purposes, only one compound can be covered under each IND application.

OPPORTUNITIES FOR EARLIER DECISION MAKING

In traditional drug development, lots of resources are spent on candidate products that will fail rather late in clinical development. In the Critical Path Initiative (also known as "Challenge and Opportunity on the Critical Path to New Medical Products") of March 2004, the FDA concluded that a new product development toolkit was urgently needed to reduce the time and resources expended on candidate products that are unlikely to succeed (U.S. Food and Drug Administration, 2004). Exploratory clinical studies are considered to be part of these efforts. In January 2006, the FDA presented a guidance document titled "Guidance for Industry, Investigators, and Reviewers—Exploratory IND Studies" (U.S. Food and Drug Administration, 2006). The document is a reminder that exploratory clinical studies outside the regulatory process for the registration of a drug product are, and have always been, accepted by the FDA. The key principle is a risk-benefit-based approach. Because exploratory clinical studies require less, or different, preclinical support than standard IND studies, a limited number of subjects can be exposed with a limited dose range for a limited period of time. Due to the limitations in number, dose range, and treatment period, exploratory clinical studies present fewer potential risks than do traditional phase I studies. The results from exploratory clinical studies can help to identify promising candidates early in the drug development process for continued development and to eliminate those lacking promise. For example, exploratory clinical studies can provide important information on PK properties and target engagement facilitating an early go-or-no-go decision for specific compounds or projects, thus improving R&D productivity with limited investments (U.S. Food and Drug Administration, 2006).

Chapter 4.3 more specifically describes different types of exploratory clinical studies and suggests when they would provide value to projects.

REFERENCES

Arrowsmith, J., Miller, P., 2013. Trial watch: phase II and phase III attrition rates 2011–2012. Nat. Rev. Drug Discov. 12, 569.

Kola, I., Landis, J., 2004. Can the pharmaceutical industry reduce attrition rates? Nat. Rev. Drug Discov. 3, 711–715.

Pammolli, F., Magazzini, L., Riccaboni, M., 2011. The productivity crisis in pharmaceutical R&D. Nat. Rev. Drug Discov. 10, 428–438.

Paul, S.M., Mytelka, D.S., Dunwiddie, C.T., Persinger, C.C., Munos, B.H., Lindborg, S.R., et al., 2010. How to improve R&D productivity: the pharmaceutical industry's grand challenge. Nat. Rev. Drug Discov. 9, 203–214.

U.S. Food and Drug Administration, 2004. Innovation or stagnation? Challenge and opportunity on the critical path to new medical products. U.S. Department of Health and Human Services, Food and Drug Administration. http://www.fda.gov/downloads/ScienceResearch/SpecialTopics/CriticalPathInitiative/CriticalPathOpportunitiesReports/ucm113411.pdf (accessed 18.06.14.).

U.S. Food and Drug Administration, 2006. Guidance for industry, investigators, and reviewers—exploratory IND studies. U.S. Department of Health and Human Services, Food and Drug Administration, Center for Drug Evaluation and Research. CDER, Rockville, MD. http://www.fda.gov/downloads/Drugs/GuidanceComplianceRegulatoryInformation/Guidances/ucm078933.pdf (accessed 18.06.14.).

Chapter 4.3

Exploratory Clinical Studies ("Phase 0 Trials")

Cecilia Karlsson

Exploratory clinical studies are also referred to as phase 0 trials, as they are conducted prior to the traditional phase I dose escalation, safety, and tolerability studies (Lappin, Noveck, and Burt, 2013). As regulatory authorities in their guidance documents use the terms *exploratory clinical studies* or *exploratory IND studies* rather than *phase 0 trials*, these terms are used throughout this chapter (ICH guideline M3(R2), 2009; U.S. Food and Drug Administration, 2006). The driver for exploratory clinical studies has been the decline in pharmaceutical productivity, leading to few, new marketed drugs. A key question has been how to generate human data earlier and cheaper, facilitating earlier closure of projects that do not hold promise and focusing resources on promising compounds. This can be done by an exploratory clinical study approach. Early evaluation of, for example, pharmacokinetics (PK) and pharmacodynamics (PD) in humans through administration of limited doses to a limited number of subjects for a limited period of time will require a shorter preclinical package to support these studies, and thus, generation of human data can be done faster and cheaper without placing study subjects at risk.

Different types of exploratory clinical studies can be conducted and are described in this chapter. The first paper to describe the microdosing concept was published in 2003, although the maximum dose was not defined at that time (Lappin and Garner, 2003). Regulatory authorities followed with a position paper from the European Medicines Agency on microdosing in 2004 (European Medicines Agency, 2004), guidance from the U.S. FDA on exploratory IND studies in 2006 (U.S. Food and Drug Administration, 2006), and guidance from the Ministry of Health, Labor and Welfare in Japan on microdose clinical studies in 2008 (Ministry of Health, Labor and Welfare, 2008). The current international standards for nonclinical safety studies for the conduct of human clinical trials and marketing authorization for pharmaceuticals, ICH guideline M3(R2), came into effect in December 2009, and includes a section on exploratory clinical trials (ICH guideline M3(R2), 2009). The microdosing position paper from the European Medicines Agency (2004) is now superseded by ICH M3(R2).

An exploratory IND can support clinical trials with several compounds with a common biological target in healthy or minimally ill subjects. The studies can demonstrate pharmacological activity but not maximum tolerated dose (MTD) levels, and they have no therapeutic or diagnostic intent (U.S. Food and Drug Administration, 2006).

Exploratory Clinical Studies Can Help to

- Determine whether a mechanism of action defined in experimental systems can also be observed in humans (proof-of-mechanism [PoM])
- Provide information on pharmacokinetics (PK) and PK/pharmacodynamics (PD)
- Select the most promising lead product from a group of candidates designed to interact with a particular therapeutic target in humans based on PK and/or PD
- Explore biodistribution using imaging technologies

TYPES OF STUDIES

Depending on the goal of the exploratory clinical study, different types of studies requiring different levels of toxicological support can be considered, including both single- and multiple (repeated)-dose studies. The ICH M3(R2) provides guidance on five exploratory clinical study approaches, with recommendations for supporting nonclinical studies. The five different exploratory clinical approaches include (1) microdosing involving not more than a total dose of 100 μg that can be administered as a single dose or divided doses in any subject; (2) microdosing involving ≤5 administrations of a maximum of 100 μg per administration (a total of 500 μg per subject); (3) single-dose trials at subtherapeutic doses or into the anticipated therapeutic range; (4) and (5)

Principles of Translational Science in Medicine. http://dx.doi.org/10.1016/B978-0-12-800687-0.00024-4

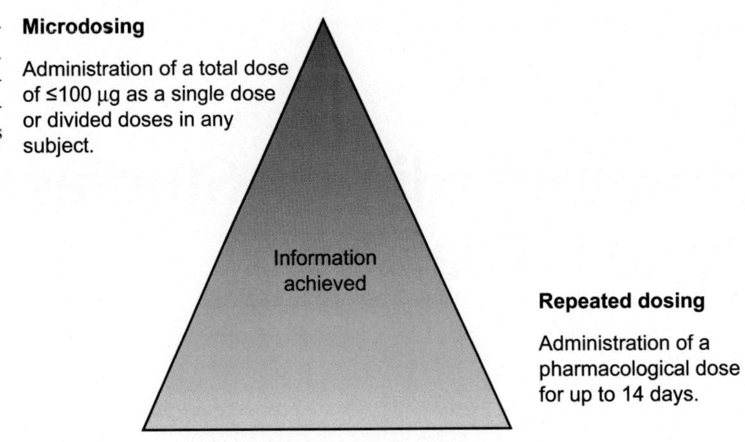

FIGURE 4.3 Depiction of different types of studies that can be performed under an exploratory IND. The two extremes are microdosing and repeated-dosing studies. The two study types will provide different information and will also require different levels of toxicological support.

multiple dose trials up to 14 days of dosing for determination of PK and PD in humans in the therapeutic dose range, with no intention to support the determination of MTD. Approaches 4 and 5 are supported by two different nonclinical approaches.

In summary, the "smallest" exploratory clinical study is microdosing, whereas the "largest" exploratory clinical study is a repeated-dose trial of a pharmacological dose for up to 14 days (Figure 4.3).

Because exploratory clinical studies involve administration of either subtherapeutic doses of a product or doses expected to produce a therapeutic but not toxic effect, the potential risk to human subjects is less than in a traditional phase I study that, for example, seeks to establish an MTD. Therefore, exploratory clinical studies can be initiated with less, or different, preclinical support than is required for traditional studies.

MICRODOSING

Microdosing is a testing paradigm that includes administration of minute amounts of the test substances to any subjects (ICH guideline M3(R2), 2009; Lappin and Garner, 2003). As described previously, for a microdosing approach, not more than 100 μg of the test substance is administered at the same time, and the total amount administered will not exceed 500 μg. The test substance is usually labeled with a highly specific radioactivity using a radionuclide that does not change the biochemical properties of the compound. The fate of the radioligand is assayed by the use of ultrasensitive detection methods, such as accelerator mass spectrometry (AMS) or positron emission tomography (PET), which provide information on PK and distribution, respectively. AMS microdosing is considered a nonradioactive method, whereas PET microdosing exposes a person to 2–4 mSv, which is in the same order as a body computed tomography investigation (Bergström, Grahnén, and Långström, 2003). Microdosing studies have also been performed using nonlabeled compounds and sensitive liquid chromatography-tandem mass spectrometry (LC-MS/MS) (Maeda and Sugiyama, 2011). The choice of analytical technique, AMS or LC-MS/MS, is to a great extent driven by the expected plasma-drug concentrations and the limit of quantification of the analytical method (Lappin, Noveck, and Burt, 2013).

Accelerator Mass Spectrometry Microdosing

Drugs enriched with ^{14}C are routinely used to examine their own metabolism and PK. Because of the long half-life of ^{14}C, a far more sensitive method is to directly detect the ^{14}C atoms themselves, and AMS is based on that principle. For a reliable, reproducible AMS measurement, only about 1000 ^{14}C atoms need to be detected (Lappin and Garner, 2003).

AMS allows administration of subpharmacological doses of ^{14}C-labeled investigational drugs with the goal of obtaining preliminary information regarding the absorption, distribution, metabolism, and excretion of test compounds (Turteltaub and Vogel, 2000). At different time points after drug administration, urine, feces, blood, and possibly also cerebrospinal fluid (CSF) samples can be collected for analysis. Urine and feces are recovered to measure the amount of drug that is excreted by these pathways, whereas blood and CSF are collected to measure total drug and/or metabolite concentrations. The samples are collected over a time period that is sufficiently long to account for several plasma half-lives expected of the drug under study.

Predictability of Accelerator Mass Spectrometry Microdosing

In an ideal situation, a compound displays linear kinetics across subpharmacological (microdose) and pharmacological dose ranges (Figure 4.4) (Lappin and Garner, 2003; Garner, 2010). However, transporter mechanisms and binding to macromolecules are processes that could lead to nonlinearity at very low doses (Buchan, 2007).

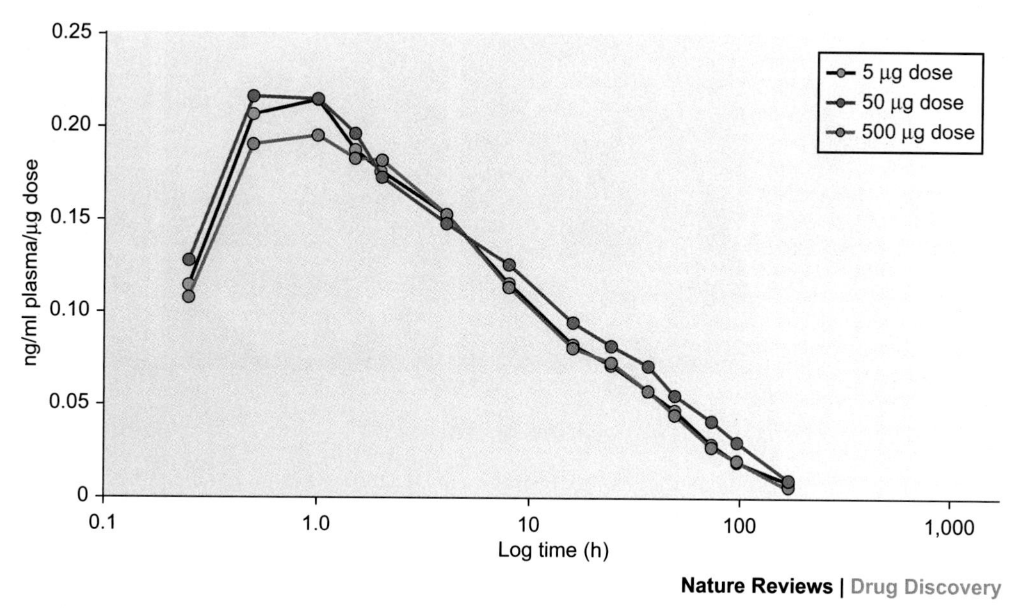

FIGURE 4.4 Semi-log plot comparing elimination of a drug from plasma following a 5, 50, or 500 μg oral dose. The plot shows linear kinetics across the dose range. *(From Lappin and Garner, 2003, reprinted by permission from Macmillan Publishers Ltd:* Nat. Rev. Drug Discov., *2, 233–240, © 2003.)*

To explore the utility of microdosing, two validation studies have been conducted in collaboration between pharmaceutical companies, universities, and research institutes known by the acronyms of the Consortium for Resourcing and Evaluating AMS Microdosing (CREAM) and EU Microdosing AMS Partnership Programme (EUMAPP). The CREAM trial was designed to assess the predictability of PK at therapeutic doses from human microdosing. Five test compounds (diazepam, midazolam, ZK253, warfarin, and erythromycin) were administered at a microdose level and at a therapeutic dose level to subjects in a cross-over design. Microdose data for three drugs (diazepam, midazolam, and ZK253) predicted the therapeutic dose PK well. For warfarin, microdose data predicted clearance for the therapeutic dose, but no linearity in the distribution PK was seen. One drug (erythromycin) could not be tested due to degradation in the stomach (Buchan, 2007; Lappin et al., 2006). The EUMAPP trial evaluated the proportionality of dose to exposure, or PK, of six drugs (paracetamol, phenobarbital, sumatriptan, propafenone, clarithromycin, fexofenadine). Of the six drugs examined, the microdose data would have predicted the PK for all molecules within a two- to three-fold factor of the pharmacological dose PK (Garner, 2010). In other words, if the EUMAPP molecules had been novel molecules, the results would have permitted a decision regarding whether or not it would be worthwhile to continue developing these molecules. Taking all data generated so far into account, about 80% of drugs tested between a microdose and a therapeutic dose have demonstrated scalable PK within a factor of two-fold for any given parameter (Lappin, Noveck, and Burt, 2013). For oral administration, about 60% of the drugs tested showed scalable PK within a factor of two. For intravenous administration, the PK for all drugs examined so far have been scalable (Lappin, Noveck, and Burt, 2013).

Positron Emission Tomography Microdosing

PET is a noninvasive tomographic imaging method, which makes it possible to determine drug distribution and concentration in vivo in humans (Bergström, Grahnén, and Långström, 2003). The investigational drugs are labeled with short-lived positron-emitting radionuclides and administered to human volunteers, mostly as intravenous injections. The positrons interact with electrons present in the atoms that make up surrounding tissues. Two oppositely directed gamma rays are then produced that can be detected by an array of detectors surrounding the volunteer. The radiotracer of major interest is ^{11}C. It has a short half-life of only 20 minutes and must therefore be manufactured in very close proximity to the PET instrument (Lappin and Garner, 2003).

There are two different approaches to PET microdosing. First, an isotope-labeled ligand can be used to displace the investigational drug. This is useful, for example, for drugs developed to treat disorders of the central nervous system, and it establishes both the entry of the drug into the brain and whether it binds to the specific receptor it is designed to target. Second, the investigational drug itself can be labeled, and its ability to reach target tissues can be studied (Lappin and Garner, 2003).

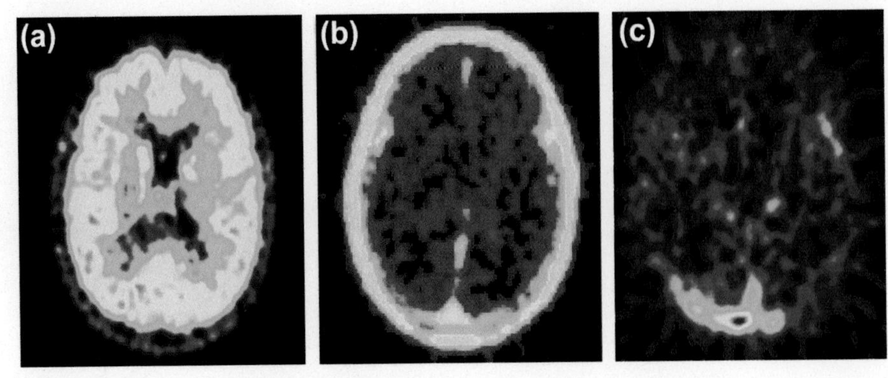

FIGURE 4.5 Examples of uptake of microdoses of ^{11}C-labeled drugs into the brain measured by positron emission tomography (PET). Figure a illustrates a drug with good brain penetration, Figure b illustrates a drug with intermediate penetration, and Figure c illustrates a drug with poor brain penetration. *(From Lappin and Garner, 2003, reprinted by permission from Macmillan Publishers Ltd:* Nat. Rev. Drug Discov., *2, 233–240, ©* *2003.)*

Predictability of Positron Emission Tomography Microdosing

Many enzyme systems and receptor-mediated transport systems can be saturated at high concentrations of drugs, and drug distribution at very low concentrations can be governed by high-affinity binding sites. This implies that it is possible that the distribution, metabolism, and elimination pattern can be different at tracer doses and at therapeutic doses. However, the exchange of drugs between tissue and plasma is usually affected little by dose (Bergström, Grahnén, and Långström, 2003). In a study by Lappin and Garner, three examples of ^{11}C-labeled drugs under development were administered at microdosing levels and at therapeutic levels (Figure 4.5). They showed the same features of brain penetration or lack of brain penetration at both levels, supporting the concept that PET microdosing can predict drug entry into the brain (Lappin and Garner, 2003).

REPEATED DOSING

Repeated dosing for up to 14 days into the therapeutic dose range but not MTD is the most informative exploratory clinical study that can be performed. Such a design would allow PK as well as biomarker and efficacy readouts after both single and multiple dosing for several disease areas.

SHOULD EXPLORATORY CLINICAL STUDIES BE PERFORMED IN ALL PROJECTS?

The answer is no. If animal models are predictive and if there are no specific issues, an exploratory clinical study will only delay development time.

Exploratory clinical studies will add value only to projects with specific issues. For example, in projects in which animal models have low or unknown predictability to humans, it would be valuable to conduct an exploratory clinical study evaluating target engagement and efficacy. This would require that readouts are available that will "respond" within 14 days of treatment. Also, in projects in which PK properties are difficult to predict, or if there are safety concerns—not at the MTD—an exploratory clinical study would be of value. If more than one compound is available in the preclinical phase, an exploratory clinical study will also provide the opportunity to rank and benchmark several compounds with a common biological target under the same exploratory IND, in order to identify the most promising candidate for continued development. All these efforts would strive to provide important human feedback to the project that could steer the "go-or-no-go" decision for a certain target or compound(s) at an early stage in the drug development process. Such information would help increase pharmaceutical productivity by earlier attrition of compounds that do not hold promise for further development.

PRACTICAL APPLICATIONS

Example

In one project, three compounds are available with a common biological target.

How can we select the most promising compound for continued development?

1. If predictive animal models are available, this would be our first choice. It will be the fastest way to identify the most promising compound and progress it to a "traditional" phase I study.
2. If predictive animal models are *not* available, an exploratory clinical study would be considered.

The need for an exploratory clinical study approach should be identified as soon as possible. Frontloading of activities such as compound supply, formulation, toxicology studies, development of biomarkers, and so on is needed, and contact with a regulatory authority must be initiated. For a successful outcome, a high level of cooperation, communication, and sharing of information between drug discovery and clinical development is needed.

In the aforementioned example, an exploratory clinical study with the aim of evaluating the PK and PD of the three compounds is performed. What kind of study would be most relevant? A repeated-dose trial during up to 14 days involving up to a pharmacological dose in minimally ill subjects (if possible) would provide an opportunity to obtain enough data to choose the most favorable compound with which to proceed.

Possible Outcomes of the Suggested Repeated-Dose Trial

1. PK: The effects of repeated dosing in the target population would be observed.
2. PD: Determination of both proof-of-mechanism (PoM) and proof-of-principle (PoP) may be possible.
3. Safety: No information would be available on the MTD, but other safety signs might be observed.
4. Important feedback to drug discovery would take place.

Only one exploratory IND is needed for all three compounds, as they have a common biological target. In this example, use of an exploratory IND approach would probably provide enough data to select the most promising compound for continued development.

Assuming the most promising compound is selected based on generated human data, can the researchers then go directly into traditional phase I studies? The answer is no. The reduced preclinical safety studies are unsuitable for registration and support of traditional phase I studies and need to be repeated. The exploratory IND also needs to be closed before a standard IND can be opened for the same compound.

Another possibility would be to perform a full safety package for all three compounds before performing the exploratory IND study. After completion and evaluation of the exploratory IND study, a standard IND could then be opened without any delays, as the supporting toxicology studies would have already been performed. However, more resources than necessary would then have been spent on two compounds for which development would be stopped.

In summary, exploratory clinical studies provide a wide range of opportunities to achieve clinical information early on in the drug development process needed for an informed decision.

REFERENCES

Bergström, M., Grahnén, A., Långström, B., 2003. Positron emission tomography microdosing: a new concept with application in tracer and early clinical drug development. Eur. J. Clin. Pharmacol. 59, 357–366.

Buchan, P., 2007. Smarter candidate selection—utilizing microdosing in exploratory clinical studies. Ernst Schering Res Found Workshop 59, 7–27.

European Medicines Agency, 2004. Position paper on the non-clinical safety studies to support clinical trials with a single microdose. CPMP/SWP/2599/02/ Rev. 1. http://www.emea.europa.eu/pdfs/human/swp/259902en.pdf (accessed 12.02.08.).

Garner, R.C., 2010. Practical experience of using human microdosing with AMS analysis to obtain early human drug metabolism and PK data. Bioanalysis 2, 429–440.

ICH guideline M3(R2) on non-clinical safety studies for the conduct of human clinical trials and marketing authorization for pharmaceuticals, 2009.

Lappin, G., Garner, R.C., 2003. Big physics, small doses: the use of AMS and PET in human microdosing of development drugs. Nat. Rev. Drug Discov. 2, 233–240.

Lappin, G., Kuhnz, W., Jochemsen, R., Kneer, J., Chaudhary, A., Oosterhuis, B., et al., 2006. Use of microdosing to predict pharmacokinetics at the therapeutic dose: experience with 5 drugs. Clin. Pharmacol. Ther. 80, 203–215.

Lappin, G., Noveck, R., Burt, T., 2013. Microdosing and drug development: past, present and future. Expert Opin. Drug Metab. Toxicol. 9, 817–834.

Maeda, K., Sugiyama, Y., 2011. Novel strategies for microdose studies using non-radiolabeled compounds. Adv. Drug Deliv. Rev. 63, 532–538.

Ministry of Health, Labor and Welfare, Japan, 2008. Pharmaceutical and Medical Safety Bureau, Tokyo. Microdose clinical studies.

Turteltaub, K.W., Vogel, J.S., 2000. Bioanalytical applications of accelerator mass spectrometry for pharmaceutical research. Curr. Pharm. Des. 6, 991–1007.

U.S. Food and Drug Administration, 2006. Guidance for industry, investigators, and reviewers—exploratory IND studies. U.S. Department of Health and Human Services, Food and Drug Administration, Center for Drug Evaluation and Research, CDER, Rockville, MD. http://www.fda.gov/downloads/Drugs/GuidanceComplianceRegulatoryInformation/Guidances/ucm078933.pdf (accessed 18.06.14.).

Chapter 4.4

Adaptive Trial Design

Martin Wehling

The scientific value of a clinical trial critically depends on the exclusion of bias, which would render the result invalid. If, for example, 100 readouts of clinical trials were available and the choice of readouts was unregulated, one would always find five significant results, on average, if the alpha error was 5%. To avoid this, a prospective trial will be designed in a very specific manner in that, for example, the main readouts ("endpoints") are prespecified and classified as primary and secondary endpoints. The same holds true for a long list of study parameters such as inclusion/exclusion criteria, number of patients/subjects to be included, duration of the intervention, blinding, statistical analysis, and interpretation of readouts.

This strict interpretation of prospective planning yields the highest possible relevance and reliability of conclusions drawn from a clinical trial. It is still the leading design for late-phase trials, such as phase IIb or phase III trials. In these trials, the main properties of new drugs or devices, which are efficacy and safety, need to be proven beyond a reasonable doubt. This is achieved best through a fixed design, prespecified at all levels of sophistication, and of course other major features of "gold-standard" trials (controlled, double-blind, randomized, prospective trials).

At earlier stages of clinical development, this static type of trial design may seem to be suboptimal because new findings from the running trial, from another trial on the same compound, or even from unrelated trials testing novel biomarkers may render the original trial design either outdated or insufficient. As a common example, one could envision the impact of dose-finding studies on running clinical trials in which the dose estimate from previous experiences, including animal results, may have been used in a prospective larger trial but found insufficient in those newer trials or even in the earlier subjects/patients of the same trial. Continuing the trial according to its original design would be a waste of time and money but, above all, would represent an ethical problem because no results of clinical importance may emerge. This implies that the subjects/patients on the test drug would have been exposed to avoidable risk that is clearly unacceptable under ethical auspices.

Thus, the adaptation of a given trial design during the course of its execution seemed highly desirable, and in 2006 the FDA (U.S. Department of Health and Human Services, Food and Drug Administration, 2006) and the EMEA (European Medicines Agency, 2006) published the first guidances on the adaptation of clinical trial designs. In 2010, the FDA published a more extensive and elaborate document on this topic (U.S. Food and Drug Administration, 2010) with the final version still pending at the time of this writing.

In this document, *adaptive design clinical study* is defined as "a study that includes a prospectively planned opportunity for modification of one or more specified aspects of the study design and hypotheses based on analysis of data (usually interim data) from subjects in the study." This definition restricts the use of the term to adaptations derived only from data of the same study. It should be mentioned that the term is also used for such modifications resulting from data of other studies or sources, as mentioned previously. Invariably, it is important that the adaptations must be planned prospectively (and should be discussed with the authorities beforehand); if it is not planned as an adaptive trial, modifications of the design during a running trial are very difficult and hard to analyze statistically.

Two levels of data impact need to be separated: design modification in response to blinded and unblinded data. Responses to unblinded data (interim analyses of clinical endpoints with intervention/control comparisons) may essentially weaken the value of results or render the entire study invalid. Responses to blinded data, such as the increase of subjects if aggregated endpoints (sum of endpoints in all study arms with blinding maintained) are below expectation, are much less troublesome and may even be imposed if not prospectively planned.

Examples for modifiable parameters proposed in the FDA document are given in Table 4.3.

Depending on the main areas of modification, different types of adaptive trials may be separated. Chow (2014) classifies adaptive trials into the following categories:

- An adaptive randomization design
- An adaptive group sequential design
- A flexible sample size re-estimation design

Principles of Translational Science in Medicine. http://dx.doi.org/10.1016/B978-0-12-800687-0.00025-6

- A drop-the-losers design
- An adaptive dose-finding design
- A biomarker-adaptive design
- An adaptive treatment-switching design
- An adaptive-hypothesis design
- A phase I/II (or II/III) adaptive seamless trial design
- A multiple adaptive design

TABLE 4.3 Possible study design modifications in adaptive design clinical studies

• Study eligibility criteria (either for subsequent study enrollment or for a subset selection of an analytic population)
• Randomization procedure
• Treatment regimens of the different study groups (e.g., dose level, schedule, duration)
• Total sample size of the study (including early termination)
• Concomitant treatments used
• Planned schedule of patient evaluations for data collection (e.g., number of intermediate time points, timing of last patient observation, and duration of patient study participation)
• Primary endpoint (e.g., which of several types of outcome assessments, which time point of assessment, use of a unitary vs composite endpoint or the components included in a composite endpoint)
• Selection and/or order of secondary endpoints
• Analytic methods to evaluate the endpoints (e.g., covariates of final analysis, statistical methodology, Type I error control)

(From U.S. Food and Drug Administration, 2010.)

Most classifiers are self-explanatory; adaptive randomization means adjustment of randomization schemes, especially in small sample size studies. In adaptive group sequential design studies, a variety of parameters may be varied (sample size, treatment arms, study endpoints, dose, and/or treatment duration). In the "drop-the-loser design," study arms without expected results may be closed and replaced by new study arms, for example, at higher doses. In the adaptive treatment-switching design, a patient may be switched from the unsuccessful study arm to a successful arm; this approach is mainly used in cancer studies. In the phase I/II (or II/III) adaptive seamless trial design, two phases of clinical development are combined in one study. In the phase I/II design, the maximal tolerable dose may be defined first and then applied to test efficacy/safety in patients in the same trial.

Adaptive design clinical studies are increasingly used in drug development. In a recent Elsevier Business Intelligence Survey, 30 biostatisticians and product development executives in pharmaceutical and clinical research firms reported that more than 75% of clinical trial practitioners had performed or considered performing an adaptive trial. However, 85% of those respondents had experience with fewer than five adaptive trials (cited from Rosenberg, 2011).

This high prevalence of adaptive trial design is the response to its obvious benefits (Chow and Corey, 2011):

1. The investigator may correct wrong assumptions made at or even before the beginning of the trial.
2. The investigator may choose the most promising option early.
3. The design allows for using emerging external information to improve the trial.
4. It opens the opportunity to cope with unexpected events.
5. It helps to speed up development process.

Conversely, there are major concerns about disadvantages because every adaptation, even if prespecified, may carry the risk of bias (selecting the good, avoiding the bad). This is particularly true for those adaptive design types considered as being "less well understood" by the FDA (U.S. Food and Drug Administration, 2010). They are:

- Adaptations for Dose Selection Studies
- Adaptive Randomization Based on Relative Treatment Group Responses
- Adaptation of Sample Size Based on Interim-Effect Size Estimates

- Adaptation of Patient Population Based on Treatment-Effect Estimates
- Adaptation for Endpoint Selection Based on Interim Estimate of Treatment Effect
- Adaptation of Multiple-Study Design Features in a Single Study
- Adaptations in Non-Inferiority Studies

Three of those less well-understood design types have been compared for benefits and challenges/obstacles (Chow and Corey, 2011).

As can be seen in Table 4.4, the downsides are mainly related to statistical problems, but also loss of information. A major concern of authorities is also that statistical power is lost and biases could be introduced by adaptive trial design, in particular if unblinded data are used for modification. Conversely, the use of blinded data is seen as much less critical.

The statistical background of adaptive design clinical trials is given in Chapter 6.

To illustrate the utility of this type of design, a simple example is given here for illustration as published by Veal et al. (2013).

The compound 13-cis-retinoic acid is used as treatment for high-risk neuroblastoma patients. It is sensitive to a pharmacogenetic variation (cytochrome P450 polymorphisms in variants 2C8, 3A5, and 3A7). Thus, fixed dosing in patients under 12 kg of body weight resulted in plasma levels below 2 μmol/L in 8 of 11 patients. The dose regimen had to be adjusted to reach sufficient exposure; this dosing by plasma levels is one of the most basic and valuable examples for an adaptive design study.

Another example for an adaptive trial design is shown in Figure 4.6. It demonstrates that accumulating data will govern patient allocation to different drug doses to improve the gain in knowledge that would not be possible if the number of subjects in the study arms were invariably predefined (Geiger et al., 2012).

This figure shows an example of an adaptive design clinical trial with individualized dosing. Adaptive seamless phase 2/3 study design. AWARD-5 is a double-blind randomized, placebo-controlled, adaptive, dose-finding, inferentially seamless phase 2/3 study. The two randomization schemes used divide the study into two stages: adaptive randomization (stage 1) and fixed randomization (stage 2). Seven doses of dulaglutide, a glucagon-like-peptide 1 analog ("dula"), will be evaluated in stage 1 along with comparators. Patients will be adaptively allocated across these doses based on accumulating data. The decision point is defined as the time when sufficient information has accrued to either select dula doses or to stop the trial. If dose selection occurs, stage 2 will begin. Only patients assigned the selected doses and comparators will continue to be followed; all others will stop further participation in the trial. Additional patients will enroll and be assigned to the comparators or dula (from Geiger et al., 2012, reprinted by permission of SAGE Publications).

TABLE 4.4 Comparison of benefits and risks for three "less well understood" adaptive trial design types

Design	Flexibility/Benifits	Challenges/Obstacles
Adaptive Randomization Design	• Unequal probability of treatment assignment • Assign subjects to more promising treatment arm	• Randomization schedule not available prior to the conduct of the trial • Not feasible for large trials or trials with long treatment duration • Statistical inference is often different, if not impossible, to obtain
Adaptive Dose Finding Design*	• Drop inferior dose group early • Modify/add additional dose groups • Increase the probability of correctly identifying the MTD with limited number of subjects	• Selection of initial dose • Selection of dose range under study • Selection criteria and decision rule • Risk of dropping promising dose groups
Two stage Seamless Adaptive Design (either phase I/II or phase II/III)	• Combine two studies into a single study • Fully utilize data collected from both stages • Reduce lead time between studies • Shorten the development time • Additional adaptations such as drop-the-loser, adaptive randomization, and adaptive hypothesis may be applied at the end of the 1st stage	• The control of the overall type I error rate • Sample size calculation/allocation • How to perform analysis based on combined data collected from both stages? • Is the O'Brien-Fleming type of boundaries feasible?

**For example, adaptive dose escalation designs for cancer trials.*
(Chow and Corey, 2011, Open Access.)

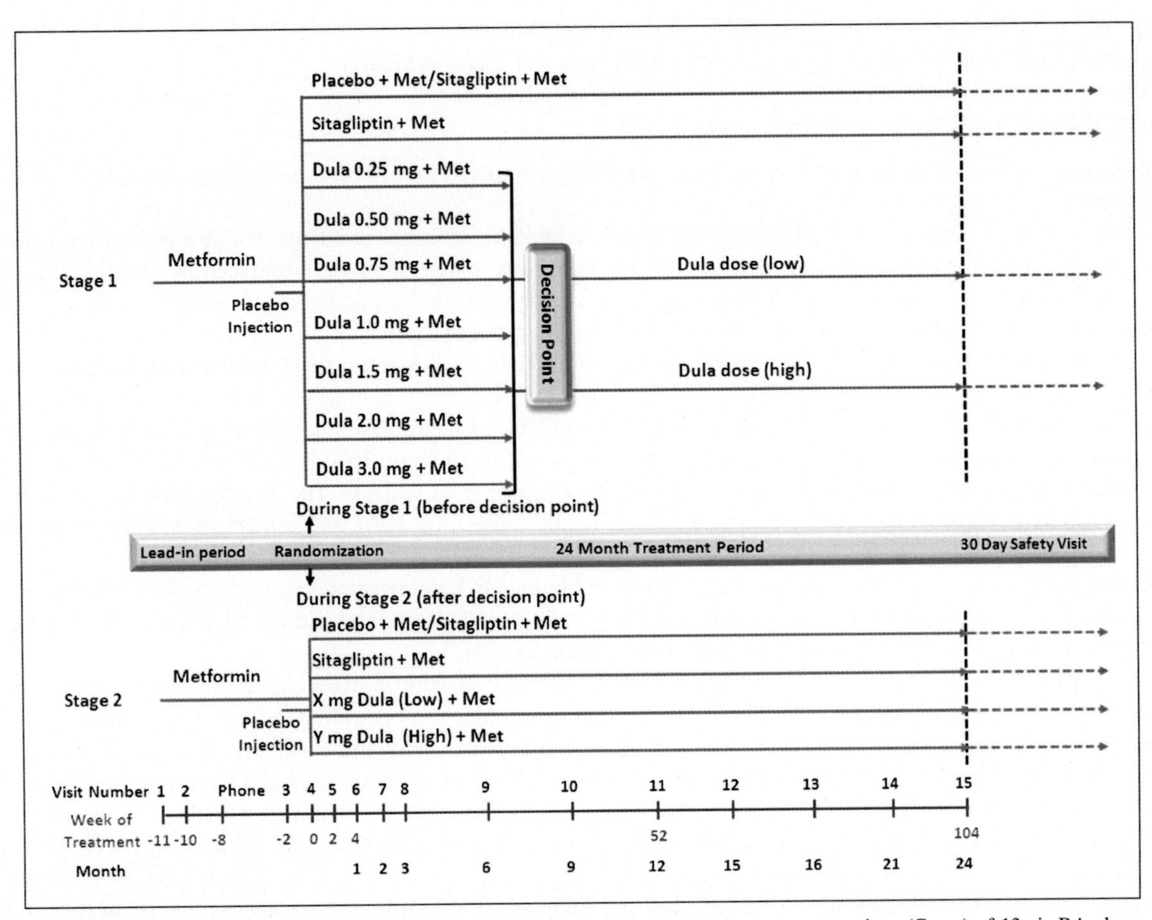

FIGURE 4.6 Example of adaptive design clinical trial with individualized dosing. Peak plasma concentrations (Cmax) of 13-cis-RA observed with protocol-based dosing and following dose increases to identify an individualized dose for all patients with initial Cmax values less than 2 μmol/L ($n = 20$, from Veal et al., 2013, by permission from American Association for Cancer Research, *Clin. Cancer Res.* **19**, 469–479).

REFERENCES

Chow, S.C., 2014. Adaptive clinical trial design. Annu. Rev. Med. 65, 405–415.

Chow, S.C., Corey, R., 2011. Benefits, challenges and obstacles of adaptive clinical trial designs. Orphanet. J. Rare Dis. 6, 79.

European Medicines Agency, 2006. Methodological issues in confirmatory clinical trials with flexible design and analysis plan. Reflection paper, Committee for Medicinal Products for Human Use. CPMP/EWP/2459/02, London, UK. http://www.ema.europa.eu/ema/pages/includes/document/open_document.jsp?webContentId=WC500003617 (accessed 03.03.14.).

Geiger, M.J., Skrivanek, Z., Gaydos, B., Chien, J., Berry, S., Berry, D., 2012. An adaptive, dose-finding, seamless phase 2/3 study of a long-acting glucagon-like peptide-1 analog (dulaglutide): trial design and baseline characteristics. J. Diabetes Sci. Technol. 6, 1319–1327.

Rosenberg, R., 2011. CenterWatch Online News, July 18 2011. Despite seeing benefits, industry awaits FDA final guidelines before adopting adaptive design trials. http://www.centerwatch.com/news-online/article/1928/despite-seeing-benefits-industry-awaits-fda-final-guidelines-before-adopting-adaptive-design-trials#sthash.EYT4P51T.dpbs (accessed 03.03.14.).

U.S. Department of Health and Human Services, Food and Drug Administration, 2006. Critical path opportunities list. http://www.fda.gov/downloads/scienceresearch/specialtopics/criticalpathinitiative/criticalpathopportunitiesreports/UCM077258.pdf (accessed 03.03.14.).

U.S. Food and Drug Administration, 2010. Draft Guidance for Industry—Adaptive Design Clinical Trials for Drugs and Biologics. U.S. Food Drug Admin, Rockville, MD. http://www.fda.gov/downloads/Drugs/.../Guidances/ucm201790.pdf (accessed 03.03.14.).

Veal, G.J., Errington, J., Rowbotham, S.E., Illingworth, N.A., Malik, G., Cole, M., et al., 2013. Adaptive dosing approaches to the individualization of 13-cis-retinoic acid (isotretinoin) treatment for children with high-risk neuroblastoma. Clin. Cancer Res. 19, 469–479.

Chapter 4.5

Combining Regulatory and Exploratory Trials

Martin Wehling

Regulatory authorities have issued clear guidances about the clinical studies that are necessary to support market approval of a new drug. Typically—as described earlier—such "regulatory" trials (meaning they are demanded by authorities) are allocated to three (four) phases of clinical development. Phase I mainly looks at pharmacokinetics in healthy volunteers (if possible; exceptions for "toxic" drugs, e.g., in oncology, which have to be explored in patients for ethical reasons) to determine the fate of the drug in the human body. This is normally done using single ascending doses and multiple ascending doses studies with major interest in plasma levels, route of excretion, metabolites, and parameters of distribution. In phase II, efficacy and safety are tested in a smaller group of patients (up to 400). In phase IIa, the very important dose-finding for clinically relevant effects is mandatory; in phase IIb, doses are applied to patients to prove efficacy and safety at a basic level (so-called proof-of-concept). In phase III (so-called pivotal trial), all previous data are corroborated and, thereby, finally safety and efficacy of a new drug in the given patient population are established; this is the ultimate prerequisite for market approval. Phase IV describes data from studies performed after market approval, which may be mandated by authorities to extend approval beyond preset time frames.

This somewhat idealistic linear sequence of clinical trials has grave limitations in drug development; it would be fine if the success of a novel compound were granted in all cases, which, however, is not true for 99% of all projects on new compounds (see Chapter 1). Late attrition of projects is costly and time-consuming, and the need to reduce it has been the major driver behind the hype about translational medicine in industry.

Therefore, early human experience with new compounds appeared as one of main strategies to cope with that challenge, and that experience was not fully achieved by the regulatory set of mandatory studies. This was the background for so-called phase 0 studies in which early orientation about human pharmacokinetics could be found, for example, by microdosing (see Subchapter 4.3). In addition to early pharmacokinetic data, understanding the mode of action, biomarker quality, or pathophysiological processes in conjunction with a new compound seems to be essential for the proper profiling of a new compound, even for the identification of a disease treatable by this compound.

Upon reflection of this demand, the term *exploratory clinical trial* (see Subchapter 4.3) was coined, with a variety of synonyms such as *physiological study* and *nonregulatory trial*. The latter synonym should not be confused with the term *nonregulated,* which is absolutely not true for these studies; they need to be performed under the same auspices of subject protection, proper statistical planning, and good clinical practice–conforming conduct as regulatory studies including the requirement of informed consent and IRB consultation.

According to the main related FDA document (U.S. Department of Health and Human Services, 2006), an exploratory investigational new drug study can help sponsors to

- Determine whether a mechanism of action defined in experimental systems can also be observed in humans (e.g., a binding property or inhibition of an enzyme)
- Provide important information on pharmacokinetics (PK)
- Select the most promising lead product from a group of candidates designed to interact with a particular therapeutic target in humans, based on PK or pharmacodynamic properties
- Explore a product's biodistribution characteristics using various imaging technologies

While early documents on exploratory studies and phase 0 studies emphasized pharmacokinetic profiling (e.g., by microdosing), other aspects have gained more interest recently, especially those relating to biomarker development and validation, mode of action, disease subclassification (= personalization), and others.

Principles of Translational Science in Medicine. http://dx.doi.org/10.1016/B978-0-12-800687-0.00026-8

TABLE 4.5 Stepwise validation of exploratory data generated by modern imaging as potential classifier biomarker

Step 1	Carry out a large phase II trial or several parallel phase II trials (100–200 patients). Collect the data provided by the new imaging tool and the data provided by the gold standard method (RECIST,* for example). Collect the response rate according to the gold standard.
Step 2	Compare in responders and nonresponders the distribution of the potential classifier biomarkers generated with the different methods of analysis: determine the optimal thresholds (receiver–operator curves).
Step 3	Determine the performance of the potential classifier biomarkers: the discrimination (sensitivity, specificity, positive predictive value, negative predictive value, accuracy) and the calibration (Brier score, for example).
Step 4	External validation with the use of a large multicenter prospective trial (clinical usefulness).

**RECIST, Response Evaluation Criteria In Solid Tumors.*
(From Cousin, Taieb and Penel, 2012, by permission of Lippincott Williams & Wilkins: *Curr Opin Oncol.* **24,** 338–344.)

It is important that such exploratory aspects could emerge from the regulatory trials if added to the study readouts and implemented in the design. A given study may measure the established outcome biomarker for proof-of concept in phase IIb, but at the same time generate valuable data on other biomarkers to be established and developed. This should be exemplified by studies on cancer treatment. In studies on solid tumors, response to treatment may be measured by the RECIST criteria (Gomez-Roca et al., 2011), which are mainly based on tumor diameters and new lesions. The rapid progress of modern imaging techniques, however, gives rise to more sophisticated and stratifying biomarkers that may be used after further exploration and refinement in the future.

Table 4.5 shows how novel cancer biomarkers may evolve from regulatory trials to which exploratory components are added (Cousin, Taieb and Penel, 2012).

Figure 4.7 demonstrates how an imaging procedure may impact on the future readouts of related studies: while RECIST criteria would indicate "stable disease," imaging clearly shows massive regression/necrosis but also a relapse of tumor growth over time (Cousin, Taieb and Penel, 2012).

With this study representing the profiling of novel imaging techniques as one example for exploratory aspects of regulatory studies, obviously the same approach can be utilized for serum biomarkers. In a regulatory study on different anti-HIV treatments, a substudy was embedded in which inflammatory biomarkers were profiled as predictors of disease response or progression. As can be seen in Figure 4.8 (McComsey et al., 2014), commonly used biomarkers of inflammation (high-sensitivity C-reactive protein, interleukin 6, and the soluble receptors for tumor necrosis factor alpha, type I and II) correlated well with AIDS-defining events (and non-AIDS-defining events). Thus, patients at high risk of developing AIDS as defined by those novel biomarkers may be stratified into more intense treatment schemes or receive second-line treatments earlier.

Another tremendously important area of exploratory clinical trials is the understanding of disease to be exploited as a source of druggable targets. In a recent study, exploratory aspects were added to a weight-loss intervention, and several adipokines and other metabolically relevant parameters were measured. A correlation between quality-of-life measures and leptin was found which may help to further understand its role in weight regulation; ultimately, the understanding of this role could guide new drug development and finally support weight loss and weight-associated diseases such as cancer (Linkov et al., 2014). Here, the exploratory aspects are not combined with a regulatory trial demonstrating that exploratory aspects may be implemented in almost all interventional studies; the given example is typical for academic research, which is still the main source of disease understanding as a prerequisite for successful drug development. Thus, the term *exploratory trial* is much wider than those concerning only drug-related trials, but in general it describes any human experiment that cannot directly be used in regulatory processes. Such trials, however, may guide drug development with great impact on speed and quality, and need to be considered especially at the earliest human experiences with new drug candidates. As there is tremendous pressure to streamline clinical development and avoid costly late attrition, the intertwining of regulatory ("must do") and exploratory trials is pivotal for success, and their integration into trials fulfilling both goals ("must do," "need to know") represents a valid and successful strategy.

(a)

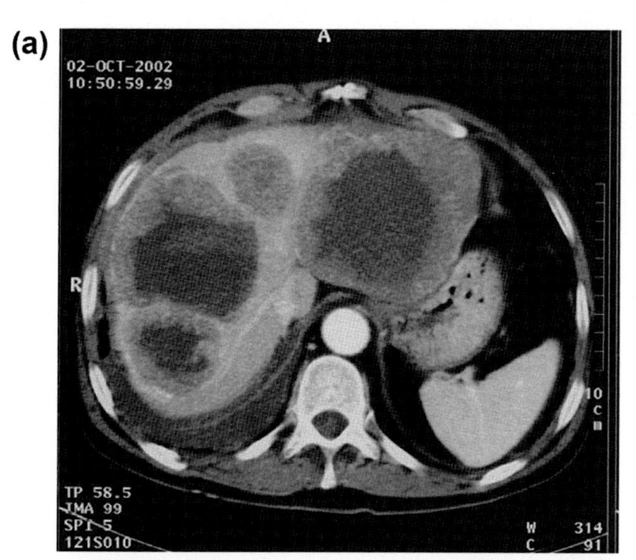

(b)

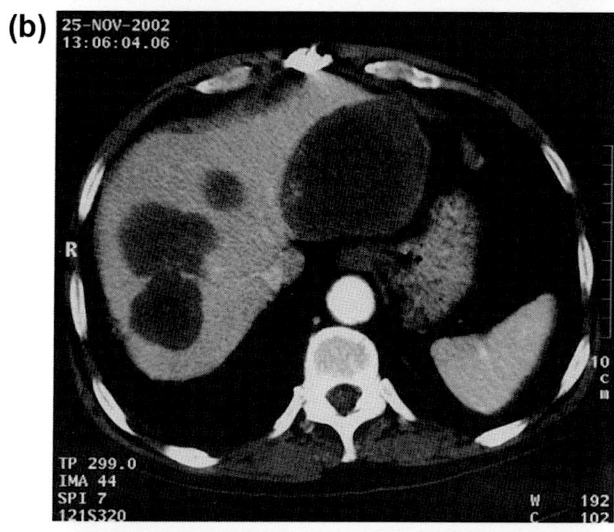

(c)

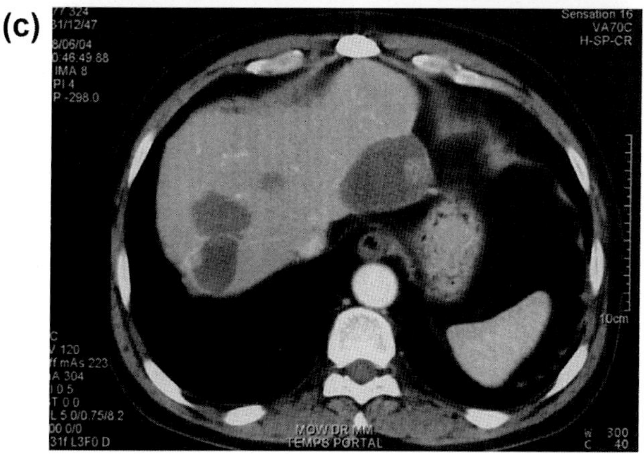

FIGURE 4.7 Assessment of liver metastasis from gastrointestinal tumor treated by imatinib mesylate. (a) Baseline assessment; (b) assessment after 6 weeks, massive necrosis of all target lesions (stable disease according to RECIST/partial response according to Choi criteria); and (c) 2 years later, appearance of nodules in the mass and increase in tumor density (stable disease according to RECIST/progressive disease according to Choi criteria). *(From Cousin, Taieb and Penel, 2012, by permission of Lippincott Williams & Wilkins:* Curr Opin Oncol. ***24****, 338–344.)*

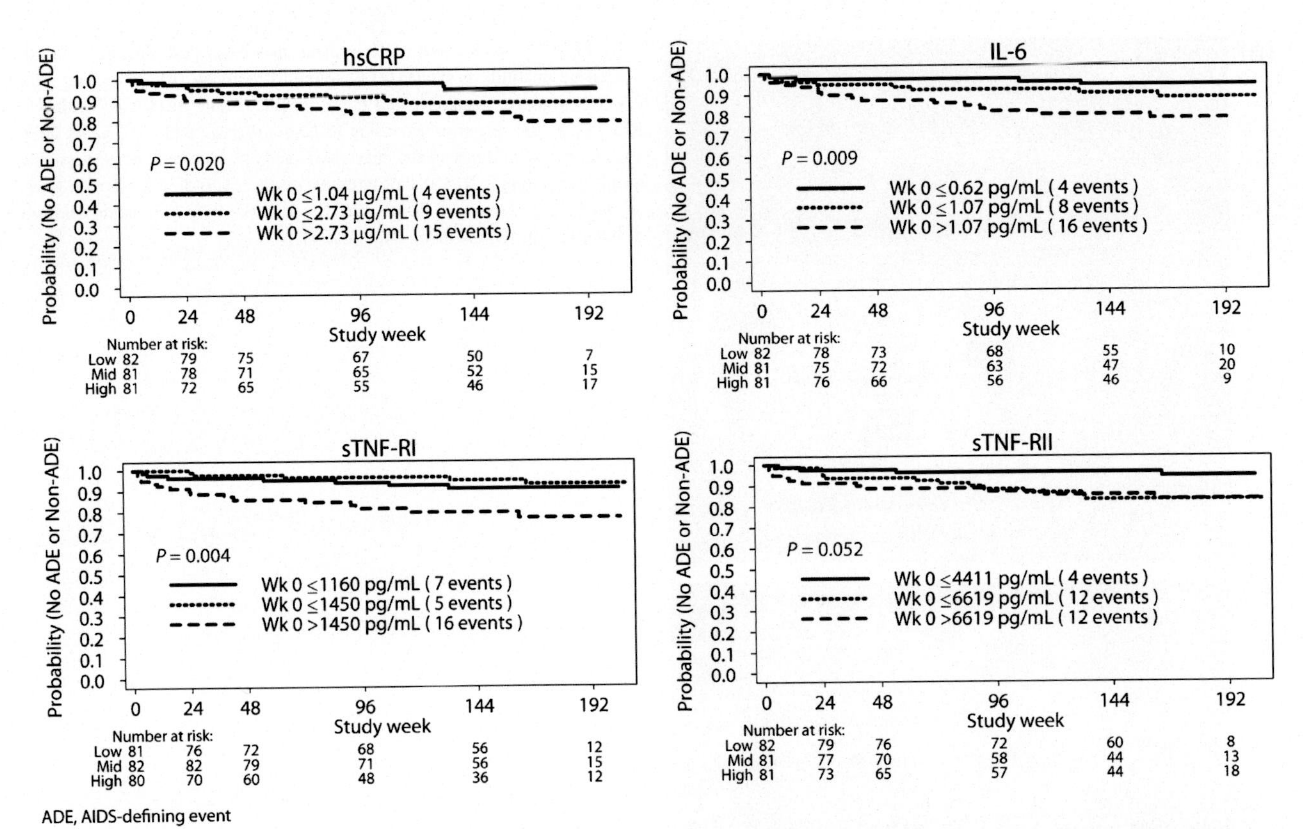

FIGURE 4.8 Kaplan–Meier plots stratified by tertiles for high-sensitivity C-reactive protein (hsCRP), interleukin 6 (IL-6), and the soluble receptors for tumor necrosis factor alpha, type I (sTNFR-I) and II (sTNF-RII). Samples were from a substudy (A5202) of AIDS Clinical Trials Group A5224s in which different antiviral regimens were tested in HIV patients. AIDS-defining events (ADE) and non-AIDS-defining events were measured. *(From McComsey et al., 2014, by permission of Lippincott Williams & Wilkins, Inc.:* J Acquir Immune Defic Syndr. ***65****, 167–174).*

REFERENCES

Cousin, S., Taieb, S., Penel, N., 2012. A paradigm shift in tumour response evaluation of targeted therapy: the assessment of novel drugs in exploratory clinical trials. Curr. Opin. Oncol. 24, 338–344.

Gomez-Roca, C., Koscielny, S., Ribrag, V., et al., 2011. Tumour growth rates and RECIST criteria in early drug development. Eur. J. Cancer 47, 2512–2516.

Linkov, F., Burke, L.E., Komaroff, M., et al., 2014. An exploratory investigation of links between changes in adipokines and quality of life in individuals undergoing weight loss interventions: possible implications for cancer research. Gynecol. Oncol. pii, S0090-8258(14)00035-3.

McComsey, G.A., Kitch, D., Sax, P.E., et al., 2014. Associations of inflammatory markers with AIDS and non-AIDS clinical events after initiation of antiretroviral therapy: AIDS Clinical Trials Group A5224s, a substudy of ACTG A5202. J. Acquir. Immune. Defic. Syndr. 65, 167–174.

U.S., 2006. Department of Health and Human Services Food and Drug Administration Center for Drug Evaluation and Research (CDER). Guidance for industry, investigators, and reviewers exploratory IND studies. http://www.fda.gov/downloads/Drugs/GuidanceComplianceRegulatoryInformation/Guidances/ucm078933.pdf (accessed 04.03.14.).

Chapter 4.6

Accelerating Proof-of-Concept by Smart Early Clinical Trials

Martin Wehling

Newer estimates about the costs of a new drug are now claimed to be at US$5 billion (Herper, 2013), reflecting not only the costs of successful but also all failed drugs. The time required to develop a new drug ranges from 10 to 15 years (Pharmaceutical Research and Manufacturers of America, 2013). The rapid increase of both costs and time has to do with many facts, including highly sophisticated regulation of clinical trials to protect participants by complex safety surveillance and reporting, strict and extensive rules regarding preclinical toxicology data, reduced margins of success in the light of competing drugs—just to name a few culprits for this development. It is obvious that the ever-shrinking remaining patent life of 5–15 years, which guarantees a high cashback after market approval, cannot generate enough funds to maintain this model of exorbitant costs and excessive development times.

This dilemma was the main reason for launching translational medicine as a potential cure. From the beginning, it was clear that a pivotal early step in clinical development is the proof-of-concept (PoC) study, typically obtained in the phase IIb clinical trial. As can be seen in Figure 2.25 (Subchapter 2.2), this is the point in time at which the highest (relative) value increment occurs because the translational risk dramatically drops if a positive (clinical) PoC is reached. This still does not guarantee market success because safety or patient stratification issues may be detected only in the very large phase III trial(s), but it reduces the risk of failure considerably.

It is only a natural consequence to adapt the development strategy for a new drug to obtain positive PoC as early as possible. This goal was one of the major objectives of the Critical Path Initiative in the United States (U.S. Food and Drug Administration, 2004) or the Innovative Medicines Initiative (IMI; Goldman, 2011) in Europe. This initiative was originally derived from the New Safe Medicines Faster initiative (European Federation for Pharmaceutical Sciences, 2014) by the European Federation for Pharmaceutical Sciences (EUFEPS). The wording already implied one of main objectives—namely, to speed up drug development.

As these compelling forces were behind the new movement called *translational medicine,* almost all efforts in the context were aimed at an accelerated PoC as a prerequisite of early, less-costly attrition and shorter timelines in development. Thus, this book detailing the manifold aspects of translational science in medicine could also be seen as supporting accelerated PoC in its entirety.

As shown in Subchapter 2.2, the translational risk critically depends not only on—among other factors—the quality of biomarkers but also on the early program of clinical studies. This chapter focuses on the translational power and its improvement in the early studies of clinical development. Clinical studies optimized under these auspices (gaining a maximum of translational power out of the smallest number of subjects/patients studied for the shortest possible duration of an intervention) are called *smart early clinical trials* here. This term is not in common use, nor has it been defined elsewhere; it should just underline the potential of shaping the early trial program in a translationally relevant way to reach the objectives mentioned previously.

The following short list comprises major goals of smart early human trials:

- To render proof-of-mechanism (PoM), proof-of-principle (PoP), proof-of-concept (PoC)
- To validate biomarkers with particular emphasis on patient stratification/responder concentration
- To find effective and safe dosing
- To gather as much information as possible in the least feasible number of patients
- To profile competitive developments early on

The terms *PoM, PoP,* and *PoC* are defined in Subchapter 4.1. Clearly, PoM is often achieved in pharmacokinetic regulatory phase I trials to which measurements of serum biomarkers will be added, which are proximal to the receptor addressed

Principles of Translational Science in Medicine. http://dx.doi.org/10.1016/B978-0-12-800687-0.00027-X

by the drug. Thus, adding a little complexity, which in most cases just means drawing few additional milliliters of blood, may demonstrate that the drug not only shows at appreciable levels in the blood, but also gets access to the target at sufficient amounts. It then may trigger the cascade of events that may eventually lead to clinical effects.

More sophisticated biomarkers could be added, e.g., imaging of radiotracers to visualize the distribution of a drug in the body, in particular its entry into the central nervous system (CNS). These are just a few examples for combining regulatory trials with exploratory aspects, thereby gaining much more information in much shorter times.

Phase 0 studies (microdosing), adaptive trial design, combining phase I and II, and exploratory trials are key components of "smart designs," which are essential for successful early human trials.

The main avenues to achieve acceleration of PoC are convincingly summarized in the session topics of a scientific meeting organized by Cambridge Healthtech Institute (CHI; http://www.healthtech.com/Proof-Of-Concept/, accessed March 6, 2014):

- Precision & Personalized Medicine Initiatives and The Effects on PoC Decision Making
- Accelerating Clinical Decision Making: Patient Stratification Using Pharmacogenomics and Protein Biomarkers
- Novel Indications Development—New Strategies for Old Drugs
- Improving PoC Study Designs & Methods
- Novel Alternate Endpoints for Clinical PoC

The latter two points are directly related to the topic of this chapter ("smart early clinical trials") and should be exemplified by a virtual case study on a new antiatherosclerotic drug. The unmet medical need is to reduce major cardiovascular events, mainly myocardial infarction, stroke, and related morbidity and mortality, by altering the biology of atherosclerotic plaques in a manner that should reduce plaque progression and render lesions more stable, less inflammatory, and less thrombogenic. Thus, stabilization of the vulnerable plaque is the goal (Tomey, Narula and Kovacic, 2014), and so far the most effective principles are statins (cholesterol-lowering drugs), which were proven to reduce plaque inflammation and clinical endpoints (major cardiovascular events, death).

In a traditional model (Figure 4.9), a two-stage set of early human trials would try to prove PoP in a 3-month trial demonstrating PoM markers (direct products of the target, e.g., in the lipoxygenase pathway), inflammatory serum markers (e.g., high sensitivity C-reactive protein), and measurements of vascular function (e.g., NO-mediated vasodilation) to change as expected. This, however, may indicate but does not prove that the vulnerable plaque is stabilized, that its inflammatory activity is reduced, which would be required as PoC (this still does not prove that major clinical endpoints including deaths are reduced; this needs to be proven in the large phase III trial). PoC would require a second study over 1–1.5 years in which plaque size and morphology would be measured by intravascular ultrasound and magnetic resonance tomography (MRT). The entire time to PoC starting from phase IIa (proper dose-finding) to IIb (PoC) would be 2 years.

In a smart study design, modern biomarkers for plaque imaging would be employed (Figure 4.10); in particular, MRT with or without molecular markers or gadolinium- or iron-containing contrast media positron emission tomography (PET) with or without CT are promising readouts to rapidly demonstrate plaque stabilization (Fleg et al., 2012). Ultrasound and histology methods (e.g., in atherectomy samples from carotid surgery) could be used as well. Because atherosclerosis is a generalized disease, easily accessible plaques, for example, in the carotid artery, may serve as indicator plaques for the disease in smaller, hidden arteries such as the coronaries.

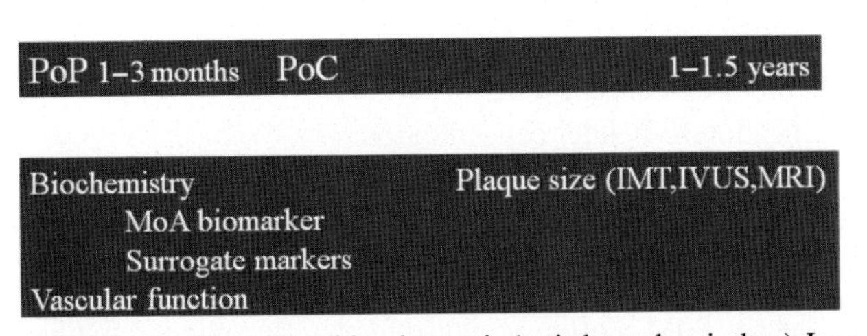

FIGURE 4.9 Phase II traditional scenario (antiatherosclerotic drug): Low risk—slow pace, two studies, PoP and PoC biomarkers separated.

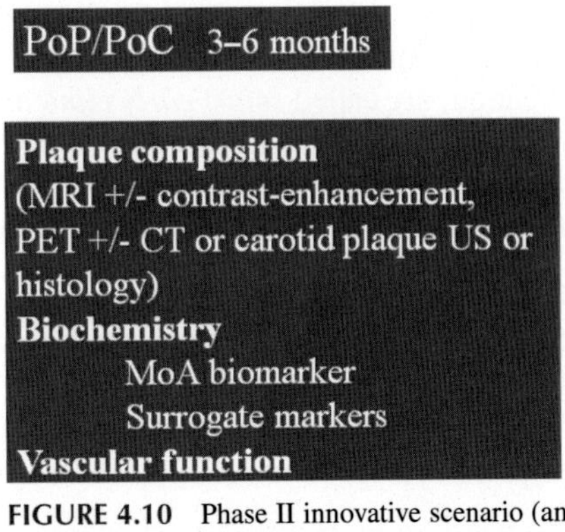

PoP/PoC 3–6 months

Plaque composition
(MRI +/- contrast-enhancement,
PET +/- CT or carotid plaque US or
histology)
Biochemistry
MoA biomarker
Surrogate markers
Vascular function

FIGURE 4.10 Phase II innovative scenario (antiatherosclerotic drug): Higher risk—high speed, one study, PoP and PoC biomarkers combined.

The combination of such biomarkers directly visualizing drug effects on the plaque morphology and stability with PoM markers in one trial would cut the time to PoC down to 3–6 months. This means a gain of 1–1.5 years in the development of such a drug.

A similar example is given in Subchapter 4.5 for cancer (Figure 4.7), in which more powerful imaging biomarkers would detect effects more rapidly and sensitively if used in future studies. An adaptive trial design would allow for using such novel biomarkers even if identified as being reliable, sensitive, and accessible just in the early segments of the same trial.

Another important area for the acceleration of PoP relates to patient stratification and responder concentration (the first two and—in part—the third points in the CMI short list shown previously). In Subchapter 2.2, a famous (or, more accurately, infamous) example for the beneficial impact of responder concentration is given for gefitinib (Table 2.13). Patients with activating mutations in the EGFR were much more likely to respond to therapy of lung cancer than those without this mutation. Obviously, there were indications for this association even before the unstratified trial was initiated, but they were not implemented in this study. Only after its failure, a new trial was designed that included this stratification; this trial was successful enough to allow for market approval. At the time those studies were designed, the principles of translational medicine were just in their infancy, and the potential of adaptive trial design was not yet fully appreciated. Today, a smart early human trial combining phase II regulatory aspects and the need to find stratifying markers very likely would be addressed by an adaptive design; it would allow for interim analyses to detect the most useful markers to predict a positive response to be used in the later stages of the same trial, eventually leading to clearcut, clinically relevant effects. As shown in Figure 4.11, the novel markers will lead to discontinuation of treatment in the putative nonresponders and to responder concentration in the remaining sample (Cousin, Taieb, and Penel, 2012).

Another evolving example relating to the development of successful treatments for malignant melanoma is given here. B-RAF mutations are present in 50% of all melanoma tumors, and specific inhibitors (e.g., vemurafenib) are successfully used in treatment alongside novel immunotherapeutics (ipilimumab). However, not all patients respond and may have other relevant mutations (Giovanni et al., 2014). In a so-called targeted approach, a finer subclassification of the individual set of relevant mutations should be utilized to "target" treatments to those mutations and use related inhibitors. Giovanni et al. (2014) propose a scheme of genetic analyses for presently known mutations (Table 4.6) as a sequential stratification approach. This scheme could be used in early clinical trials with adaptive design, which would not test a single intervention but find the optimal markers and clinical effects of related inhibitors. After approval, the same approach would have to be used in individual patients as a mandatory combination of diagnostics ("companion diagnostics") and treatment.

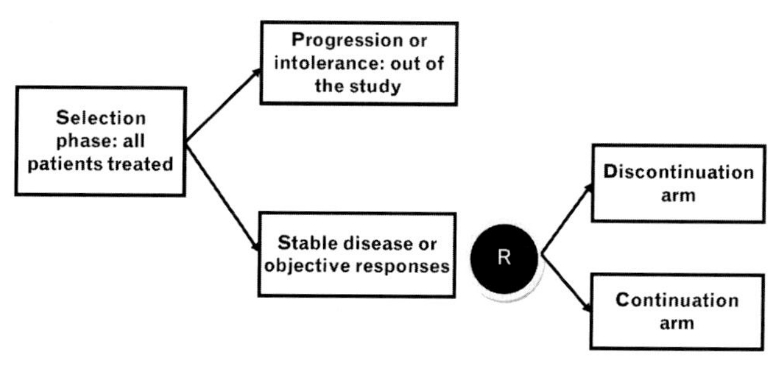

FIGURE 4.11 Discontinuation scheme for an adaptive design clinical trial exploring (early segment) and utilizing (later segments after interim analyses) a novel biomarker for stratification. *(From Cousin, Taieb, and Penel, 2012, with permission from Lippincott Williams & Wilkins:* Curr. Opin. Oncol. ***24****, 338–344.)*

TABLE 4.6 Stepwise model to characterize the mutational status of melanomas

(1) Screen for V600E *BRAF** mutation in melanoma patients with advanced disease (i.e., unresectable stages III and IV) as well as those at high risk of disease progression (stages IIIB and IIIC).
(2) In case of negative-V600E *BRAF* mutation, look for other non-V600E *BRAF* mutations (i.e., K, M, R, D).
(3) Melanomas not showing *BRAF* mutations should be investigated for *N-RAS* mutations.
(4) Double-negative *BRAF* and *N-RAS* melanomas should be further explored for *KIT* mutations or amplifications. This is even more relevant for acral and mucosal melanomas that should be investigated for both *BRAF* and *KIT* mutations at the first step.
(5) Triple-negative melanomas may benefit from *GNAQ* mutation evaluation, especially for uveal melanoma.

**For abbreviations, see the Human Genome Organization (http://www.genenames.org/).*
(From Giovanni et al., 2014, Open Access.)

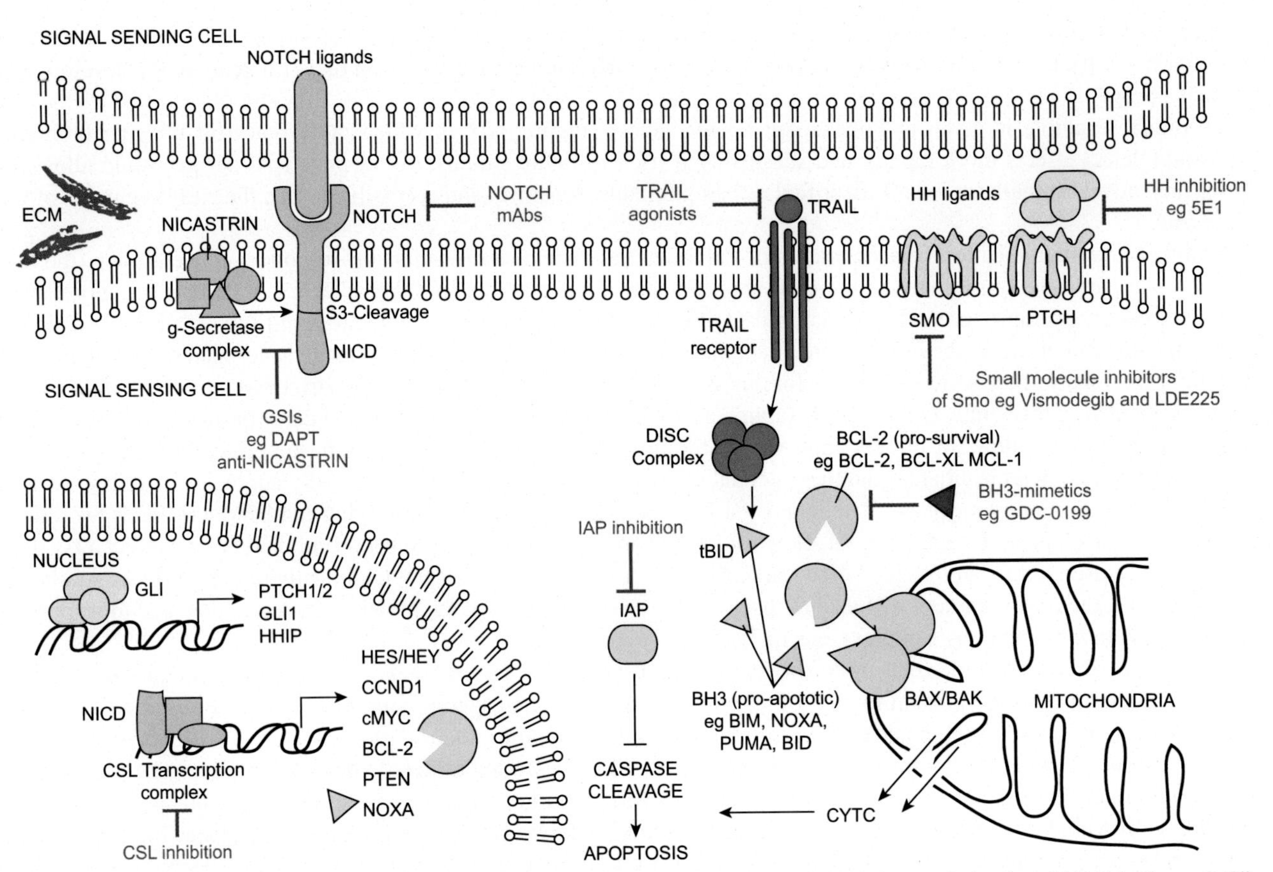

FIGURE 4.12 Novel therapeutic targets in triple negative breast cancer. *(From O'Toole et al., 2013, by permission from BMJ Pub.* Group: J. Clin. Pathol. *66, 530–542.)*

Breast cancer has the longest history of classification approaches by tissue biomarkers, with the earliest one being estrogen or progesterone receptors (ER and PR) as stratifiers. Meanwhile, many other markers have been proven to correlate with outcomes and treatment effects, with HER2 being firmly established, and maybe others such as the PI3K-AKT-mTOR pathway or the FGFR family to come. Yet, if ER, PR, and HER2 are negative ("triple negative breast cancer"), this population is still called "targetless" (O'Toole et al., 2013). Figure 4.12 depicts a current plot of potential drug targets in "triple negative" breast cancers. It is easily conceivable that early human trials on novel breast cancer drugs could test some of the markers concomitantly; the study design could be changed as the trial progresses in that nonresponding patients who are stratifiable by those novel markers may be discontinued.

If taken together, the proper stratification of patients and related responder concentration is one of the most promising approaches in reducing development times, in particular those up to clinical PoC. As a commentary, it may be noted that this approach, however, is all but new; in the past, it was simply called *disease classification*. The main difference between disease classification 50 years ago and today is the range and power of biomarkers applied, mainly reflecting the breakthrough discoveries in genetics, "omics," but also and maybe even more importantly, imaging and serum biomarkers.

REFERENCES

Cousin, S., Taieb, S., Penel, N., 2012. A paradigm shift in tumour response evaluation of targeted therapy: the assessment of novel drugs in exploratory clinical trials. Curr. Opin. Oncol. 24, 338–344.

European Federation for Pharmaceutical Sciences, 2014. New safe medicines faster—on how to rethink and accelerate drug development. http://www.eufeps.org/NSMF (accessed 05.03.14.).

Fleg, J.L., Stone, G.W., Fayad, Z.A., Granada, J.F., Hatsukami, TK.S., Kolodgie, F.D., et al., 2012. Detection of high-risk atherosclerotic plaque: report of the NHLBI Working Group on current status and future directions. JACC Cardiovasc. Imaging 5, 941–955.

Giovanni, P., Giovanni, P., Aldo, T., Pietro, L., Gabriele, L., Fabio, G., Caterina, L., 2014. Molecular targeted approaches for advanced BRAF V600, N-RAS, c-KIT, and GNAQ melanomas. Dis. Markers 2014, 671283.

Goldman, M., 2011. Reflections on the Innovative Medicines Initiative. Nat. Rev. Drug Discov. 10, 321–322.

Herper, M., 2013. The cost of creating a new drug now $5 billion, pushing big pharma to change. FORBES Pharma & Healthcare 8/11/2013. http://www.forbes.com/sites/matthewherper/2013/08/11/how-the-staggering-cost-of-inventing-new-drugs-is-shaping-the-future-of-medicine/ (accessed 05.03.14.).

Pharmaceutical Research and Manufacturers of America, 2013. Biopharmaceutical Research Industry Profile 2013. http://www.phrma.org/sites/default/files/pdf/PhRMA%20Profile%202013.pdf (accessed 05.03.14.).

O'Toole, S.A., Beith, J.M., Millar, E.K., West, R., McLean, A., Cazet, A., et al., 2013. Therapeutic targets in triple negative breast cancer. J. Clin. Pathol. 66, 530–542.

Tomey, M.I., Narula, J., Kovacic, J.C., 2014. Year in review: advances in understanding of plaque composition and treatment options. J. Am. Coll. Cardiol. pii, S0735–1097(14)01088–2.

U.S. Food and Drug Administration, 2004. Challenge and opportunity on the critical path to new medical products. http://www.fda.gov/downloads/ScienceResearch/SpecialTopics/CriticalPathInitiative/CriticalPathOpportunitiesReports/ucm113411.pdf (accessed 07.02.14.).

Chapter 5

Pharmaceutical Toxicology

Steffen Ernst
Updated from 1st edition, former authors: Ernst S (corresponding), Boyer S, & Platz S

INTRODUCTION

Toxicology is one of the oldest, truly translational scientific disciplines among the life sciences. As illustrated in a picture from the sixteenth century (*Regnerus de Graaf De Succo Pancreatico*), the formal investigation of disease and the quest for appropriate treatment has always involved testing of therapeutic remedies in animals to explore possible effects and side effects (Figure 5.1). In fact, numerous examples illustrate that many new pharmacological insights and opportunities for therapy were derived from unexpected side effects discovered in toxicological studies.

FIGURE 5.1 A sixteenth-century depiction of medical research. Note the sick patient in the background. The physicians are dissecting a human body in the middle, and various alive and dead animals can be seen in the foreground that were (presumably) used for experimentation. *(From Amberger-Lahrmann and Schmähl, 1993.)*

Principles of Translational Science in Medicine. http://dx.doi.org/10.1016/B978-0-12-800687-0.00028-1

Toxicology is the scientific discipline of studying the adverse effects of xenobiotics. The term *xenobiotics* is taken from the Greek words *xenos* ("stranger" or "foreign") and *bios* (life). *Xenobiotics* refers to any possible chemical or biological substance from any origin, such as industrial chemicals, pharmaceuticals, and plant- or animal-derived substances, and to all sorts of industrial or occupational by-products or pollutants. Traditionally, the ultimate goal of toxicology, as a scientific discipline, is to prevent harm to humans by poisonous substances. However, especially during the Italian Renaissance, this principle was reversed, and some toxicological experimentation was focused on the opposite—the discovery of particularly toxic, forensically undetectable substances as a means to eliminate or disable enemies.

Today, toxicology has evolved into a broad scientific area with numerous subdisciplines and specialty sections. The original, almost exclusive focus on humans has been extended to the investigation of adverse effects on livestock, wild animals, habitats, and even entire ecosystems. The study of xenobiotics of chemical or biological origin was extended to other harmful entities such as electromagnetic radiation, magnetic fields, and even sound waves.

Given the breadth and depth of the field, this chapter covers only the basic aspects of pharmaceutical toxicology that are considered relevant for the identification and development of new pharmacologically based therapies. This area of toxicology as practiced today has emerged in the pharmaceutical industry following the thalidomide disaster in the 1960s. The first part of this chapter provides a general overview of the principles of pharmaceutical toxicology, followed by an introduction to traditional, regulatory toxicology, and finally, the much newer area of drug discovery toxicology.

BASIC PRINCIPLES OF TOXICOLOGY

Elements of Toxicity

Any macroscopic, toxicological response depends on at least three basic interrelated aspects: the test substance, the dose or exposure, and the duration of exposure (Figure 5.2).

Test Substance (Xenobiotic)

Each chemical or biological xenobiotic has the ability to interact with a large array of molecular structures. In theory, any element within a biological system can be a target for interaction by an external compound. The external compound can affect classical pharmacological receptors, enzymes, transporters, carrier proteins, or structural elements in a direct or indirect manner. Often each interaction between a test compound and an internal molecule or structure causes one particular side effect. The entirety of all possible biological interactions of a given substance may be called a compound-inherent biological profile—a mental thought construct that can never be characterized or explored in total. The group of elements within such a biological profile may be called the toxicological profile. However, such profiles are rather abstract constructs. Because the biological profile includes all possible molecular interactions in all biological systems, it is generally impossible to explore and characterize all aspects. In addition, the biological and toxicological profiles differ from organism to organism, from species to species, and from individual to individual, as well as over time. Both profiles are absolute only in one individual at one point in time. For example, the principal ability to interact with a particular receptor may be exactly the same in two different individuals; however, other aspects in their biological profiles may be different enough to elicit significantly different overall responses.

Many in vitro tests aim to identify the principal ability of test compounds to interact with selected molecular targets deemed to mediate certain adverse effects. However, there is usually no proof that compounds identified in such a way actually elicit the side effect expected. This depends critically on other elements of a toxicological response, as illustrated in the following two sections.

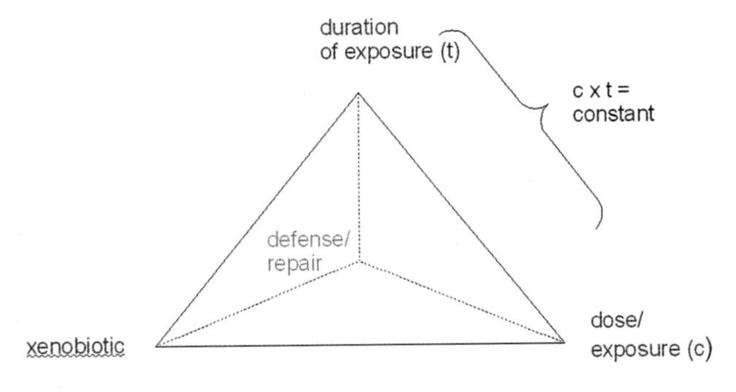

FIGURE 5.2 Depiction of the relationship between the three basic elements of toxicity. The corners of the triangle illustrate that a test compound (often called a xenobiotic) must be given at a high enough dose for a sufficient period of time.

TABLE 5.1 Approximate acute mean lethal doses (LD50) of some chemical agents

Agent	LD50 (mg/kg)*
Ethanol	10,000
Sodium chloride	4,000
Ferrous sulfate	1,500
Morphine sulfate	900
Phenobarbital sodium	150
Picrotoxin (plant)	5
Strychnine sulfate	2
Nicotine	1
d-Tubocurarine	0.5
Hemicholinium-3	0.2
Tetrodotoxin (puffer fish)	0.1
Dioxin (TCDD)	0.001
Botulinum toxin	0.00001

**LD50 is the dosage (mg/kg body weight) causing death in 50% of exposed animals.*
(From Klaassen, 2013. Casarett and Doulls Toxicology. 8th ed., with kind permission by McGraw-Hill Education)

Dose or Exposure

It is almost superfluous to point out that a xenobiotic has to be given at a sufficiently high dose to elicit a biological or toxicological response. The relationship of dose and principal ability to cause toxicity was recognized by Paracelsus (1493–1541), who stated the oldest and most fundamental paradigm of toxicology: "What is there that is not poison? All things are poison and nothing [is] without poison. Solely the dose determines that a thing is not a poison." In other words, any possible xenobiotic is poisonous if given at a high enough dose. This is the case for typical toxicants like dioxin, botulinum toxin, or insecticides, as it is for all components of our daily life, including table salt and even drinking water. However, the concept of focusing on dose alone is certainly outdated, as the dose given is only a crude correlate to estimating the actual exposure of the biological system (e.g., the test animal). Thus, in addition to the dose, it is essential to assess how much of a given dose is actually reaching the systemic circulation and the concentration of the substance in certain organs, tissues, or other compartments. Previously, the assessment of test compound concentrations posed a major challenge if nonradiolabeled substances were concerned. New methodologies like *mass spectrometry imaging* simplified this task significantly. Once such information is assessed and the toxicological profile of a compound is established, the dose alone may be an appropriate means of communication to, for example, compare the relative toxicity of different compounds (Table 5.1). In this context, chemicals with the ability to cause harm at comparatively small doses are considered toxic, whereas other chemicals that require large, often unrealistic amounts to cause harm are considered nontoxic.

Duration of Exposure

Finally, in addition to a dose resulting in sufficient exposure, a toxic compound needs to be present in the organism, at the target structure, for a sufficient period of time to cause an adverse effect. Each side effect or toxicity may require different exposure durations (and doses). For example, benzene causes sedation at short, acute exposure durations, whereas longer ones cause bone marrow toxicity. For each type of toxicity, there is also a close relationship between the dose or exposure and the duration of exposure. For many (if not all) adverse effects, there exists a window in which the dose and duration of exposure of a toxicant are in a linear relationship. The same toxicological response may be triggered by a low dose given for a long period of time or by a high dose given for a proportionally shorter duration. The controversial German Nobel laureate Fritz Haber (1868–1934) was the first to describe such linear relationships for different warfare gases in lethality experiments. He formulated the so-called Haber's Law of Toxicology:

$$c \times t = \text{constant}$$

where c is the concentration of a gas in the inhaled air and t is the duration of inhalation treatment. Karl Rozman at the University of Kansas City demonstrated the applicability of the principle to routes of administration other than inhalation, longer treatment periods (even lifetime studies), and a variety of side effects other than lethality, including cancer. However, it is important to stress that the linear relationship of dose and time for a given toxicity requires explicit knowledge and experimental control of exposure. Confounding factors like changing pharmacokinetics over time (exposure) due to induction of metabolizing enzymes may easily lead to loss of required control. In addition, the relationship reveals itself only under fairly stable biological conditions. For example, significant defense and repair mechanisms in response to a toxicant may render it difficult to demonstrate the linear relationship of dose and time.

Repair and Defense

In addition to the simplified concept of test substance, exposure, and exposure duration, it is important to take into account the body's significant ability to defend itself from toxicological challenges and to repair possible damages. In principle, all defense mechanisms focus on altering one of the three basic elements of a toxicological response. For example, barriers may hinder the transfer of a compound into the system or into a particular macroscopic or molecular compartment, thus changing the exposure and exposure duration at the target site. Specialized transporters may move a compound away from a certain site or from the body as such. Modifications of the toxicant (metabolism) may change the compound to make it less harmful or to facilitate increased elimination, thus changing the biological and toxicological properties. Furthermore, the body can react to a challenge by actually changing the biology and, thus, the biological or toxicological profile of a compound. Harihara M. Mehendale at the University of Louisiana at Monroe demonstrated that the severity of acute toxicity can be significantly changed by inducing competent repair mechanisms, for example, the mechanisms found in the liver. Rats receiving a nonlethal dose of the liver toxicant carbon tetrachloride became resistant and survived a subsequent, normally lethal dose of the same compound. Thus, especially in acute and subacute time frames, aspects of defense and repair are important with regard to the onset and extent of an adverse reaction. At longer treatment durations, such mechanisms probably become less dynamic, with a considerably lower impact on the nature and severity of toxicity. This excludes incomplete attempts to repair damage, which may even be part of the toxicological profile after longer exposure. For example, liver fibrosis (cirrhosis) after long-term ethanol exposure can be considered an incomplete and inadequate attempt to deal with the constant damage to the organ.

Differences in the ability to fight off or repair toxicological damage are also quite important from a translational standpoint. Many diseases—for example, diabetes mellitus—have been shown or are suspected to reduce defense mechanisms. Because preclinical, toxicological studies are conducted in young, healthy animals, they do not address the possibility of disease-related increased susceptibility.

The basic elements of a toxicological response described in the preceding sections appear quite simple and appealing. However, it cannot be stressed enough that these three elements are hardly ever under complete control in a toxicological study. The biological and toxicological profile of any xenobiotic is never really known to its full extent. In many instances, it may not be known whether the given compound is actually the cause of a deleterious effect at a molecular target. The actual damage may be caused by a metabolite of the parent compound, a secondary molecule such as a reactive oxygen species, or even an induced endogenous compound such as TNFalpha. The dose may be under less than full control due to variable degradation of the chemical substance, for example, the degradation of a substance with a low pH in the stomach or chemical instability as such. The usual measurement of blood concentrations may not necessarily correlate with the concentration in the target organ, tissue, or cellular compartment. For the same reason, the exposure duration at the target site may be unknown and highly variable.

Some old pharmaceuticals like acetylsalicylic acid (Aspirin®) demonstrate that even after more than a century of clinical use, new, hitherto unknown biological aspects are revealed almost every year. Thus, a researcher may be unable to gain full control of all three basic elements of the toxicological response of even one single compound in his or her lifetime!

Dose–Response Relationship

Once a toxicity of a compound is identified in an animal model, it is important to put it into perspective according to the actual risk this particular effect may have on the human target population. A basic analytic tool in this respect is the analysis of the relationship between the dose or exposure and the response to that dose or exposure.

The relationship of dose and spectrum of effects is one of the most fundamental concepts in toxicology. Such dose–response relationships can be looked at from two principal angles. One describes the gradually increasing severity of a particular toxicological response relative to the dose in one individual. The other describes the variability of such an effect in a target population. Because the measured effects of an individual dose–response relationship are continuous over a range of doses, an individual relationship is called a graded dose–response relationship.

Examples of graded dose–response relationships include the dose-dependent increase of sedation by an anesthetic and the inhibition of an enzyme by a respective inhibitor. The variability of sedation or enzyme inhibition in a population receiving the same doses would constitute the second type of dose response. Based on empirical evidence, dose–response relationships are often expressed on a log-normal scale.

The analysis of dose–response relationships yields the doses needed to elicit a minimal and maximal toxicological response. The highest dose at which no side effect is observed is often referred to as the no observed adverse effect level (NOAEL). This dose is often used as a basis for comparison with other parameters. For example, a large quotient derived from the division of the NOAEL of a relevant toxicological endpoint by the dose required for a desired, therapeutic effect would indicate a relatively safe compound with large safety margins. A small quotient would indicate the opposite. The slope of the dose response indicates the difference between the minimal and maximal doses with considerable implications as to the safety of the compound. For example, a toxicant with a steep slope indicates a comparatively small difference between the dose required for a minimal effect and that required for a maximal effect. If the curve illustrates the rate of lethality relative to dose, the steep slope indicates that even small dose increases result in many more deaths. In contrast, a shallow slope indicates that much larger doses are required to yield a measurable increase in mortality. Thus, the slope of the dose–response curve of the second type of dose response indicates the relative susceptibility of the study population to the side effect under investigation (Figure 5.3).

Figure 5.3 illustrates the most common dose–response curve—a sigmoid-shaped curve—but others are possible as well. For example, essential nutrients like vitamins express a U-shaped graded dose response in which doses below the daily requirements and above a toxicity threshold are associated with side effects. Thresholds—doses or exposure concentrations below which no side effects are detected—are observed for many types of toxicities. In fact, thresholds probably exist for all types of toxicity. However, they may vary greatly among individuals or may be located at very low concentrations. For some classes of compounds and effects—for example, chemical carcinogens—conservative models reject the concept of a threshold and assume that even a single molecule may be able to cause cancer.

Appropriate use of dose–response relationships implies a number of aspects:

- The effect is related to exposure to the test compound.
- The extent of the effect is related to the dose.
- There is a molecular or other target with which the molecule interacts to produce the effect.
- The response as such and the degree of the response are related to the concentration of the agent at the target site.
- The concentration at the site is, consequently, related to the dose administered.

There may be exceptions with regard to the need for a molecular target when the toxicity is related to the mechanistic impairment of a particular biological function, for example, lung toxicity due to large amounts of inert dust particles exceeding the lung's self-cleaning capacity or the fallout of compounds in the urine leading to kidney and/or bladder lesions.

For the reasons outlined previously, it is clear that a single compound may have different dose responses for each type of toxicological effect. These different effects may even occur at different points in time. With regard to the evaluation of a given compound, great care must be spent on the selection of the most appropriate side effect. This is often, but not always, the toxicity elicited by the lowest dose applied (Figure 5.4).

Differences among Species

One of the key textbooks in toxicology states that "experimental results in animals, when properly qualified, are applicable to humans" (Klaassen, 2001, p.25). However, it needs to be recognized that even fundamental differences in the spectrum

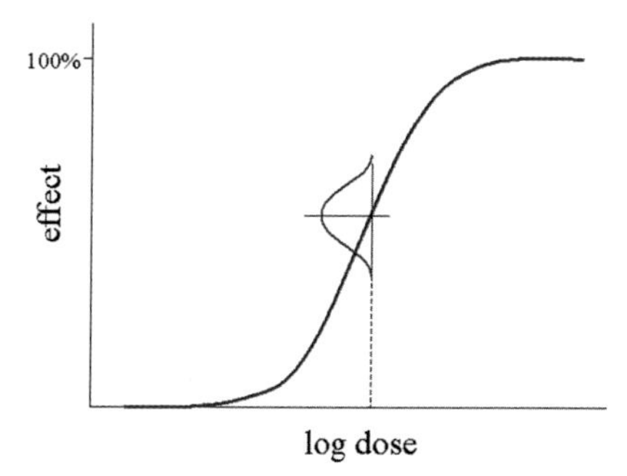

FIGURE 5.3 Example of a graded dose–response curve (for a single individual). If such curves were obtained from a number of individuals, the effect would vary in a normally distributed fashion at any given dose.

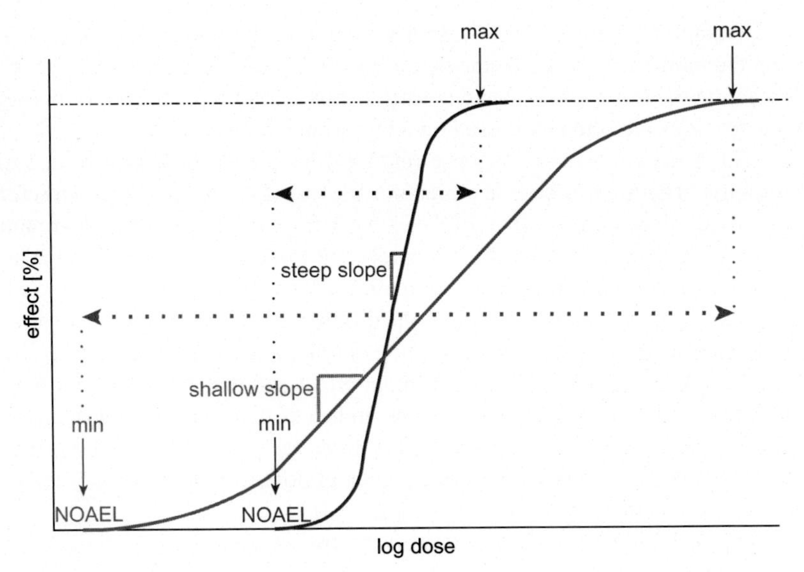

FIGURE 5.4 Illustration of two dose–response curves with different slopes and NOAEL. Assuming the two curves represent lethality by two different compounds, it is noteworthy that when all animals represented by the blue curve are dead, only little more than half of the animals represented by the red curve have been affected, even though the red curve has a lower NOAEL.

and nature of toxicity may occur in different animal species. An often-highlighted case is the thousandfold difference in susceptibility to the acutely lethal effects of 2,3,7,8-tetrachlorodibenzi-*p*-dioxin between hamsters and guinea pigs. Thus, the selection of an animal test species alone may lead to over- or underestimation of the risks of a given compound for humans.

In principle, there are only two ways to compensate for species differences in risk assessment. One is to increase the number of species included in animal tests in the hope that the increased variability of responses will cover humans as well. This is why two species, one rodent and one nonrodent, must be included in the safety studies prior to testing of new pharmaceuticals in humans. However, there are obvious practical and ethical limitations to the increase of the number of animal species in toxicological studies. Before testing in different species, researchers should perform a concise assessment of the mechanisms leading to species differences. In many cases, comparative assessment of the exposure and exposure duration in various species and knowledge of the molecular mechanisms responsible for the differences can explain variable toxicological responses in these species. In addition, investigation of the molecular mechanism responsible for a particular side effect may help significantly in the extrapolation of toxicological information across species. However, despite a vast array of modern technologies, concise investigation of a particular toxicity can be extremely challenging and time consuming and may not be successful at all.

Viewpoints of Toxicity

Unwanted side effects and toxicity may be looked at from various angles.

- *Time:* In order for one to predict the relative risk of a given compound, time plays a critical factor. For example, side effects occurring after several weeks of dosing may be rather irrelevant for a remedy given only once to an individual.
- *Effect:* The type of toxicological response is quite important. Blunt organ damage and tissue damage carry a different weight in general compared to, for example, impairment of fertility with a substance intended to be given only to the elderly.
- *Target organ or tissue:* For example, lung toxicity found with a compound intended to treat pulmonary diseases is more critical than that of a compound for treating a urological disease.
- *Mechanism:* As outlined previously, mechanistic knowledge may be important for translating toxicological information. For example, knowledge of some types of toxicity can clearly indicate that the toxicity in question is irrelevant in humans because of the lack of similar mechanisms.
- *Susceptibility:* This is important with regard to the intended size of the patient population. It is evident that a drug intended to be given to millions of individuals without a prescription requires more than one trial performed on only a few individuals under close clinical surveillance.
- *Route of administration:* Because the route of administration has a significant impact on the exposure, exposure duration, and even the biological and toxicological profile of a substance, toxicological studies have to use the same route of administration as will be used in future clinical applications.
- *Reversible effects versus irreversible effects:* It is obvious that irreversible effects carry significantly more weight than those which heal and resolve over time.

TABLE 5.2 Examples of hazards and their effects

Type of Workplace Hazard	Example of Hazard	Example of Harm Caused	Factors Influencing Risk
Thing	Knife	Cut	Sharpness of knife
Substance	Benzene	Leukemia	Workplace concentration
Material	Asbestos	Mesothelioma	Use of filter mask
Source of energy	Electricity	Shock, electrocution	Presence of insulation
Condition	Wet floor	Slips, falls	Smoothness of surface
Process	Welding	Metal fume fever	Presence of controlled ventilation
Practice	Hard rock mining	Silicosis	Avoidance of dust

(From Canadian Centre for Occupational Health and Safety, 2008, reprinted with permission.)

Risk Assessment

The purpose of preclinical safety studies is to prevent harm and injury to healthy volunteers and patients exposed to a hitherto unknown chemical substance in clinical trials and practice. The key principle in toxicology is to first identify the principal hazards of a given compound and then estimate how likely it is that humans may experience such hazards under the predefined conditions of a clinical study or therapeutic use, and, finally, take appropriate steps to reduce the risks as much as possible. In the United States and Canada, the term *hazard* defines the principal, intrinsic toxic properties of a xenobiotic and is thus comparable to the abstract term *biological and toxicological profile,* as outlined previously. *Risk* is defined as the likelihood or probability of an adverse outcome (the term *hazard* is often, incorrectly, used in this context). Table 5.2 illustrates the relationship of hazard and risk.

The principal steps of risk assessment for investigational drugs are outlined in Figure 5.5. They include

- Identifying the principal adverse effects in two animal species
- Quantitative and mechanistic evaluation as to whether such adverse effects are likely to occur in humans under the defined conditions of clinical studies or therapeutic use (risk)
- Application of uncertainty factors
- Balancing the risk against the expected therapeutic benefit
- Removing the compound from the clinical test or redefining the conditions of clinical studies and therapeutic use as necessary

REGULATORY TOXICOLOGY

Introduction

Regulatory toxicology in the most simplistic view describes the use of toxicology in the regulatory process with the ultimate goal of protecting public health. Within the context of this book, the information related to regulatory toxicology is largely pertinent to the toxicity testing of small molecules or synonymously new chemical entities. Obviously, there are also other human therapeutics, including therapeutic proteins, vaccines, and most recently siRNA and mRNA, etc., but they differ in the toxicity testing.

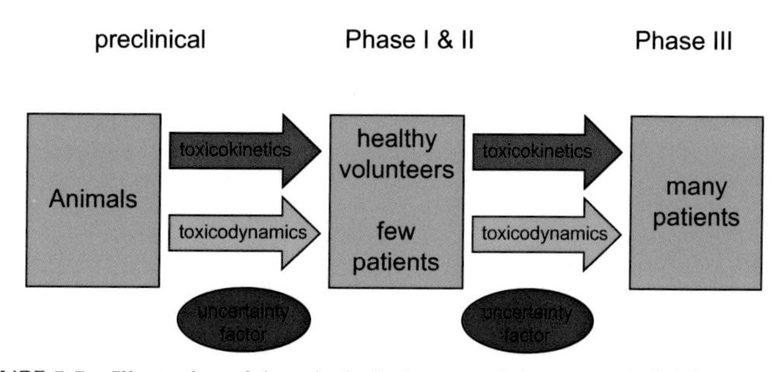

FIGURE 5.5 Illustration of the principal elements of pharmaceutical risk assessment.

Regulatory toxicology as defined previously is intimately linked to risk assessment. Risk assessment typically includes an estimate of the probability of harm, such as the probability of liver toxicity after the use of a particular drug, and a clear description of the various assumptions and uncertainties that go into the risk assessment. Risk assessment is used as part of the decision making of regulatory agencies and industry to allow the use of potential new medicines whose benefits outweigh the risks associated with their use.

As described by the National Research Council of the National Academy of Sciences (Lehman et al., 1949), risk assessment involves four components:

1. Hazard identification: An evaluation of the adverse health effects the agent is capable of causing. Examples might include the capacity of an agent to cause liver or nervous system damage or to cause cancer.
2. Dose–response assessment: A determination of how much of an agent is required to cause a toxic effect, and prediction of the exposure levels at which risk is likely to be negligible or nonexistent.
3. Exposure assessment: A determination of how much of an agent people might be exposed to under various conditions, such as use of a drug or a consumer product and environmental exposure at a hazardous waste site.
4. Risk characterization: An integration of the pertinent information from the preceding steps to characterize the risks to the exposed population. For example, what is the likelihood of liver toxicity if an individual uses a particular drug? The risk characterization also includes an explicit description of the assumptions and uncertainties that go into the risk assessment and the overall confidence in the results of the analysis.

For the overall process of risk assessment, the integration of metabolism, toxicity, pathology, and mechanism is playing an increasing role, and a better understanding of these areas is essential for creating proper regulations.

Historically, regulatory toxicology was not seen as a very dynamic field, and it is true that the principles of safety evaluation delineated by Lehman and colleagues in 1949 still describe relatively accurately the current regulatory toxicology practice as it applies to the evaluation of general organ and tissue damage. However, there is significant momentum in a combined effort of the pharmaceutical industry and regulatory agencies to evaluate opportunities for improved approaches to toxicological assessment. The Critical Path Institute (C-Path) and the Innovative Medicines Initiative are documenting this effort.

Regulatory Toxicity Studies

With the development of the pharmaceutical industry, it has become important to have an independent evaluation of medicinal products before they are allowed on the market. In the United States, a tragic mistake in the formulation of a children's syrup in the 1930s was the trigger for setting up the product authorization system under the U.S. Food and Drug Administration. In Japan, government regulations requiring all medicinal products to be registered for sale started in the 1950s. In many countries in Europe, the trigger was the thalidomide tragedy of the 1960s, which revealed that the new generation of synthetic drugs, which were revolutionizing medicine at the time, had the potential to harm as well as heal. For most countries, the 1960s and 1970s saw a rapid increase in laws, regulations, and guidelines for reporting and evaluating the data on the safety, quality, and efficacy of new medicinal products. Although the different regulatory systems were based on the same fundamental obligations to evaluate the quality, safety, and efficacy, until recently, the detailed technical requirements had diverged over time to such an extent that the industry found it necessary to duplicate many time-consuming and expensive test procedures, in order to market new products internationally. Harmonization of regulatory requirements was pioneered by the European Community in the 1980s, as the EC (now the European Union) moved toward the development of a single market for pharmaceuticals. At the same time, there were bilateral discussions between Europe, Japan, and the United States on possibilities for harmonization. This finally resulted in the birth of the International Conference on Harmonisation of Technical Requirements for Registration of Pharmaceuticals for Human Use (ICH) in 1990, a joint regulatory and industry project to improve, through harmonization, the efficiency of the process for developing and registering new medicinal products in Europe, Japan, and the United States. The ICH Topics are divided into four major categories, and ICH Topic Codes are assigned according to these categories. This includes Safety Topics—that is, those relating to regulatory in vitro and in vivo toxicity studies (Table 5.3).

Despite these efforts in harmonization, national guidance documents still exist, mainly due to a difference of opinion and the possibility of an accelerated response to new developments. Examples are the exploratory IND guidance document or the emphasis on male fertility testing in Japan, which requests that male fertility data be provided prior to the inclusion of men in clinical trials.

TABLE 5.3 ICH guidance SAFETY documents, summarized by the FDA

Carcinogenicity studies	
S1A	Need for Carcinogenicity Studies of Pharmaceuticals
S1B	Testing for Carcinogenicity of Pharmaceuticals
S1C(R2)	Dose Selection for Carcinogenicity Studies of Pharmaceuticals
Genotoxicity studies	
S2(R1)	Guidance on Genotoxicity Testing and Data Interpretation for Pharmaceuticals Intended for Human Use Guidance on Specific Aspects of Regulatory Genotoxicity Tests for Pharmaceuticals Genotoxicity: A Standard Battery for Genotoxicity Testing of Pharmaceuticals
Toxicokinetics and pharmacokinetics	
S3A	Note for Guidance on Toxicokinetics: The Assessment of Systemic Exposure in Toxicity Studies
S3B	Pharmacokinetics: Guidance for Repeated Dose Tissue Distribution Studies
Toxicity testing	
S4	Single Dose Toxicity Tests Duration of Chronic Toxicity Testing in Animals (Rodent and Nonrodent Toxicity Testing)
Reproductive toxicology	
S5(R2)	New title: Detection of Toxicity to Reproduction for Medicinal Products & Toxicity to Male Fertility Previous title: Detection of Toxicity to Reproduction for Medicinal Products Maintenance of the ICH Guideline on Toxicity to Male Fertility: An Addendum to the Guideline on Detection of Toxicity to Reproduction for Medicinal Products (in S5(R2))
Biotechnological products	
S6	Preclinical Safety Evaluation of Biotechnology-Derived Pharmaceuticals
Pharmacology studies	
S7A	Safety Pharmacology Studies for Human Pharmaceuticals
S7B	The Nonclinical Evaluation of the Potential for Delayed Ventricular Repolarization (QT Interval Prolongation) by Human Pharmaceuticals
Immunotoxicology studies	
S8	Immunotoxicity Studies for Human Pharmaceuticals
S9	Nonclinical Evaluation of Anticancer Pharmaceuticals
S10	Photosafety Evaluation of Pharmaceuticals
Joint safety/efficacy (multidisciplinary) topic	
M3(R2)	Guidance on Nonclinical Safety Studies for the Conduct of Human Clinical Trials and Marketing Authorization for Pharmaceuticals

(From U.S. Food and Drug Administration, 2008.)

Following the overall drug development pathway, toxicity testing can be separated into three phases:

1. Preclinical testing: This includes in vitro and in vivo studies to screen for potentially harmful effects before any new medicine is administered for the first time to humans.
2. Testing during phases I and II: The information is intended to provide toxicity data following subchronic and chronic treatment in animal models prior to longer clinical trials, as well as an assessment of effects on fertility and embryo-fetal development.
3. Testing during phase III: Data from toxicity studies are normally available for registration, including carcinogenicity in rodent models as well as peri- and postnatal development.

See also Figure 5.6 for a pictorial representation of these phases.

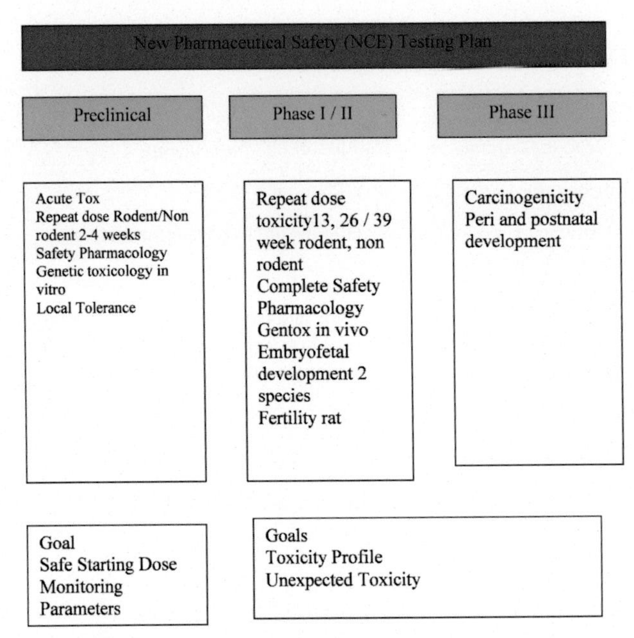

FIGURE 5.6 Illustration of the phases of toxicity testing.

The Importance of Good Laboratory Practice

As part of the risk assessment, regulatory authorities evaluate nonclinical studies as part of a submission to support an application for a product license or clinical trial. To ensure the quality, consistency, and reliability of data, the control system of good laboratory practice (GLP) was developed. GLP applies to nonclinical data and therefore includes toxicology (in vitro and in vivo), safety pharmacology, pathology, and pharmaco- or toxicokinetics, including bioanalytics.

As defined by the Organisation for Economic Co-Operation and Development (2008), the primary goal of GLP is "to ensure the generation of high quality and reliable test data related to the safety of industrial chemical substances and preparations in the framework of harmonising testing procedures for the Mutual Acceptance of Data." The detailed rules for GLP in the United States are described in the code of federal regulations 21CFR58 by the U.S. Food and Drug Administration, which addresses good laboratory practice for nonclinical laboratory studies, and those of the European Union are described in the two basic directives 2004/10/EC and 2004/9/EC (on the harmonization of laws, regulations, and administrative provisions relating to the application of the principles of good laboratory practice and the verification of their applications for tests on chemical substances) of the European Parliament and Council. As a consequence, any preclinical research that is not conducted according to these guidelines might not support an IND in the United States or a clinical trial application (CTA) in Europe.

In general, GLP embodies a set of principles that provide a framework within which laboratory studies are planned, performed, monitored, recorded, reported, and archived. It assures regulatory authorities that the data submitted are a true reflection of the results obtained during the study and can therefore be relied on when making risk or safety assessments. GLP monitoring units are part of regulatory authorities and are involved in the inspection of GLP compliance.

Animal Models

Before new medicines are administered to humans, their safety must be evaluated by screening for potentially harmful effects. This screening requires in vitro and in vivo animal models. Despite significant progress in the evaluation of new in vitro models in compliance with the 3R principles (reduce, refine, replace)—for example, EPISKIN® and EpiDerm®, which are both artificial skin models for skin irritation—it is widely accepted that in vivo models still provide the best prediction of what might happen in humans. In a concordance analysis of the toxicity of pharmaceuticals in humans and animals, the results showed a positive human concordance rate of 71% for rodent and nonrodent species, with nonrodents being predictive for 63% of cases and rodents for 43% of cases (Olson et al., 2000). The highest incidence of overall concordance was seen in hematological, gastrointestinal, and cardiovascular human toxicities, and the least was seen in cutaneous toxicities. Obviously, effects not predicted by the animal studies include subjective effects such as headaches or nausea.

Appropriate animal toxicity testing is a mandatory legal requirement. The requirements for repeated-dose toxicity studies as described by the ICH include testing in two mammalian species—usually a rodent and a nonrodent—in which the duration of treatment should equal or exceed the duration of the human clinical trial. Laboratory-bred animals such as rats, mice, and beagles are preferred because of the wealth of background data available. The researcher's confidence in the animal's background data minimizes the number of animals used and increases the likelihood that the studies will produce reliable scientific results. Whereas rats are routinely used as the rodent species, the nonrodent species is chosen based on its similarity to humans with regard to its primary pharmacodynamic response, pharmacokinetic profile, and metabolite pattern. Therefore, alternative nonrodent species like mini-pigs or nonhuman primates also have to be considered, although the use of nonhuman primates should be seen as an exception. In a normal study design, an equal number of male and female animals are tested in three doses with a route of administration identical to that intended for humans, and there is one control group. The number of animals used in any particular study is generally linked directly to the number indicated in the published regulatory guidelines. Where appropriate, the number of animals is set after consultation with professional statisticians. The husbandry of the animals has to follow the strict animal welfare rules described by the USDA and the European Council directives.

By definition, the objective of a repeated-dose toxicity study is to identify the target organ toxicity or other nonspecific toxicity of a high dose of a compound that has no adverse effects in a low dose.

As mentioned previously, the large amount of historical data available for laboratory-bred animals allows early identification of toxicity. However, due to the relatively homogenous geno- and phenotype of these animals, as well as the fact that generally only young, healthy animals are used, rare toxic effects or effects related to certain diseases or ages will rarely be picked up. At first glance, this provides an opportunity for a broader use of disease models or genetically modified animals and may be warranted if, for example, the pharmacodynamic effect by itself will cause toxicity. In general, however, the lack of extensive control data and the fact that even the best model does not reflect human disease 100% poses a challenge for the process of risk assessment that needs to be considered prior to testing.

New Approaches in Regulatory Toxicology: The Exploratory Investigational New Drug Approach

The preclinical testing strategy described in the chapter on regulatory toxicity studies presents the most common approach currently used in the pharmaceutical industry. However, in an attempt to reduce the time and resources expended on candidate products that are unlikely to succeed, the revised ICH M3 (R2) document (International Conference on Harmonisation, 2008) as well as the FDA guidance titled "Guidance for Industry, Investigators, and Reviewers—Exploratory IND Studies," published in January 2006 (U.S. Food and Drug Administration, 2006), offer alternative approaches to testing drug candidates prior to phase I clinical trials in humans and to accelerating drug development. Because of the limited preclinical safety data for supporting the clinical trial, the human exposure is limited and has no therapeutic or diagnostic intent. The value of these early clinical trials is related to exploring the pharmacokinetic profile—for example, microdosing—and the readout of specific safety or pharmacodynamic biomarkers. The preclinical safety package very much depends on the duration and extent of the human exposure.

The microdose approach is as follows:

- ≤ 100 μg: An extended single-dose toxicity study in one species, usually rodents
- ≤ 5 administrations of a maximum of 100 μg: A 7-day toxicity study in one species with structure–activity relationship (SAR) assessment of the genotoxic potential

The approach for single-dose studies up to the subtherapeutic or intended therapeutic range is as follows:

- Extended single-dose toxicity studies in rodent and nonrodent species
- Assessment of genotoxic potential (Ames mutagenicity test)
- Safety pharmacology core batteries

The approach for multiple-dose studies that last up to 2 weeks and do not test the maximum-tolerated dose is as follows:

- Approach 1: Two-week repeated-dose toxicity studies in rodent and nonrodent species with exposure multiples of anticipated human AUC, safety pharmacology, and genotoxicity
- Approach 2: Two-week repeated-dose toxicity studies in rodents testing up to a maximum tolerated dose and confirmatory studies in nonrodent species for a minimum of 3 days at the exposure intended to be the NOAEL in the rodent, safety pharmacology, and genotoxicity

However, since the first edition of this book, it became apparent that only a few large pharmaceutical companies are actively using alternative regulatory approaches to a relevant degree. In contrast, small start-up companies often use such limited approaches to save on costs prior to the application for additional funding or selling of the respective drug project or product.

BIOMARKERS

As the limitations of currently available safety biomarkers become increasingly evident—for example, the lack of sensitivity of common renal biomarkers—joint efforts have begun to take place between agencies and the industry, resulting in the establishment of collaborations like the Predictive Safety Testing Consortium (PSTC), an organization formed under the umbrella of the C-Path Institute with the goal of improving the biological evaluation of preclinical and clinical bridging biomarkers of drug safety.

One recent success is the approval by the FDA and EMEA of seven proteins or biomarkers of drug-induced kidney injury in animal studies that were found in urine and that can provide additional information about drug-induced damage to kidney cells (Kim-1, albumin, total protein, B2-microglobulin, cystatin C, clusterin, and trefoil factor-3). Previously, the FDA and EMEA had required drug companies to submit data from blood tests to demonstrate that potential therapies will not cause kidney damage. Although the blood tests accurately measure toxicity, they do so in delayed fashion: by the time the blood tests indicate toxicity—after a week of treatment—the kidneys may have sustained severe damage. The new tests for kidney biomarkers offer more immediate results and greater sensitivity. Instead of yielding results after days of treatment, the biomarker tests will indicate whether kidney cells are being damaged only a few hours after treatment begins, and the tests will pinpoint precisely which kidney cells a drug is impacting. Other groups of the PSTC are working on novel biomarkers for liver, muscle, and blood vessel toxicity, as well as carcinogenicity. However, novel insights can even be gained from old, established biomarkers, for example, cardiac troponin where new, ultrasensitive detection methods can now reliably measure baseline values in healthy animals and man.

LINKS

Regulatory Agencies and Testing Guidelines: Pharmaceuticals

- U.S. Food and Drug Administration (FDA): www.fda.gov
- European Medicines Agency (EMEA): www.emea.europa.eu
- Medicines and Healthcare Products Regulatory Agency (UK): www.mhra.gov.uk
- International Conference on Harmonisation of Technical Requirements for Registration of Pharmaceuticals for Human Use (ICH): www.ich.org

Societies

- U.S. Society of Toxicology (SOT): www.toxicology.org
- Eurotox Association of European Toxicologists & European Societies of Toxicology: www.eurotox.com
- U.S. Teratology Society: www.teratology.org
- U.S. Society of Toxicologic Pathology: www.toxpath.org
- British Toxicology Society: www.thebts.org
- European Teratology Society (ETS): www.etsoc.com
- Behavioral Toxicology Society: www.tox-portal.net

Related Sites

- ToxPortal (a gateway to several useful toxicology sites): www.toxportal.com
- Toxnet (a cluster of databases on toxicology, hazardous chemicals, and related areas): www.toxnet.nlm.nih.gov
- European Centre for the Validation of Alternative Methods (ECVAM): https://eurl-ecvam.jrc.ec.europa.eu/
- NTP Interagency Center for the Evaluation of Alternative Toxicological Methods: http://ntp.niehs.nih.gov/pubhealth/evalatm/index.html
- International Union of Toxicology (IUTOX): www.iutox.org
- Drug Information Association (DIA): www.diahome.org
- World Health Organization (WHO): www.who.int
- Association of the British Pharmaceutical Industry (ABPI): www.abpi.org.uk
- Pharmaceutical Research and Manufacturers of America (PhRMA): www.phrma.org
- Association of the Research-Based Pharmaceutical Companies in Germany: www.vfa.de
- Organisation for Economic Co-Operation and Development (OECD): www.oecd.org

THE PRACTICE OF DISCOVERY SAFETY ASSESSMENT

Safety assessment in the early drug discovery process is essentially an exercise in risk mitigation. Risk mitigation in this context means that lessons from previous safety studies should be used to influence future studies so that past problems can more easily be avoided using a combination of information and experimentation. However, most drug discovery projects actually take safety endpoints into account. In practice, as soon as a candidate compound is dosed to living systems, be they in vitro or in vivo, a safety study takes place. The combination of using lessons from previous safety studies to design specific new studies and monitoring all other studies for signs of safety issues is the basis of early safety assessment in drug discovery.

Early in the drug discovery process, when medicinal chemistry is most active and intellectual property around a therapeutic target is being claimed, the quality of the molecules is increasing. This process relies on a combination of rules, focused in vitro and in vivo experiments, as well as ad hoc studies. The benefits of this process are very clear in the areas of absorption, distribution, metabolism, and elimination (ADME); project failure due to poor ADME properties has fallen dramatically in the years following the introduction of such measures. The development of ADME sophistication continues and permeates essentially all early drug discovery projects. Rapid optimization of ADME properties has become a reality for many projects.

Replicating the relative success of ADME optimization in the area of safety is of critical importance if higher-quality drug discovery projects are to become a reality. The introduction of structured approaches to using safety or toxicology knowledge and experimentation in the earliest phases of drug discovery is gaining pace, and most discovery teams welcome input from toxicologists as long as it is timely, results in effective project decisions, and is based on sound toxicological principles and data. Given these conditions, even modest efforts at increasing the safety quality of drug candidates can pay large dividends in avoidance of project delays due to known liabilities. The key to both ADME and safety impact on early discovery projects is an ever-increasing science base supporting rapid assessment of properties relevant to the endpoints of interest at a pace that is compatible with the decision-making process. This assessment can, and does, take the form of computational, in vitro, and in vivo studies. Most safety studies are protracted; hence, this science base is slow to evolve, and learning must be extracted carefully at every opportunity. One of the primary opportunities for evolving the science base of safety in drug discovery is in problem-solving activities.

The failure of a drug or clinical candidate is invariably associated with a substantial amount of basic science that goes into problem-solving activities. This effort, if captured and integrated into the discovery process, can contribute to the development of better and more sophisticated approaches to discovery safety assessment and of meaningful refinements in the regulation of drug development and approval. Thus, the resources and momentum behind a problem-solving effort, particularly around a late-stage clinical candidate, present a unique opportunity to develop a more stable and valid science base on which to build a more rational approach to discovery safety.

One basic question involves what this increase in the sophistication of a safety science base looks like and to whom it should be directed. The choice of a therapeutic target is one critical area, and many advances have been made in recent years in the characterization of the specific role(s) of a protein in various tissues. This target safety aspect is not a one-time exercise to be carried out at the beginning of a project, but a constant vigil, a continuous endeavor to relate all aspects of the complicated life of a therapeutic target to the adverse event signals coming from preclinical and clinical studies.

Target-Based Safety Assessment

The choice of a therapeutic target is seldom made from a safety perspective. However, some concrete measures can be taken to anticipate the potential toxicological consequences of a given therapeutic intervention. The most obvious among these is an assessment of the overall function and tissue distribution of the target. The presence of the target in a nontarget tissue is a concern, and it is worthwhile to note this distribution and to assess differential functions of the target in the various tissues. Direct phenotypic information from knockout studies is invaluable in this regard and provides a clear early indication of potential concerns, not only for normal function but also for eventual reproductive disturbances. Rat and mouse knockout phenotype databases contain data that may give indications of potential pathologies if a target is inhibited. Such “soft” knowledge is, in itself, not enough to exclude or include a target for consideration but rather serves as a sound hypothetical basis for further, focused studies to address the potential for target-specific pathologies. These studies can be undertaken as soon as in vivo studies start for the assessment of efficacy in preclinical models.

An independent toxicological view of target biology needs to be a continuous part of the safety contribution to a discovery project. A thorough knowledge of the distribution and function of the target is critical in the interpretation of observations from both safety and efficacy studies. An unfortunately spectacular example of this is the role of Cox2 in coagulation. Cox2 inhibitors show very little tissue selectivity in their distribution and thus would be expected to affect essentially all Cox2 activities in the body. The pharmacological mechanism outlined for the primary beneficial therapeutic effect of Cox2 inhibitors is complex and reviewed elsewhere (Warner and Mitchell, 2004), but mostly restricted to the target effects of Cox2-inhibition in the prostanoid cascade.

This view cannot be that adopted by the toxicologist. Figure 5.7 illustrates the number of associations between Cox2 (in the center) and other proteins (orange and red symbols), small molecules (green symbols), and physiological processes (yellow symbols). Within this picture are the associations that would give indications of imbalances in coagulation cascades induced on inhibition of Cox2.

Together with observations from preclinical or clinical studies, association maps like these can serve to help toxicologists form testable hypotheses from large volumes of inconclusive data.

Assessment of target biology, as a continuous process, has a few quite basic components:

- It is necessary to monitor the developing biology of the target, not only in the target tissue. The function may not be the same in all tissues.
- Tissue distribution studies add tissues or species continuously. It is probably rare for a protein to be expressed in a tissue for no reason; if it is expressed, it is important to the functioning of that particular tissue.
- Use of these approaches seldom gives concrete answers. The best result is the development of a testable hypothesis.

Safety-Directed Drug Design

It is in the area of chemical design that perhaps the most value can be gained from translating safety data into real, tangible decisions. This is a long, tedious process, but several successful examples have been identified in which a valid set of decision-making tools can be used to warn researchers about chemical designs that contain inherent safety liabilities. In some cases this can lead to decisions before synthesis is even undertaken. In other cases the level of confidence is lower, but the tool can spark the decision to do further experiments in more sophisticated model systems to investigate the probability of a real safety

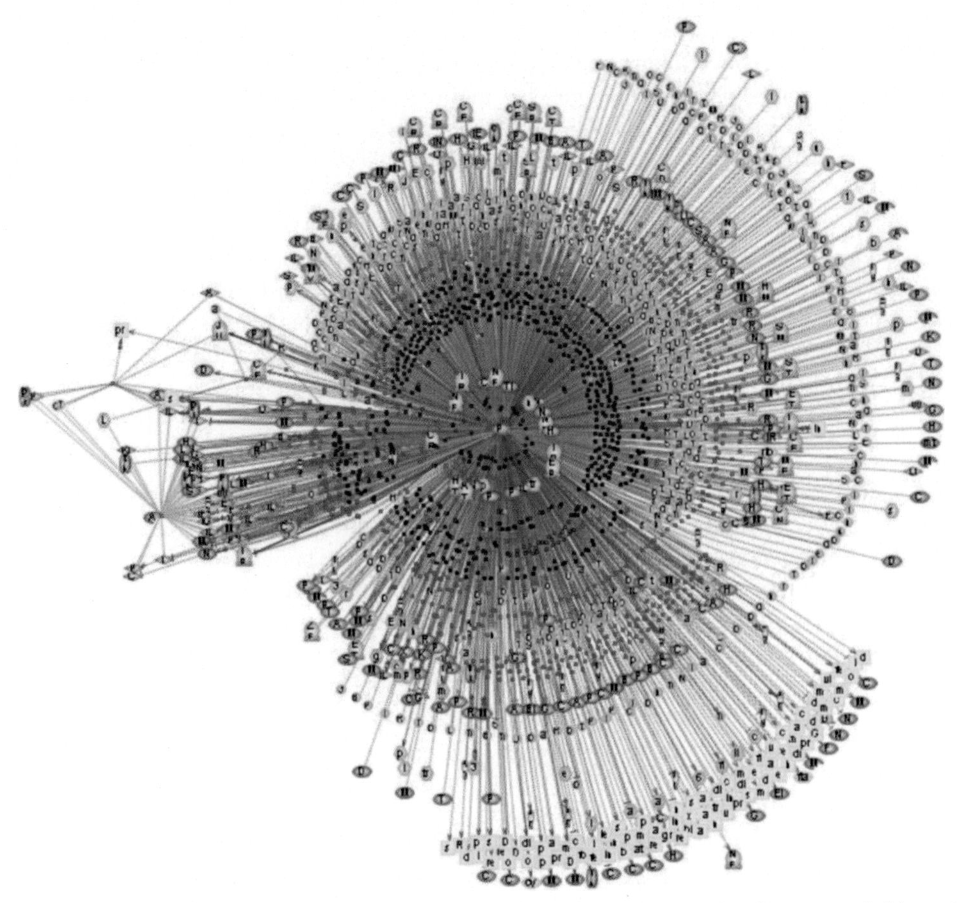

FIGURE 5.7 Depiction of connections found in the literature between Cox2 (center) and proteins (red and orange symbols), small molecules (green symbols), and physiological processes (yellow symbols). *(From Apic et al., 2005, reprinted by permission from Elsevier.)*

problem. The key to either of these scenarios is to have, first, a clear strategy behind the decision-making tools such that results from a simple test (computational, in vitro) can be followed up and confirmed in a relevant in vivo test and, second, adequate testing throughput to facilitate iterative design. Without this confirmation mechanism, very little about the *real* risk of a compound or series can be concluded, and without adequate throughput, the medicinal chemist will be mired in indecision.

What kinds of safety-related decisions can be made by the medicinal chemist? There are actually many. First and foremost, one should always be cognizant of the area of chemistry in which one is working, the therapeutic class of compounds, and any similarities they may have, both in their basic structure and in their pharmacophoric makeup, with compounds with known pharmacologies. Polypharmacology, while desirable in some special cases, carries the potential to add an unnecessary burden to the drug development project and, in many cases, can be identified and avoided by early searching of pharmacology datasets like PubChem, GVKBio, or BioPrint or by performing in vitro secondary pharmacology screening on a large number of targets. This sort of characterization hardly ever leads to the decision to stop a chemical series but, rather, aids in the identification of potential problems early enough to guide confirmatory experimentation (monitoring of blood pressure in early preclinical species for compounds active at the alpha1 adrenergic receptor, for example) and allows chemistry to be changed such that liabilities can be designed out if they are judged to be serious.

A specific case of unwanted pharmacology is activity at the cardiac hERG K+ channel. Potent blockers of the hERG channel can be found in most therapeutic classes, and thus few industrial medicinal chemists have escaped this problem entirely. Fortunately, there have been many advances in our understanding of hERG specificity and the physicochemical properties that drive hERG blockade. Protein models of the channel are now available, and scores of papers have been published on hERG structure–activity relationships. Designing away from hERG blockade while retaining all other properties remains a challenge; however, the combination of better foundation tools such as SAR and protein models, combined with enhanced screening technologies and a constantly improving picture of the preclinical and clinical testing systems required to assess the ultimate risk of fatal arrhythmias, has dramatically improved our ability to successfully manage this once-difficult problem. As mentioned previously, high-resolution preclinical and clinical studies on the causal mechanisms of safety problems allow for identification of the chemical events underlying these mechanisms. The blockade of hERG is one of these cases, and now medicinal chemists, armed with the right tools, can in most cases successfully build out hERG liabilities. From the multitude of tests and the tens of thousands of compounds characterized for this liability, chemists can now use a few rather simple rules to avoid, if not eliminate, blockades at the hERG channel.

The following are a few simple rules for avoiding hERG activity:

- Avoid extended cylinder-shaped molecules with hydrophobic features at the ends.
- Avoid aliphatic amines with a pK_a above 7.
- Avoid compounds with log P or log D values over 2.

In most cases, potent activity at the hERG channel is a showstopper for the project, and it must be addressed.

Another such endpoint is genetic toxicity. As with hERG, genetic toxicity testing is part of a mandated set of tests that must be performed before humans are exposed to a new drug candidate. As such, it must be effectively dealt with in the early phase of a project to avoid a significant issue later in the more mature project. Several very effective tools are available to the medicinal chemist that range from structural warnings to SAR models with reasonable levels of predictivity. Screens for genetic alterations also come in a number of types, with the most common for drug development being the salmonella reverse mutation, or Ames test, and the mouse lymphoma assay. As with hERG, many thousands of compounds have been characterized for their genetic toxicity liability, and from these data, design rules can be derived.

The following are a few simple rules for avoiding genetic toxicity:

- Avoid reactive species.
- Avoid aromatic amines or compounds that give aromatic amine metabolites.
- Avoid compounds that can produce other forms of reactive metabolites.

As the last point in this list illustrates, reactive metabolites are a problem for drug development. Reactive metabolites can either be relatively silent, for example, although present in significant amounts, they do not cause overt problems in preclinical and early clinical studies; or they can be more direct and thereby be clearly associated with organ-based toxicities. Silent reactive intermediates like those associated with halothane have the potential to bloom into serious problems in the clinic as more and more patients are exposed; they may also become apparent only after years of use in patients. These reactions are sometimes labeled "idiosyncratic," although in reality the reactive metabolite exists in all patients and the manifestation of a problem happens in those susceptible patients. Direct, organ-based toxicity by compounds producing reactive intermediates, such as paracetamol, is also possible and is usually easier to detect early because the compounds cause a more dose-related lesion that is seen across species and in nearly all individuals exposed.

Thus, with regard to reactive intermediates, prudence is necessary. Avoiding placing unnecessary burdens on the project as well as acting responsibly toward future patients would dictate that functional groups that can give rise to reactive metabolites be identified and at the very least be discussed and preferably avoided. Awareness can also aid the interpretation of safety studies when unexpected pathologies are observed. A catalog of in vivo pathologies that have been associated with reactive intermediates is a very useful tool and can be used in combination with a thorough library of the functional groups and substructures that can give rise to reactive intermediates.

The following are a few simple rules for avoiding reactive intermediates:

- Actively use a "warning" substructure list.
- Characterize biotransformation as early as possible.
- Look for warning signs: overt cell toxicity in genetic toxicity assays using metabolic activation, time-dependent inhibition of P450 activity, cytotoxicity in P450-expressing cell lines.
- Assess the degree of binding in combination with the intended dose. The higher the dose, the greater the reactive metabolite burden.

For those wishing to use other toxicity data to make drug design decisions, online structure toxicity resources such as ChemIDPlus, IUCLID, TOXNET, and DSSTox can complement the commercial information or model tools such as MultiCASE and DEREK. A recent, potentially useful addition to this group is the Mechanism Based Toxicity Database from GVKBio, a hand-annotated, structure-searchable database of reported toxicities. The current version of this database contains more than 13,000 entries.

In vitro tests are too numerous to list here, but several new and useful tests have recently been developed. The general trend is to move away from gross cytotoxicity assays in the target cell or tissue type and to instead focus on either nonlethality measures in target cell types or measurements of general interference in basic cellular processes via the monitoring of several endpoints simultaneously. Additionally, recent examples of promising applications of toxicogenomics give hope that this technique will finally begin to live up to its initial promise. There is an almost universal desire to enhance the predictivity of early toxicology screens, and this has led to a number of products and initiatives to improve the general area of predictive toxicology.

SUMMARY

As the researchers of today focus on integrating as much sophistication into target liability assessment and chemical design as possible, the potential for reducing safety-related failures is becoming increasingly accessible, but it is far from certain. We can be certain, however, that promising projects and drug candidates will fail less often for the reasons of the past. The problems of the future will be qualitatively different, but if we have done a conscientious job in addressing the problems of the past and translating this learning into early drug discovery project decision making, we will free up more time and resources within our safety groups for solving tomorrow's problems.

PRECLINICAL SAFETY FROM A TRANSLATIONAL PERSPECTIVE

As outlined in the beginning of this chapter, pharmaceutical toxicology or safety assessment is, through and through, a translational discipline. All aspects of preclinical safety activities aim at identifying potential hazards, predicting and quantifying risks for healthy trial subjects or patients, and, if necessary, assessing suitable means for eliminating or reducing unacceptable risks. However, there are many uncertainties in this approach. For practical reasons, only a limited number of aspects can be addressed in preclinical animal and in vitro studies. Furthermore, considerable species differences leading to changes in biological activity or pharmacokinetic properties as well as individual variations in susceptibility to certain effects and compounds add to uncertainties in the risk-assessment process. Thus, any toxicological risk-assessment process for pharmaceutical or other compounds is always incomplete and associated with uncertainties. Such uncertainties are compensated for by empirically derived safety factors. It is evident from the prevailing incidence of pharmaceutical treatment-related side effects that such factors are occasionally not sufficient to protect patients from expected and, particularly, unexpected side effects. This aspect has recently been investigated in a thorough analysis of common reasons for drug project attritions in a major pharmaceutical company (Cook et al., 2014; Figure 5.8).

Traditional regulatory toxicology has developed sophisticated methods and strategies to characterize the possible hazards and associated risks of potential drug candidates. However, this comparatively retrospective and descriptive approach has been deemed unsustainable in view of unacceptable attrition and issue rates in preclinical safety testing and clinical studies as well as ever-increasing safety expectations and requirements. Thus, based on expert knowledge, steps have been taken to actively design safer drug candidates. This is mostly done by trying to avoid known structural or pharmacokinetic

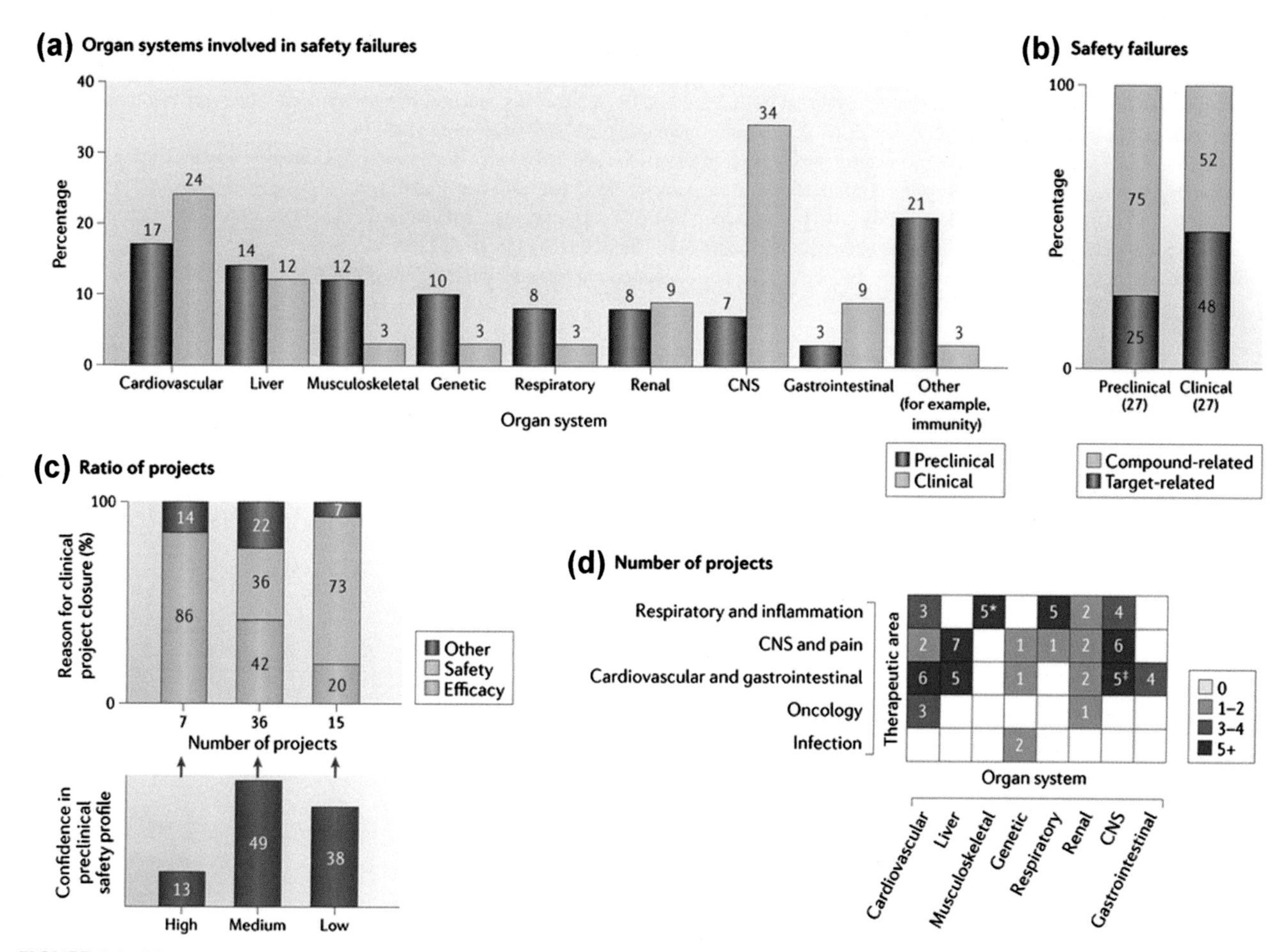

FIGURE 5.8 The overall translatability of preclinical safety issues relative to adverse reactions in the same organ or functional system in man. *(From Cook et al. [2014], reprinted by permission from Macmillan Publishers Ltd:* Nat. Rev. Drug Discov. *13, 419–431.)*

properties leading to certain side effects. The situation is comparable to the crash testing of a finished automobile (regulatory toxicology), which leads to a completely new area of expertise in which researchers are charged with designing air bags and safety crush zones (investigative or discovery toxicology). However, in contrast to the car industry, it remains to be seen if early, active safety efforts lead to the same level of improvement.

Despite the many activities in the regulatory and investigative toxicology area, one aspect remains unaddressed. There is no element in the preclinical safety package focusing directly on patients. All studies are conducted in healthy young animals and are, thus, primarily risk-assessment tools for healthy human subjects. However, many diseases are known to increase patients' susceptibility to side effects. This is certainly a relevant area within pharmaceutical toxicology that requires further development.

REFERENCES

Amberger-Lahrmann, M., Schmähl, D., 1993. Gifte-Geschichte der Toxikologie. Fourier Verlag, Wiesbaden.

Apic, G., Ignjatovic, T., Boyer, S., Russell, R.B., 2005. Illuminating drug discovery with biological pathways. FEBS Letters 579, 1872–1877.

Canadian Centre for Occupational Health and Safety, 2008. Hazard and risk. http://www.ccohs.ca/oshanswers/hsprograms/hazard_risk.html#_1_2 (accessed 11.09.14.).

Cook, D., Brown, D., Alexander, R., March, R., Morgan, P., Satterthwaite, G., Pangalos, M., 2014. Lessons learned from the fate of AstraZeneca's drug pipeline: a five-dimensional framework. Nat. Rev. Drug Discov. 13, 419–431.

International Conference on Harmonisation of Technical Requirements for Registration of Pharmaceuticals for Human Use, 2008. ICH M3 (R2) document. http://www.ich.org/fileadmin/Public_Web_Site/ICH_Products/Guidelines/Multidisciplinary/M3_R2/Step4/M3_R2__Guideline.pdf (accessed 15.09.14.).

Klaassen, C. (Ed.), 2013. Casarett and Doulls Toxicology. 8th ed. McGraw Hill, New York, p. 17. Table 2–1.

Lehman, A.J., Lang, E.P., Woodard, G., Draize, J.H., Fitzhugh, O.G., Nelson, A.A., 1949. Procedures for the appraisal of the toxicity of chemicals in foods. Food Drug Cosmet. Law 4, 412–434.

Olson, H., Betton, G., Robinson, D., Thomas, K., Monro, A., Kolaja, G., et al., 2000. Concordance of the toxicity of pharmaceuticals in humans and in animals. Regul. Toxicol. Pharmacol. 32, 56–67.

Organisation for Economic Co-Operation and Development (OECD), 2008. Good laboratory practice. http://www.oecd-ilibrary.org/environment/oecd-series-on-principles-of-good-laboratory-practice-and-compliance-monitoring_2077785x (accessed 15.09.14.).

U.S. Food and Drug Administration, 2006. Guidance for Industry, Investigators, and Reviewers—Exploratory IND Studies. http://www.fda.gov/downloads/Drugs/GuidanceComplianceRegulatoryInformation/Guidances/ucm078933.pdf (accessed 15.09.14.).

U.S. Food and Drug Administration, 2008. ICH Guidance SAFETY Documents, summarized by FDA. http://www.fda.gov/CDER/GUIDANCE/#International%20Conference%20on%20Harmonisation%20-%20Safety (accessed 11.09.14.).

Warner, T.D., Mitchell, J.A., 2004. Cyclooxygenases: new forms, new inhibitors, and lessons from the clinic. FASEB J. 18, 790–804.

Chapter 6

Translational Science Biostatistics

Georg Ferber and Ekkehard Glimm

STATISTICAL PROBLEMS IN TRANSLATIONAL SCIENCE

Drug development can be described as a series of experiments that span drug discovery, preclinical research, and clinical research. During these experiments, knowledge about the drug and the disease under study is accumulated and structured. Each experiment has a component of learning and one of confirming, and later experiments are based on what has been inferred from previous ones. More specifically, the decisions as to whether to stop or continue drug development or which experiment to perform next, as well as many details of each individual experiment, are based, to a large extent, on the information gathered up to that point. However, traditionally, the step from one experiment to the next is done rather informally.

Statistics can be used to describe and explore data and thus generate hypotheses. The ultimate goal of many statistical analyses, however, is to support or refute a hypothesis about the effect of a treatment. Traditionally, this is often done in the context of statistical tests of hypotheses. Within one experiment, statistical tests of hypothesis are set up to control the *type 1 error,* that is, to keep the probability of rejecting a true null hypothesis below a predefined threshold, as is discussed in more detail in "Design and Interpretation of an Experiment." Statistical tests of hypotheses are widely used in clinical research, as evidenced by the ubiquity of p-values in the medical literature, but, in fact, they are not always optimal for an individual step in a development process. Confirmation of a hypothesis of the efficacy and, as much as possible, of the safety of a potential medicine is essential in phase III of clinical development, in which the public and regulators need to be convinced. The same approach may be inappropriate in preclinical and early clinical phases. In these situations, support for the choice between different development strategies is needed, with the option of stopping development.

This point of view can be carried even further, if we do not limit ourselves to looking at various development strategies for one compound but ask questions like the following: Given our current knowledge, which of a number of compounds with similar indications or mechanisms of action should we concentrate on for further development? A fully decision-theoretic approach would have to specify a cost–benefit relation, uncertainty about potential treatment effects, available resources, prior knowledge about treatment effects, treatment populations, side effects, and the like, and then combine these elements into a statistical model. This may often be overambitious for an entire drug development program, but elements from this approach can be implemented into parts of the program, as is discussed further later in this chapter.

In summary, we should keep in mind that for preclinical and early clinical development more appropriate methods than classical hypothesis testing can and should be developed. In particular, this includes models that can take into account the fact that in early development phases, the information we can expect from the trials performed so far is rather limited.

Other fields in which statistics can be helpful are inference and description of relationships, such as pharmacokinetics and kinetic–dynamic relationships, as well as other approaches to dose–response modeling. For biomarkers, inference and confirmation of the relationship with clinical endpoints are important. In all these areas, a mathematical model that describes the relationship between the scientific entities is needed. Such models might be mechanistic (e.g., describing the relation in terms of a biologically motivated set of differential equations) or empirical (e.g., a smoothing spline), but they will almost invariably have a stochastic component. Statistics, thus, has a role in estimating the parameters of the model, quantifying the uncertainty of this estimation, and selecting the most adequate model. Again, in traditional applications, selection and estimation are performed within one experiment, but statistics can also provide support that spans a series of experiments. Furthermore, in safety assessments, we need to demonstrate the absence of an effect of a certain size. The specification of an upper tolerance limit for an effect indicating harm is a paramount prerequisite of any reasonable statistical treatment of drug safety, but in general this specification cannot be obtained by statistical means.

Principles of Translational Science in Medicine. http://dx.doi.org/10.1016/B978-0-12-800687-0.00029-3

In the subsequent sections, we first introduce the fundamentals of statistical modeling and mention some of the most common statistical models. We then turn to the statistical principles of the design of an experiment and introduce some designs that are particularly relevant to translational medicine. We devote a special section to multiplicity, because multiplicity is ubiquitous in translational science and because an appropriate understanding of related problems and potential solutions is very important. A section on biomarkers follows, which is centered on the concepts of sensitivity and specificity for diagnostic tests. These concepts can also be applied to biomarkers used as predictors for a clinical outcome. A section on biological modeling concludes the chapter. These concepts are illustrated with examples from pharmacokinetics and pharmacodynamics. Our goal is to convey the basic ideas and to give references for those who want or need to understand the details.

STATISTICAL MODELS AND STATISTICAL INFERENCE

As mentioned previously, statistics is concerned with the analysis of data that are subject to random variation, that is, measurement error or any other source of uncertainty that prevents perfect reproducibility of an experimental result. An important basic distinction is that between exploratory statistics, which aims at simply describing the data without any explicit attempt at drawing conclusions, and inferential statistics. The distinction can be blurred; for example, linear regression is sometimes viewed as a tool to merely summarize the relation between two measurements descriptively, but sometimes the objectives may comprise testing for statistical significance of the regression parameters. Means, variances, standard deviations, standard errors, and other measures of the location and shape of the distribution of available data are the mainstay of any statistical analysis. They are reported to summarize and highlight the predominant or most relevant features of an experiment's outcome. Many graphical tools, such as the boxplot, serve the same purpose (see Hoaglin, Mosteller, and Tukey, 1985; Maindonald and Braun, 2007; Tufte, 2001; Weiss, 2008) and have been implemented in statistical software packages like the free software R (R core team 2014, http://www.r-project.org/).

Inferential statistical modeling should always start with a clear statement about the questions to be answered. In translational medicine, the primary focus of late-stage preclinical and early clinical trials will typically be a decision about continuing a compound into full clinical development (proof-of-concept [PoC]). The predominant question may be, "Should we do this?" The decision will rest on one or two pieces of information with different weight:

1. The selection of one primary or several coprimary endpoints: Usually, these will be measurements of treatment efficacy. In translational medicine, these will often be biomarkers used as surrogates for clinical endpoints—for example, CD4-cell counts, hormone levels, tumor shrinkage, and the like. Safety parameters may also be considered, in particular if they represent "red lights" that prevent the further development of an otherwise efficacious drug.
2. The selection of secondary endpoints: It is usually advisable to restrict the number of primary endpoints to one or a few. Otherwise, multiplicity (see the following discussion) might pose a challenge. However, as Senn and Bretz (2007) convincingly argue, concerns over multiplicity should not prevent researchers from acquiring more measurements. All questions of scientific interest that are not covered by the primary analysis may be relegated to the secondary endpoint analyses.

Of course, if the decision is to continue into full development, a number of follow-on questions arise. Although it may not be possible to address all of them in a translational medicine trial, it is important to consider them as early as possible. They are as follows:

- Selecting a formulation and the type of application (oral, intravenous, subcutaneous, etc.)
- Selecting a dose and a regimen
- Assessing safety and the safety–efficacy trade-off
- Determining the patients to be treated and subpopulations of special interest

In preclinical phases, the focus may be more on selecting from a number of candidate substances than on a go-or-no-go decision for a single one. Decision making in this setting will similarly be based on one or a few primary endpoints and supportive evidence from secondary endpoints. In toxicology, however, a subset of trials has confirmatory character, because their purpose is to demonstrate to the public that the substance is safe enough to be administered to humans.

Inevitably, these questions require a stochastic model to describe the relationship between a treatment and its effect on the intended patient population. It would be hopeless to attempt to provide here an exhaustive overview of the approaches to model clinical data—a model may be extremely simplistic (the percentage of responders to a treatment defined by a threshold on a single lab parameter) or very complicated (a multivariate nonlinear mixed-effects model adjusting for several covariates).

It may be purely empirical or motivated by biological considerations, or it may be a model that is a hybrid between the two. It may be parametric or nonparametric, longitudinal or cross-sectional, a parallel or a cross-over design, and Bayesian (i.e., including prior information) or frequentist (assuming there is nothing outside this trial). However, all such models have two aspects in common:

1. Their stochastic nature: The response to treatment is not deterministic, but varies within a range of possibilities (whether this is inherent or just due to the fact that we do not know all influences is a philosophical question that does not concern us here).
2. Their approximate character: A model does not need to be "true"; however, it must be a reasonable approximation of reality. How close to the truth this approximation is depends on the situation at hand and also on the primary object of interest. For example, if we are primarily interested in the effect of a treatment, and if patients have been randomized to one of two treatment arms, we do not usually try to find many additional sources of variation between patients to put into, say, the model covariates X. We may, however, consider covariates that are already known to exert an important influence. The objective of including these in the model then is to reduce variation in the treatment effect estimate or to explain differences in the treatment effects between individuals, but the interest is not primarily in these covariates themselves.

As a third aspect, one may add that the focus of interest is usually on a population, rather than on the individual, but there can be exceptions.

The generic form of a statistical model will in many cases be

$$y = f(X, \beta, \varepsilon),$$

where y is called the response, X is a set of covariates or factors influencing the response, β is a set of unknown parameters that relate the factors to the response, and ε is an error term inducing variation in the individual responses. *Response* is a synonym for *endpoint* and denotes the primary quantity of interest, highlighting its dependence on the factors, in particular the ones that are deliberately modified in the experiment. The factors X might be categorical—and usually treatment allocation (e.g., new treatment or placebo) will be one of them—or continuous, for example, the patient's age, the baseline (pretreatment) blood pressure, or the dose of an administered drug. Traditionally, the error term was often considered additive, such that the models are $y = f(X,\beta) + \varepsilon$. The classical interpretation of ε is that of a measurement error. Modern developments have generalized this concept to mixed-effects models in which the error structure can be incorporated into the model (e.g., Diggle et al., 2002; Fitzmaurice, Laird, and Ware, 2011). This is of particular interest when repeated measures on a patient are taken. For readers interested in statistical methodology, we have added a short overview of the statistical models used in translational medicine at the end of this chapter in "Statistical Models." This section uses more technical terminology than the rest of this chapter and contains further reading that requires some knowledge of statistical sciences.

The selection of an appropriate model should be made by a team of experts that includes statisticians. However, as is discussed in "Biological Modeling," the decision about the right model is one that a statistician cannot—or at least should not—make alone.

Major research in statistical inference is predominantly focused on estimation of the model parameters β. Just as there are many different models, there is a multitude of estimation methods, and presenting them all is certainly beyond the scope of this text. The most important general principle is probably that of maximum likelihood (Cox and Hinkley, 1974). More important than knowing the technical details, however, is having an appreciation of the fundamental challenges of any modeling attempt:

- The unknown parameters β are estimated from the available data y and X. Hence, the estimates are subject to some uncertainty, quantified by standard errors and confidence intervals.
- A proper statistical analysis will involve some diagnostic checks for goodness of fit, that is, a quantitative assessment of how well the data match the assumed model. The general idea is usually to calculate a measure of deviance between the observed values and those predicted by the model. In addition, the individual contributions to the deviance, which are called residuals, should be checked for poorly predicted outcomes and also for nonrandom patterns.
- A complex model with many parameters may seem to fit the observed data well, but it will often work poorly when applied to predict future responses. This phenomenon is called overfitting. Several techniques are available to assess this. Traditionally, goodness-of-fit criteria that included a penalty for the number of parameters (e.g., Akaike's AIC) have been used. With the advent of powerful computers, more complex models have gained importance. These call for more sophisticated model-checking procedures like cross-validation, splits into training and test datasets, and resampling methods. See chapter 7 in Hastie, Tibshirani, and Friedman (2009) for an overview of model assessment techniques.

Although at several points in the drug development process, it is certainly necessary to fit stochastic models to given data, another challenge lies in the acquisition of the data, that is, the design of the experiment. Because the researchers usually have a certain model in mind before the experiment is started, they should set it up in such a way that the model parameters can be estimated as precisely as possible. Hence, the stochastic model needs to be considered even before any data are acquired. This entails sample size estimation, collection of covariate information, randomization (simple or stratified), and allocation of total sample size to strata and treatment groups and, possibly, to dose categories and the number of dose levels and their values.

DESIGN AND INTERPRETATION OF AN EXPERIMENT

As stated previously, when planning an experiment, one should have in mind the way it is going to be analyzed. Once the experiment is performed, it is usually difficult or impossible to make up for deficiencies in design. In addition, any change made to an analysis plan after the data are known undermines the credibility of a confirmatory analysis.

The quality of an experiment can be seen as the degree to which the researchers indeed get correct answers to the questions asked. Lack of quality in this sense can be attributed to two components: systematic error (bias) and random variability. Stated the other way round, the quality of an experiment has two components: reliability measures the degree to which similar results would be obtained if the same experiment was repeated, whereas validity describes the degree to which the experiment actually measures what the researchers want to measure. When designing an experiment, one should take measures to maximize validity and reliability under the conditions given by ethical and financial as well as by time constraints. Among the most efficient measures for ensuring validity in clinical and preclinical trials is the use of a control group, the use of randomization for the assignment of subjects to the experimental groups, and the blinding of subjects as well as experimenters. In many cases, further precautions need to be taken to avoid influences from different sources being confounded. As an example, in a cross-over design, such as an experiment, where each subject receives a sequence of experimental treatments (including control) over a number of periods, researchers run the risk of confounding the effect of time (i.e., a period effect) with a treatment effect.

The key measure to ensure reliability in an experiment is standardization. This is a way to reduce variability and thus increase reliability. Standardization, however, has its limits, and there may be influential factors that cannot be standardized. For example, the effect of a drug may depend on body weight, and performing the experiment with subjects whose body weights are in a narrow range may not be reasonable. Such a factor may then be a candidate for inclusion as a covariate in the statistical model. This augmented model is likely to show reduced variability compared to the model without the factor. The reliability of an experiment can be further increased and the variability of the outcome decreased by increasing the sample size, that is, the number of independent experimental units included. Determining a sufficient sample size is a key step in planning an experiment.

There is an obvious conflict between standardization and the validity of inference. An experiment performed under very strict conditions will produce results that can only be generalized to situations with a similarly rigid setting. In a phase III clinical trial, in which we want to obtain results that generalize to the population of all patients with a given disease, every ethically justifiable effort should be made to include a representative sample of this population in the experiment, and restrictions of age, concomitant diseases, and so on should be avoided unless they affect patient safety. However, in early clinical development, such as in a PoC study, the subjects included in a trial will often serve as models rather than representatives of a larger population. In such cases, it often is acceptable to impose restrictions on the populations under investigation in order to increase homogeneity. This, of course, applies a fortiori to preclinical settings.

The most straightforward design of an interventional experiment is the parallel group design. Subjects are randomly assigned to two or more experimental groups. Subjects in all groups are treated in an identical manner, apart from the difference in the intervention to be tested. Because in such a design the only difference between what happens to the subjects in different treatment groups is the intervention under study, there is relatively little room for confounding, and the interpretation of the outcome is usually straightforward. The drawback of a parallel group design is that comparisons between treatments are always comparisons between groups of different subjects. If there is heterogeneity in the subjects under study, this heterogeneity will affect the reliability of the between-group comparison. One way to reduce this variability is to look at changes from a pretreatment baseline value. As a rule, if, in the population under study, the correlation between baseline values and observations at the endpoint of the study is at least 0.5, looking at the change from baseline will reduce the variability of the measurement, compared to looking at the value at the endpoint only. In some cases and depending on the nature of the measurement, percent change from baseline

$$\frac{x_e - x_b}{x_b} \cdot 100$$

(with x_e being the observation at the endpoint and x_b the one at baseline) or percent of baseline may be an alternative function to look at (see, however, Vickers, 2001).

In some cases, cross-over designs that compare different treatments within one subject can be used. The simplest among them is the two-period cross-over design, in which subjects are treated with the two treatments under study in two subsequent periods, with a sufficient washout in between. In cross-over designs, researchers must take care to distinguish time-related effects from treatment effects. In a two-period cross-over design, assigning half of the subjects to the sequence *a–b* (assuming that *a* and *b* are the treatments to be compared) and the other half to *b–a* allows period effects (e.g., habituation to the experimental situation) to be separated from treatment effects, provided there is an additive relationship between the two. If, however, the effect of the treatment in the first period carries over in any way to the second period, this will introduce a bias into the estimates of both the period and the treatment effect. A carryover effect cannot be identified in a two-period cross-over design.

If $k > 2$ treatments are to be compared, the two-period cross-over can be extended to a k-period design. In this case, each subject is assigned to one of a number of sequences. In a Latin square, k sequences of k periods are used in such a way that each treatment appears in each period exactly once. This gives the best efficiency for estimating treatment and period effects. It is possible to achieve even more balance: in a Williams design, every treatment will appear in each period the same number of times, and, in addition, every treatment will be preceded by every other treatment. For even k, k sequences can achieve this balance; for odd k, $2k$ sequences are needed.

The presence of carryover will always be a problem in a cross-over design. If one makes rather unrealistic assumptions regarding the carryover (e.g., an additive effect that only affects the subsequent period), this effect can be estimated in Latin square designs for $k > 2$. However, we follow Senn (1993) and recommend that a cross-over design should be considered only if a relevant carryover can be excluded a priori. In addition, stability over time is another prerequisite for a cross-over design. Therefore, a cross-over trial should rarely be used in progressive diseases or if seasonal effects are to be expected.

Estimation of the dose–response relationship is a common objective in many trials in translational medicine. We will usually use $k + 1$ treatment groups using k ascending doses and a placebo group. (We use the term *placebo*, although our considerations apply to any form of inactive control. This is to avoid confusion with an active control, which will be dealt with later.) Parallel groups or, if the prerequisites are fulfilled, Latin square cross-over designs could be used. A simple way of analyzing such a trial is to compare each dose to the placebo group. If we expect a monotone dose–response relationship, this expectation can be incorporated into the analysis procedure (Peng, Lee, and Liu, 2006; Williams, 1972). In particular, as a general rule, the most efficient design is one with treatment groups of equal size; however, a design with multiple comparisons to the placebo group can be more efficient if the placebo group is larger. This might be considered in particular in trials with healthy volunteers. The methods cited previously assume a monotone dose–response relationship and do not address details of its shape. If we want to estimate the shape of the dose–response relationship and, in addition, if we want to be able to deal with nonmonotone relationships, we can attempt to model the shape and identify the best-fitting model. The MCPmod method (Bretz, Pinheiro, and Branson, 2005; Pinheiro et al., 2014) combines a modeling approach with a many-to-one comparison of dose levels and placebo in a very efficient way.

In pharmaceutical research, there is a trend toward trials that include active control groups in addition to a placebo group. There are several reasons why an active control could be used. If the primary purpose is to compare the test treatment to a standard active treatment, researchers must guard against a situation in which the standard treatment does not show an effect for whatever reason. Having a placebo arm in the trial in addition to an active control is the simplest way to achieve this. The placebo arm is even more useful if the researchers are looking for noninferiority rather than superiority of the test treatment in comparison to the active control. Note that in such a case, it may be sufficient to assign only a small fraction of the subjects to the placebo treatment (Pigeot et al., 2003). Even if the goal is a comparison of the test treatment with the placebo, inclusion of an active control may be valuable. In this case, the active control will be used to show assay sensitivity. In particular, in early trials exploring a biomarker, an active control can help distinguish a failed experiment (in which neither the active control nor the test is different from the placebo) from an inactive treatment (in which the active control differs from the placebo, but the test treatment does not). The demonstration of assay sensitivity using an active control is particularly important in trials set up to demonstrate safety. As an example, the Thorough QT/QTc Study described in the ICH E14 guidance (International Conference on Harmonisation, 2005) explicitly recommends the use of an active control for this purpose.

Traditionally, hypothesis testing has played a central role in the inferential analysis of experiments. This paradigm is based on three elements:

1. A statistical model of the experiment: This model describes how the data are generated. It may, in addition, also describe relationships between measurements or experimental conditions within the experiment. A test statistic that is a function of the data as described in the model is a central part of the model.
2. A null hypothesis H_0 that is related to the test statistic in the model.
3. Data observed.

The test statistic is computed based on the data observed, and this value is compared with the predictions of the model. Based on this comparison, we either reject the null hypothesis in favor of an alternative, or we cannot reject it.

The most common use of hypothesis tests is probably to show the existence of a drug effect. In a placebo-controlled trial, a null hypothesis would state that there is no difference between the test drug and placebo. To take a more specific example, assume that we are measuring an endpoint Y that we can assume to be normally distributed as $N(\mu_i, \sigma)$ for $i = 0$ (placebo) and $i = 1$ (test) with mean values μ_i and a common standard deviation σ. In this case the null hypothesis would be

$$\mu_1 \leq \mu_0$$

and the alternative hypothesis would be

$$\mu_1 = \mu_0 + \Delta$$

with some relevant difference $\Delta > 0$. However, there are other situations. Let us assume that we are interested in the safety of the drug. In this case, the endpoint may be a parameter describing an undesired effect. We essentially exchange the role of our null and alternative hypotheses. Our null hypothesis would be

$$\mu_1 \geq \mu_0 + \delta$$

where $\delta > 0$ is the margin of increase in Y that would be of no concern. Our alternative hypothesis could be

$$\mu_1 = \mu_0$$

if this is what we hope to be able to show. Similar situations can arise in a test for efficacy if we compare the test drug with an active control, and we cannot expect the test to be more efficacious than the control. This is often done in so-called non-inferiority trials in which a new drug may not be more efficacious than some standard treatment but has other advantages such as increased tolerability, a better safety profile, or easier administration.

Two types of errors might arise in hypothesis testing: the error of rejecting a null hypothesis that is true (type I error) and the error of not rejecting a null hypothesis if an alternative hypothesis is true (type II error). The usual way to proceed is to fix an upper limit α for the probability of a type I error and to derive a threshold for the test statistic that corresponds to this limit. Depending on the choice of an alternative hypothesis, on the variability of the data, and on the sample size, we can then compute the probability of a type II error. When one is planning an experiment, this relationship can be used to determine the sample size that is needed to make sure that the risk of missing a relevant alternative hypothesis—for example, an effect size of at least Δ—is kept below a predefined threshold β. Whereas the null hypothesis and the alternative hypothesis should be based on scientific—and thus rarely statistical—arguments and the question to be investigated, the choice of the accepted error probabilities is rather arbitrary. It is very common to set α to 0.025 or to 0.05, whereas for β, values as large as 0.1 or even 0.2 are common. Larger values for β have been criticized as unethical (Halpern, Karlawish, and Berlin, 2002). These choices may be reasonable in a setting in which evidence comes from a number of similar trials, and the final conclusion may depend on a formal or informal meta-analysis (Sutton et al., 2007). In a PoC setting, however, in which the outcome of a trial determines the fate of a new drug or therapy, such a choice seems inadequate. A naïve approach could be to look at the expected costs that the decision to stop or continue development entails. On the one hand, there are the costs of the PoC trial to be planned, which increase with sample size. On the other hand, there are the costs that are consequences of a type I or type II error, which need to be weighted with the probability for such an error to calculate the expected loss. In a PoC setting, the costs of a type I error would essentially be the costs of a phase II program; the costs associated with a type II error are those of a lost opportunity to develop and market an effective drug or those entailed by a switch to a less effective alternative compound in the pipeline of a development portfolio. These costs may be substantially higher than the costs of a phase II program, let alone those of the PoC trial itself. If so, it would seem logical to keep α approximately at the range that is currently being used, or even to relax the requirements on type I error control, but to require a substantially smaller acceptable type II error probability. The increased costs of the PoC trial might often be compensated by the reduced risk of missing an opportunity.

This reasoning is very crude. In particular, it considers only one alternative hypothesis, and it does not take into account prior knowledge. Decision theory based on Bayesian statistics offers a much more refined framework. A number of authors have addressed this problem for various settings (Cheng, Su, and Berry, 2003; Halpern, Brown, and Hornberger, 2001; Mayo and Gajewski, 2004; Pezeshk and Gittins, 2002, 2006). Yin (2002) specifically addresses the question of sample size calculation for a PoC study using a biomarker.

A related concept is that of the expected value of a drug development project (Senn 1996). This concept has been developed mainly to determine the optimal timing of studies in drug development, but it can also be applied when designing individual studies with respect to their size and error probabilities.

The considerations made here for a clinical PoC study apply also to earlier situations, if one experiment decides the fate of a drug. We will never know which potent drugs have been lost because of the insufficient power of an early decisive study.

Alternative ways to determine the sample size of a trial include a method that ensures that a confidence interval of at most a certain width is obtained with a given probability. A more interesting concept is that of predictive power. This is a hybrid of frequentist and Bayesian concepts. Essentially, predictive power is the Bayesian probability that a future planned trial will be significant in the frequentist sense. The attractiveness of the concept comes from the fact that it spans across more than one trial in a development program (see, e.g., chapter 6 in Spiegelhalter, Abrams, and Myles, 2004).

MULTIPLICITY

The issue of multiplicity arises if a single trial is intended to answer more than one research question. For example, if several hypotheses about the effect of a treatment on different biomarkers are investigated simultaneously, then the probability of making a wrong decision about at least one of them increases with the number of hypotheses. Likewise, if we look at a number of subgroups within the same experiment, multiplicity comes into play.

As an illustration of the effect of multiplicity, consider independent hypotheses, all investigated with statistical tests at the "local" level $\alpha = 0.05$. If 10 out of all investigated hypotheses are true, then the probability of at least one false rejection is

$$1-(1-0.05)^{10} \approx 40\,\%$$

A closely related problem is that of selection bias. If a trial involves the observation of several biomarkers regarding their response to treatment, and if only the results of the ones displaying the strongest reaction are considered, then the treatment effect estimates will overestimate the true treatment effects on these selected biomarkers. This is a particular challenge in gene expression analyses in which thousands of genes are screened for differential expression.

The traditional methods of dealing with this problem have been developed in the context of statistical hypothesis testing. In this context, strong control of the familywise error rate (FWER) is a frequent requirement: the probability of rejecting *any* true hypothesis is controlled at a fixed level α, usually 5%. Among the many methods of achieving this, probably the best known is Bonferroni's method, which consists of simply doing k tests, each at level α/k. This method has a tendency to "stick" with the null hypothesis; that is, it is a conservative method. Other widespread approaches that are less conservative include those of Scheffé, Tukey, and the Dunnett many-to-one comparison method (Dunnett,1955) for normally distributed data. More general approaches to correcting for multiplicity that can be used with almost any statistical test are the closed test principle (Marcus, Peritz, and Gabriel, 1976) and the partitioning principle (Finner and Straßburger, 2006; Takeuchi, 1973; see Hsu, 1996 for an overview).

FWER control is a very strict requirement, especially when considering many hypotheses. It places a lot of emphasis on the type I error, invariably at the expense of the type II error, that is, at the probability of detecting true differences. Hence, it is not advisable to attempt strong FWER control with all endpoints of a clinical trial. It is perfectly acceptable to label most of the clinical trial objectives as "exploratory" and analyze the corresponding data without multiplicity adjustment. Obviously, any findings of this exploratory analysis need to be confirmed in an independent trial, where they are tested as a primary hypothesis, that is, with control of multiplicity if applicable.

In spite of this, multiplicity adjustment is important if the number of biomarkers is very large. The advances in "omics" technology have multiplied the number of biomarkers assessed in translational medicine. Thus, recent years have seen a revival of multiplicity adjustment methods. Advances have been achieved mainly in two directions:

(1) Resampling methods make use of modern computer power. To illustrate the general idea, assume a treatment is to be compared with a placebo and that there are many endpoints (e.g., several biomarkers that all measure treatment effect) per patient. The resampling approach starts by assuming that no one knows to which group (treatment or placebo) the patients belong. It then generates a complete list of all possible, hypothetical allocations of the patients to the two groups and, for each generic allocation, obtains the corresponding value of a predefined test statistic, for example, the minimum of all p-values of the single endpoint comparisons. If the observed value of an endpoint (i.e., the value that is obtained from the real observed patient-group allocation) is in the "extreme tail" of the empirical distribution of this quantity, we have proof of a significant effect of the treatment in this endpoint with strong control of the FWER (Westfall and Young, 1993; Westfall, 2011). If it is not possible to generate a complete list due to the computational burden, a random sample can be taken instead. For sparse data, rotation approaches are available (Langsrud, 2005; Läuter, Glimm, and Eszlinger, 2005). The advantage of these methods is that they automatically adjust for correlations between the endpoints, whereas older methods like Bonferroni or Simes are based on worst-case scenarios.

(2) The concept of false discovery rate (Benjamini and Hochberg, 1995) relaxes the strict FWER requirement. It requires that the probability of falsely rejected hypotheses does not exceed a designated percentage of all considered hypotheses, typically 5%. Hence, it is now accepted that, for example, among 100 true hypotheses, 5 may be rejected. Most research in gene expression data analysis is focusing on this definition (Efron, 2010; Storey, 2003; Storey and Tibshirani, 2003).

Views on the extent to which formal multiplicity adjustment is necessary in medical research differ in the statistical community (Senn and Bretz, 2007). In the context of translational medicine, where any findings will have to pass the thorough test of a phase III development, this is usually an issue only when selecting from a large number of biomarkers. In that case, however, some care is required, in particular if the PoC is declared based on a selected endpoint.

BIOMARKERS

The U.S. National Institutes of Health Director's Initiative on Biomarkers and Surrogate Endpoints (Biomarkers Definition Working Group, 2001) defines a biomarker as "a characteristic that is objectively measured and evaluated as an indicator of normal biological processes, pathogenic processes, or pharmacological responses to a therapeutic intervention." As a rule, biomarkers are not clinically relevant by themselves but are used as predictors for clinically relevant outcomes. In translational medicine, the ultimate outcome to be predicted is usually the clinical efficacy or safety of a treatment when used in a certain indication. In many cases, only one aspect of efficacy is covered by a biomarker. For example, a necessary condition for a drug to act in the central nervous system is that it crosses the blood–brain barrier. Any unspecific measure of brain function, such as the resting electroencephalogram, can be used to ascertain this, but measuring the action of the drug on its target in the brain may be more valuable. Another use of biomarkers is in diagnostic tests. In this case, biomarker expression levels are often compared to a threshold value to decide on the presence or absence of a disease. In phase III trials, the efficacy of a drug is often also measured by a binary outcome, such as remission (the definition of which will, of course, depend on the indication). This setting is somewhat similar to that of a diagnostic test; therefore, the concepts introduced in the following have been applied to both situations.

Sensitivity and specificity are to diagnostic tests what type I and type II error are to statistical tests of hypotheses. Sensitivity is the probability of correctly identifying a condition, typically the presence of a disease, by a positive diagnostic test result. Hence, it is the equivalent of power = 1−type II error probability. Specificity is the probability of correctly identifying the absence of the disease by a negative test and is thus equivalent to 1−type I error probability.

Obviously, sensitivity and specificity depend on the threshold to which the biomarker is compared. If, for example, high values of the biomarker indicate the presence of a disease, we can increase the specificity of the test by increasing the threshold. This is done at the cost of a decrease in sensitivity. An extremely useful tool is the receiver operating characteristic (ROC) curve. In this context, this is a plot of sensitivity versus 1−specificity. For a perfect diagnostic test, the ROC curve would have an inverted L-shape; a useless diagnostic test (i.e., the equivalent of tossing a coin) would be described by a straight line from (0,0) to (1,1).

Of course, diagnostic tests should have both high sensitivity and specificity, but the right balance between them can be tricky. For example, consider an extremely rare disease that afflicts only 1 in 10,000 people. Suppose that the sensitivity of the test is 1 and the specificity is 0.999. If 100,000 patients are screened, $10 + 0.001 \times 99{,}990 \approx 110$ is the expected number of people identified as diseased, but in fact only $10/110 \approx 9\%$ of these do suffer from it! The ratio of true positive results among all results is called *positive predictive value*. Together with the corresponding negative predictive value (true negatives/all negatives), it is an additional measure of the quality of a diagnostic test supplementing sensitivity and specificity.

If we can increase the specificity to 1 by decreasing the sensitivity to 0.8, we would expect to find 8 diseased people and miss 2 while not having any false alarms. Depending on how grave the disease is, this might be the preferable option. (Of course, in this example, the right answer is probably not to do mass screening but to test on suspicion only.) The example illustrates that if an assay is developed and a threshold is picked, there is no off-the-shelf solution like simply picking the point at which sensitivity + specificity is at a maximum.

There is much discussion about the validation of biomarkers. Roughly speaking, a biomarker is validated if the accuracy of the predictions derived from it is known and is sufficiently good. If a biomarker is used in a confirmatory context, validation is very important. In such a context, we would speak of a surrogate endpoint. Strictly speaking, if a biomarker is validated for the prediction of the efficacy or safety of a therapy, then this validation applies only to the exact conditions that were used when validity was established. These conditions include the patient population, therapy used, and so on. In particular, any new therapeutic approach would require a revalidation of a biomarker. If a new therapeutic approach has an effect on a biomarker that is similar to that of a standard reference therapy, we thus may not automatically conclude that it will have a comparable effect on the clinical endpoint. To be valid, such a conclusion must be strongly supported by biological and clinical arguments. Blood pressure is one such mechanistically validated biomarker that is generally accepted as a surrogate endpoint. However, the fact that, for new antihypertensive therapies, large morbidity and mortality studies are conducted shows the limitations of even this well-established and well-understood surrogate endpoint. In translational medicine, such an overly strict requirement on validation is not appropriate. If we are looking for efficacy in this setting, it is important to have reasonably sensitive markers so that drugs that have therapeutic potential can be identified. If we are looking for safety, on the other hand, the biomarkers should be reasonably specific. In this way, the most promising drugs can be selected for further development, but, obviously, some risk of failure in later phases will still remain. If, however, the development of a useful drug is discontinued due to an erroneous biomarker, this is usually an irreversible decision that incurs the cost of a lost opportunity in terms of patient and financial benefit.

Formally, a biomarker used to predict a binary outcome can be validated by determining the ROC of the marker. If a biomarker is used to predict another continuous outcome, the simplest measure for validation would be the correlation of the biomarker and the clinical endpoint. There is extensive discussion about more elaborate methods for validating surrogate endpoints, but this is of limited relevance to the field of translational science. For an in-depth discussion, see, for example, Alonso and Molenberghs (2007).

If we have a set of biomarkers with moderate sensitivity and specificity, we may attempt to construct a composite marker with better quality. This approach is promising, if the various markers measure different entities, so that the correlation between them is at most moderate. For example, several combinations based on logic can be constructed. If we start with two biomarkers, we can derive a test that is positive only if both biomarkers are positive. In this case, specificity will increase over each of the individual markers, at the cost of sensitivity. Alternatively, we might look at a test that requires only one of the two markers to be positive. This obviously would increase sensitivity.

We can also look at linear combinations, that is, weighted sums of a set of biomarker values. This leads to linear discriminant analysis or logistic regression. Under the model of multivariate normality for each of the two groups (e.g., normal and diseased), and assuming that the within-group covariance matrix is the same for both groups, the optimal separation of the two groups is accomplished by a linear discriminant function that maximizes the distance between the groups relative to the within-group variance.

A word of caution is appropriate here: in order to fit a discriminant function that can be used for future patients, that is, one that is a predictor beyond the sample of data from which it is derived, it is important that the number of observations is much larger than the number of biomarkers considered. If this is not the case, the resulting situation is similar to overfitting (see the next section). As a rule of a thumb, some authors recommend that for linear discriminant analysis the number of observations per group should be at least 3 times the number of biomarkers considered, but recommendations of up to 10 times can also be found (Ferber, 1981; Romeder, 1973).

BIOLOGICAL MODELING

As mentioned previously, stochastic models can be purely empirical or motivated by the subject matter. A purely empirical model is not based on any reasoning about underlying biological mechanisms; it simply tries to emulate a factor–response relationship under some, generally mild, restrictions on the smoothness of this relationship. Hastie, Tibshirani, and Friedman (2009) give an excellent overview of this topic in section 5 of their book.

If we have knowledge of the underlying biological mechanisms, such empirical modeling is inefficient. For example, if we know that the concentration of a substance in the blood of a patient, considered as a function of time after intake, follows a one-compartmental model, we should restrict attention to time–response relationships that conform to this model.

Knowledge of such underlying mechanisms is problem specific and ranges from very precise to very vague. Transformation to linearity is a very common example of the latter. In fact, this case is so common that it is usually not considered a case of biological modeling. However, use of a logarithmic transformation is based on the general observation that growth is often exponential such that a logarithmic transformation renders growth data linear. This observation is then extended to deal with phenomena that reveal a skewed distribution. Many lab parameters like creatinine and serum ferritin are examples of responses that are routinely log-transformed before analysis.

From the statistician's point of view, a biological phenomenon that can be linearized in this fashion is relatively easy to analyze, since standard linear model theory (analysis of variance [ANOVA], analysis of covariance [ANCOVA], and linear regression) can be applied to the data. Phenomena that are nonlinear in nature *and* cannot be linearized easily are much more challenging. Rather than attempting an exhaustive overview of the types of nonlinear models used in the various fields, let us consider two illustrative and important examples and discuss some salient features. More complete overviews are given by Davidian (2009), Ratkowsky (1990), and Seber and Wild (1989).

Example 1: Pharmacodynamics

Studies have investigated the effect of metformin on the level of hemoglobin A_1C (HbA_1C) in patients suffering from diabetes mellitus. Metformin lowers HbA_1C. In addition, within the range of doses investigated, the lowering effect increases as the dose increases, but with a diminishing return. Relatively little is known beyond these rather vague notions. In this situation, the so-called Emax model is very popular. It assumes that the response (lowering of HbA_1C) has an expected value given by the Michaelis–Menten equation $E_0+(E_{max}\cdot d)/(ED_{50}+d)$, where d denotes dose. The three parameters of the model are E_{max}, the maximum attainable effect; ED_{50}, the dose at which half of the maximum effect is achieved; and a placebo effect E_0 (this can be set to 0 in many applications). Figure 6.1 shows the shape of Emax–dose–response relationships. The shape is primarily determined by the ED_{50}. The smaller the ED_{50}, the more bent the curve.

The Emax model can be derived from enzyme kinetics following the Michaelis–Menten model. The idea behind this is, however, rather heuristic. Such models are sometimes called semimechanistic.

Example 2: Pharmacokinetics

In pharmacokinetic analyses, more sophisticated approaches are common. This field deals with the modeling of concentration–time profiles. One- and two-compartment models are most frequently used for this purpose (Rowland and Tozer, 1995). The most popular model is the one-compartment model with first-order absorption and elimination:

$$C(t) = d_0 \cdot \frac{f}{V}\frac{k_e k_a}{(k_a - k_e)}(\exp(-k_e t) - \exp(-k_a t))$$

Here, $C(t)$ denotes the concentration at time t; k_e and k_a represent the elimination and absorption rate constant, respectively; d_0 represents the initial dose; f represents the fraction of the drug absorbed; and V represents the volume of distribution. This model is derived from the assumption that the rate of absorption (into the compartment) is proportional to the amount available for absorption and that the rate of elimination (from the compartment) is proportional to the concentration. Figure 6.2 shows a typical concentration–time profile derived from this model.

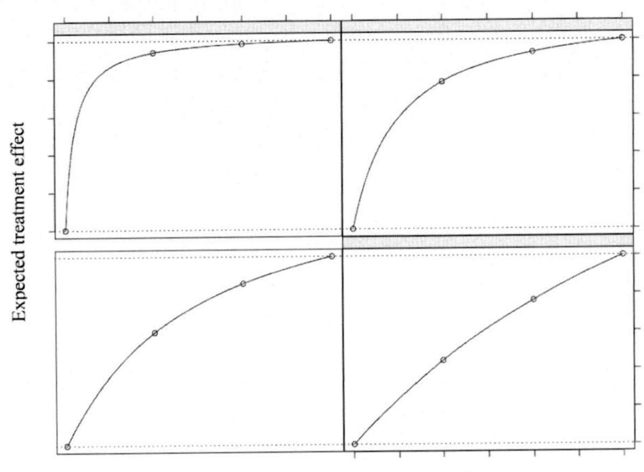

FIGURE 6.1 Emax model shapes for different ED_{50}.

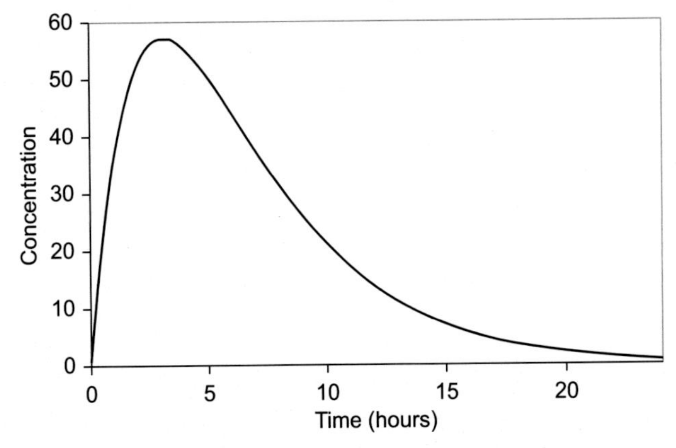

FIGURE 6.2 A one-compartment model.

Models like these are deterministically based on assumptions about how a biological mechanism will work. In example 1, one of the key assumptions is that the mechanism follows a law of mass action. The assumptions in example 2 include not only the proportionality of rates but also the key assumption of dose proportionality—that is, the assumption that doubling the dose, for example, will shift the entire concentration–time curve upward by a factor of two. In reality, these assumptions may or may not be fulfilled. In addition, even if we believe that the models are "true," the parameters of the models have to be estimated. If we have some concrete data, like concentration profiles from the plasma samples of a number of test animals, these data will not follow our assumed model perfectly. Hence, just like with linear regression, we are forced to find the best fit.

There is much room for sophistication in both the deterministic and the stochastic parts of the model. In the deterministic part, we may want to generalize from one-compartment models to multicompartment models (Reddy et al., 2005). In the pharmacodynamic example 1, a Hill parameter can be added if the effect of the dose on the HbA_1C level is below the prediction from the Emax model at low doses.

Regarding the stochastic part, the simplest approach would be to use, for example, the method of least squares to find the Emax curve that shows minimum quadratic deviation from the observed HbA_1C values. This approach requires no specific assumption about the distribution of an error term in the model, but if we assume a normally distributed additive measurement error, then the least-squares solution coincides with the maximum-likelihood (ML) estimate of the Emax profile. If patient i receives dose d_i, the model is

$$\text{reduction in HbA}_1\text{C}\,(i) \;=\; E_0 + (E_{\max}\cdot d_i)\,/\,(\text{ED}_{50} + d_i) \;+ \varepsilon_i$$

with independent, identically normally distributed errors ε_i.

If we desire something closer to "reality," we might want to generalize this model. For example, it might be assumed that the errors of repeated measures from the same patient are correlated, or we might regard the model parameters themselves as random variables with an associated random error. The latter assumption leads to so-called (random effects) nonlinear mixed models.

Various software packages are available to fit such models. NONMEM (http://www.iconplc.com/technology/products/nonmem/) allows users to specify a pharmacokinetic model in terms of differential equations. The software solves the differential equations numerically and estimates the parameters by ML (or, to be more precise, by approximations of the ML solutions). In other software packages like SAS PROC NLMIXED (http://support.sas.com/documentation/) or the R-procedure nlme (http://cran.r-project.org/web/packages/nlme), the model must be specified in terms of the model equation. SAS PROC NLMIXED and R nlme do not allow users to specify the model in terms of differential equations, but they use fitting algorithms that are superior to those implemented in NONMEM. Combining nlme and nlmeODE (Tornøe et al., 2004) in R allows the same functionality as NONMEM.

Some words of caution are in order. As mentioned previously, there is a trade-off between the complexity—or rather its absence—of a model and its "closeness to the truth." In translational medicine, there are usually relatively few observations available to fit a model. With sparse data, however, it is impossible to check the correctness of model assumptions, and problems with overfitting quickly outweigh an apparent "realism." Ratkowsky (1990) calls this "the trap of overgenerality" and applies the principle of Occam's razor: "Other things being equal, parsimony is more likely to reflect the truth than its opposite."

To give a rather simplistic example with observations at just two dose levels, we cannot estimate an Emax model, because the number of parameters is larger than the number of dose levels, and the model is overparametrized. Even if we have more dose levels, the data may be such that ML estimates do not exist, are not unique, tend to infinity, or are on the boundary of the parameter space, thus invalidating asymptotic properties, which are usually relied on to generate confidence intervals and statistical tests.

If a numerical optimization algorithm fails to converge, the reasons for this can be manifold. There may be numerical issues with the algorithm itself (like poor start values or inappropriately wide or small steps within the iterations). Often, however, the issue is with the model assumptions (overparametrization or "near" overparametrization). It can even happen that a model describes the data well, but, due to an unfortunate constellation of observations, it cannot be fit, although there is no indication that it is "wrong." Figure 6.3 gives a simple illustrative example in which the observed response at four dose levels is such that the slope of a logistic (S-shaped) fit through the points becomes infinity and the ED_{50} becomes nonunique, although there is little indication of any lack of fit.

If convergence of the fitting algorithm is successful, overparametrized models might still suffer from a tendency toward instability: minor changes in the observed data might cause dramatic changes in the model fit with corresponding uncertainty about the conclusions derived from the model. To less-than-careful modelers, this will not be obvious from measures of variation such as standard errors and confidence intervals. These are usually derived under the assumption that the model is correct and are thus far too narrow in these situations.

In pharmacokinetics, the focus of interest sometimes is on a specific parameter rather than on the entire concentration–time profile. The most popular ones are the area under the curve (AUC), the maximum concentration (C_{max}), the time of maximum concentration (t_{max}), and the compound's half-life ($t_{1/2}$). These parameters can be either derived from one of

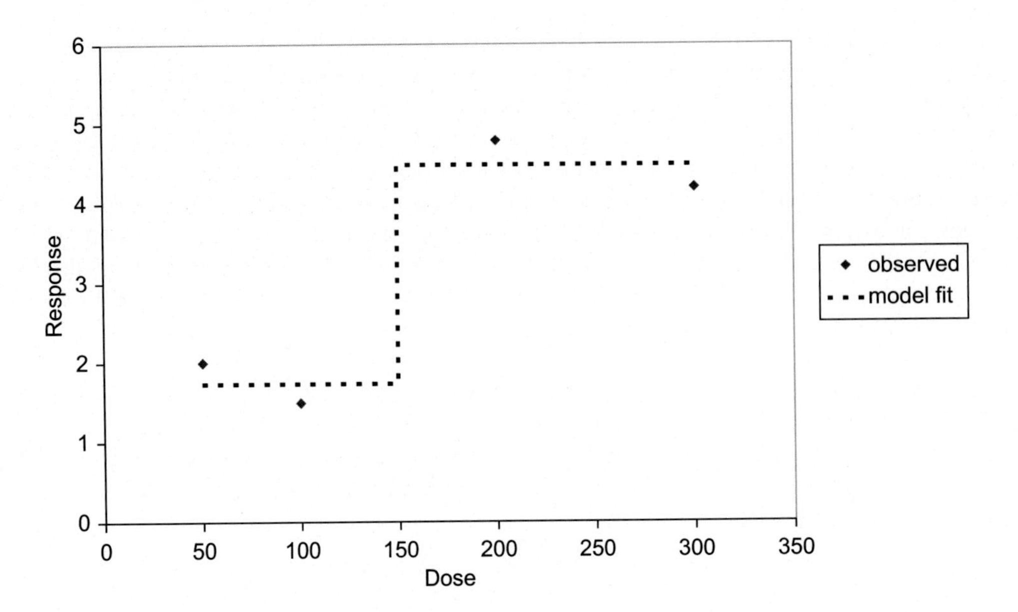

FIGURE 6.3 A logistic model for dose–response. The results at the four dose levels are such that the ED_{50} cannot be uniquely estimated and the slope parameter becomes infinity.

the parametric models described so far (e.g., the one-compartment model) or estimated nonparametrically from observed data. The well-known trapezoidal rule for estimating the AUC is an example of the latter approach. The advantages of such nonparametric methods include less specific model assumptions, hence more robustness. Apart from the limited scope, the disadvantages include less precision if the model is (approximately) correct and some uncertainties regarding the estimated quantities (e.g., Is C_{max} really close to one of the points in time where we have measurements? Which time points "in the tail" in Figure 6.2 should be excluded from calculating AUC?).

The reasons for the ubiquity of nonparametric estimates are in part historical. Today's computer power enables us to perform numerical optimizations and, if everything else fails, simulations that a generation ago were inconceivable. In the past decades, a lot of work went into the algorithms to fit more sophisticated parametric models, and many advances were made (chapter 20 in Molenberghs and Verbeke, 2005; Pinheiro and Bates, 2000).

Still, the balance between simplicity and complexity is a delicate one. The availability of powerful computer packages is no safeguard against their misuse, and fitting complex nonlinear models remains an art that requires an in-depth knowledge of biology, numerical optimization methods, and mathematical statistics. The combination of these skills in a single individual is rare. In accord with our background in biostatistics, we feel it is appropriate to highlight one specific danger: the illusion that all attention can be focused on development of a "realistic" model because some black box optimizer implemented in a software package will churn out the parameter estimates with confidence intervals and significance tests. The reality is not as simple as that. However, a close collaborative effort of teams bringing together experts from all fields can achieve much.

STATISTICAL MODELS

Almost all statistical models used in translational medicine fall into one of the following categories:

- Models for a continuous and normally distributed response: linear model (regression, ANOVA, and ANCOVA) (McCulloch, Searle, and Neuhaus, 2008): such models can be written as

$$y = X\beta + \varepsilon,\ \varepsilon \sim N(0, \Sigma),$$

  where y is the vector of responses, X is the matrix of covariates, and ε is an error vector assumed to be normally distributed with the variance–covariance matrix Σ. The t-test is one of the simplest examples of a test for the significance of a parameter in such a model. The category includes linear mixed-effects models (the mixed-effects model occurs when more than one component of Σ is unknown and must be estimated). The model is often justified asymptotically by the central-limit theorem.

- Models for a response that is continuous but not normally distributed:
  - Transformation to normality (e.g., Box–Cox; see, e.g., Box, Hunter, and Hunter, 2005)
  - Nonparametrics or transformation to ranks or rank-based scores (e.g., Conover, 1999)
  - Nonlinear modeling (e.g., compartmental models, Emax) (Seber and Wild, 1989)
- Models for a categorical (often binary) response (Agresti, 2013):
  - Generalized linear model (GLM), usually logistic regression, sometimes Poisson regression, or other GLM link functions: cloglog, negative binomial.
  - Many tests of contingency tables; e.g., the chi-square test, Fisher's exact test, and the Cochran–Mantel–Haenszel test are special cases of parameter tests in these models.

REFERENCES

Agresti, A., 2013. Categorical data analysis, 3rd ed. Wiley, New York.

Alonso, A., Molenberghs, G., 2007. Surrogate marker evaluation from an information theory perspective. Biometrics 63, 180–186.

Benjamini, Y., Hochberg, Y., 1995. Controlling the false discovery rate: a practical and powerful approach to multiple testing. J. R. Stat. Soc. (B) 57, 289–300.

Biomarkers Definition Working Group, 2001. Biomarkers and surrogate endpoints: preferred definitions and conceptual framework. Clin. Pharmacol. Ther. 69, 89–95.

Box, G.E.P., Hunter, J.S., Hunter, W.G., 2005. Statistics for experimenters: design, innovation, and discovery, 2nd ed. Wiley, New York.

Bretz, F., Pinheiro, J.C., Branson, M., 2005. Combining multiple comparisons and modeling techniques in dose-response studies. Biometrics 61, 738–748.

Cheng, Y., Su, F., Berry, D.A., 2003. Choosing sample size for a clinical trial using decision analysis. Biometrika 90, 923–936.

Conover, W.J., 1999. Practical nonparametric statistics, 3rd ed. Wiley, New York.

Cox, D.R., Hinkley, D.V., 1974. Theoretical statistics. Chapman & Hall/CRC, London.

Davidian, M., 2009. Non-linear mixed-effects models. In: Fitzmaurice, G., Davidian, M., Verbeke, G., Molenberghs, G., (Eds.), Longitudinal data analysis. Chapman & Hall/CRC Press, Boca Rotaon, pp. 107–141. Chapter 5.

Diggle, P.J., Heagerty, P., Liang, K.Y., Zeger, S.L., 2002. Analysis of longitudinal data, 2nd ed. Oxford University Press, Oxford.

Dunnett, C.W., 1955. A multiple comparison procedure for comparing several treatments with a control. J. Amer. Statist. Assoc. 50, 1096–1121.

Efron, B., 2010. Large-scale inference. Cambridge University Press, London.

Ferber, G., 1981. Automatic description of EEG background activity. Meth. Inf. Med. 20, 32–37.

Finner, H., Straßburger, K., 2006. On δ-equivalence with the best in k-sample models. J. Am. Stat. Assoc. 101, 737–746.

Fitzmaurice, G.M., Laird, N.M., Ware, J.H., 2011. Applied longitudinal data analysis, 2nd ed. Wiley, New York.

Halpern, J., Brown, B.W., Hornberger, J., 2001. The sample size for a clinical trial: a Bayesian-decision theoretic approach. Stat. Med. 20, 841–858.

Halpern, S.D., Karlawish, J.H., Berlin, J.A., 2002. The continuing unethical conduct of underpowered clinical trials. JAMA 288, 358–362.

Hastie, T., Tibshirani, R., Friedman, J., 2009. The elements of statistical learning, 2nd ed. Springer, New York.

Hoaglin, D.C., Mosteller, F., Tukey, J.W. (Eds.), 1985. Exploring data tables, trends and shapes. Wiley, New York.

Hsu, J.C., 1996. Multiple comparisons: theory and methods. Chapman and Hall, London.

International Conference on Harmonisation of Technical Requirements for Registration of Pharmaceuticals for Human Use, 2005. E14: the clinical evaluation of QT/QTc interval prolongation and proarrhythmic potential for non-antiarrhythmic drugs. http://www.ich.org/fileadmin/Public_Web_Site/ICH_Products/Guidelines/Efficacy/E14/E14_Guideline.pdf (accessed 28.07.14.).

Langsrud, Ø., 2005. Rotation tests. Stat. Comp. 15, 53–60.

Läuter, J., Glimm, E., Eszlinger, M., 2005. Search for relevant sets of variables in a high-dimensional setup keeping the familywise error rate. Stat. Neerlandica. 59, 298–312.

Maindonald, J., Braun, J., 2007. Data analysis and graphics using R: an example-based approach, 2nd ed. Cambridge University Press, Cambridge.

Marcus, R., Peritz, E., Gabriel, K.R., 1976. On closed testing procedures with special reference to ordered analysis of variance. Biometrika 63, 655–660.

Mayo, M.S., Gajewski, B.J., 2004. Bayesian sample size calculations in phase II clinical trial using informative conjugate priors. Controlled Clin. Trials 25, 157–167.

McCulloch, C.E., Searle, S.R., Neuhaus, J.M., 2008. Generalized, linear, and mixed models, 2nd ed. Wiley, New York.

Molenberghs, G., Verbeke, G., 2005. Models for discrete longitudinal data. Springer, New York.

Peng, J., Lee, C.I., Liu, L., 2006. Max-min multiple comparison procedure for comparing several dose levels with a zero dose control. J. Appl. Stat. 33, 549–555.

Pezeshk, H., Gittins, J., 2002. A fully Bayesian approach to calculating sample sizes for clinical trials with binary responses. Drug Inform. J. 36, 143–150.

Pezeshk, H., Gittins, J., 2006. A cost-benefit approach to the amount of experimentation in clinical trials. Drug Inform. J. 40, 407–411.

Pigeot, I., Schäfer, J., Röhmel, J., Hauschke, D., 2003. Assessing non-inferiority of a new treatment in a three-arm clinical trial including a placebo. Stat. Med. 22, 883–899.

Pinheiro, J.C., Bates, D.M., 2000. Mixed effects models in S and Splus. Springer, New York.

Pinheiro, J.C., Bornkamp, B., Glimm, E., Bretz, F., 2014. Model-based dose finding under model uncertainty using general parametric models. Stat. Med. 33, 1646–1661.

Core Team, R., 2014. R: A language and environment for statistical computing. R Foundation for Statistical Computing. URL Austria, Vienna. http://www.R-project.org/ (accessed 06.08.14.).

Ratkowsky, D.A., 1990. Handbook of nonlinear regression models. Marcel Dekker, New York.

Reddy, M., Yang, R.S., Andersen, M.E., Clewell III, H.J., 2005. Physiologically based pharmacokinetic modeling: science and applications. Wiley-Interscience, New York.

Romeder, J.M., 1973. Méthodes et programmes d'analyse discriminante. Dunod, Paris.

Rowland, M., Tozer, T.N., 1995. Clinical pharmacokinetics: concepts and applications, 3rd ed. Williams & Wilkins, Baltimore.

Seber, G.A., Wild, C.J., 1989. Nonlinear regression. Wiley, New York.

Senn, S., 1993. Cross-over trials in clinical research. Wiley, Chichester.

Senn, S.J., 1996. Some statistical issues in project prioritisation in the pharmaceutical industry. Stat. Med. 15, 2669–2702.

Senn, S., Bretz, F., 2007. Power and sample size when multiple endpoints are considered. Pharm. Stat. 6, 161–170.

Spiegelhalter, D.J., Abrams, K.R., Myles, J.P., 2004. Bayesian approaches to clinical trials and health-care evaluation. Wiley, Chichester.

Storey, J.D., 2003. The positive false discovery rate: a Bayesian interpretation and the q-value. Ann. Stat. 6, 2013–2035.

Storey, J.D., Tibshirani, R., 2003. Statistical significance for genomewide studies. Proc. Natl. Acad. Sci. USA 100, 9440–9445.

Sutton, A.J., Cooper, N.J., Jones, D.R., Lambert, P.C., Thompson, J.R., Abrams, K.R., 2007. Evidence-based sample size calculations based upon updated meta-analysis. Stat. Med. 26, 2479–2500.

Takeuchi, K., 1973. Studies in some aspects of theoretical foundations of statistical data analysis. Toyo. Keizai. Shinpo-sha, Tokyo (in Japanese).

Tornøe, C.W., Agersø, H., Jonsson, E.N., Madsen, H., Nielsen, H.A., 2004. Non-linear mixed-effects pharmacokinetic/pharmacodynamic modelling in NLME using differential equations. Comput. Methods Programs Biomed. 76, 31–40.

Tufte, E.R., 2001. The visual display of quantitative information, 2nd ed. Graphics Press, Cheshire, CT.

Vickers, A.J., 2001. The use of percentage change from baseline as an outcome in a controlled trial is statistically inefficient: a simulation study. BMC Med. Res. Methodol. 1, 6.

Weiss, N.A., 2008. Introductory statistics, 8th ed. Addison-Wesley, Boston.

Westfall, P.H., 2011. On using the bootstrap for multiple comparisons. J. BioPharm. Stat. 21, 1187–1205.

Westfall, P.H., Young, S.S., 1993. Resampling-based multiple testing: examples and methods for p-value adjustment. Wiley, New York.

Williams, D.A., 1972. The comparison of several dose levels with a zero dose control. Biometrics 28, 519–531.

Yin, Y., 2002. Sample size calculation for a proof of concept study. J. BioPharm. Stat. 12, 267–276.

Chapter 7

Intellectual Property and Innovation in Translational Medicine

Palmira Granados Moreno and Yann Joly[1]

INTRODUCTION

The large sums of private and public money devoted to research and development and the scientific and technological advancements in medicine have not yet translated into many clinical products (Bornstein and Licinio, 2011; FitzGerald, 2005). Translational medicine intends to find ways to remedy these delays and encourage a faster, less expensive, and more effective translation from basic research done in the labs to clinical applications that benefit patients. Intellectual property can contribute to this process. In order to provide an overview of the way in which intellectual property can be of assistance in the medical translation process, this chapter describes details about patents both in general and in relation to specific medical tools and products. Since access to information and to research tools, as principles of open science exalt, can also be instrumental in translational medicine to further medical innovation, this chapter also refers to principles of open science that promote such access. Finally, it also discusses the importance of collaboration and partnerships and of combining practices of patenting with principles of open science to achieve that collaboration and those partnerships.

CONTEXT

General Description of Translational Medicine

Translational medicine (TM) is the branch of medical research whose purpose is to facilitate a faster, easier, and less expensive transition between preclinical medical research (also known as basic research) and clinical applications that could cure diseases in humans, thus allowing the achievement of the social value of science (i.e., to be the source of useful products that improve people's lives) (Joly, Livingstone, and Dove, 2012; Wehling, 2010). The translation process can be understood as comprising three stages. The first stage translates from basic science (laboratory—in vitro and in animals) to potential clinical applications. The second stage involves the translation of the proposed human applications into proven clinical applications. In this stage, drug development and safety and efficacy evaluations (e.g., clinical trials) take place. The third stage translates the proven clinical applications into the implementation and adoption of clinical practices (Drolet and Lorenzi, 2011). The list of issues that need to be addressed in the execution of TM includes scientific research (e.g., biological, genetic, medical, biomedical), funding and business decisions, regulatory systems (including clinical trials, among other issues), legal and ethical frameworks, communication and networking among the different parties, product development, and intellectual property (Albani and Prekken, 2009; Prestwich et al., 2012).

One of the most common barriers to the translation of biomedical research is that many researchers, particularly those in academia, tend to focus on areas or projects in which they are interested for their scientific value (*l'art pour l'art*) and on writing papers with a strong scientific impact. This research activity does not necessarily result in clinical applications (Wehling, 2010). TM could influence the focus of some researchers by suggesting areas or projects with higher probabilities of translation into clinical applications or with higher social value. Another problem is the high costs associated with

1. The authors would like to acknowledge the editorial assistance of Katie Saulnier from the Centre of Genomics and Policy, McGill University, and the financial support of the Fonds de Recherche du Quebec-Santé (FRQ-S).

Principles of Translational Science in Medicine. http://dx.doi.org/10.1016/B978-0-12-800687-0.00030-X

the development of innovative drugs, diagnostic tests, and medical devices. On the one hand, academic research is expensive and usually publicly funded. The high costs associated with academic research are likely to make exclusive public funding unsustainable. On the other hand, among the sources of the highest costs taken on by the industry are typically the clinical trials that must be undertaken to determine the efficacy and safety of a particular product on human beings. TM can help to broaden the sources of funding by facilitating early collaboration and partnerships between academia and industry, making the process less burdensome for the parties involved.

Strategies and practices of intellectual property (as part of a broader scheme of commercialization[2]) and principles of open science can facilitate collaborations and partnerships among research groups, and thereby contribute to the achievement of TM tasks and objectives. In addition to the obvious scientific and technological background required, execution of the aforementioned combination involves knowledge and concepts from health policy, health economics, law, and social marketing communication (Glasgow, 2009). Researchers, government, biotechnology and pharmaceutical industries, physician scientists, regulators, policymakers, investors, business developers, nongovernmental organizations, and funding agencies are some of the parties that would benefit from these practices and strategies (Translational Medicine, 2014; Bubela, Fitzgerald, and Gold, 2012; Wehling, 2010).

Intellectual Property and TM

One of the most successful ways to translate basic research into clinical applications is through collaboration and partnerships from the early stages of basic research (bench) to the stages of production of clinical applications (bedside). Early partnerships of biotechnology and pharmaceutical industries with universities and public research centers could enable all the parties to jointly make decisions on which research projects to develop and the steps that should be taken, despite their well-known differences in incentives and viewpoints concerning innovation. This could lessen the problem of having many academics focusing on research that does not necessarily translate into clinical applications (Bornstein and Licinio, 2011). Early collaborations between basic and clinical researchers and the sharing of knowledge and findings obtained from new and old R&D processes in the in vitro or animal and human stages of the research may permit researchers to implement shorter, less expensive, and more successful clinical trials (Wehling, 2006). Collaborations and partnerships could also enable the sharing of costs and risks associated with the R&D process between basic researchers (e.g., universities) and clinical researchers (e.g., biotechnology and pharmaceutical companies), which would eventually make it less burdensome for both. Partnerships and commercialization strategies can encourage continued public funding and attract other sources of funding, including private ones.

Forms of intellectual property (e.g., patents, licenses, patent pools[3]) developed to protect the findings and resulting inventions can facilitate collaborations and partnerships (Caulfield, Harmon, and Joly, 2012; Gold, 2008; Joly, Livingstone, and Dove, 2012). Intellectual property rights also help to assign market value to different products resulting from research. Market value can favor market expectations regarding the outcomes of research, thereby incentivizing actors to create and join collaborations and partnerships. The market value assigned to the various products also facilitates negotiations during the creation and development of collaborations and partnerships by determining the value of the collaborators' and partners' contributions[4] (Bornstein and Licinio, 2011; FitzGerald, 2005).

Basic Functioning of Patents[5]

Patents, along with trade secrets and licensing agreements, are the dominant form of intellectual property and the most commonly used tools associated with biotechnology and pharmaceutical research, development, and commercialization (Bubela, Fitzgerald, and Gold, 2012; Joly, Livingstone, and Dove, 2012). Patents grant their holders a 20-year[6] set of

---

2. *Commercialization* is defined as "the process of extracting economic value from new products, processes and knowledge through the use of intellectual property rights, licensing agreements, and the creation of spin-off companies" (Joly et al., 2010).
3. A *patent pool* is a set of agreements between two or more patent holders to license one or more of their patented inventions to one another or to third parties (Andrews, 2002; van Zimmeren et al., 2011).
4. The drawbacks of certain practices of intellectual property are introduced at the end of the following section.
5. The overview herein described covers the way patents are regulated in the United States, Canada, Mexico, and Europe. It makes references to the Canadian Patent Act (2013), the European Patent Convention (2013), the Mexican Industrial Property Law (2012), the U.S. Patent Act (2014), the Canadian Manual of Patent Office Practice (Canadian Patent Office, 1998), the European Directive 98/44/EC on biotechnological inventions (European Parliament, 1998), and the European Patent Office Guidelines (European Patent Office, 2012).
6. This 20-year period starts on the date on which the patent application is filed (Canadian Patent Act, 2013; European Patent Convention, 2013; Mexican Industrial Property Law, 2012; U.S. Patent Act, 2014).

exclusive rights to make, use, sell, and import the invention they protect. Inventions[7] can be patented provided that they are new, they result from an inventive activity, they are useful (or have a utility or industrial application in some jurisdictions), and they fall within what constitutes patentable subject matter (Canadian Patent Act, 2013; Canadian Patent Office, 2010; European Patent Convention, 2013; Mexican Industrial Property Law, 2012; U.S. Patent Act, 2014; Vaver, 2011).

An invention is considered new if it has not been described, disclosed, or predicted in the national or international prior art by the day on which the patent application is filed or within the 12 months prior to that filing date. Prior art in medical research includes written documents (e.g., scientific/technical articles, books, newspaper's or magazine's commentaries, and teaching materials), patent applications, oral presentations, conferences, and Internet posts available to the public. In order to maintain the invention's novelty, it is recommended that the researchers make information describing the invention available to the public only when the patent application has been filed or within the 12 months prior to such a filing. It is also recommended that they use confidentiality agreements to protect the novelty of the invention by contractually preventing those individuals with access to the information related to the invention from disclosing it.

An invention results from an inventive activity[8] (or is nonobvious[9]) when the inventor employed some creativity or ingenuity while devising and producing the invention and an unimaginative person skilled in the art does not see the invention as the next obvious thing to happen in the technical or scientific field. In this context, a person skilled in the art would be a fictional researcher, physician, or clinician with ordinary scientific expertise and access to publicly available information associated with the invention. As in the case of novelty, this requirement is evaluated at the date on which the patent application is filed.

An invention is considered useful[10] (or susceptible to having an industrial application[11]) when it has a practical utility or it can be produced or used for the purposes described in the patent application. The utility, usefulness, or industrial application is either demonstrated or can be soundly predicted at the date on which the patent application is filed. For instance, a drug is patentable as a product and its active ingredient as a composition of matter provided that they treat or are predicted to be able to treat a specific disease in a new and nonobvious way. This association to a specific disease or disorder constitutes the utility or industrial application.

Four categories of innovations relevant to medical innovation are commonly considered outside the patentable subject matter, and are therefore excluded from patent protection in most jurisdictions (Canadian Patent Office, 2010; European Patent Office, 2012; Mexican Industrial Property Law, 2012; World Trade Organization, 1994; Vaver, 2011). The first category of innovations excluded from patentability includes essentially biological processes for the production, reproduction, and propagation of plants and animals.[12] Higher life forms are the second category. This exception, however, does not exist in all jurisdictions. Whereas Canada considers higher (multicellular) forms live organisms, such as the oncomouse,[13] unpatentable [*Harvard College v. Canada* (Commissioner of Patents) 2002]; in the United States, nonnaturally ocurring, nonhuman multicellular living organisms, including animals such as the oncomouse, are patentable.

The third category of innovations excluded from patentability includes biological and genetic material as found in nature and discoveries that merely reveal something that exists in nature, regardless of whether they are known. The exclusion from patentability in these two cases is explained by the lack of human inventiveness or ingenuity in materials and discoveries that already exist in nature. As we discuss later, this category of innovations is relevant in translational medicine because the main issue debated in cases involving gene patents has been whether isolated gene sequences fall within this exception.

The last category of exceptions includes surgical and therapeutic methods[14] used in human beings or animals.[15] Methods involving the excision of tissue, organ, or tumor samples from the body are considered forms of surgery and therefore

7. *Invention* is defined as any human creation (e.g., art, process, machine, manufacture, composition of matter) that transforms matter or energy that already exists in nature for the benefit of men so they can satisfy their needs (Canadian Patent Act, 2013; Mexican Industrial Property Law, 2012; U.S. Patent Act, 2014).

8. European Patent Convention, 2013.

9. Canadian Patent Act, 2013; Mexican Industrial Property Law, 2012; U.S. Patent Act, 2014.

10. Canadian Patent Act, 2013; U.S. Patent Act, 2014.

11. European Patent Convention, 2013; Mexican Industrial Property Law, 2012.

12. Plants Breeders' Rights (another form of intellectual property) usually protect this type of process. This exclusion prevents multiple intellectual property rights on the same advance or invention.

13. The oncomouse was a genetically modified mouse created at Harvard University to carry an activated gene that significantly increases the mouse's susceptibility to cancer used in cancer research.

14. Some jurisdictions, such as Europe, also exclude diagnostic methods practiced on the human body from patentability. This exception, however, does not extend to diagnostic tests performed on body tissues or fluids after they have been removed from the human body, provided that those tissues or fluids are not returned to the human body (European Patent Office, 2012; European Patent Convention, 2013).

15. For purposes of this exception, a therapeutic or medical method cures, prevents, or ameliorates an ailment or pathological condition or treats a physical abnormality or deformity. The reason behind this exception is to prevent having patent rights interfering with saving a human or animal life.

unpatentable. This exception, however, does not extend to drugs and products for treating human or animal illnesses, which are patentable, or to purely cosmetic treatments.

Patents are applied for on a country-by-country basis, and they have only national coverage of protection. In each country, the application has to comply with national legal and administrative requirements. This is true even in the specific case of a European patent.[16] Even though there are no patents with international coverage, a PCT application[17] filed at the moment of filing the first patent application allows all the patent applications filed in different countries to have the same filing (priority) date (Patent Cooperation Treaty, 2001; Vaver, 2011).

Four types of patents are associated with medical research: patents on products (e.g., drugs, diagnostic tests), patents on processes or methods (e.g., methods that test genetic alterations or abnormalities, methods of extraction, or manipulation of genes), patents on compositions of matter (e.g., inventions that combine different types of genetic material, inventions concerning genes and the role they play in the development of diseases or in regulating bodily functions), and patents on improvements (Cook-Deegan, 2008; Resnik, 2004; Vaver, 2011).

A patent holder can be a person, a group of people, or a company/organization/institution. Patent holders have the right to exclusively make, use, sell, and/or import their patented invention. Patent rights may be assigned[18] or licensed[19] to third parties, thus allowing patent holders to profit from and commercially exploit their patented inventions.

Users (e.g., researchers) are allowed to use patented inventions without needing to obtain the patent holder's authorization (known as users' rights) in the following cases. Researchers can use a patented invention for experiments and research; however, the concept of "research" in this context is vague, and the extent of use allowed changes from jurisdiction to jurisdiction.[20] Generic manufacturers can use the patented invention to apply for a generic drug's regulatory approval to enter the market and to submit supporting samples before the patent of the original drug expires.[21] Developed countries are allowed to export and less-developed countries are allowed to obtain a compulsory license to import patented drugs.[22] The purpose of users' rights in general is to balance the interests of both parties (patent holders and users), limit the extent of control granted to the patent holders, and ensure the promotion of innovation. Additionally, in the area of medicine, users' rights also intend to promote access to health care. In addition to the above-mentioned users' rights, the patent system provides two cases in which a compulsory license might be granted: when the government needs to address a national emergency or extreme urgency (e.g., public health crisis) and when the patent holder is considered to be abusing his or her patent.[23]

---

16. A European patent is filed before the European Patent Office (EPO) in accordance with the European Patent Convention. A European patent does not grant international protection. For a European patent to have effect and be enforceable in any European country, it is required that a translation of the European patent granted be filed, evaluated, and granted in accordance with national law by each of the European countries where patent protection is sought.

17. A PCT patent application is a patent application filed in accordance with the Patent Cooperation Treaty. For more information, see PCT FAQ at World Intellectual Property Organization online: http://www.wipo.int/pct/en/faqs/faqs.html.

18 Assignments transfer ownership on the patented invention (and all the rights granted by the patent) to a third party. They may be partial or total and must be in writing.

19. Licenses grant a third party consent or permission to use, make, sell, or import the patented invention. They may be sole, exclusive, or nonexclusive. Exclusive licenses need to be in writing, whereas others may be oral. Licenses may be gratuitous or for a royalty.

20. The reason behind this right is to allow and encourage the use of the information and knowledge disclosed in the patent application to promote further innovation. This exception is valid provided that it does not conflict with the patent holder's right to commercialize his or her invention.

The uncertainty on the definition of "research" for purposes of this exemption is reflected in the different approaches taken in different jurisdictions. In Canada, fair research, development, and experimentation using a patented invention is allowed without the need of the patent holder's authorization until the moment on which the product is ready to be shown to a prospective customer or the product is approved to be commercialized (which is when commercialization is considered to begin). In the United States on the other hand, university research does not fit in this category of allowed use because faculty members' research is considered to be furthering the university's business (Madey v. Duke University, 2002; Organisation for Economic Co-operation and Development, 2004; Resnik, 2004; Vaver, 2011).

21. The purpose of this user's right or exception is to avoid extending the monopoly that a patent grants beyond the stated term of 20 years. With this user's right, the generic manufacturer can start marketing the generic version of the patented drug right after the patent expires.

22. The purpose of this right is to address problems in accessing health care in least developed countries (Agreement on Trade Related Aspects of Intellectual Property Rights 1995; Doha Declaration on the TRIPs Agreement, 2001; Vaver, 2011).

23. A patent holder is considered to be abusing his or her patent when he or she fails to meet the local demand for a patented product, when he or she unreasonably refuses to grant a license, or when he or she imposes abusive terms in a license. In both cases, the government could grant a nonexclusive compulsory license to use the patented invention without the patent holder's authorization for the scope and duration of the purpose for which it was authorized (e.g., health national emergency), and give the patent holder a reasonable fee (Agreement on Trade Related Aspects of Intellectual Property Rights, 1995; Andrews, 2002; Vaver, 2011).

The patent application comprises a description (also known as disclosure) of the invention[24]; drawings, graphics, or blueprints that complement the above-mentioned description; a summary or abstract of the description; and one or more claims that define and delimit the scope and extent of the protection requested. In patent applications where gene sequences are involved, it is necessary to include a list of the relevant nucleotide or amino acid sequences. In case the invention refers to or includes biological material that cannot be described, a deposit of such biological material with the International Depository Authority is required by some patent offices (Canadian Patent Office, 2010; European Patent Office, 2012).

Importance of the Patent System

The importance of the patent system lies in the innovation that it is expected to promote. First, the patent system is said to promote innovation by giving researchers and investors economic incentives to engage in innovative research projects. For investors, the economic incentive is having the exclusive right to make, use, sell, or import the patented invention or to authorize others to do so in exchange for a royalty, which allows them to recoup their investments. For researchers, the incentive is having the exclusive right to make, use, sell, or import the patented invention and thereby being economically rewarded for their work (Castle, 2009; Jensen and Murray, 2005; Sumikura, 2009; Vaver, 2011). Second, the patent system is said to promote innovation by creating a state of the art or knowledge commons from which to build up innovation. It does so by requiring the patent holder, as part of the patent application, to disclose the description, details, components, uses, and functionality of the invention to the extent that it allows any person skilled in the art to reproduce the invention (Andrews, 2002; Jensen and Murray, 2005). Third, the patent system is said to promote innovation by facilitating information markets. Commodities such as the information and knowledge related to inventions can be integrated into patents (i.e., as exclusive economic rights with a recognized value assigned). This acquired form of "packaging" makes it easier for such commodities to be exchanged and to circulate in markets (Hope, 2008).

Despite the benefits that the patent system may bring, there are concerns about the negative impact that patents may have on research and development and, particularly with respect to medicine, on access to health care[25] (Canadian HIV-AIDS Legal Network et al., 2010; Gold, 2008; Heller and Eisenberg, 1998; Hope, 2008; Stiglitz, 2006). These concerns have contributed to a renewed interest in open science approaches.

Open Science

Definition and Principles

Open science is a movement characterized by principles of sharing and quickly disseminating knowledge in order to foster scientific process, maximize the impact of research, and meet humanitarian goals (Caulfield, Harmon, and Joly, 2012). It originated in the late sixteenth and early seventeenth centuries to make scientific research, data, and dissemination accessible to all, departing from the regime of secrecy of "nature's secrets," whose norm was to withhold knowledge from the vulgar multitude (David, 2004). According to the open science movement, seeking truth is a communal activity: "The only way in which a human being can make some approach to knowing the whole of a subject is by hearing what can be said about it by persons of every variety of opinion, and study all modes in which it can be looked at by every character of mind" (Mill, 1879).

The main principles of open science include the open sharing of data, fast dissemination of knowledge, the building of research based on the accumulation of previous findings (i.e., building on each other's work), and cooperation among scientists (Hope, 2008). These principles subscribe to Merton's norms of science (communalism,[26] universalism,[27] disinte

---

24. The description gives details about the invention and how to put it to use. It includes a recount of its components, its use, and the best-known method to reproduce the invention. All the information included in the description should be clear and thorough so as to allow a person skilled in the art to fully and unambiguously understand the invention, know its utility, and be able to reproduce it, if necessary or desired. In order to comply with this part of the description, it is recommended that the researcher keep a journal of the invention documenting every step followed throughout the whole process that led to the final outcome of the invention and the reasons for asserting the alleged industrial application. References to the current state of the art (e.g., articles, books, commentaries, patents) associated with the invention help to show that the invention is in fact new and nonobvious.

25. Some of these concerns are described more thoroughly in the following sections.

26. Communalism emphasizes the importance of commonly owned knowledge.

27. Universalism encourages keeping information open to all persons, regardless of their personal characteristics (e.g., race, class, gender, religion, nationality, profession, etc.).

restedness,[28] originality,[29] and skepticism[30]), which represent ideal institutional values that science should tend to (Hope, 2008; Merton, 1973).

Many of these principles have been integrated into a number of international policies and guidelines, such as the Organisation for Economic Cooperation and Development (OECD) Principles and Guidelines for Access to Research Data from Public Funding, Organisation of the Human Genome (HUGO) Statement on Human Genomic Databases, Genome Canada, and the U.S. National Institutes of Health Final Statement on Sharing Research Data, among others.[31]

Benefits of Open Science

Some of the arguments in favor of open science mention that sharing and making data and knowledge openly available leads to better data quality, the formation of good practice, and a sharing culture (Hope, 2008). For instance, a large network of researchers from a variety of backgrounds focusing on a specific topic from different angles and through different lenses is more likely to reach new findings, identify more mistakes, and discover more or better ways to reach the same results than a small homogeneous group of people. Open science is also helpful in avoiding unnecessary duplication of research. Openly sharing research tools and results creates an information commons, which everyone can use and build on. Fast dissemination allows this commons to be created, expanded, and updated in short periods of time. Since researchers can access what others are doing and reuse what they have discovered, they can start where others left off and devote their resources to further advancement in innovation (Caulfield, Harmon, and Joly, 2012). Furthermore, the more that information, platforms, and data are used, openly shared, and collaboratively exchanged, the more open, creative, and useful said information, platforms, and data will become (Giustini, 2006).

Renewed Interest in Open Science

With the emphasis given to the intellectual property system as the main source of incentives for researchers to engage in research projects and for investors to devote resources to those projects, the open science model and its principles have been demoted, or at least restricted. For instance, certain practices with patents are deemed contrary to principles of scientific production and diffusion for the following reasons. Given that patents assign individual exclusive property rights to use and control the patented invention (and the associated scientific and technological knowledge), patent holders have the possibility of preventing others from using the invention and the knowledge associated with it. This goes against the principle that science should be shared, cumulative, and cooperative. Moreover, researchers may delay publications related to the invention until the patent application is filed, thus withholding important information about the invention. This action is contrary to the principle that scientific products should be rapidly shared (Hope, 2008).

The open science model, however, regained popularity and its support became stronger with the occurrence of two events. The first was the development of technological tools (i.e., advanced computers, Internet, storage tools, etc.) that allow faster and easier ways to share and disseminate knowledge, findings, and data (Rentier, 2014). The second was the rise of the open source movement in the late twentieth century.[32] The open source movement highlighted the potential negative effects that the proliferation of intellectual property rights could bring to innovation[33] and therefore questioned the need and convenience of excessive individual intellectual property rights (Hope, 2008). Particularly in the field of medical research, important world health issues, such as the HIV-AIDS South African crisis,[34] have also influenced the return

---

28. Disinterestedness refers to the scientists' selfless approach to advance science and research.
29. Initially not included, originality alludes to the novelty of the contribution.
30. Skepticism encourages subjecting all knowledge to rigorous and structured trials of scrutiny, replication, and verification.
31. For more examples, see Caulfield, Harmon, and Joly (2012).
32. Open source is a production and development model based on open, free, and universal access. It stemmed from the Free Software movement at the end of the twentieth century. While it originally focused on the software industry, as it exalts the importance of maintaining the source code of a software program free (as in freedom) for everyone to use, develop, and build on, it has also been proposed in other areas such as health and agricultural biotechnology. The model is implemented by asserting intellectual property rights over the creative work and granting an open license that authorizes the nonexclusive use of the copyrighted (and sometimes patented) creative work, either to review the information associated with the work (e.g., source code), to use the work or invention, to modify it and adapt it to one's needs, or to build on it and create derivative works. As part of this open source license, the work that results from using the copyrighted/patented work (derivative work or improvement) needs to be licensed under these same terms of openness and nonexclusivity. The benefits suggested by this model include large and assorted collaboration, decentralized open production, low engagement costs, and peer review and feedback loops that will potentially lead to high-quality outcomes (Hope, 2008; Zittrain, 2004).
33. This issue is developed more thoroughly below.
34. In 1998, pharmaceutical companies who held patents over HIV-AIDS antiretroviral drugs sued the government of South Africa for importing and manufacturing their patented drugs without their permission in order to fight the AIDS epidemic (Gold, 2008).

to principles of open science and a different approach to intellectual property. The foregoing conditions have favored the adoption and implementation of different models of collaboration and partnership as new ways to combine principles of open science and forms of intellectual property, with the purpose of achieving greater levels of innovation and better access to new products and services (Gold, 2008). Open innovation is one example.

Open Innovation

Open innovation is a model of collaboration that recognizes that innovation is both more efficient and effective with the participation and collaboration of different parties. The term *open innovation* is based on the assumption that an efficient R&D process incorporates parties, ideas, resources, and strategies internal and external to the company. The purpose of the collaboration suggested is not only to accelerate the R&D process, but also to expand the markets and to maximize revenues. The commercial aspect of these two last purposes constitutes the main difference with open science, as the latter usually focuses only on the advancement of science and the R&D process. In order to achieve its purposes, the model proposes the use and trade of intellectual property rights (i.e., copyrights, patents, and trademarks) to establish collaborations and partnerships throughout the different stages of the R&D process. Through these collaborations and partnerships, the cooperating parties are able to share ideas, information, research tools, and strategies, which is likely to allow for a more cumulative and collective perspective of the R&D process. This could result in a more efficient and faster R&D process because duplication of efforts and work could be avoided and different stages of the R&D could be worked on simultaneously. Furthermore, the use and trade of intellectual property rights capitalizes the R&D process through the royalties generated by the use of the protected inventions. The open innovation model, nonetheless, has some important drawbacks. The first drawback is the complexity in determining the ownership on all the parts that comprise the invention, when it is created in collaboration with multiple parties. Likewise, practices of licensing among numerous parties can become complicated and confusing as the many terms and conditions may be difficult to fully understand and follow. A third drawback is the risks of disclosing essential and commercially valuable information while sharing information as part of the collaboration. One example of the open innovation model in medical research is the collaboration between AstraZeneca and MRC Laboratory of Molecular Biology created to understand the mechanisms of human diseases and to develop new therapies for Alzheimer's disease, cancer, and rare diseases (Astra-Zeneca Global, 2012; Chesbrough, Vanhaverbeke, and West, 2006; Joly, 2014).

Public–Private Partnership Models

Definition and Objectives

A public–private partnership (PPP) is a cooperative arrangement between the public and the private sectors,[35] where both sectors work together, sharing their resources, risks, and responsibilities in order to achieve a common or compatible objective (Kwak, Chih, and Ibbs, 2009). PPPs can be national and international. PPP models emerge from the idea that neither the public nor the private sector is able to successfully and efficiently see the innovation process from bench to bedside and sustain it in the long term by itself. The R&D process developed solely in the public might be too slow and ineffective (Kwak, Chih, & Ibbs, 2009). The process maintained only in the private sector might be too burdensome for the parties involved and bring about unequal distribution of the final products and services. Hence, PPPs aim to facilitate the translation of basic research into clinical products, including new treatments, drugs, and vaccines (Mittra, 2013; Reich, 2000). They also intend to fulfill a social and moral obligation to improve the health of people in developing countries (Reich, 2000).

Benefits of Public–Private Partnership Models

The main benefit of a PPP is that the public and private sectors share resources (e.g., economic funds, knowledge, data and samples, clinical expertise, patient base, infrastructure), risks, responsibilities, and rewards as partners. They also create greater networks of experts with different insights. Not only can a PPP alleviate the burdens inherent to the innovation process and make the process more efficient, less costly, and faster, but more importantly, it can facilitate and make the R&D process more fluid and holistic (Kwak, Chih, and Ibbs, 2009; Mittra, 2013). The main problems encountered by PPPs are the inherent complexities of basic and clinical science, the high and sometimes unrealistic expectations, and the different objectives and interests originating in both the public and the private sectors (Mittra, 2013).

---

35. In the TM context, the public sector comprises publicly funded or not-for-profit organizations, research centers, and universities, and it usually focuses on basic research. The private sector refers to for-profit organizations, such as biotechnology and pharmaceutical companies, and it usually centers on clinical research (Reich, 2000).

In the area of medical research and health, PPPs have proved to be more efficient, faster, and capable of achieving better outcomes than the industry alone. One example of a successful PPP is in the area of neglected diseases (Bubela, Fitzgerald, and Gold, 2012).

TRENDS IN TRANSLATIONAL INTELLECTUAL PROPERTY

Patents are the main form of intellectual property used in TM to protect inventions and commercialize products. Patents can be obtained on research tools, on new and repositioned drugs, on genetic tests, on risk prediction models, and on personalized therapies.

Patents and Research Tools

Genes (gene sequences), cDNA,[36] and biomarkers[37] (e.g., genes, proteins, metabolites, enzymes), as research tools, can be the basis for the development of genetic tests, new and repositioned drugs, gene therapy, and personalized medicine[38] when they are found to be linked to a disease or disorder (Chandrasekharan and Cook-Deegan, 2009; Sumikura, 2009). The role that these research tools play in the development of clinical products has lured investors (private and some public) and some researchers to pursue patents on research tools, not only to commercialize the inventions associated with them, but also to maintain a competitive position in the field by excluding their competitors from using those research tools (Kesselheim and Shiu, 2014).

For the purposes of this patent analysis, we will differentiate gene sequences and biomarkers from cDNA. Whereas gene sequences and biomarkers are found in the human body and subsequently isolated from it, cDNA is created in a laboratory. The composition of gene sequences and biomarkers also differs from those of cDNA. The differences among these research tools impact the way in which they meet the patent requirements and therefore also their patentability.

Gene sequences and biomarkers found in the human body are not patentable. Merely identifying, locating, and correlating gene sequences and biomarkers to a disease, disorder, or reaction is considered a discovery. The reason for this assertion is that, regardless of the complexity and hard work involved in identifying, locating, and finding that correlation, those identified, located, and correlated genes and biomarkers have not been transformed into anything new; they are exactly the same as found in nature, and therefore they are excluded from patentability.

The patentability of isolated gene sequences, however, is a more complex issue. A gene isolated from the human body has been commonly considered patentable for the following two reasons. First, an isolated gene's chemical qualities, characteristics, and structure are different from those of a gene in the human body because the process of isolation breaks the natural bonds of that segment from the rest of the genome in the human body. This difference was deemed by the courts and patent manuals and regulations of most jurisdictions sufficient to make isolated genes different from how they are found in nature and therefore patentable, provided that they met the other requirements (e.g., new, nonobvious, and utility or industrial application). Second, this "new" isolated gene was "created" by a human-invented process of isolation. The process of isolating a gene was considered to require human ingenuity, and therefore, the resulting gene was also considered to be a product of human ingenuity.

However, in June 2013, the U.S. Supreme Court stated that isolated gene sequences should be considered discoveries and therefore not patentable. The U.S. Supreme Court justified its decision by explaining that the isolated genes maintain the same molecules as the genes found in the human body, and that the differences found at the ends of the segments of the isolated genes are merely ancillary to the isolation (*Association for Molecular Pathology v. Myriad Genetics, Inc.*, 2013; Shanahan, 2014).

---

36. Complementary DNA (cDNA) is DNA that complements or matches mRNA. cDNA is synthesized by copying an mRNA chain using the enzyme reverse transcriptase. As mRNA is formed after all introns have been removed, cDNA contains only the exons (portion of a gene that codes for amino acids) of DNA, omitting the introns. Scientists use cDNA to express a particular protein in a cell that does not normally express such a protein (*Association for Molecular Pathology v. Myriad Genetics, Inc.*, 2013; Bradley, Johnson, and Pober, 2006; Human Genome Project, 2014).

37. A biomarker (or biological marker) is defined as a "characteristic that is objectively measured and evaluated as an indicator of normal biological processes, pathogenic processes, or pharmacologic responses to a therapeutic intervention." There are different types of biomarkers. A prognostic biomarker provides information on the likely course of a disease in untreated individuals. A predictive biomarker advises on the subpopulations of patients who are most likely to respond to a given therapy. A diagnostic biomarker indicates the existence of a disease. A drug-related biomarker helps to determine whether a drug will likely be effective in a specific patient or how the patient's body will likely process the drug. Examples of biomarkers include genes, proteins, metabolites, and enzymes (FitzGerald, 2005; Kesselheim and Shiu, 2014; Strimbu & Tavel, 2010; Wehling, 2008; Wehling, 2010).

38. Personalized medicine is a practice of medicine that takes into account the individual's genetic profile to influence medical decisions concerning the prevention, diagnosis, and treatment of a disease (National Human Genome Research Institutes. National Institutes of Health, 2014).

Other jurisdictions, such as Europe, Canada, and Mexico, have not pronounced on their patentability criteria with respect to the patentability of isolated gene sequences since the U.S. Myriad Case decision. European patent regulation[39] continues to state that whereas discoveries of natural substances, such as the sequence or partial sequence of a gene, are not considered patentable, biological material that has been isolated from its natural environment, however, is patentable. Canada and Mexico have a similar approach. In these jurisdictions, isolated gene sequences are patentable as compositions of matter (Bradley, Johnson, and Pober, 2006; Canadian Patent Act, 2013; European Parliament, 1998; European Patent Office, 2012; Mexican Industrial Property Law, 2012).

cDNA is held patentable because it is synthetic DNA created in the laboratory with human ingenuity and is considered different from the DNA existing in nature. Whereas the natural human DNA has exons and introns, the cDNA includes only the exons and excludes the introns (*Association for Molecular Pathology v. Myriad Genetics, Inc.*, 2013; European Patent Office, 2012).

Isolated gene sequences and biomarkers (in some jurisdictions) as well as cDNA can be patented as part of the drugs, therapies, and genetic tests in which they are used. This association is what provides the element of utility, usefulness, or industrial application. They can be patented as an essential part in a process or as compositions of matter with the purpose of being used in a specific genetic test, drug, or therapy and be mentioned as such in the claims that comprise the drug, therapy, or genetic test patent. The description or inclusion of gene sequences, biomarkers, or cDNA in these terms could provide the patent holder exclusive rights not only over the invention (e.g., the process followed in determining the presence of an abnormal or mutated gene), but also over the research activities associated with the isolated gene sequence, biomarker, and/or cDNA (e.g., any research activities associated with an already identified disease and with the discovery of any other disease that could be potentially linked with that gene sequence and/or cDNA) (Andrews, 2002; European Patent Office, 2012).

Patents on Genetic Tests and Personalized Therapies

Genetic Tests

Genetic tests allow physicians and clinicians to detect in patients the existence or the likelihood of developing a chromosome or gene abnormality or mutation, and therefore of developing genetic diseases or disorders. Genetic tests can also help to detect the existence, effect, and function of biomarkers responsible for the processing of drugs or treatments, thereby contributing to the development of personalized medicine. There are four steps required to perform a diagnosis test. The first step is to identify and locate the specific chromosomes, genes, or biomarkers associated with the disease, disorder, or reaction. The second step is to determine the standard sequences or state of those genes or biomarkers to be used as a reference. The third step is to examine whether the individual has those genes, mutations, or biomarkers. The fourth step is to analyze the likelihood of the individual developing the associated genetic disease, disorder, or reaction (Andrews, 2002; Bradley, Johnson, and Pober, 2006; Sumikura, 2009).

Genetic tests can be patented as processes or as products. A genetic test can be patented as a process if it describes a new and nonobvious method for detecting a genetic abnormality or mutation in association with a genetic disease or disorder or as a process to detect a biomarker to determine the reaction to a drug or treatment (Canadian Patent Office, 2010). In order for the method to be patentable, it cannot consist merely of observing, comparing, and stating laws of nature; it needs to go beyond simply stating correlations between genetic variants and clinical phenotypes and beyond comparing an individual's DNA sequence with a reference sequence. The way in which the steps of the genetic test are executed, the instruments are used, and the results and correlations are interpreted and disclosed needs to be new, nonobvious, and specific to a utility or industrial application for the test to be patentable (*Association for Molecular Pathology v U.S. Patent and Trademark Office and Myriad Genetics Inc.*, 2012; Bakshi, 2014; Klein, 2007; *Mayo Collaborative Services v. Prometheus Laboratories, Inc.*, 2012; *PerkinElmer Inc. v. Interma Ltd., 2011*). Some jurisdictions, such as Europe and Mexico, exclude diagnostic methods practiced on the human body from patentability. This exclusion, however, does not apply to methods that require that the diagnosis be performed outside the body; therefore, gene tests performed in vitro, for instance, are considered patentable (European Patent Office, 2012; Mexican Industrial Property Law, 2012). A genetic test can be patentable as a product when it is embodied as a kit. Instruments, apparatuses, or compositions used in these tests are also patentable as products (Canadian Patent Office, 2010; European Patent Office, 2012; European Patent Office, 2014).

39. For the purposes of this chapter, the term *European patent regulation* comprises the European Patent Convention, the Guidelines for examination in the European Patent Office, and the Directive 98/44/EC of the European Parliament and of the Council of 6 July 1998 on the legal protection of biotechnological inventions.

Personalized Therapies

Personalized medicine has the potential to assist clinicians in making diagnostic or therapeutic decisions specific to a particular individual's genetic profile by predicting whether a newly prescribed drug will likely be effective and/or safe in the treatment of a disease. Personalized therapies will often combine three main elements. The first consists in using a genetic test to detect whether the patient has the specific genes responsible for the processes of production of the biomarkers (e.g., enzymes) that regulate the analysis and processing of drugs (e.g., chemotherapy and heart or antidepressant medications). The second is predicting the likely effect of a drug, dosage, therapy, or treatment, taking into account and interpreting the results obtained from the test. The last element consists of making a decision to prescribe a specific drug, dosage, or treatment based on the prediction made with the genetic test. Since any change or abnormality in any of these genes may impact the effect that a drug may have on the patient, screening for these genes and knowing the potential effects in that particular individual can help to better prescribe a treatment (Kesselheim and Shiu, 2014).

The human ingenuity of the invention, which makes it patentable, is found in the combination of the three elements. Therefore, in this case, the invention related to personalized therapies is patentable if, in addition to being new, nonobvious, and having a specific utility or industrial application, it detects a specific gene or biomarker, determines the likelihood of developing a particular reaction, and, based on this, also suggests the proper treatment considering the specific genetic conditions of the patient (Kesselheim and Shiu, 2014; *Mayo Collaborative Services v. Prometheus Laboratories, Inc.*, 2012). The patent would grant protection over the process that integrates the genetic screening and the specific administration of a treatment. A kit that allows the aforementioned analysis to be performed is often manufactured and patented as well as a product.

Patents on Risk Prediction Models

Risk prediction models help to determine an individual's risk of developing an adverse outcome (e.g., cardiovascular disease) based on statistical analyses and the existence of certain characteristics or predictors. Biomarkers, along with family history, age, sex, and lifestyle, are among the predictors used for risk prediction models. Risk prediction models will help to determine the need for preventive strategy (e.g., changes in the individual's lifestyle) or even treatment. They can also monitor the success of a specific treatment in subjects with different predictors (Ahmed et al., 2014).

The ingenuity in risk prediction models lies in the multi-element statistical analysis to determine the likelihood of developing an adverse outcome or the success in a specific treatment, which goes beyond merely stating the presence of selected biomarkers or any laws of nature. The patent obtained in this case is on the process for predicting the likelihood of the outcome. In addition to the method, the invention often comprises a kit for which a product patent can also be obtained (Ahmed et al., 2014).

Risk prediction models may also be protected by copyright. In this case, the computer program created and used for the analysis of predictors to anticipate a condition or response is protected as a literary work by copyright. The protection granted by copyright covers the source code (i.e., selection and arrangement of instructions and algorithms used) as well as the program's screen display. The program's concepts and ideas are not protected by copyright. The term of protection extends for the life of the author plus 50 years[40] after the author's death. Since registration is not required, protection starts automatically once the work has been generated. The copyright owner has the exclusive right to distribute, reproduce, rent, modify, and license (i.e., authorize others to exercise these rights) the program.

Patents on New and Repositioned Drugs

Repositioned drugs are known drugs and compounds (patented or not) paired with new uses, diseases, or conditions. They can also consist of a different formulation or combination of compounds. The advantage of drug repositioning over regular drug development is that the drug being repositioned has already passed a number of tests for safety and efficacy. This can make the costs and process of R&D lower and the safety approval shorter (Chong and Sullivan, 2007).

Drugs (new and repositioned) can be patented as products and their active ingredients as compositions of matter. The drug and the active ingredient have to comply with the patent requirements: new, nonobvious, and have a utility or industrial application.[41]

40. This term of protection fluctuates between 50 (Canada), 70 (U.S. and Europe), and 100 (Mexico) years after the author's death depending on the specific jurisdiction (Canadian Copyright Act, 2012; European Council Directive 93/98/EEC, 1993; European Directive 2009/24/EC, 2009; European Patent Convention, 2013; Mexican Copyright Law 2014; U.S. Copyright Law, 2011).

41. The patents granted to new drugs are independent from the exclusive marketing rights granted by the U.S. Food and Drug Administration in the United States after the drug has been approved (U.S. Food and Drug Administration. U.S. Department of Health and Human Services, 2014).

For repositioned drugs to be patented, the new use, disease, or condition (industrial application or utility), besides being different from the one associated with the known drug or compound, has to be new and nonobvious. This means that the new use, disease, or condition cannot have been described in the state of the art or cannot have been obvious to a person skilled in the art. Furthermore, given that the repositioned drug or compound is related to or derives from a potentially already patented drug or compound, in order for the repositioned drug to be patentable, it is necessary to verify that it does not infringe on any third party's patents. Therefore, it is required to determine whether there are patents over any of the parts or components of the repositioned drug or compound held by any third party. In the case that there are patents, it is necessary to obtain a license to make, use, and/or sell that known drug or compound from its patent holder. These actions are not only required to obtain a patent, but also to commercialize the repositioned drug or compound (Ashburn and Thor, 2004; European Patent Office, 2012).

Secrecy

Finally, in addition to patenting, academic researchers and the industry have used trade secrets and confidentiality agreements (also known as nondisclosure agreements) in order to protect certain information (e.g., formulas, designs, patterns), know-how (e.g., processes, practices), instruments, compilation of information, and data associated with collected samples. This information has to comply with three requirements in order to be considered protectable by trade secrets: generally unknown to the public, confer economic benefit, and its owner must make reasonable efforts to keep its secrecy.

Secrecy is particularly useful not only as a means of protection per se, but also as a preliminary phase of patents. As mentioned previously, two of the requirements for an invention to be patentable are that said invention is (a) new and (b) nonobvious at the moment the patent is filed (or within the 12 months previous to that filing date). While an invention is being developed, a lot of information is created and a lot can be disseminated through publications, collaborations, academic presentations, and more. This dissemination could destroy the invention's novelty. Confidentiality agreements can maintain the invention's novelty by requiring that the parties involved in the development process enter into a contractual obligation to not disclose any of the information related to the invention until the patent is filed (Andrews, 2002; Caulfield, 2010; Joly, Livingstone, and Dove, 2012).

Secrecy is an attractive option while protecting information and inventions in medical research because the requirements are simpler, the time of protection is unlimited, it can cover anything that has not been made available to the public (as opposed to only patentable subject matter), and the costs associated are less than those associated with patents (e.g., there are no registration fees). More on the advantages and disadvantages of secrecy is described in the following section.

DISCUSSION

A Perspective on the Future of Genetic Patents

There are two major debates in association with the patenting of gene sequences that have a direct bearing on TM and intellectual property.[42] The first refers to the patentability of gene sequences: whether they constitute mere discoveries or they involve human ingenuity. The second focuses on the impact of gene patents on the promotion of innovation. They are explained in the following sections.

Patentability of Gene Sequences: Myriad Case

The debate around the patentability of gene sequences centers on whether they are patentable subject matter. As mentioned previously, patent systems, in most jurisdictions, exclude genetic material as found in nature and discoveries that merely reveal something that exists in nature because they lack human ingenuity. Gene sequences found in the human body fall within this unpatentable category; they are genetic material as found in nature. cDNA, on the other hand, is artificially created in laboratories, and it has structural and molecular differences with the DNA in the human body, which makes it patentable. Whether isolated gene sequences are excluded from patentability, however, has changed over time and from jurisdiction to jurisdiction.

---

42. There are two other debates regarding gene patents. The first debate focuses on the impact of gene patents on access to health care services. The second debate revolves around the morality of attributing property rights on genes, as they are considered life. These two debates will not be discussed here because, although important, they do not have a direct relevance to TM or intellectual property.

Before *Association for Molecular Pathology v. Myriad Genetics, Inc.*, 2013 (Myriad Case),[43] in the United States, an isolated gene sequence could be patented due to the fact that isolating a gene sequence or DNA segment from the human body was enough to allege that it was different from how it exists in nature.[44] The reasoning behind this criterion was that such a gene sequence or DNA segment existed in nature only inside the human body. When it was isolated and removed from the human, it was stripped from the conditions that surround it, changing some of its chemical qualities, characteristics, and structure. These new conditions were not created by nature itself, but by techniques created by human ingenuity.

The patentability of isolated gene sequences, however, changed with the Myriad Case. In June 2013, the U.S. Supreme Court refuted the isolation criteria, asserting that isolated DNA sequences are not patentable because they are the same as the ones naturally found in the human body and the differences between them are just ancillary to the isolation.

The impact of this decision varies. In the United States, this decision lays some groundwork for patentable subject matter for future genetic inventions. Given that the U.S. Supreme Court ruled Myriad's isolated gene sequences of BRCA1 and BRCA2 unpatentable and therefore, invalid, subsequent patent applications on isolated gene sequences filed in the United States shall be deemed equally unpatentable. With respect to already granted U.S. gene patents, notwithstanding that this decision does not automatically invalidate them, as it only dealt with and ruled on the facts and matters of the Myriad Case, it does set a precedent concerning how isolated gene sequences shall be interpreted in terms of patentability for purposes of the patent system in that country. This decision makes existing patents on isolated gene sequences practically unenforceable. Subsequent cases addressing patent infringement issues related to isolated gene sequences shall be decided in favor of the defendant and declare no patent infringement, because for the latter to exist, a valid patent is required. The Myriad Case decision may also indirectly influence other jurisdictions in their future gene patent policies.

Outside the United States, however, the U.S. Supreme Court decision in the Myriad Case does not seem to have influenced court ruling yet. European patent regulation still states that biological material that has been isolated from its natural environment is patentable. Canada and Mexico follow a similar policy (Canadian Patent Office, 2010; Mexican Industrial Property Law, 2012; European Parliament, 1998; European Patent Office, 2012). These jurisdictions still have gene patents validly granted over isolated gene sequences, and they seem to still consider isolated gene sequences patentable as long as they have a utility or industrial application (e.g., diagnostic tests, drugs and personalized medicine, and gene therapies), they are new and have a certain level of human ingenuity, and they are nonobvious or involve an inventive step[45] (Canadian Patent Office, 2010; Mexican Industrial Property Law, 2012; European Patent Office, 2012; European Patent Office, 2014). In Australia, patents over isolated genes have recently been upheld (Shanahan, 2014).

Patenting of Gene Sequences and Promotion of Innovation

The purpose of the patent system has been stated as the promotion of innovation for the benefit of society. However, some literature and even some cases have raised a concern that the current proliferation of patents could, in fact, do the opposite of what patents are supposed to do, and hamper innovation.

The concern, known as the tragedy of the anticommons,[46] is that excessive patent rights in genetic research, and particularly in gene sequences, will lead to underuse and/or wasteful use of resources, thus hindering innovation. This concern is explained by the high or burdensome transaction costs involved in clearing all the intellectual property rights

---

43. *Association for Molecular Pathology v. Myriad Genetics, Inc.*, 569 U.S. 12-398 (2013). As a mere reference, this case involved the U.S. patents (1) 5,747,282, (2) 5,693,473, and (3) 5,837,492 granted to Myriad over the BRCA1 and the BRCA2 genes, whose mutations are associated with an increase risk (50%–80% risk for breast cancer and 20%–50% risk for ovarian cancer as opposed to the 12%–13% risk of an average risk) of developing breast and ovarian cancer. Myriad discovered the exact location of the BRCA1 and BRCA2 in chromosomes 17 and 13. With this information, it determined the "normal" nucleotide sequence of the BRCA1 and the BRCA2 genes and the specific mutations associated with breast and ovarian cancer. The Myriad patents' claims asserted ownership on isolated DNA sequences of BRCA1 and BRCA2 genes, a list of 15 "normal" nucleotides therein contained, and the specific mutations in those genes associated with breast and ovarian cancer. They also asserted ownership on cDNA nucleotide sequences (synthetic DNA) that code for the BRCA amino acids. These patents would give Myriad the exclusive right to isolate a person's BRCA1 and BRCA2 genes or any strand that contains the 15 nucleotides listed or of any other within those genes, and to synthetically create BRCA cDNA (*Association for Molecular Pathology v. Myriad Genetics, Inc.*, 2013).

44. Biotechnological inventions such as inventions involving human genes or DNA sequences are patentable if the biological material (human genes or DNA sequences) is isolated from its natural environment or produced by means of a technical process even if it previously occurred in nature (European Patent Office, 2012).

45. The European patent regulation provides an additional requirement for an invention to be patentable: not contrary to public order. In order for an invention to be patentable, it cannot be against public order. This makes, for instance, inventions for human cloning or that use human embryos unpatentable (European Patent Office, 2012).

46. Anticommons can be understood as the underuse of a resource because said resource has multiple owners, each of which has a right to exclude others from using it (Heller and Eisenberg, 1998).

(i.e., ensuring the invention is not infringing any third party's intellectual property rights) over the necessary research tools and resources in a given R&D process. For instance, in a research field (e.g., medical genetics) where there are multiple patents associated with research resources and tools (e.g., gene sequences), identifying which patents are relevant to the development of a particular product or technology, what is the scope of ownership of those patents, and who owns them can be highly technical, ambiguous, time-consuming, and costly (Chandrasekharan and Cook-Deegan, 2009; Heller and Eisenberg, 1998; Hope, 2008). Innovation is also said to be hindered because, in the case in which the negotiations to obtain licenses to work on a patented product or technology do not prosper and a license is not obtained, the network of researchers and resources focused on that product or technology is likely to be smaller. This reduction of researchers and resources would delay further development or improvement of current innovation. The third concern is associated with patent thickets.[47] Areas densely populated by patents require researchers and innovators to engage in burdensome negotiations and high transaction costs. Researchers whose topics relate to such multilevel patent thickets may be consequently discouraged from continuing their original projects and opt to shift their focus to a less patent populated topic in order to avoid such a hassle. These three concerns are particularly strong in the biomedical community given the cumulative nature of biomedical research and the importance that gene sequences play in the field (Hope, 2008).

However, up to today, regardless of the extensive theoretical and academic debate about whether the patent system truly promotes or hinders innovation, the empirical evidence available is ambiguous. There is little empirical evidence supporting the idea that patents have actually promoted innovation in the biotechnology industry or that patents have harmed innovation (Attaran, 2004). Furthermore, it has been observed that the aforesaid tragedy of the anticommons can be avoided if transactions costs are kept low by, for instance, developing infrastructures or frameworks where intellectual property owners grant each other licenses, such as patent pools or intellectual property clearinghouses[48] (Caulfield, 2009; Hope, 2008; Merges, 2001).

Trade Secrecy as an Option (Pros and Cons)

Trade secrecy has been used as an option to protect an invention when patents do not protect the invention or when there is an improvement or alteration of it not covered by the existing patent. A trade secret is defined as information (e.g., formulas, know-how, processes, information, patterns, designs, instruments, practices, or compilation of information) with economic value that its owner keeps confidential to maintain a competitive advantage (Trade Secret Definition). The owner of the trade secret needs to take all the necessary steps to keep the information confidential in order for it to be considered a trade secret. The main instrument used for these purposes is a confidentiality agreement (or nondisclosure agreement). A confidentiality agreement can include a thorough description of the information that shall be considered secret and the penalties provided in the case of breach of confidentiality. In general, violation of trade secrets is punished by a fine, imprisonment, or both.

The main advantage of using trade secrets as an option to protect an invention and/or its improvements is the low cost involved in acquiring protection. Since there are no registration fees, the only cost involved is taking the necessary measures to keep the information and the invention confidential, which is very low, particularly in comparison to the cost required to obtain a patent.

There are nonetheless some significant disadvantages associated with using trade secrets to protect inventions. The first disadvantage of trade secret protection is that, regardless of the measures that the owner of the trade secret may take to protect his or her trade secret, the information he or she is trying to protect may still be discovered by independent research or reverse engineering. This leads to the second disadvantage, which is that the protection is usually short-lived and difficult to preserve (Resnik, 2004). A third disadvantage is precisely the negative impact that the secrecy involved in a trade secret may have on innovation. Disclosure and sharing of knowledge have social value per se. They are the basis and source of innovation. Without disclosure of knowledge, there is no basis to build on; there is no rich source of inspiration (Williams, 2013). Furthermore, keeping the state of the art of research secret and undisclosed may lead to duplication of research and a waste of economic resources, time, and effort.

---

47. A patent thicket is defined as "a dense web of overlapping intellectual property rights [in this case, patents] that must be hacked through in order to actually commercialize new technology" (Shapiro, 2001).

48. A clearinghouse, in the context of the patent system, is a neutral platform that brings together and matches patent holders and licensees by facilitating license agreements between them. The benefits of a patent clearinghouse for patent holders and licensees include low negotiation costs as standard licenses are frequently used, maximized dissemination of the patent holders' inventions, and low search costs for licensees. A clearinghouse can also help to monitor and enforce the invention's terms of use, to collect and distribute royalties, and to mediate and arbitrate in the case of disputes (van Zimmeren et al., 2011).

Toward Balanced Innovation Environment

Both commercialization and open science can play a key role in translational medicine. On the one hand, commercialization is a very useful, and in some cases even essential, way to fund basic research and to translate it into clinical applications. Patents are the most commonly used strategy to achieve commercialization in the biotechnology and the pharmaceutical industries (Kaye, Hawkins, and Taylor, 2007; OECD, 2004). On the other hand, principles of open science are the basis for further medical innovation, given the cumulative nature of science (Caulfield, Harmon, and Joly, 2012; David, 2003; OECD, 2004). Open data sharing, fast dissemination of information, common ownership of knowledge, and cumulative research building are already integrated not only in day-to-day scientific practices, but also in national and international policies and guidelines. However, strong and exclusive emphasis on commercialization has been shown to lead to data withholding, delayed publication and, in some cases, to the impediment of collaboration (Caulfield, 2010; Heller and Eisenberg, 1998; Jensen and Murray, 2005; Joly, Livingstone, and Dove, 2012; Regenberg and Mathews, 2011; Shapiro, 2001). Similarly, exclusive practices of open science can lead to problems with incentives and sustainability, particularly considering the substantial investment required for lab equipment and clinical trials as well the long-lasting processes. Consequently, it seems advisable to design infrastructures that balance the use of both strategies.

A framework that balances commercialization and access to information and to research tools could involve the combination of patenting practices with tools for collaboration such as the creation of patent pools, cross-licensing,[49] and patent clearinghouses. This combination could help to ease negotiations, lessen the problems that patent thickets may cause to research, and keep transaction costs low while still promoting commercialization (Andrews, 2002; Gold, 2008; Sumikura, 2009). Additionally, the aforesaid balance could incorporate principles and practices of open science (e.g., open sharing and rapid dissemination) that would favor knowledge exchange, knowledge translation, partnerships, and collaboration[50] (Joly, Livingstone, and Dove, 2012).

In addition to structuring a framework that balances commercialization and open science, TM strategies should include properly trained individuals (experts) throughout the whole process from bench to bedside. These experts should be knowledgeable not only in terms of health and safety regulation, government-related administrative processes, science, and patents (Mankoff et al., 2004), but also in terms of how to best implement the balancing framework suggested previously. Particularly regarding patents, the translation process would benefit from having, as in-house experts, lawyers and patent agents with training in medical science (in the particular research field at hand) and medical scientists with training in intellectual property. Their knowledge and contribution could help to create a more effective patent landscape that would inform researchers and those in charge of commercializing the invention about existing patents and advise on filing for new ones. Likewise, these in-house experts need to be knowledgeable in principles, models, frameworks, and strategies of open science and open innovation. This expertise would allow them to design and implement frameworks that combine obtaining patents and protecting the economic aspects of the inventions with enabling fast and open access to information and research tools and promoting collaborations and partnerships, thus attaining the aforementioned balance. For instance, these experts could inform researchers in the preclinical and clinical stages about the existing patents that would need to be negotiated and cleared in order to avoid future litigation. They could also advise them on which patents it could be advantageous to obtain and help to prepare the applications by collecting information all throughout the R&D process and identifying the relevant state of the art. These experts could also advise on the creation of patent pools, cross-licensing, patent clearinghouses, and partnerships. Policies of rapid sharing that simultaneously promote access to information, maintain medical inventions protected, and encourage building collaborations could also be developed. Finally, their expertise could be helpful in patent infringement proceedings in which researchers may be involved in any stage of the translational process (Kaye, Hawkins, and Taylor, 2007).

CONCLUSION

TM, commercialization (patents), and open science (access to information and research tools) aim to promote and enable the advancement of medical science and innovation. TM advances medical science by facilitating the translation of basic research into clinical applications that would ultimately benefit patients. This translation requires rigorous scientific research, sufficient funding, timely and appropriate business decisions, compliance with regulatory systems and legal and

49. Cross-licensing is a form of bilateral license in which two patent holders grant each other a license for the use and exploitation of their patented inventions (van Zimmeren et al., 2011).

50. In some places, such as Quebec, these strategies, combined with people exchange, entrepreneurship, and commercialization, are encompassed in the term *valorization*. This term is considered broader than *commercialization* because it is not motivated primarily by profit, as the latter is (Joly, Livingstone, and Dove, 2012).

ethical frameworks, extensive communication and networking among the different parties, and product development. Patents, being the most common form of intellectual property used as a strategy of commercialization in the field, advance science in three ways. First, patents may entice researchers and investors into engaging in medical research by providing economic rights that allow them to recoup their investment and to profit from the medical invention. Second, patents, used as commodities, help to assign a market value to the medical inventions they protect, therefore facilitating their trade and commercialization in the information market. Third, patents favor the creation of a state of the art or creative commons by requiring a thorough description of the invention that enables any person skilled in the art to reproduce it and by releasing all that information into the public domain after the 20-year period granted of protection. Principles of open science (e.g., open sharing of data, fast dissemination of knowledge, and the importance of commonly owned knowledge to build cumulative research) promote innovation by facilitating access to information and research tools. However, strong and exclusive emphasis on each one of them has its drawbacks.

Developing and implementing a TM framework that combines practices of commercialization, such as patents, with principles of open science can maximize the benefits and mitigate the unintended consequences of intellectual property strategies and of open science principles. This proposed framework should be designed and executed in ways that encourage and favor collaboration and partnership between preclinical (academics) and clinical (industry) R&D, thereby getting the optimal social and economic value of current medical research (Caulfield, Harmon, and Joly, 2012). Some of the strategies that the proposed framework could include in order to achieve collaborations and partnerships are, on the one hand, cross-licenses, patent pools, and clearinghouses, and on the other hand, policies of fast dissemination and open sharing among the parties to such partnerships. The proposed framework would also benefit from training and employment of in-house experts that implement the aforementioned framework and integrate practices and principles of open science and of patents in the relevant medical research area. Implementing this framework all throughout the R&D process, beginning with its early stages, could favor a more efficient, fair, and productive collaboration among the different parties and actors involved in the R&D process (Williams, 2013).

REFERENCES

Legal Citations

Agreement on Trade Related Aspects of Intellectual Property Rights (1995). Uruguay.

Association for Molecular Pathology v. Myriad Genetics, Inc., 569 (U.S. Supreme Court, June 2013).

Association for Molecular Pathology v U.S. Patent and Trademark Office and Myriad Genetics Inc., 689 (U.S. Federal Circuit 2012).

Canadian Copyright Act (2012). Canada.

Canadian Patent Act (2013). Canada.

Doha Declaration on the TRIPs Agreement (2001). Qatar.

European Council Directive 93/98/EEC (1993). Europe.

European Directive 2009/24/EC (2009). Europe.

European Patent Convention (2013). Germany.

European Parliament, 1998. Directive 98/44/EC of the European Parliament and of the Council of 6 July 1998 on the legal protection of biotechnological inventions.

Harvard College v. Canada (Commissioner of Patents) 28155 (SCC 2002).

Madey v. Duke University, 307/1531 (U.S. Fed. Cir. 2002).

Mayo Collaborative Services v. Prometheus Laboratories, Inc., 566 (U.S. Supreme Court, March 20, 2012).

Mexican Copyright Law (2014). Mexico.

Mexican Industrial Property Law (2012). Mexico.

Patent Cooperation Treaty (2001). United States.

PerkinElmer Inc. v. Intema Ltd., 2011–1577 (U.S. Court of Appeals, November 2012).

U.S. Copyright Law (2011). United States of America.

U.S. Patent Act (2014). United States.

Scientific Citations

Ahmed, I., Debray, T.P., Moons, K.G., Riley, R.D., 2014. Developing and validating risk prediction models in an individual participant data meta-analysis. BMC Med. Res. Methodol. 14, 1–15.

Albani, S., Prekken, B., 2009. The advancement of translational medicine—from regional challenges to global solutions. Nat. Med. 15, 1006–1009.

Andrews, L.B., 2002. Genes and patent policy: rethinking intellectual property rights. Nat. Rev. 3, 803–808.

Ashburn, T.T., Thor, K.B., 2004. Drug repositioning: identifying and developing new uses for existing drugs. Nat. Rev. Drug Discov. 3, 673–683.

Astra-Zeneca Global, 2012. Alzheimer's, cancer and rare disease research to benefit from landmark MRC-AstraZeneca compound collaboration. Downloaded from AstraZeneca Media: http://www.astrazeneca.com/Media/Press-releases/Article/20121131-astrazeneca-MRC-collaboration-disease-research.

Attaran, A., 2004. How do patents and economic policies affect access to essential medicines in developing countries. Health Aff. (Millwood) 23, 155–166.
Bakshi, A.M., 2014. Gene patents at the Supreme Court: Association Molecular Pathology Inc. v. Myriad Genetics. J. Law Biosci. 1, 183–189.
Bornstein, S.R., Licinio, J., 2011. Improving the efficacy of translational medicine by optimally integrating health care, academia and industry. Nat. Med. 17, 1567–1569.
Bradley, J., Johnson, D., Pober, B., 2006. Medical Genetics. Blackwell Publishing, United Kingdom.
Bubela, T., Fitzgerald, G.A., Gold, R., 2012. Recalibrating intellectual property rights to enhance translational research collaborations. Sci. Transl. Med. 4, 1–6.
Canadian HIV-AIDS Legal Network et al., 2010. Publications. Canadian HIV-AIDS Legal Network. Downloaded from Canadian HIV-AIDS Legal Network http://www.aidslaw.ca/publications/publicationsdocEN.php?ref=1151.
Canadian Patent Office, 2010. Manual of Patent Office Practice.
Castle, D., 2009. Introduction. In: Castle, D., Castle, D. (Eds.), The role of intellectual property rights in biotechnology innovation. Edward Elgar, United Kingdom, pp. 1–15.
Caulfield, T., 2009. Human gene patents: proof of problems? Chicago-Kent Law Rev. 84, 133.
Caulfield, T., 2010. Stem cell research and economic promises. J. Law, Med. Ethics 38, 303–313.
Caulfield, T., Harmon, S.H., Joly, Y., 2012. Open science versus commercialization: a modern research conflict. Genome Med. 4, 1–11.
Chandrasekharan, S., Cook-Deegan, R., 2009. Gene patents and personalized medicine—what lies ahead? Genome Med. 1, 1–4.
Chesbrough, H., Vanhaverbeke, W., West, J., 2006. Open innovation: researching a new paradigm. Oxford University Press, United Kingdom.
Chong, C.R., Sullivan, D.J., 2007. New uses for old drugs. Nature 448, 645–646.
Cook-Deegan, R., 2008. Gene patents. In: Crowly, M. (Ed.), From birth to death and bench to clinic: The Hastings Center Bioethics briefing book for journalists, policymakers, and campaign. The Hastings Center, New York. 69–72.
David, P., 2003. The economic logic of "open science" and the balance between private property rights and the public domain in scientific data and information: a primer. In: Esanu, J.M., Uhlir, P.F. (Eds.), The role of scientific and technical data and information in the public domain. National Academies Press, Washington, DC, pp. 19–34.
David, P.A., 2004. Understanding the emergence of 'open science' institutions: functionalist economics in historical context. Ind. Corp. Change 13, 571–589.
Drolet, B.C., Lorenzi, N.M., 2011. Translational research: understanding the continuum from bench to bedside. Transl. Res. 157, 1–5.
European Patent Office, 2012. Guidelines for examination in the European Patent Office. European Patent Office, Munich.
European Patent Office, 2014. Patenting Issues. Patents in Biotechnology. Downloaded from European Patent Office: http://www.epo.org/news-issues/issues/biotechnology.html.
FitzGerald, G.A., 2005. Anticipating change in drug development: the emerging era of translational medicine and therapeutics. Nat. Rev. 4, 815–818.
Giustini, D., 2006. How Web 2.0 is changing medicine. Br. Med. J. 333, 1283–1284.
Glasgow, R.E., 2009. Critical measurement issues in translational research. Res. Social Work Prac. 19, 560–568.
Gold, R., 2008. Toward a new era for intellectual property: from confrontation to negotiation. International Expert Group on Biotechonology. Innovation, and Intellectual Property, Montreal.
Heller, M.A., Eisenberg, R.S., 1998. Can patents deter innovation? The anticommons in biomedical research. Science 280, 698–701.
Hope, J., 2008. BioBazaar. The open source revolution and biotechnology. Harvard University Press, Cambridge, MA.
Human Genome Project (HGP), 2014. cDNA (Complementary DNA). http://humangenes.org/cdna-complementary-dna.
Jensen, K., Murray, F., 2005. Intellectual property landscape of the human genome. Science 310, 239–240.
Joly, Y., 2014. Propriete intellectuelle et modeles de collaboration ouverte. In: Rousseau, S., Moyse, P.-E. (Eds.), Propriete intellectuelle. JurisClasseur Quebec. Collection Droit des Affaires. LexisNexis, Canada, pp. 31–39.
Joly, Y., Caulfield, T., Knoppers, B.M., Harmsen, E., Pastinen, T., 2010. The commercialization of genomic research in Canada. Health Policy 6, 24–32.
Joly, Y., Livingstone, A., Dove, E.S., 2012. Moving beyond commercialization: strategies to mazimize the economic and social impact of genomics research. GPS-Genome Canada, Ottawa.
Kaye, J., Hawkins, N., Taylor, J., 2007. Patents and transaltional research in genomics. Nat. Biotechnol. 25, 707–739.
Kesselheim, A., Shiu, N., 2014. The evolving role of biomarker patents in personalized medicine. Clin. Pharmacol. Ther. 95, 127–129.
Klein, R.D., 2007. Gene patents and genetic testing in the United States. Nature Biotechnol. 25, 989–990.
Kwak, Y.H., Chih, Y., Ibbs, C.W., 2009. Towards a comprehensive understanding of public private partnerships for infrastructure development. CMR 51, 51–78.
Mankoff, S.P., Brander, C., Ferrone, S., Marincola, F.M., 2004. Lost in translation: obstacles to translational medicine. J. Transl. Med. 2, 1–5.
Merges, R., 2001. Institutions for intellectual property transactions: the case of patent pools. In: Dreyfuss, R.C., Zimmerman, D., & First, H. (Eds.), Expanding the boundaries of intellectual property: innovation policy for the knowledge society. Oxford University Press, Oxford, pp. 123–165.
Merton, R.K., 1973. The normative structure of science. In: Storer, N.W. (Ed.), The sociology of the science. University of Chicago Press, Chicago, pp. 267–278.
Mill, J.S., 1879. The collected works of John Stuart Mill. Volume XVIII. Essays on politics and society. Part 1. Downloaded from Online Library of Liberty http://oll.libertyfund.org/titles/233.
Mittra, J., 2013. Exploiting translational medicine through public–private partnerships: a case study of Scotland's Translational Medicine Research Collaboration. In: Mittra, J., Milne, C.-P. (Eds.), Transational medicine: the future of therapy? Pan Stanford Publishing, Boca Raton, FL, pp. 213–232.
National Human Genome Research Institute. National Institutes of Health, 2014. Personalized Medicine. Talking Glossary of Genetic Terms. https://www.genome.gov/glossary/index.cfm?id=150.

Organisation for Economic Co-operation and Development, 2004. Patents and Innovation: Trends and Policy Changes. OECD, Paris.
Prestwich, G.D., Bhatia, S., Breuer, C.K., Dahl, S.L., Mason, C., McFarland, R., et al., 2012. What is the greatest regulatory challenge in the translation of biomaterials to the clinic? Sci. Transl. Med. 4, 1–6.
Regenberg, A., Mathews, D.J., 2011. Promoting justice in stem cell intellectual property. Regen. Med. 6, 79–84.
Reich, M.R., 2000. Public–private partnerships for public health. Nat. Med. 6, 617–620.
Rentier, B., 2014. Rebirth of science. https://www.youtube.com/watch?v=OKDxsAkqVh4.
Resnik, D.B., 2004. Owning the genome: a moral analysis of DNA patenting. State University of New York Press, Albany, NY.
Shanahan, L., 2014. Court rules that breast cancer gene can be patented. Downloaded from *The Australian*: http://www.theaustralian.com.au/news/health-science/court-rules-that-breast-cancer-gene-can-be-patented/story-e6frg8y6–1227048900037?nk=9653a33aaca7342bbe24bf814dfa7d83.
Shapiro, C., 2001. Navigating the patent thicket: cross licenses, patent pools, and standard setting. In: Jaffe, A.B., Lerner, J., Stern, S. (Eds.), Innovation policy and the economy, vol. 1. MIT Press, Cambridge, MA, pp. 120–150.
Stiglitz, J.E., 2006. Scrooge and intellectual property rights. Br. Med. J. 333, 1279–1280.
Strimbu, K., Tavel, J.A., 2010. What are biomarkers? Curr. Opin. HIV AIDS 5, 463–466.
Sumikura, K., 2009. Intellectual property rights policy for gene-related inventions—toward optimum balance between public and private ownership. In: Castle, D., Castle, D. (Eds.), The role of intellectual property rights in biotechnology innovation. Edward Elgar, United Kingdom, pp. 73–97.
Trade Secret Definition. Downloaded from *Black's Law Dictionary*: http://thelawdictionary.org/trade-secret/.
Translational Medicine. Science Translational Medicine. Definition. Downloaded from *Science Magazine*, August 15, 2014. http://www.sciencemag.org/site/marketing/stm/definition.xhtml.
U.S. Food and Drug Administration. U.S. Department of Health and Human Services, 2014. Development and Approval Process (Drugs). United States.
Vaver, D., 2011. Intellectual Property Law. Irwin Law, Canada.
van Zimmeren, E., Vanneste, S., Matthijs, G., Vanhaverbeke, W., Van Overwalle, G., 2011. Patent pools and clearinghouses in the life sciences. Trends Biotechnol. 29, 569–576.
Wehling, M., 2006. Translational medicine: can it really facilitate the transition of research "from bench to bedside"? Eur. J. Clin. Pharmacol. 62, 91–95.
Wehling, M., 2008. Translational medicine: science or wishful thinking? J. Transl. Med. 6, 31.
Wehling, M., 2010. Introduction and definitions. In: Wehling, M. (Ed.), Principles of translational science in medicine. From bench to bedside. Cambridge University Press, New York, pp. 1–17.
Williams, H.L., 2013. Intellectual property rights and innovation: evidence from the human genome. J. Polit. Econ. 121, 1–27.
World Trade Organization, 1994. Agreement on trade-related aspects of intellectual property rights. World Trade Organization, Geneva.
Zittrain, J., 2004. Normative principles for evaluating free and proprietary software. U. Chi. L. Rev. 71, 1–21.

Chapter 8

Translational Research in the Fastest-Growing Population: Older Adults

Kevin P. High and Stephen Kritchevsky

INTRODUCTION

Why Study Aging?

Aging is not a disease state, so why devote a chapter to aging in a textbook on translational science in medicine? There are a number of convincing arguments to do so, but perhaps most important is that aging research has the potential to dwarf the "return on investment" for research focused on any single disease. Age is such a strong risk factor for multiple illnesses across many organ systems that addressing aging itself has a greater potential benefit for extending lifespan than curing any single disease (Miller, 2002). The average lifespan of a 50-year-old human in the developed world is about 31 additional years (i.e., median life expectancy of about age 81). It is estimated that curing cancer would add only 4–5 years of life; a similar number of years would be gained from curing heart disease. Even curing cancer and heart disease and diabetes and stroke would add only an estimated 15 years to the current status (i.e., median life expectancy would increase to about 95 years). However, if one could retard aging to a degree first identified in the 1930s through caloric restriction studies in animals, it is estimated that median life expectancy at age 50 would increase to nearly 60 years—adding about 30 years to the current total life expectancy (Miller, 2002). Thus, investing in aging research is likely to at least rival, if not exceed, the impact of research aimed at cure or prevention of any specific disease.

A second reason to study aging is that older adult populations throughout the world are growing rapidly. Median life expectancy in humans has increased markedly in the last two centuries and is now nearly 80 years in most industrialized nations. Most of this gain has come from improved sanitation, public health measures that have reduced maternal and traumatic deaths, and advances in treating/preventing infections. However, since the middle of the last century, there have been tremendous medical advances to manage common chronic diseases that are causes of middle-aged mortality, such as cardiovascular disease and many types of cancer that have further contributed to median longevity. As a result, there are rapidly rising populations of older adults in essentially all industrialized nations including the United States (Figure 8.1a; U.S. Census Bureau, 2014). This trend is occurring in developing countries as well, and UN Aging statistics suggest more people age 80+ years will live in developing than developed countries by 2025 (United Nations, 2014). This segment of the population accounts for the majority of health care expenditures in the United States and other developed countries. The average yearly expenditure for health care of those age 65–74 years is almost $10,000, and more than double that figure for those age 85 years and over (Figure 8.1b; U.S. Census Bureau, 2014).

Lifespan versus Healthspan

Despite marked increases in median lifespan in the last two centuries, maximum human lifespan has remained essentially constant at about 110 years. Extending this may not be feasible or, at the very least, a problem of enormous scientific and ethical consideration. Further, with improved survival of acute illnesses that previously resulted in early/middle-aged adult deaths, there has been a concomitant increase in those living, and aging, with accumulating chronic conditions (Figure 8.1c; Centers for Disease Control and Prevention, 2014). This realization has led to the concept of "healthspan" (reviewed in Kirkland, 2013)—i.e., the period of time one spends in healthy, active life before the occurrence of functional limitation and dependence. Extending healthspan rather than concentrating just on lifespan has become a major goal of aging research. Focusing on healthspan may also have a greater impact on health care cost than disease-focused or lifespan extension. Surviving one illness

Principles of Translational Science in Medicine. http://dx.doi.org/10.1016/B978-0-12-800687-0.00031-1

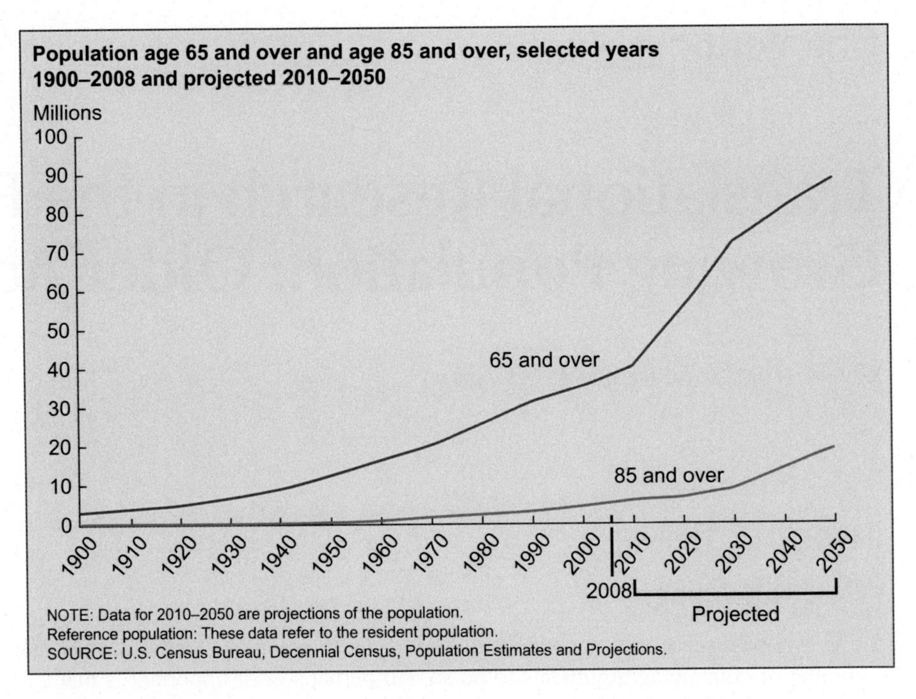

FIGURE 8.1a Actual (1900–2008) and projected (2010–2050) number of persons in the U.S. age 65+ or 85+ years.

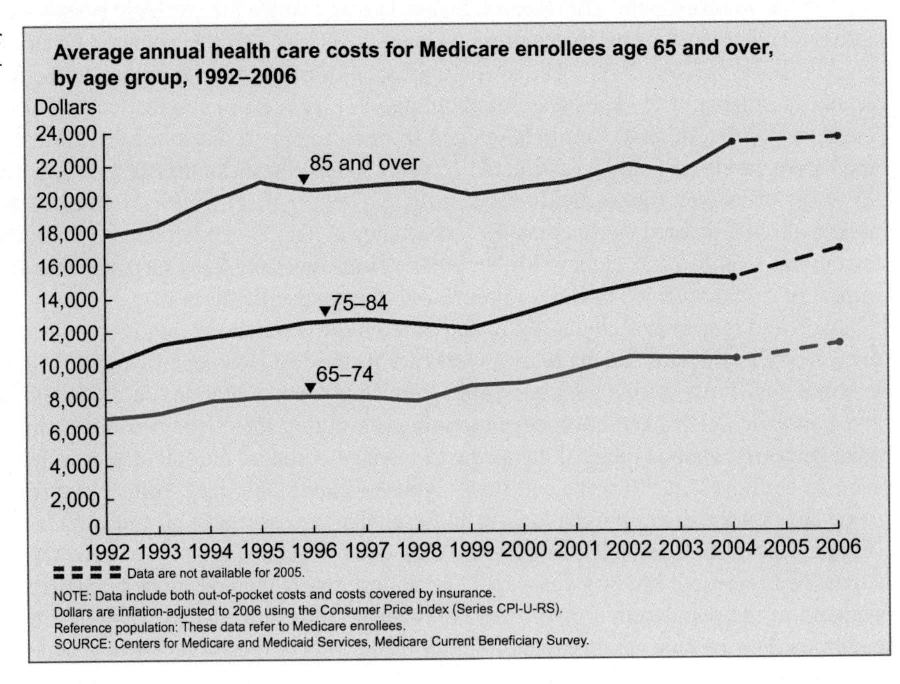

FIGURE 8.1b Mean health care costs per Medicare enrollee age 65–74, 75–84, and 85+ years from 1992 to 2006.

essentially means we live to experience another illness and another and another, increasing the lifetime cost of health care for an individual. However, improving healthspan has the potential to be markedly cost-saving by pushing severe, debilitating illnesses to the very end of life, reducing the time one requires high cost, labor-intensive care and support.

GERONTOLOGY VERSUS GERIATRICS

Though an oversimplification, aging research can be broadly divided into *gerontology,* which is the study of aging, and *geriatrics,* which is the study of caring for older adults. Older patients typically experience multimorbidity, the presence of three or more concurrent chronic illnesses. In addition, many health problems of older patients are not referable to a specific organ or disease process (e.g., fatigue, slowness, and weakness). As researchers new to the field of aging become aware of gerontology and the benefits of including gerontologic approaches, they frequently ask, "What's the biomarker of aging I should

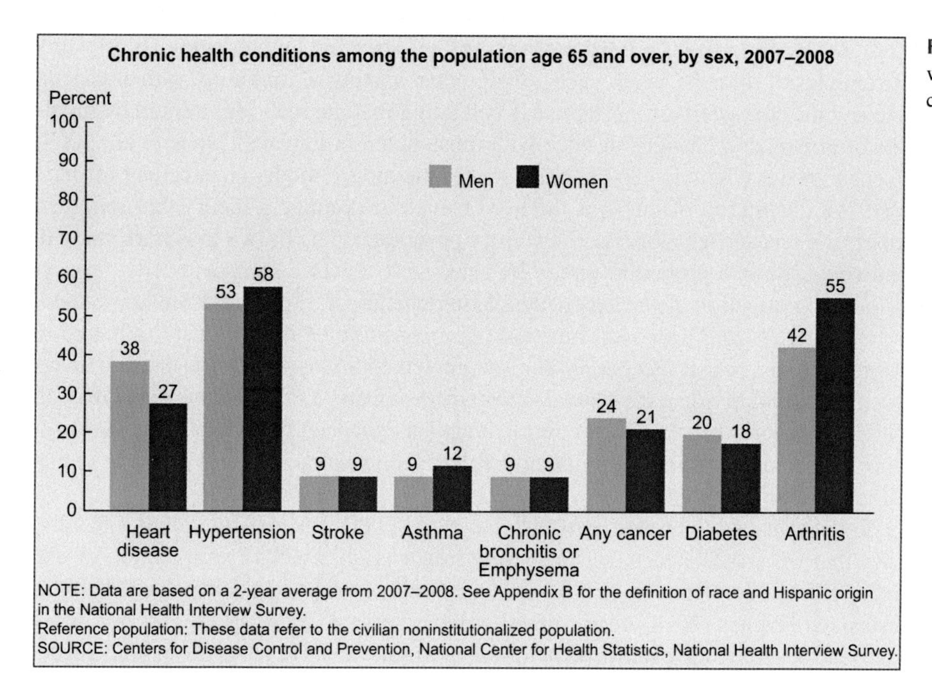

FIGURE 8.1c Percent of U.S. men or women age 65 years and older with the indicated health condition.

follow to see if age affects *XYZ* process?" This is a naïve question; the aging "process" itself is a multifaceted, multi-organ phenomenon in which physiologic reserve—the capacity to recover from insult—gradually wanes. There have been a number of excellent recent reviews (Burch et al., 2014; Lopez-Otin et al., 2013) that outline the biologic "hallmarks of aging"; they include genomic, biochemical, cellular, and systems biology indicators of age. However, no single factor is clearly "causative" or in the causal pathway for aging, just like most complex illnesses where there is no single cause, but "risk factors" that increase the likelihood of disease. Aging appears to be best represented by a web of causation; genetics, nutrient sensing and energetics, senescence, proteostasis, and cell–cell communication all appear to play a role (Burch et al., 2014).

Geriatrics focuses on the prevention and treatment of disease in older adults. This differs from that of young adults in several important ways. First, risk factors for illness in older adults may differ markedly from those in younger adults. For example, elevated cholesterol is a strong risk factor for cardiovascular disease in young and middle-aged adults. But as one ages beyond 70, cholesterol becomes a debated risk factor. This may be due to a number of important changes with age. It may be due to a "healthy survivor effect"; those for whom hypercholesterolemia was going to be a risk factor have already experienced disease or did not survive to old age, so in the remaining seniors there appears to be no additional risk. Many older adults have a substantial burden of subclinical vascular disease. These persons are often treated by analysts as being disease free, the consequence of which is to dampen risk factor relationships. In addition, many physiologic measurements serve as indicators of overall health in advanced age. The loss of the ability to maintain a cholesterol level, blood pressure, or weight—all are hallmarks of a high risk for mortality in older populations. Thus, it is difficult to tease apart whether a low level represents an indicator of low future disease risk or an indicator of high risk of homeostatic decompensation. Furthermore, other age-critical risk factors (e.g., arterial stiffness, change in body composition) may dwarf the effects of cholesterol as one ages. Second, as noted previously, older adults rarely have a single disease, but almost always experience illness in the context of other chronic conditions—arthritis and heart disease, cognitive impairment and diabetes—often multiple diseases. This mandates that appropriate treatments be examined in common comorbid conditions. An excellent example is heart failure in a patient who also has renal dysfunction. Treatment with an angiotensin-converting enzyme inhibitor (e.g., lisinopril), while beneficial for heart failure, may lead to unacceptable hyperkalemia or decreased glomerular filtration in those with advanced kidney disease; thus, inclusion of common dyads and triads of disease is a critical factor for translational research in aged populations. Finally, the hardest endpoint in nearly all basic science or clinical trial models of disease is survival. It is unambiguous and unarguably relevant. However, for many older adults, death is not the worst outcome. On the contrary, life in pain or dependent on others may be a fate "worse than death." Thus, patient preference for outcomes is a critical factor for translational research in geriatric research.

ANIMAL MODELS OF AGING

Animal models play a critical role in translational research, and there are important considerations when such models are applied to aging and diseases of older adults. The most commonly used animal model is certainly the mouse model.

There are distinct reasons to use, and not use, mice for translational research for investigating human aging. To illustrate the relative pros and cons, we will use examples of immune senescence—the gradual waning of immunity with advancing age. Maue and colleagues recently reviewed the characteristics of human T-cell aging that are well represented by mouse models, demonstrating relative parallels of age-related changes in mice with those noted in humans (Maue et al., 2009). This general conservation of immune senescence across mice and humans makes the mouse model an excellent resource for experiments only achievable in mice. An illustration of this was the use of mice to examine a theory that senescent cells hypersecrete inflammatory cytokines (the senescence-associated secretory phenotype, SASP) in a landmark study by Baker et al. (2011). Elimination of senescent cells in a progeroid mouse by selectively deleting senescent cells (targeted by their expression of p16) resulted in marked mitigation of the accelerated aging phenotype. The effect was presumably achieved through elimination of the senescent cells *and* improved function of surrounding cells previously impaired by cytokines produced by cells that had "aged" into the SASP (Peeper, 2011). A complementary approach was used to inactivate p16 in T or B lymphocytes *at a specific age and in a specific immune compartment* using a *Lck–Cre* system (Liu et al., 2011). This strategy maintained a "youthful" phenotype in the T-cell compartment but enhanced lymphoid malignancies in B cells. These types of experiments show the potential to which animal models have contributed, and will continue to play a critical role, for translational research in immune senescence.

However, rodent models frequently do not translate to human aging (Akbar et al., 2000; Fauce et al., 2008), including failure in humans of vaccine approaches that were validated in rodents (MacGregor et al., 1998). Extrapolation of data between species must therefore be made with caution. In a review titled "Of Mice and Not Men: Differences between Mouse and Human Immunology," Mestas and Hughes (2004) detail the differences between these species. Major differences include expression level, diversity and response of pattern recognition receptors (e.g., toll-like receptors; further explored by Barreiro et al., 2009; Copeland et al., 2005; Munford, 2010), distinct differentiation signals for T-cell subtypes, and a large number of variances in cell surface marker and/or receptor-ligand specificity that dictate migration and function of immune cells. Further, humans live much longer than mice, and the memory immune cells of humans must be maintained for a substantially greater time. Persistence of human memory T-cell pools is likely subjected to additional constraints that may not apply to murine experimental systems (Akbar et al., 2000). Another key difference between mice and men is their prior exposure history. All humans age in the presence of persistent viral infections, but the combination of viruses present is as unique as the number of people on earth (Virgin, Wherry, and Ahmed, 2009). Mice are raised in "specific pathogen-free conditions," to ensure that viruses like murine hepatitis virus do not confound experimental design. Variable exposure and immune activation to persistent viral infections carried by laboratory mice are rarely if ever included in reports of animal models of immune senescence, and it is commonly assumed, almost certainly falsely, that viral exposure and immune response to chronic virus infection in older mice is similar to that of young adult mice. In such experiments, changes attributed to age may reflect the impact of chronic viral infection as the true underlying cause.

To address some of these issues, researchers have transplanted human bone marrow in immunocompromised mice to reconstitute human immune cells or to accept fully differentiated human peripheral blood cells. These "humanized" mouse models have promise for immunology studies (reviewed in Shultz, Ishikawa, and Greiner, 2007), and some age-related illnesses (e.g., Alzheimer's disease; Bruce-Keller et al., 2011) have been substantially investigated using humanized mouse models. However, a note of caution must be sounded with regard to this model. While intrinsic immune cell aging may be relatively well represented by humanized mice, extrinsic forces (human physiology of the liver and kidneys, species variation in response to stressors, diet, etc.) are not represented in this model. The importance of this was shown in a recent study by Warren and colleagues using serum or mononuclear cells from different species (Bruce-Keller et al., 2011; Warren et al., 2010). In that study, inflammation in response to lipopolysaccharide (LPS) was better predicted by the species from which serum was derived than the species of origin for the cell, suggesting there are circulating factors that may influence responses more than intrinsic cell properties.

HUMAN APPROACHES TO TRANSLATIONAL AGING RESEARCH

The goal of biomedical research focusing on the mechanisms underlying aging is to increase the human healthspan. A common conceptual framework for approaching this goal is presented in Figure 8.2. Many physiologic parameters decline with age including skeletal muscle mass, bone mineral density, maximum heart rate, forced expiratory volume in 1 second (FEV_1), and glomerular filtration rate. These changes contribute to predictable declines in functional performance reflected in things like usual walking speed, peak exercise capacity, psychomotor speed, and working memory. The reduction in performance is called "functional impairment." Over time "functional impairments" may become large enough to affect how an older person interacts with his or her environment, making tasks more difficult, a condition referred to as "functional limitation." If these declines continue, a person may lose the ability to meaningfully interact with his or her

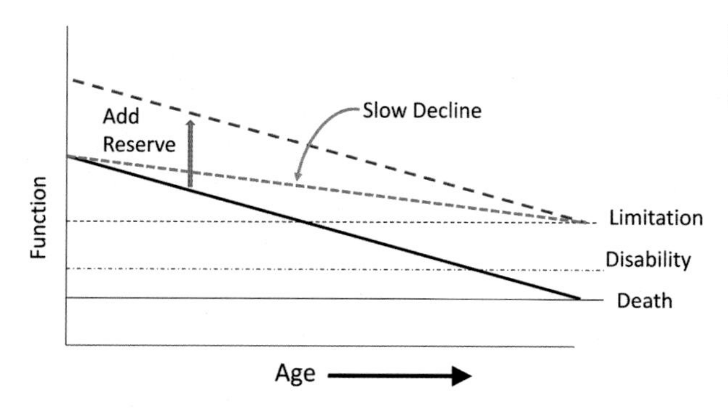

FIGURE 8.2 A model for how age-related physiologic changes can affect healthspan and lifespan. Alternative strategies to delay the onset of disability and death include adding physiologic reserve or slowing physiologic decline.

environment without help. This person is said to be "disabled." Ultimately, declines may be so great that they are incompatible with life. Healthspan in human populations is commonly defined as the length of life lived without disability.

The conceptual model suggests two primary strategies for extending healthspan: building functional reserve and slowing rates of decline (Figure 8.2). Reserve reflects "spare" physiologic capacity beyond that required to fully interact with the environment. Depending on the plasticity of the organ system, one can bring reserve to old age, or one can build reserve even in old age. Muscle weakness is a major risk factor for physical disability. On average, strength declines about 1% each year from about age 40 and accelerates after about age 70. A 50-year follow-up of 2,775 British infants born in 1946 showed that participants who were larger babies had stronger grips as middle-aged adults (Kuh et al., 2002). Data from a cohort of Hawaiian men of Japanese descent have shown that grip strength at age 50 was a strong risk factor for disability 25 years later (Iwaoka and Tannebaum, 1976). On the other end of the age spectrum, Fiatarone and colleagues showed that even in nonagenarians a weight-training regimen increased strength by an average of 174% and improved gait speed by 48%, indicating reserve can be gained even at very old age (Fiatarone et al., 1990).

Studies of the rate of decline of physiologic systems and functions are less common because they require multiple measures of the physiologic system measured far enough apart to be able to distinguish true change from measurement variability. The best established ongoing study of the aging process is the Baltimore Longitudinal Study on Aging (BLSA), which has followed some participants for more than 40 years (BLSA, 2014; Shock et al., 1984). The BLSA shows that while on average many physiologic systems decline with age, there is substantial person-to-person variability in how these systems change through time with some persons declining very quickly and others maintaining a level of physiologic function with little evident decline over decades. Behaviors like strength training and intense aerobic exercise are associated with increased strength and aerobic reserve, but they do not seem to slow the rate of decline of these capacities. Tobacco use accelerates decline in lung function. Poor diet is associated with faster declines in cognitive function. Declines in physiologic function can be accelerated by disease events and associated periods of physical inactivity/bed rest. HIV/AIDS and/or its treatment accelerate declines in physiologic function; a myocardial infarction can permanently impair the heart's ability to deliver blood. Interestingly, rates of change do not appear to be strongly correlated across physiologic systems. The CHS study has looked at longitudinal changes in several physiologic measures representing different organ systems, and the correlation in these changes over time is low (Kim and Guengerich, 1990).

Identifying Determinants of Longevity and Healthspan Using Epidemiologic Approaches

The most rudimentary approach to identifying age-related process is to compare young and old persons. This is a frequently used approach but has proven to be unreliable. It is often difficult to assure the representativeness of the two groups, so results can be difficult to replicate. It is also difficult to separate age-related changes from the effects of common comorbidities and disease conditions common among older adults. In an attempt to remedy this problem, some investigators have identified samples of older persons with no identifiable clinical conditions. This is difficult to do primarily because many older persons without a clinical diagnosis may have substantial subclinical pathology. For example, about one-third of older persons have clinically manifest cardiovascular disease, and another one-third have substantial subclinical disease using extensive noninvasive imaging. Thus, the definition of "disease free" will vary depending on the intensity of disease detection. Furthermore, young–old comparison studies have frequently identified differences that have not panned out. Older adults tend toward lower levels of insulin-like growth factor-1 (IGF-1), growth hormone, and DHEA-S, but repletion experiments have generally shown no functional benefits and some unexpected harms.

Long-lived persons and populations have been studied in the attempt to identify physiologies and pathways associated with longevity. Several populations across the globe have an unusually large number of persons living to advanced age: Okinawans and Seventh-day Adventists are two examples. The common factors across these populations include life-long physical activity, meaningful societal engagement of elders, and a diet that is rich in fruits and vegetables and sparse in animal flesh. While these populations can provide clues to habits that may enhance longevity, they cannot be used to test mechanistic hypotheses. Many research groups have studied centenarians to try to identify the basis for their extreme longevity. This approach has met with limited success, and no clear risk factor profile has emerged. However, it is clear that longevity runs in families, so many groups are employing genomic approaches to identify molecular pathways important to human longevity. Many candidate loci have been identified within given populations, but few have been consistently replicated. Current best evidence implicates genetic variability at the *APOE* and *FOXO3* loci (Broer et al., 2014). The *APOE4* genotype is strongly associated with Alzheimer's disease, and the *APOE2* allele may exert a protective effect. *FOXO3* is involved in IGF-1/insulin signaling. It has a homologue in the *C. elegans* model *Daf-16* gene, where its expression is important to lifespan regulation. The issues and complexity in conducting and interpreting genetic studies of human aging and longevity are well summarized elsewhere (Brooks-Wilson, 2013).

Longitudinal cohort studies, like the Health Aging and Body Composition Study, and the Cardiovascular Health Study have proven to be productive for addressing new hypotheses regarding the aging process, healthspan, and lifespan. These studies enumerate and carefully phenotype large samples of older persons who are then followed and re-examined at intervals over extended periods of time. The studies also typically have biospecimen repositories to allow the efficient examination of new hypotheses, as well as identification of persons of similar ages and overall health statuses to test factors predicting the age-related outcomes. The repeated measures approach allows investigators to parse out physiologic and molecular differences that are associated with longer life from those that may change in the anticipation of death. One can look at early deaths (within the first few years) separately or look at risk factor trajectories until the time of the event. Such an analysis suggests that many of the cognitive changes underlying Alzheimer's disease begin to manifest themselves 8–10 years before a diagnosis is made.

TESTING TREATMENTS TO EXTEND HEALTH- AND LIFESPAN

The National Institute on Aging (NIA) sponsors a program to support the systematic evaluation of treatments with the potential to extend lifespan and delay disease in mice (National Institute on Aging, 2014a, 2014b). Basic science has identified a number of targets for new and existing therapies to extend life, and a testing program was put in place to assure that the initial evaluations of new therapies was conducted in a systematic manner and collected a uniform set of measures beyond simply noting mean and maximal lifespans of treated animals. The program has identified the drug rapamycin as having life-extending properties in both male and female mice (Miller et al., 2014), while acarbose, 17-α-estradiol, and nordihydroguaiaretic acid extend lifespan more effectively in male mice (Harrison et al., 2014). The typical design involves both male and female mice with treatment initiation between 4 and 10 months of age. Mice are followed until all have died. This is roughly equivalent to starting these treatments in humans at between 16 and 40 years of age and following them until age 100. The interventions testing program paradigm is appropriate for determining whether a compound has life-extending properties in a model mammalian system, but it would be very difficult to translate this approach directly into humans. The mean life expectancy of a 40-year-old living in the United States is about 40 years. It is logistically and financially infeasible to conduct a 40–60 -year experiment in a sufficient number of persons to evaluate an intervention's effect on life expectancy. There are also serious ethical considerations. Participants would be asked to take a test compound for the next 40–60 years. The intervention would not treat any particular complaint or disease, so the immediate benefits to those receiving the test intervention would be expected to be small. The study would be ethical only if the long-term safety of the intervention was virtually certain because the risk of potential harm would outweigh uncertain long-term benefits.

Biomarkers provide a strategy for evaluating longevity treatments in humans. Biomarkers are substitute study endpoints when the desired endpoint cannot be measured for either ethical or practical reasons (Figure 8.3). Biomarkers have a strong statistical link to the desired outcome; although not necessarily in the causal pathway, they move in a direction after interventions that predict clinical outcomes. For example, serum high LDL-cholesterol is a biomarker for atherosclerotic heart disease. Its elevation is related to higher coronary heart disease risk, and interventions that lower LDL-C also lower the risk of heart disease. The link is so well established that new drugs can be approved on the basis of their cholesterol-lowering properties rather than proving a reduction in heart disease risk. The strength of LDL-C as a biomarker for heart disease is based on its causal role in the promotion of atherosclerosis. The search for biomarkers of longevity is complicated by the fact that a single pathway or mechanism that leads to long life has not been identified. Nevertheless, it is possible to propose criteria for biomarkers such that if they were positively affected by a new treatment, one would expect life extension with some confidence.

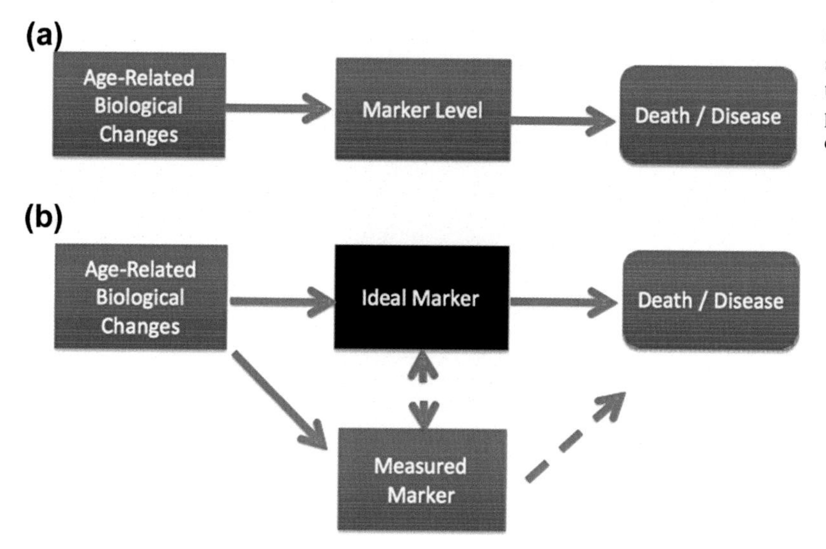

FIGURE 8.3 Biomarkers are measurements used to substitute for an outcome that may be rare or take a long time to develop. Biomarkers may be (a) on the causal pathway or (b) correlated with causal marker or the outcome itself.

Box 8.1 An Ideal Longevity Biomarker Would

1. *Correlate with chronological age.* If a marker does not change with age, it is unlikely to change in response to an intervention that affects the aging process.
2. *Have a strong and consistent relationship predicting long-term mortality across a variety of populations, disease conditions, and comorbidities independent from age.* Measures of acute physiologic distress may predict short-term mortality but may not predict long-term survival; the relationship should be seen regardless of underlying disease-specific pathologies.
3. *Be responsive to change.* If the marker is unchangeable, it cannot be used to evaluate an intervention.
4. *Show a monotonic relationship with mortality.* For some physiologic measures, the relationship with mortality is curvilinear: pushing levels too low might lead to increased mortality (e.g., blood pressure).
5. *Be objectively measured.* Subjective measures can be powerful measures of outcome, but their interpretation across studies, populations, and cultures is problematic.
6. *Changes in the biomarker should be associated with differences in mortality risk.* This increases the confidence that changes caused by an intervention would be reflected in differences in long-term mortality.

A scan of the epidemiologic literature identifies a number of candidate longevity biomarkers including Interleukin-6 (IL-6) levels, muscle strength, usual gait speed, FEV_1, Peak $V0_2$, Cystatin-C, and the Digit Symbol Substitution Test (DSST). Each satisfies most of the criteria listed in Box 8.1: each changes with age, each is consistently related to all-cause mortality, and the mortality relationship is monotonic across their ranges. Each is objectively measured. IL-6, muscle strength, gait speed, and Peak $V0_2$ are sensitive to change. It is unclear whether FEV_1, Cystatin-C, or DSST can be improved with an intervention. However, there are proven interventions that can slow the rate of change of both FEV_1 and Cystatin-C. Table 8.1 shows how some of these markers predict 12-year mortality in the Health Aging and Body Composition Study, a study of 3,075 well-functioning community-dwelling older adults aged 70–79. There is a strong gradient in all-cause mortality across each of these measures. A composite score based on sum of ranks of each of the components shows even a stronger gradient with an over threefold higher mortality risk in those doing most poorly across all the components compared to those with the most favorable biomarker profile. Treatments showing a significant relative improvement in the composite would be expected to extend life expectancy in humans. It should be noted that several of the proposed biomarkers reflect functions that are desirable in and of themselves—improved gait and walking speed, for example—and thus benefits to weaker or slower older adults would tip the risk–benefit ratio allowing for the ethical testing of novel interventions over shorter time frames in persons limited in these domains.

LIMITATIONS FOR BOTH ANIMAL AND HUMAN MODELS

Some issues of translational work in the field of aging are inadequately addressed in either current animal models or human studies. The usual investigative paradigm employed and rewarded by grant/manuscript reviewers is that of focusing on a single, isolated experimental variable (e.g., specific gene knockouts) while controlling as many other potentially

TABLE 8.1 Twelve-year mortality risk in the Health ABC cohort by gender-specific quantiles (7th) of candidate biomarkers

Marker	Worst Category					Best Category
Interleukin-6	67%	62%	55%	50%	44%	41%	35%
Grip Strength	61%	51%	55%	53%	49%	48%	39%
Gait Speed	70%	56%	55%	48%	48%	44%	33%
% Predicted FEV_1	68%	56%	49%	45%	40%	46%	44%
Crystatin-C	73%	51%	54%	47%	48%	44%	40%
Digit-Symbol Substitution Test	72%	60%	53%	50%	46%	42%	30%
Composite Score	78%	69%	52%	49%	40%	34%	23%

influencing factors as possible. This approach has been applied not only to mouse models as outlined previously, but also to humans; again we will use immune senescence as an example. The SENIEUR criteria (Ligthart et al., 1984) were originally derived to highly select aged adults with little/no comorbid illness for comparison to young adults with no illness to isolate the factor of "age." Only individuals who have aged successfully are enrolled in such studies, but these are not at all "typical" seniors; they are the "elite" aged. This point was made by several researchers who argue for a broader paradigm (Castle, Uyemura, and Makindan, 2001). The greatest value for human studies using criteria like SENIEUR is when the "elite" group is compared to older adults with comorbidity (Trzonkowski et al., 2009). The same is true of mice purchased from aging colonies. If only the very aged are used in experiments and compared to young adults, the senior group represents only those animals that have aged in an elite fashion and survived to old age. The strategy of using an evolved, broader model for animal studies in which old "well" versus old "diseased" comparisons are made (rather than just old/young comparisons) is virtually never employed. Further, assessment of physical and cognitive function in animal models that mimic human functional outcomes of interest (e.g., disability) are critical to develop further (e.g., Shively et al., 2012) if animal models are to translate to humans.

The explicit exclusion of animal and human studies that include comorbidity in order to keep the models "simple" is perhaps the biggest disconnect in the translational science of aging. *In disease-focused research, single-hit animal models or predominant illness cohorts may be appropriate experimental models, but translational geriatrics demands that the investigative model* ***embrace complexity*** *as a foundational element.* About half of all Medicare patients have three or more chronic conditions (e.g., co-existent heart disease, diabetes, and arthritis). The effects of multimorbidity are not additive but multiplicative (Boyd et al., 2008). For example, the relative risk for disability is increased 2.3-fold in those with heart disease, 4.3-fold for those with arthritis, but 13.6-fold for those with both (Boyd et al., 2008). It is impossible to ignore these issues when conducting translational geriatric research without almost assuring failure when applying research from simple, age-focused animal or human models to the vastly heterogeneous senior population.

EXAMPLE OF TRANSLATIONAL RESEARCH IN AGING: CALORIE RESTRICTION

Since the 1930s scientists have known that reducing caloric intake can have a profound effect on lifespan. This is among the most robust interventions in all of science with the observation holding true in yeast, flies, worms, and rodents (Fontana, Partridge, and Longo, 2010) (Figure 8.4). There is conflicting literature with regard to the effect of calorie restriction (CR) in nonhuman primates (NHPs) with increased median lifespan [*ad lib* fed animals 25.9 years (IQR 7.10) vs 28.3 years (IQR 7.29) in calorie-restricted animals] in one study (Colman et al., 2014). Another study showed no difference in lifespan whether CR was initiated early or late in adulthood, but there was a trend in the CR group toward a lower incidence of age-related disease (cardiovascular, diabetes, neoplasia, $p = 0.06$) (Mattison et al., 2012). Extensive investigation of the mechanisms underlying this observation using both CR and mutational approaches has shown that conserved nutrient sensing [particularly IGF and target of rapamycin (TOR) (Figure 8.4) (Mirzaei, Suarez, and Longo, 2014)], and redox pathways (Walsh, Shi, and Remmen, 2014) are very likely implicated.

Of course, starvation—extreme calorie restriction—has severe, lifespan-limiting effects, so optimally restricting calories in a "modest" range—25%–30%—seems to have the greatest effect in animal models (Fontana, Partridge, and Longo, 2010).

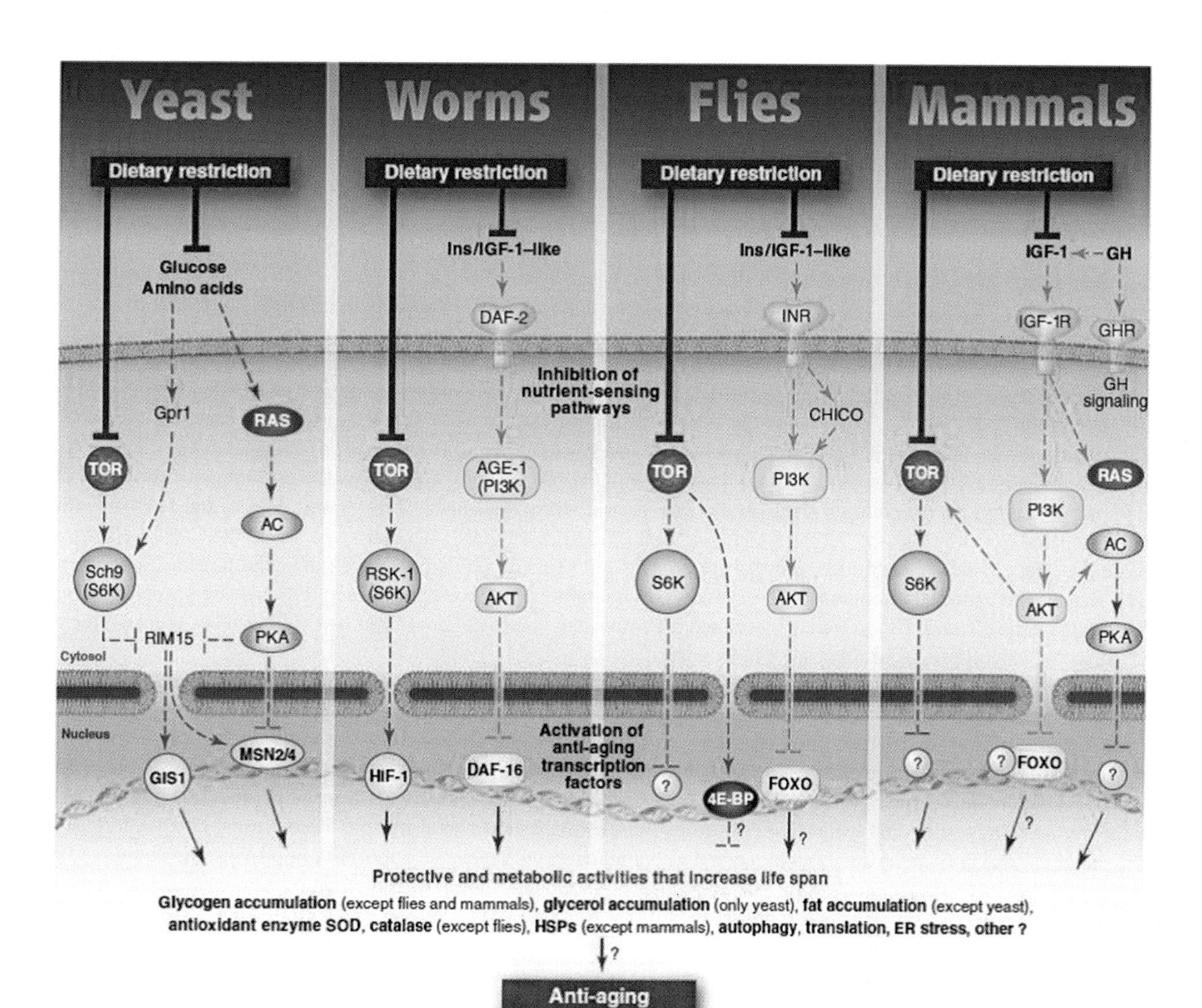

FIGURE 8.4 The pathways implicated as likely mechanisms by which dietary restriction exerts anti-aging effects appear to be conserved from yeast to worms to flies and even mammals. *(Reprinted from Fontana, Partridge, & Longo, 2010, with permission.)*

Similarly, the timing of caloric restriction is critical, even with only modest CR. There are data that suggest the earlier one starts calorie restriction, the more effective it is; however, there are significant negative effects of even moderate CR on reproductive capacity (Fontana, Partridge, and Longo, 2010). CR not only extends lifespan but also mitigates disease development (Anderson and Weindruch, 2012; Colman et al., 2014), and maintains physical function in rhesus macaques (Yamada et al., 2013), thus increasing healthspan as well.

Small molecule CR "mimetics" may hold the most real promise for human interventions, and once pathways were elucidated, at least one such compound, rapamycin, was tested in the NIA's interventions testing program and shown to extend median lifespan in both male (10%–23%) and female (18%–26%) mice (Miller et al., 2011, 2014) and slow age-related declines in physical activity (Wilkinson et al., 2012). However, there do appear to be specific organ (testicular degeneration, cataracts) and molecular alterations (insulin sensitivity, IGF changes) that differ between CR and rapamycin (Miller et al., 2014; Wilkinson et al., 2012); thus, not surprisingly, the pleotropic effects of CR are not completely recapitulated by a single small molecular mimetic.

Evaluating the effects of CR in humans is complex. Chronic CR as determined by low body mass index(BMI) is associated with poor health in humans. There are few "natural experiments" because most undernourished human populations are also typically impoverished with high burdens of communicable disease and high mortality rates. In developed countries, while there tends to be an excess of mortality in older adults with either very high or very low BMIs (weight / height^2; BMIs), there is little relationship between BMI and mortality across a wide range of intermediate BMIs. Very low BMIs in older persons are usually a sign of poor health and ongoing unresolved health issues. These observations do not directly address the question of whether voluntary and nutritionally adequate caloric restriction has benefits in terms of either healthspan or lifespan in older adults. The question has been approached from a number of directions building on the substantial knowledge base built upon CR animal model systems.

Mercken and colleagues identified 15 (BMI = 19.2 +/– 1.1 kg/m^2) middle-aged individuals who had voluntarily restricted their energy intake by 30% for an average of 9.6 years (Mercken et al., 2013). Their muscle gene expression profile was compared to that of nonobese age-matched controls (BMI = 25.3+/– 2.3 kg/m^2) consuming a typical Western diet. The transcriptional profile of the calorically restricted group showed downregulation of IGF-1/insulin signaling pathways compared to the controls. The pattern of transcriptional changes was remarkably homologous to that seen within the muscles of calorically restricted rats with downregulation of genes related to inflammation and an upregulation of genes associated with mitochondrial biogenesis, autophagy, and antioxidant enzymes. Investigators from the Pennington Biomedical Research Center evaluated three different strategies to achieve 10% weight loss in nonobese adults in a 6-month pilot study (Heilbronn et al., 2006). Compared to the control group, weight loss was associated with lower insulin levels but not lower glucose levels. DHEAS and protein carbonyl levels were not affected, but caloric restriction did significantly lower core body temperature and lean body mass-adjusted resting metabolic rate. Cells derived from humans after they have undergone a weight loss intervention show an increase in heat stress resistance, Sirt1 Protein levels, and PGS-1α mRNA levels compared to cells collected before the intervention (Allard et al., 2008). The NIA sponsored the CALERIE study to explore the effect of a 25% restriction in nonobese adults for 2 years on physiologic and molecular targets (Rochon et al., 2011). Results are expected imminently. These studies do not target older populations, and thus provide proof-of-principle information on the effect of CR on longevity-related pathways, but would not provide strong evidence in support of the idea that CR would extend life- or healthspan in older adults.

Using a biomarker approach, one can evaluate the likelihood that energy restriction would reduce mortality rate in older persons. There have been a handful of clinical trials that evaluated weight reduction in older overweight and obese adults defined as a mean baseline age of ≥ 65 years. These studies show that short-term (12–18 month) weight reduction significantly increases gait speed, increases peak VO_2, and reduces IL-6 levels (Nicklas et al., 2004; Rejeski et al., 2011; Villareal et al., 2011). While muscle mass is lost when total mass is reduced, it does not appear that weight loss interventions affect muscle strength, suggesting that weight loss improves muscle quality defined as muscle force/muscle cross-sectional area (Villareal et al., 2011). Weight loss both alone and in conjunction with exercise appears to improve some cognitive functions in older adults (Napoli et al., 2014). The effect of weight reduction on renal function or pulmonary function has not been evaluated in older adults. However, data from the Look AHEAD trial, which examined the effect of long-term weight reduction and physical exercise on cardiovascular events in persons with Type II diabetes, showed a 30% reduction in the onset of very-high-risk-chronic kidney disease after 8–10 years of follow-up (Look AHEAD Research Group, 2014).

There have been some studies of the effect of CR on mortality in humans despite the challenges of doing so. The Swedish Obesity Study followed 2010 persons receiving gastric banding or by-pass surgery and compared their mortality experience to the experience of a matched control (Sjostrom et al., 2007). The surgical procedures *generated a sustained weight* loss of between 14% and 25% and were associated with a 24% reduction ($p = 0.04$) in all-cause mortality over 10.9 years of follow-up. However, as a nonrandomized study, it is unclear if selection bias or confounding contributed to the observed mortality advantage. In addition, the study did not include older persons. The best evidence for a benefit from caloric restriction comes from randomized trials evaluating weight loss interventions compared to control or non-weight loss interventions. There are three reports in the literature of randomized trials of weight loss that followed persons for more than 5 years: the ADAPT study of obese and overweight older persons with knee osteoarthritis, the TONE study of overweight and obese older adults with mild hypertension, and the aforementioned Look AHEAD trial (Shea et al., 2010, 2011; Wing et al., 2013). Only ADAPT and TONE targeted older persons with mean baseline ages > 65. All three studies found that persons randomized to weight reduction had lower all-cause mortality with follow-up times ranging from 8 to 10 years. The relative reductions in risk were 50% ($p = 0.01$), 12% ($p > 0.05$), and 15% ($p = 0.11$) for ADAPT, TONE, and Look AHEAD, respectively. If one pools the published data, a significant reduction in total mortality is observed. The mortality benefits from the human studies may be due to the treatment of obesity rather than biologic changes associated with CR. The ADAPT study was an 18-month intervention, but the mortality advantage persisted throughout the follow-up period. The Look AHEAD and SOS study survival curves are instructive in that in both studies, the survival advantage for the weight-losing groups did not begin to manifest until 4–5 years after randomization. This delay in benefit provides a benchmark for understanding how long it may take for changes on the cellular level to play out as reductions in mortality risk. In contrast, beneficial changes in many biomarkers are evident after 6 months of intervention.

TRANSLATIONAL AGING RESOURCES

Animals and Animal Tissues

There are important animal resources for aging research primarily assembled by the NIA (National Institute on Aging, 2014d). A number of rodent resources are available including mice and rats across the age spectrum of various strains, F1 crosses, and calorie-restricted animals. In addition, a tissue bank of specimens is available. All investigators with an aging-focused grant are eligible to access these resources, but pricing and access priority vary by specimens/strain and funding source (NIA grantee vs other). Other options for rodents include retired breeders from several manufacturers; however, investigators are cautioned that retired males often fight when re-paired in housing and may require individual cages. Females that have produced multiple litters have substantial variation from nulliparous females for some biologic systems (e.g., immunology).

NHP resources are also available (University of Wisconsin–Madison, 2014). When compared to rodents, NHPs better model a number of critical issues in aging research: NHPs have more human-like joints and movement that may better simulate arthritis, functional decline, and sarcopenia than rodent models. Recently, functional assessments that mimic physical performance measures in humans have been demonstrated to decline with age in NHPs (Shively et al., 2012).

Cohorts/Populations

The NIA and other NIH institutes have made decades-long investments in a number of large cohorts either directly to follow aging (e.g., BLSA, Health ABC), disease-focused in an aging-related area [e.g., Multi-Ethnic Study on Atherosclerosis (MESA), Cardiovascular Health Study (CHS)], or in specific groups that were initially middle aged but have now entered "geriatric" status [e.g., Women's Health Initiative (WHI)]. Other important surveys provide critical data in aging cohorts (e.g., Health and Retirement Survey). These are summarized in a number of websites with access information and specimen repositories (National Archive of Computerized Data on Aging, 2014; National Institute on Aging, 2014c; National Institutes of Health, 2014a).

Tools and Toolboxes

The NIH Toolbox (National Institutes of Health, 2014b) is a recently released, multidimensional set of brief measures assessing cognitive, emotional, motor, and sensory function across the lifespan. The NIH Toolbox was created to address the need for a standard set of measures that could be used across diverse study groups and designs in various settings to assess neurological and behavioral function facilitating the study of changes across the lifespan. Many resources in the toolbox are free, but fees may apply for use of the NIH Toolbox in projects not funded by the NIH.

The Patient Reported Outcomes Measurement Information System (PROMIS®) (National Institutes of Health, 2014c) is a free resource developed to provide validated patient-reported outcomes for clinical research and practice. PROMIS® uses computer adaptive testing and traditional paper instruments in global health, physical function, fatigue, pain, sleep/wake function, emotional distress, and social health. PROMIS® provides resources in four key areas: (1) Comparability—measures standardized across common domains, metrics, and conditions; (2) Reliability and validity—metrics for each domain have been rigorously reviewed and tested; (3) Flexibility—can be administered in a variety of ways or formats; and (4) Inclusiveness—designed to span all study populations regardless of literacy, language, physical function, or life course.

CONCLUSION

Translational research in aging has the potential to influence lifespan and healthspan beyond the boundaries of traditional disease-based research. There are many animal models and human cohorts that are useful in studying aging, but important limitations include marked differences among them that make translating animal findings to humans difficult, a number of key limitations that affect longitudinal studies in both humans and animals, and the need to study multimorbidity and complexity for translational research to be generalizable to most older adults. Focusing on healthspan rather than lifespan by including functional outcomes as endpoints is critical and requires experimental design and analysis plans unique to aging research. A number of resources are available from the NIA, the Centers for Disease Control and Prevention, and other agencies to assist translational research in this area.

REFERENCES

Akbar, A.N., Soares, M.V., Plunkett, F.J., Salmon, M., 2000. Differential regulation of CD8+ T cell senescence in mice and men. Mech. Ageing Dev. 121, 69–76.

Allard, J.S., Heilbronn, L.K., Smith, C., Hunt, N.D., Ingram, D.K., Ravussin, E., et al., 2008. In vitro cellular adaptations of indicators of longevity in response to treatment with serum collected from humans on calorie restricted diets. PLoS One 3, e3211.

Anderson, R.M., Weindruch, R., 2012. The caloric restriction paradigm: implications for healthy human aging. Am. J. Hum. Biol. 24, 101–106.

Baker, D.J., Wijshake, T., Tchkonia, T., LeBrasseur, N.K., Childs, B.G., van de Sluis, B., et al., 2011. Clearance of p16Ink4a-positive senescent cells delays ageing-associated disorders. Nature 479, 232–236.

Barreiro, L.B., Ben-Ali, M., Quach, H., Laval, G., Patin, E., Pickrell, J.K., et al., 2009. Evolutionary dynamics of human toll-like receptors and their different contributions to host defense. PLoS Genet. 5, e1000562.

BLSA, 2014. Baltimore Longitudinal Study of Aging. http://www.blsa.nih.gov/.

Boyd, C.M., Ritchie, C.S., Tipton, E.F., Studenski, S.A., Wieland, D., 2008. From bedside to bench: summary from the American Geriatrics Society/National Institute on Aging Research Conference on Comorbidity and Multiple Morbidity in Older Adults. Aging Clin. Exp. Res. 20, 181–188.

Broer, L., Buchman, A.S., Deelen, J., Evans, D.S., Faul, J.D., Lunetta, K.L., et al., 2014. GWAS of longevity in CHARGE consortium confirms APOE and FOXO3 candidacy. J. Gerontol. A. Biol. Sci. Med. Sci. pii: glu166. [Epub ahead of print].

Brooks-Wilson, A.R., 2013. Genetics of healthy aging and longevity. Hum. Genet. 132, 1323–1338.

Bruce-Keller, A.J., Gupta, S., Knight, A.G., Beckett, T.L., McMullen, J.M., Davis, P.R., et al., 2011. Cognitive impairment in humanized APPxPS1 mice is linked to Abeta(1-42) and NOX activation. Neurobiol. Dis. 44, 317–326.

Burch, J.B., Augustine, A.D., Frieden, L.A., Hadley, E., Howcroft, T.K., Johnson, R., et al., 2014. Advances in geroscience: impact on healthspan and chronic disease. J. Gerontol. A. Biol. Sci. Med. Sci. 69 (Suppl. 1), S1–S3.

Castle, S.C., Uyemura, K., Makinodan, T., 2001. The SENIEUR Protocol after 16 years: a need for a paradigm shift? Mech. Ageing Dev. 122, 127–130.

Centers for Disease Control and Prevention NHS, 2014. http://www.cdc.gov.

Colman, R.J., Beasley, T.M., Kemnitz, J.W., Johnson, S.C., Weindruch, R., Anderson, R.M., 2014. Caloric restriction reduces age-related and all-cause mortality in rhesus monkeys. Nat. Commun. 5, 3557.

Copeland, S., Warren, H.S., Lowry, S.F., Calvano, S.E., Remick, D., 2005. Acute inflammatory response to endotoxin in mice and humans. Clin. Diagn. Lab Immunol. 12, 60–67.

Fauce, S.R., Jamieson, B.D., Chin, A.C., Mitsuyasu, R.T., Parish, S.T., Ng, H.L., et al., 2008. Telomerase-based pharmacologic enhancement of antiviral function of human CD8+ T lymphocytes. J. Immunol. 181, 7400–7406.

Fiatarone, M.A., Marks, E.C., Ryan, N.D., Meredith, C.N., Lipsitz, L.A., Evans, W.J., 1990. High-intensity strength training in nonagenarians. Effects on skeletal muscle. JAMA 263, 3029–3034.

Fontana, L., Partridge, L., Longo, V.D., 2010. Extending healthy life span—from yeast to humans. Science 328, 321–326.

Harrison, et al., 2014. Aging Cell 13(2), 273–282. http://dx.doi.org/10.1111/acel.12170.

Heilbronn, L.K., de Jonge, L., Frisard, M.I., DeLany, J.P., Larson-Meyer, D.E., Rood, J., et al., 2006. Effect of 6-month calorie restriction on biomarkers of longevity, metabolic adaptation, and oxidative stress in overweight individuals: a randomized controlled trial. JAMA 295, 1539–1548.

Iwaoka, W., Tannebaum, S.R., 1976. Photohydrolytic detection of N-nitroso compounds in high-performance liquid chromatography. IARC Sci. Publ. 51–56.

Kim, D.H., Guengerich, F.P., 1990. Formation of the DNA adduct S-[2-(N7-guanyl)ethyl]glutathione from ethylene dibromide: effects of modulation of glutathione and glutathione S-transferase levels and lack of a role for sulfation. Carcinogenesis 11, 419–424.

Kirkland, J.L., 2013. Translating advances from the basic biology of aging into clinical application. Exp. Gerontol. 48, 1–5.

Kuh, D., Bassey, J., Hardy, R., Aihie, S.A., Wadsworth, M., Cooper, C., 2002. Birth weight, childhood size, and muscle strength in adult life: evidence from a birth cohort study. Am. J. Epidemiol. 156, 627–633.

Ligthart, G.J., Corberand, J.X., Fournier, C., Galanaud, P., Hijmans, W., Kennes, B., et al., 1984. Admission criteria for immunogerontological studies in man: the SENIEUR protocol. Mech. Ageing Dev. 28, 47–55.

Liu, Y., Johnson, S.M., Fedoriw, Y., Rogers, A.B., Yuan, H., Krishnamurthy, J., et al., 2011. Expression of p16(INK4a) prevents cancer and promotes aging in lymphocytes. Blood 117, 3257–3267.

Look AHEAD Research Group, 2014. Effect of a long-term behavioural weight loss intervention on nephropathy in overweight or obese adults with type 2 diabetes: a secondary analysis of the Look AHEAD randomised clinical trial. Lancet. Diabetes Endocrinol. 2, 801–809.

Lopez-Otin, C., Blasco, M.A., Partridge, L., Serrano, M., Kroemer, G., 2013. The hallmarks of aging. Cell 153, 1194–1217.

MacGregor, R.R., Boyer, J.D., Ugen, K.E., Lacy, K.E., Gluckman, S.J., Bagarazzi, M.L., et al., 1998. First human trial of a DNA-based vaccine for treatment of human immunodeficiency virus type 1 infection: safety and host response. J. Infect. Dis. 178, 92–100.

Mattison, J.A., Roth, G.S., Beasley, T.M., Tilmont, E.M., Handy, A.M., Herbert, R.L., et al., 2012. Impact of caloric restriction on health and survival in rhesus monkeys from the NIA study. Nature 489, 318–321.

Maue, A.C., Yager, E.J., Swain, S.L., Woodland, D.L., Blackman, M.A., Haynes, L., 2009. T-cell immunosenescence: lessons learned from mouse models of aging. Trends Immunol. 30, 301–305.

Mazzoni, M.M., Giannaccini, G., Luccacchini, A., Bazzichi, L., Ciompi, L.M., Pasero, G., et al., 1986. Urate binding globulin: specific antibody preparation. Adv. Exp. Med. Biol. 195 Pt A, 387–392.

Mercken, E.M., Crosby, S.D., Lamming, D.W., JeBailey, L., Krzysik-Walker, S., Villareal, D.T., et al., 2013. Calorie restriction in humans inhibits the PI3K/AKT pathway and induces a younger transcription profile. Aging Cell 12, 645–651.

Mestas, J., Hughes, C.C., 2004. Of mice and not men: differences between mouse and human immunology. J. Immunol. 172, 2731–2738.

Miller, R.A., 2002. Extending life: scientific prospects and political obstacles. Milbank Q. 80, 155–174.

Miller, R.A., Harrison, D.E., Astle, C.M., Baur, J.A., Boyd, A.R., de Cabo, C.R., et al., 2011. Rapamycin, but not resveratrol or simvastatin, extends life span of genetically heterogeneous mice. J. Gerontol. A. Biol. Sci. Med. Sci. 66, 191–201.

Miller, R.A., Harrison, D.E., Astle, C.M., Fernandez, E., Flurkey, K., Han, M., et al., 2014. Rapamycin-mediated lifespan increase in mice is dose and sex dependent and metabolically distinct from dietary restriction. Aging Cell 13, 468–477.

Mirzaei, H., Suarez, J.A., Longo, V.D., 2014. Protein and amino acid restriction, aging and disease: from yeast to humans. Trends Endocrinol. Metab. pii: S1043–2760(14)00127-1. doi: 10.1016/j.tem.2014.07.002. [Epub ahead of print].

Munford, R.S., 2010. Murine responses to endotoxin: another dirty little secret? J. Infect. Dis. 201, 175–177.

Napoli, N., Shah, K., Waters, D.L., Sinacore, D.R., Qualls, C., Villareal, D.T., 2014. Effect of weight loss, exercise, or both on cognition and quality of life in obese older adults. Am. J. Clin. Nutr. 100, 189–198.

National Archive of Computerized Data on Aging, 2014. http://www.icpsr.umich.edu/icpsrweb/NACDA/.

National Institute on Aging, 2014a.

National Institute on Aging, 2014b. Interventions Testing Program (ITP). http://www.nia.nih.gov/research/dab/interventions-testing-program-itp.

National Institute on Aging, 2014c. Publicly Available Databases for Aging-Related Secondary Analyses in the Behavioral and Social Sciences. http://www.nia.nih.gov/research/dbsr/publicly-available-datasets.

National Institute on Aging, 2014d. Scientific Resources. http://www.nia.nih.gov/research/scientific-resources.

National Institutes of Health, 2014a. National Institute on Aging Population Studies Database. http://nihlibrary.ors.nih.gov/nia/ps/niadb.asp.

National Institutes of Health, 2014b. NIH Toolbox for the Assessment of Neurological and Behavioral Function. http://www.nihtoolbox.org/Pages/default.aspx.

National Institutes of Health, 2014c. Patient Reported Outcomes Measurement Information System. http://www.nihpromis.org/.

Nicklas, B.J., Ambrosius, W., Messier, S.P., Miller, G.D., Penninx, B.W., Loeser, R.F., et al., 2004. Diet-induced weight loss, exercise, and chronic inflammation in older, obese adults: a randomized controlled clinical trial. Am. J. Clin. Nutr. 79, 544–551.

Peeper, D.S., 2011. Ageing: old cells under attack. Nature 479, 186–187.

Rejeski, W.J., Brubaker, P.H., Goff Jr., D.C., Bearon, L.B., McClelland, J.W., Perri, M.G., et al., 2011. Translating weight loss and physical activity programs into the community to preserve mobility in older, obese adults in poor cardiovascular health. Arch. Intern. Med. 171, 880–886.

Rochon, J., Bales, C.W., Ravussin, E., Redman, L.M., Holloszy, J.O., Racette, S.B., et al., 2011. Design and conduct of the CALERIE study: comprehensive assessment of the long-term effects of reducing intake of energy. J. Gerontol. A. Biol. Sci. Med. Sci. 66, 97–108.

Shea, M.K., Houston, D.K., Nicklas, B.J., Messier, S.P., Davis, C.C., Miller, M.E., et al., 2010. The effect of randomization to weight loss on total mortality in older overweight and obese adults: the ADAPT Study. J. Gerontol. A. Biol. Sci. Med. Sci. 65, 519–525.

Shea, M.K., Nicklas, B.J., Houston, D.K., Miller, M.E., Davis, C.C., Kitzman, D.W., et al., 2011. The effect of intentional weight loss on all-cause mortality in older adults: results of a randomized controlled weight-loss trial. Am. J. Clin. Nutr. 94, 839–846.

Shively, C.A., Willard, S.L., Register, T.C., Bennett, A.J., Pierre, P.J., Laudenslager, M.L., et al., 2012. Aging and physical mobility in group-housed Old World monkeys. Age (Dordr) 34, 1123–1131.

Shock, N.W., Greulich, R.C., Andres, R., Arenberg, D., Costa, J.P., Lakatta, E.G., et al., 1984. Normal human aging: The Baltimore Longitudinal Study of Aging. Government Printing Office, Washington, DC.

Shultz, L.D., Ishikawa, F., Greiner, D.L., 2007. Humanized mice in translational biomedical research. Nat. Rev. Immunol. 7, 118–130.

Sjostrom, L., Narbro, K., Sjostrom, C.D., Karason, K., Larsson, B., Wedel, H., et al., 2007. Effects of bariatric surgery on mortality in Swedish obese subjects. N. Engl. J. Med. 357, 741–752.

Trzonkowski, P., Mysliwska, J., Pawelec, G., Mysliwski, A., 2009. From bench to bedside and back: the SENIEUR Protocol and the efficacy of influenza vaccination in the elderly. Biogerontology 10, 83–94.

United Nations, 2014. World Population Ageing 1950–2050. http://www.un.org/esa/population/publications/worldageing19502050/.

University of Wisconsin–Madison, National Primate Research Center, 2014. Internet Primate Aging Database. http://ipad.primate.wisc.edu.

U.S. Census Bureau, 2014. http://www.census.gov.

Villareal, D.T., Chode, S., Parimi, N., Sinacore, D.R., Hilton, T., Armamento-Villareal, R., et al., 2011. Weight loss, exercise, or both and physical function in obese older adults. N. Engl. J. Med. 364, 1218–1229.

Virgin, H.W., Wherry, E.J., Ahmed, R., 2009. Redefining chronic viral infection. Cell 138, 30–50.

Walsh, M.E., Shi, Y., Van, R.H., 2014. The effects of dietary restriction on oxidative stress in rodents. Free Radic. Biol. Med. 66, 88–99.

Warren, H.S., Fitting, C., Hoff, E., Adib-Conquy, M., Beasley-Topliffe, L., Tesini, B., et al., 2010. Resilience to bacterial infection: difference between species could be due to proteins in serum. J. Infect. Dis. 201, 223–232.

Wilkinson, J.E., Burmeister, L., Brooks, S.V., Chan, C.C., Friedline, S., Harrison, D.E., et al., 2012. Rapamycin slows aging in mice. Aging Cell 11, 675–682.

Wing, R.R., Bolin, P., Brancati, F.L., Bray, G.A., Clark, J.M., Coday, M., et al., 2013. Cardiovascular effects of intensive lifestyle intervention in type 2 diabetes. N. Engl. J. Med. 369, 145–154.

Yamada, Y., Colman, R.J., Kemnitz, J.W., Baum, S.T., Anderson, R.M., Weindruch, R., et al., 2013. Long-term calorie restriction decreases metabolic cost of movement and prevents decrease of physical activity during aging in rhesus monkeys. Exp. Gerontol. 48, 1226–1235.

Chapter 9

Translational Medicine: The Changing Role of Big Pharma

C. Simone Fishburn

INTRODUCTION

The growth of translational research has coincided with ground-shifting changes in the landscape of drug development, seen in new structures, relationships, and attitudes of the key players involved.

Twenty years ago, there was one key player: big pharma. Now, big pharma sits alongside successful biotech companies, start-ups, academic organizations, and venture firms and exists in an innovation ecosystem that includes multiple other service providers that play important roles in drug development.

This chapter discusses how pharma is waking up to the changes around it and adapting its operational strategy to find new ways of bringing more drugs to market.

HISTORY: HOW DID WE GET HERE?

The oldest pharmaceutical companies date back to the mid-19th century, when they began to industrialize the generation of chemicals for medical use. Indeed, all the corporations currently recognized as "big pharma" have their roots in companies founded in the second half of the 19th century or early 20th century (see Table 9.1).

By contrast, the biotech industry was launched in the latter part of the 20th century with the founding of Genentech in 1976, followed by Amgen in 1980. Those companies were formed to capitalize on the newly discovered DNA technology and provide an efficient means of making highly targeted protein-based drugs. The result was the creation of new recombinant protein drugs including, significantly, antibody-based drugs that added a completely new therapeutic modality to address unmet medical needs (see "Pharmas and Biotechs: What's in a Name?").

Pharmas and Biotechs: What's in a Name?

As the drug development landscape has grown and the organizations have changed, the names have stuck but the old definitions for a pharmaceutical company or a biotech company have started to lose relevance. Traditionally, pharmaceutical companies produced small molecule drugs and were fully integrated drug development companies with their own sales forces. Biotechs, by contrast, produced biologics or therapeutics derived from living organisms, such as cell-based or DNA-based therapies, and were often barely bigger than an R&D organization, with no sales force. Pharmas, on the one hand, were conceived of as large, high market-cap organizations with a powerful political lobby. Biotechs, on the other hand, were thought of as small, nimble, generally lower market-cap organizations that used innovative approaches to solve medical problems. The attitudes persist today—although the reality belies the definitions. As Table 9.1 shows, five of the biggest biotech companies have market caps in the range of—and sometimes greater than— traditional pharmas. Genentech is, in fact, now owned by Roche but operates independently and is still often considered a biotech. Medimmune, now owned by AstraZeneca, has a similar status. However, these big biotechs—including Genentech before its acquisition—are fully integrated companies with full sales divisions and multiple marketed products. In addition, while there are many smaller biotechs, including over 400 publicly listed ones and over 1,500 privately held ones (Hodgson, 2006; Lähteenmäki and Lawrence, 2005), many of them are developing therapeutics that are traditional small molecule drugs rather than biotechnology products. Likewise, pharmas are now developing biotechnology products in addition to small molecules.

This chapter defines the sectors based on their origins, since the behaviors and attitudes align most closely with the history of the company and its foundation rather than with its structure today or the technology basis of its products.

Principles of Translational Science in Medicine. http://dx.doi.org/10.1016/B978-0-12-800687-0.00032-3

TABLE 9.1 Big pharma and top biotech companies*

	Company Name	Year Founded	Number of Marketed Products (Originated in the Company)	Market Cap†	Country
Big Pharma	Abbott Laboratories	1900	58	$64 B	U.S.
	AbbVie Inc.	2013	33	$89 B	U.S.
	AstraZeneca plc	1999	108	$93 B	U.K.
	Baxter International Inc.	1931	44	$41 B	U.S.
	Bayer AG	1863	84	$84 B	Germany
	Boehringer Ingelheim GmBH	1885	27	privately held	Germany
	Bristol-Myers Squibb Co.	1887	82	$83 B	U.S.
	Eli Lilly and Co.	1876	87	$68 B	U.S.
	GlaxoSmithKline plc	2000	225	$117 B	U.K.
	Johnson & Johnson	1887	101	$291 B	U.S.
	Merck & Co. Inc.	1891	134	$174 B	U.S.
	Novartis AG	1996	164	$218 B	Switzerland
	Novo Nordisk A/S	1923	44	$121 B	Denmark
	Pfizer Inc.	1849	205	$187 B	U.S.
	Roche	1896	129	$249 B	Switzerland
	Sanofi	1973	159	$145 B	France
	Takeda Pharmaceutical Co. Ltd.	1781	59	$36 B	Japan
	UCB Group	1928	18	€14.5 B	Belgium
Big Biotech	Amgen Inc.	1980	61	$105 B	U.S.
	Biogen Idec Inc.	1978	17	$81 B	U.S.
	Celgene Corp.	1986	23	$76 B	U.S.
	Genentech Inc.	1976	35	-	U.S.
	Gilead Sciences Inc.	1987	46	$162 B	U.S.

**The definition of pharma and biotech companies has changed over time (see box). Today's big pharmas trace their roots to the late 19th and early 20th century. Abbvie Inc. was formed from a split of Abbott Labs in 2013. Novartis, GlaxoSmithKline, AstraZeneca, and Sanofi were formed from pharma mergers in the 1990s. The biggest biotech companies have market caps that are comparable to those of the big pharmas, and numbers of in-house originated products that are in the same range as the pharmas. Genentech is now owned by Roche but operates independently and is still thought of by many as a biotech. Market capitalization value in billion dollars, based on stock price in August 2014.*
Source: Google Finance, Yahoo Finance.

Until then, pharmaceutical companies had been the sole source of inventing, manufacturing, and selling new drugs. In the decades in which they dominated the industry, the corporations became large, unwieldy organizations that relied on their own employees for new ideas and whose only real competition was from the few other pharmaceutical companies that existed. The pharma companies created huge campus-type properties, removed from city centers, that did not foster—let alone depend on—outside connections or collaborations and that appeared to offer long-term security and stability to employees.

The result was that the large inward-looking organizations were slow to notice the changes going on around them in biomedical research, technology, and even workforce attitudes.

Causes of Change

The most significant change was the biotech boom, sparked largely by huge advances in not only biology, but also computers and software engineering. New possibilities opened up for rapid data processing, compound generation, and gene manipulation that led to the development of transformative technologies such as polymerase chain reaction, sequencing, gene chips, high throughput screening, and transgenic mice. Together, these created opportunities to address the causes of disease in ways that were unimaginable a decade or two before. Those advances also enabled—and were partly a response to—the human genome project, which itself sparked a "genomics bubble" in which investors and entrepreneurs looked to the power of genomics to unlock knowledge about diseases that had hitherto been black boxes to biologists, and to the potential for population genomic studies to yield medicines tailored to individual patients (Collins, 1999).

Arguably the biggest stimuli for the changed environment—including changed attitudes—were two important pieces of legislation that democratized innovation and opened the door for individuals to pursue their own ideas of how to create new drugs and technologies, and how to capitalize on their own ideas.

The first was the Bayh–Dole Act of 1980 that allowed universities, nonprofit organizations, and small businesses to patent and essentially own inventions arising from federally funded research (Markel, 2013). The aim of the act was to stimulate growth in the sluggish economy by allowing unexploited research discoveries from universities and hospitals to be commercialized.

The second was the Hatch–Waxman Act of 1984 that defined patent exclusivity terms in order to spur innovation for both generic and brand name drugs (Kesselbaum, 2011). For innovators of new molecular entities (NMEs), one of the key provisions of that law was the extension of a drug's patent lifetime after Food and Drug Administration (FDA) approval to compensate for some of the time lost during clinical testing and the FDA review period (Kesselbaum, 2011). By providing a period of patent exclusivity, the act aimed to enable drug developers to recover their investment and secure financial returns that justified their risk. In short, the act provided another incentive for venture firms and other investors to promote entrepreneurship in drug discovery.

These developments also caused a tremendous shift in the outlook for scientists themselves, and led to a marked increase in entrepreneurship among academics in life sciences. Thus, a Ph.D. no longer suggested a choice between a path forward in academia or a career in a big corporation, but opened doors to financial and intellectual independence with the option of founding one's own company or joining a start-up. This diverted top talent away from pharma companies, which had once offered a stable and steady income, and created a marketplace of opportunity and ideas in which researchers could take control of their own careers.

Business-Focused Solutions: M&A and Licensing

As pharma companies started to notice the changes and looked to increase their productivity, they turned first to business-focused—rather than science-focused—solutions. Primary among those were mergers and acquisitions (M&A), which provided a way to obtain new pipeline compounds invented elsewhere and to cut costs by eliminating duplicate administrative, management, and operational roles as two companies became one (LaMattina, 2011).

During the 1990s, described as "the era of mergers" by La Mattina (2011), a flurry of activity saw consolidations of major pharmaceutical companies that reduced the number of big pharma companies and laid the groundwork for the landscape that exists today.

- In 1989, Bristol-Myers merged with the Squibb Corporation and combined their R&D divisions, forming Bristol-Myers Squibb Co.
- In 1989, SmithKline Beckman and The Beecham Group plc merged to form SmithKline Beecham plc, which then merged with Glaxo Wellcome plc in 2000 to form GlaxoSmith Kline plc (GSK).
- In 1996, Ciba-Geigy Ltd. merged with Sandoz Ltd. to form Novartis AG.
- In 1999, the Swedish company Astra AB merged with the U. K.-based Zeneca group to form AstraZeneca plc (AZ).
- In 1999, Pharmacia & Upjohn Inc. merged with Monsanto Co. to form Pharmacia Corp. (The agricultural chemical division was spun off in 2000 under the name Monsanto.)

In addition, over the following decade, many pharma companies acquired rather than merged with other pharma companies to bolster their pipelines.

- Pfizer Inc. acquired Warner-Lambert Co. in 2000, Pharmacia Corp. in 2003, and Wyeth in 2009.
- Sanofi-Synthelabo S.A. acquired Aventis S.A. in 2004. In 2011 the company changed its name to Sanofi.
- Bayer AG acquired Schering AG in 2006.
- Merck & Co. Inc. acquired Schering-Plough Corp. in 2009.

Concurrently, the biotech world was undergoing enormous expansion as the venture community rushed to create new start-ups that rapidly formed a significant market of small- and medium-sized enterprises (SMEs) (Hodgson, 2006; Lähteenmäki and Lawrence, 2005). During the 1980s and 1990s, the number of small companies that produced NMEs doubled from 78 to 145, reflecting the new reality that this sector of the industry was now squarely in play as competitors for big pharma (Munos, 2009).

Pharma companies did not ignore this boom, and in fact stood behind many of the biotech-generated products that made it to market. Indeed, several pharma companies created their own venture firms that operated independently from the company's business division, had separate dedicated venture funds, and went to scout the latest biotechnologies and fund start-ups (Fletcher, 2001).

More significantly though, as pharma companies sought to bolster their pipelines, they adopted a strategy of in-licensing and created business development structures and staff to support it (Papadopoulos, 2000). They turned to biotech companies to license compounds or platform technologies with the promise to yield multiple compounds.

This move created a symbiotic relationship between biotech companies and pharma. The majority of biotech companies did not have the funds or the expertise to carry out large or complex clinical trials, or to negotiate their way through the regulatory process all the way to approval. In addition, they did not have the sales force to fully commercialize the assets.

Thus, licensing deals gave biotech companies a much-needed influx of money and a way to have their products reach patients, while they allowed pharma companies to boost to their pipelines, brought significant revenue, and partly solved their declining ability to innovate.

The Productivity Crunch

By 2000, the big pharma companies still dominated the industry in terms of revenues from drugs sold, market value, dollars invested, and employee numbers. But they now faced significant competition for ideas and access to early innovation from biotech companies, academic spinouts, and the growing biotech venture community. That became apparent within the next decade when drugs invented during the 1990s had made their way through clinical development to FDA approval. Between 1998 and 2007, only 44% of the innovative drugs (Kneller, 2010) approved by the FDA originated in big pharma, whereas 25% originated in biotech companies and 31% in academia. Innovative drugs were defined as "scientifically novel" if they had a novel mechanism of action or were first in a distinct class of compounds at the time of approval (Kneller, 2010). In addition, three-quarters of the drugs from academia were first licensed to biotech companies and only one-quarter to pharma companies (Kneller, 2010).

The wake-up call for the pharma companies was the plateau in productivity that became apparent in the early 2000s, and that flew in the face of the dramatic increases in R&D investment over the previous years (Munos, 2009; Pammoli, Magazzini, and Riccaboni, 2011; Paul et al., 2010).

From 2000 to 2004, an average of 25 NMEs per year received FDA approval, the same number that had been approved a decade earlier between 1990 and 1994 (Center for Drug Evaluation and Research, 2014). By 2010, the pharmaceutical industry was spending over $65 billion a year on R&D (PhRMA, 2011), but still only 21 NMEs drugs were approved that year. The bottom line was that the pharma companies' strategy of M&A and in-licensing did not increase the number of drugs reaching the market.

Several reviews have explored the reasons behind the falling productivity, often defined as the cost per NME, and have highlighted failures in methods for target validation, project scrutiny, processes for making go-or-no-go decisions, and even internal reward systems within pharma companies (Booth and Zemmel, 2004; Cuatrecasas, 2006; Cook et al., 2014; Munos, 2009; Paul et al., 2010; Williams, 2011).

The genomics revolution had promised a slew of new drugs that many assumed would follow inevitably from the explosion of information available that could yield new therapeutic targets. On the premise that investment would be proportional to output, the industry poured money into R&D. However, that premise ignored the importance of innovation and the many elements that contribute to the successful conversion of an idea into a drug—a process now encompassed by the term *translational research.*

TRANSLATIONAL SOLUTIONS

The productivity crunch has given rise to the much-used term *the valley of death*, which describes the gulf between the tremendous breakthroughs being made in cellular and molecular biology and the static levels of new treatments, diagnostics, and preventative tools reaching the market (Butler, 2008; Meslin, 2013).

The pharmaceutical industry has not been the only stakeholder to suffer or to blame for the valley of death. First among its victims are patients, who have not seen the medical advances that should have come from the large public investment in biomedical research. In the last decade, NIH and academia have also stepped up to re-examine their role and have begun to participate more actively in the drug development ecosystem.

Role of the National Institutes of Health (NIH)

The early 2000s arguably represent the dawn of the era of translational research, which was partly a response to the fact that biological research had drifted away from its historical focus on finding treatments for diseases. Over the previous decades, NIH-funded and other academic research became entrenched in details of basic research, exploring how cells and proteins work, and what drives biological processes. When NIH leaders gathered with top clinicians and scientists to discuss the status, the conclusion was that while basic research is a fundamental building block to creating therapies, not enough of the building was going on (Butler, 2008).

In response, Elias Zerhouni, then director of the NIH—the biggest U.S. public funder of biomedical research—announced in 2003 the Roadmap for Medical Research to chart a path toward drug development by funding programs to share knowledge, resources, and infrastructure for biomedical researchers within academia (Check, 2003). In 2006, the NIH launched the clinical and translational sciences award (CTSA) program that reached over 60 academic medical institutions across the United States. In 2012, under Zerhouni's successor, Francis Collins, the NIH opened its newest institute, the National Center for Advancing Translational Sciences (NCATS).

Overall, the changes have started to bring researchers closer to clinicians, and many hospital-based researchers have welcomed the opportunity to work with laboratory scientists who can contribute molecular expertise to their clinical projects.

Integrated Discovery Nexuses

As the field moves toward dissolving the silos that have separated the various pools of funding, expertise, and knowledge, a growing trend has emerged of integrated discovery nexuses that foster innovation by bringing academic researchers together with experts from industry, venture communities, and contract research organizations (CROs) (Figure 9.1) (Fishburn, 2013a).

The essential feature of a discovery nexus that distinguishes it from a simple bilateral alliance is that it incorporates most of the players in the drug development ecosystem. The nexuses are partly a consequence of the layoffs from pharma during the 1990s and 2000s and the growth of entrepreneurism in the field. Together they caused a migration of talent to small businesses and created a pool of independent consultants that provide services, expertise, and funding to help innovators translate their discoveries to tractable programs or commercially viable opportunities.

In some cases, such as Stanford University's SPARK program, these programs are housed within and run by a university. In others, such as the California Institute of Quantitative Biosciences (QB3), they involve multiple universities and are partly state-funded (May, 2011; Fishburn, 2014a). QB3 is a nonprofit organization with a commercialization arm, the Innolab, and research facilities at three University of California campuses—UC Berkeley, UC San Francisco, and UC Santa Cruz. Pharma companies have also begun to generate nexuses—such as J&J's Janssen Labs, which has a no-strings-attached model and provides facilities, resources, and education to give entrepreneurial scientists access to expertise in translational science from across the ecosystem.

In addition, several state and local governments have created science parks and biotech incubators that provide laboratory space and facilities to spur business creation in their communities (Fishburn, 2013a).

PUBLIC–PRIVATE PARTNERSHIPS: THE NEW MANTRA

One important consequence of this changing landscape is increasingly industry-friendly attitudes among academic researchers who recognize not only new options for funding, but also opportunities for converting their discoveries into meaningful products.

Big pharma is responding to the changes in academia and the innovation landscape by creating independent external discovery units. In their previous business models, pharma companies' interactions with academia were primarily aimed at licensing intellectual property and then developing the assets in-house. Now, these companies are acknowledging they cannot simply put a bigger "welcome" sign on the door; they need to invest and actively foster innovation in the other sectors of the ecosystem. Thus, different models of public–private partnerships (PPPs) have emerged that are not simply licensing deals, but involve ongoing interactions between pharma companies and academic organizations. (Tralau-Stewart et al., 2009).

Pharma and Academia

Several pharma companies have launched discovery centers designed to foster ongoing collaboration between academic researchers and company scientists and to complement the pharma companies' internal R&D activities.

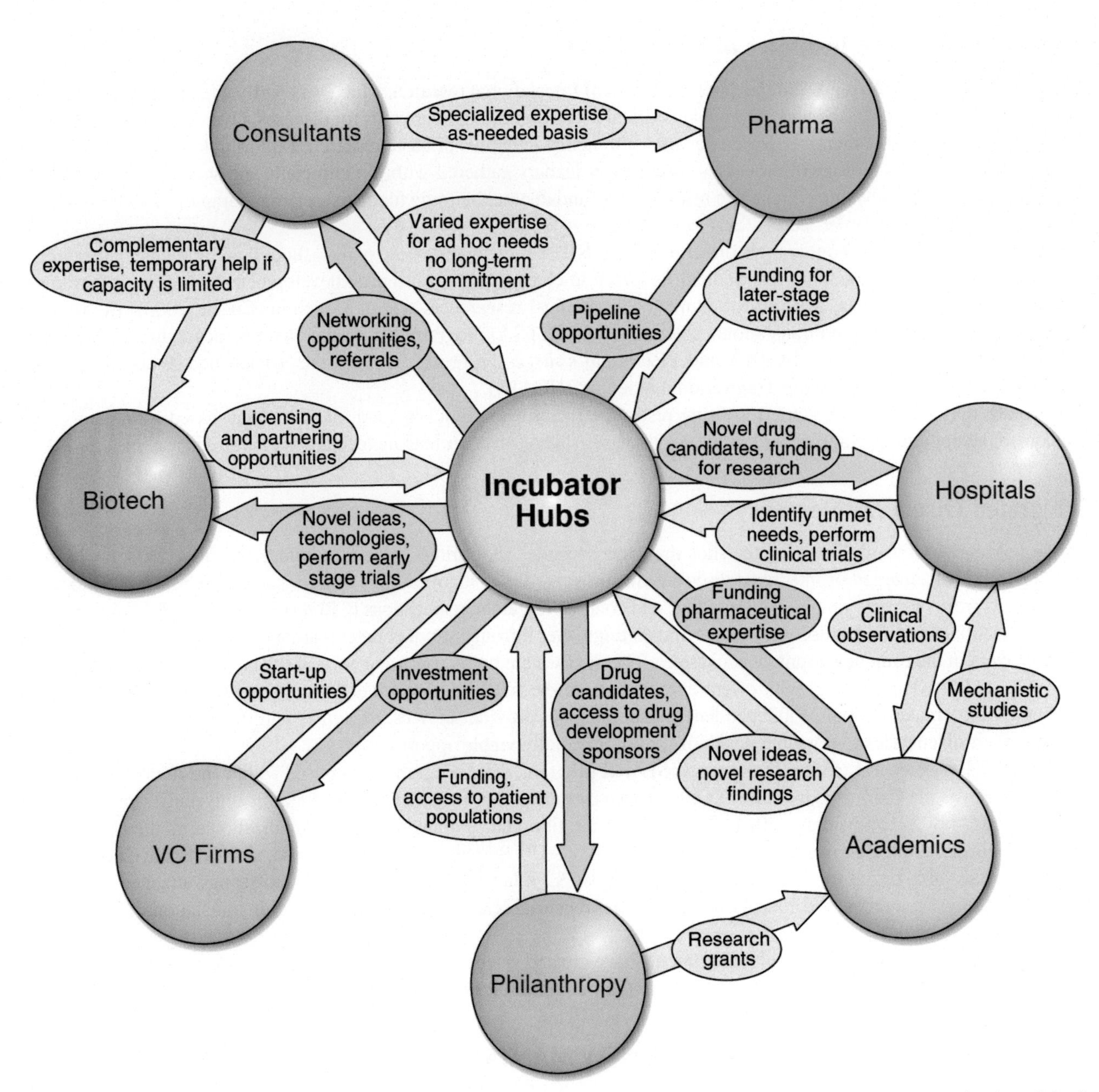

FIGURE 9.1 The integrated discovery nexus. Translational programs or centers, often based in universities or hospitals, serve as incubator hubs for helping commercialize and advance laboratory discoveries. They create a network of interactions with participants from many parts of the drug discovery ecosystem, each of which has something to gain from their involvement in the network. Examples include Stanford University's SPARK program, University of California's QB3 program, CTSI program at UCSF, and J&J's Janssen Labs (Fishburn, 2013a; Fishburn, 2014a). *(Figure is reproduced from Fishburn, 2013a with permission from Elsevier.)*

Pfizer launched its Global Centers for Therapeutic Innovation (CTI) in 2010 to create collaborations with academia and help bridge the gap between discovery and the clinic. To date, CTIs have been founded in four cities: Boston, New York, San Diego, and San Francisco. Together, they encompass 23 academic institutions and have a portfolio of 25 projects. Pfizer funds the preclinical and clinical development programs and provides access to compound libraries, screening capabilities, antibody development technology, and other expertise. The CTI model allows researchers to publish their work, and the IP is jointly owned by the university and Pfizer. However, the pharma has an exclusive commercial option to the compounds if they move to the clinic, and the university receives milestone payments if the option is exercised (Cain, 2010; Ratner, 2011).

Similarly, Johnson & Johnson (J&J) announced in 2012 that it would form four innovation centers in academic and biotech hot spots: Boston, Massachusetts; California; London; and Shanghai. The goal is to form alliances and collaborations with academics and entrepreneurs that can take different forms—such as early stage funding, creation of start-ups, or straightforward licensing deals. The J&J model involves about 25 J&J scientists in a center, mixed with corporate investors and business development executives (Senior, 2013). These innovation centers operate independently from the discovery nexus model of J&J's Janssen Labs.

GSK has taken a different approach to externalizing innovation that has a similar goal of fostering ongoing relationships with academic scientists. Rather than set up a discovery nexus, GSK created the Discovery Partnerships with Academia (DPAc) program to partner with individual academic researchers on early stage projects (Haas, 2011).

Thus, GSK started to compete with VC firms to provide an alternative for researchers to forming a start-up if they wanted to commercialize their discoveries. In the program, the academic researcher partners with a GSK team and gains access to the company's preclinical capabilities to carry out ADMET, PK, toxicology, and other studies, while the company obtains exclusive commercial rights to drugs that might result from the alliance (Haas, 2011).

In addition to these structured alliances, pharma companies have started to partner with universities in ongoing projects to identify targets or tap into new technologies to develop specific therapeutic product candidates (Table 9.2).

These partnerships aim to benefit both parties as pharma companies obtain access to new science at low risk and researchers receive funding, expertise, and training to help advance their translational projects. From the pharma companies' perspective, they solve their inherent problem of having limited internal innovation because of their size, structure, and culture (Fishburn, 2013a; Munos, 2010; Tralau-Stewart, 2009). By forming ongoing partnerships, however, rather than simply licensing the assets, they defray the risk, as there is less money up front, and they have longer to assess the molecules or technology, or to select the optimal compounds. Perhaps more importantly, they increase probability of success for early stage projects by providing expertise about the key factors to consider for translating drugs—an area in which academics frequently have little training. Such factors encompass the likely medical use of a compound in the relevant patient population, in addition to what would be required for an FDA filing—such as dosing, route, and frequency of administration; PK/PD; formulation; and appropriate toxicology assessment.

Biotech Companies and Academia

As the ecosystem grows and academia become more translationally savvy, partnerships are also emerging between biotech companies and academia (Kling, 2011). In these cases, the main driver is not the externalization of innovation, since many of these companies still have a strong entrepreneurial drive and internal cultures that foster risk taking. But because biotech companies have limited funding and an obligation to shareholders to drive programs to the clinic, they often cannot afford to continue groundbreaking research into new technologies or targets that could bolster or complement their pipeline programs. Partnering with academia offers them a way to access cutting-edge science at a lower financial and human resource cost than performing those studies in-house.

Between 2011 and 2013, about 300 partnerships per year were formed between biotech companies and academia, including licensing deals and ongoing alliances (BioCentury, 2014). U.S. institutions led the trend and were involved in the majority of deals in 2011 and 2012. However, the trend to partner has spread to Europe and beyond, and by 2013, the non-U.S. deals surpassed those made with U.S.-based academic organizations (Figure 9.2).

Several biotech companies incorporated academic partnering into their core strategy. Evotec AG is a German company that has deliberately pursued partnering with academic organizations to advance therapeutic candidates to complement its original business of fee-for-service operations (Lou, 2013). Since 2011, the company has started 11 programs within its EVT innovate arm and has signed at least seven deals with public institutions. Regulus Therapeutics Inc. is another company that includes academic organizations in its partnering strategy. Regulus works with over 37 academic laboratories to expand its microRNA-based therapeutic pipeline (Menzel and Xanthopoulos, 2012). Other biotech companies that are active in forming academic alliances include Astellas Pharma Inc., Biogen Idec Inc., and Foundation Medicine Inc.

For academic organizations, partnering with biotech companies offers welcome funding but carries risks that a company will lock up the IP and not move forward on a project due to strategic business considerations or simply to block a competitor (Kling, 2011). The solution lies in careful deal structures that university technology transfer offices are becoming more adept at negotiating.

PPPs are coming to define this phase of the industry's evolution and have been created to help solve the innovation gap. It will take several more years to see if PPPs can significantly increase the number of drugs that reach the market and can serve as a real solution rather than just a bubble.

TABLE 9.2 Partnerships between pharma companies and academic labs in 2012 and 2013

Pharma	Academic Institution	Date Announced	Aim of Collaboration	Therapeutic Area or Technology
AstraZeneca plc	Karolinska Institute	Jan-12	Develop diagnostic imaging tools to aid in drug development for neurological diseases, such as chronic pain, AD, and PD	Neurology; Imaging
	Cornell University; Feinstein Institute for Medical Research; University of British Columbia; Washington University	Jul-12	Drug discovery partnership focused on apolipoprotein E (ApoE) epsilon 4 in AD	Neurology
	Broad Institute of MIT and Harvard	Sep-12	Discover antibacterials and antivirals	Infectious disease
	Fudan University	Dec-12	Research on mechanism of the herbal compound leonuri in cardiovascular disorders such as chronic heart failure, ischemic stroke, and atherosclerosis	Cardiovascular disease
	Vanderbilt University	Jan-13	Discover and develop positive allosteric modulators (PAMs) of muscarinic acetylcholine receptor M4 to treat psychosis and other neuropsychiatric symptoms	Neurology
	Cancer Research UK; University of Manchester	Jun-13	Discover and develop cancer therapeutics involved with DNA damage response and other pathways	Cancer
	Cancer Research U.K.; University of Cambridge	Jul-13	Research tumor mutatons and therapies in prostate, pancreatic, and other cancers	Cancer
	Tufts University	Jul-13	Research brain diseases including Alzheimer's disease (AD), Parkinson's disease (PD), neurodevelopmental and autism spectrum disorders (ASD)	Neurology
	Hadassah University Hospitals	Sep-13	Discover and develop compounds for cancer, respiratory diseases and diabetes	Cancer; respiratory diseases; metabolic diseases
	University of Maryland	Sep-13	Research collaborations on cancer, cardiovascular, metabolic, respiratory, inflammatory autoimmune, and infectious diseases	Cancer; cardiovascular, metabolic, respiratory, autoimmune, infectious diseases
	Wyss Institute for Biologically Inspired Engineering at Harvard University	Oct-13	Develop and validate animal-cell versions of Organs-on-Chips for preclinical drug safety testing	Organ screening technology
	Johns Hopkins University	Dec-13	Research on immune system cells in tumor growth; targets and mechanisms in rheumatoid arthritis (RA); validation of therapeutic targets; assessment of mAb combinations for clinical candidate selection; research on new ways to manufacture complex next-generation biologic drugs for cancer	Autoimmune, cancer
	University of California San Francisco	Feb-14	Discover and develop small molecules and biologics in a range of indications	Multiple

TABLE 9.2 Partnerships between pharma companies and academic labs in 2012 and 2013—cont'd

Pharma	Academic Institution	Date Announced	Aim of Collaboration	Therapeutic Area or Technology
Bayer AG	Children's Hospital & Research Center Oakland	Mar-13	Identify new therapies for rare blood diseases with an emphasis on sickle cell anemia	Hematological disorders
	Broad Institute of MIT and Harvard	Sep-13	Discover and develop therapeutics that target genomic changes in cancer	Cancer
Boehringer Ingelheim GmbH	Swiss Federal Institute of Technology (ETH) Zurich	Sep-12	Investigate three undisclosed cell types involved in metabolic homeostasis that might contribute to obesity and diabetes	Metabolic diseases
	Duke Clinical Research Institute	Feb-14	Research in idiopathic pulmonary fibrosis to understand disease progression and treatment approaches, and establish biomarker bank	Respiratory diseases
Boehringer Ingelheim GmbH; GWT-TUD GmbH	Technische Universitat Dresden	Feb-12	Research the causes of diabetes and the link between excessive blood glucose and complications in other organs	Metabolic diseases
Bristol-Myers Squibb Co.	Tsinghua University	May-12	Identify and validate targets in cancer and immunology, with a focus on structural biology and mapping the 3D structure of molecular targets	Cancer; immunology
	Scripps Research Institute	Jun-12	Prepare synthetic intermediates and analogs against BMS targets using Scripps' chemistry methodologies	Chemistry
	Vanderbilt University	Oct-12	Discover, develop, and commercialize PAM of metabotropic glutamate receptor subtype 4 (mGluR4) to treat PD	Neurology
Eli Lilly and Co.	University of Edinburgh	Oct-13	Research on cancer mechanisms	Cancer
GlaxoSmith-Kline plc	Yale University	May-12	Design and develop proteolysis targeting chimeric molecules that are cell permeable, and small molecules that promote degradation of a target protein via the ubiquitin-proteasome pathway	Cancer; other
	Vanderbilt University	Feb-13	Discover, develop, and commercialize PAMs of the melanocortin 4 receptor for severe obesity	Metabolic diseases
	Cancer Research U.K.; University of Manchester	Dec-13	Develop cancer drugs against an undisclosed epigenetic regulator	Cancer
	The Wellcome Trust; University of Dundee	Mar-12	Develop treatments for Chagas disease, leishmaniasis, and African sleeping sickness	Rare diseases
Johnson & Johnson	University of Queensland	Feb-12	Develop spider venom peptides against a human ion channel for treatment of chronic pain	Neurology
	The Scripps Research Institute	May-13	Develop therapies and vaccines for infectious diseases, including influenza, using Scripps' structural biology methods	Infectious disease
	Catholic University of Leuven (KU Leuven); The Wellcome Trust	Sep-13	Discover and develop antivirals to prevent dengue infection	Infectious disease

Continued

TABLE 9.2 Partnerships between pharma companies and academic labs in 2012 and 2013—cont'd

Pharma	Academic Institution	Date Announced	Aim of Collaboration	Therapeutic Area or Technology
Novo Nordisk A/S	University of Oxford	Apr-12	Identify and develop biomarkers and targets for RA and other autoimmune inflammatory diseases	Autoimmune
Pfizer Inc.	Cold Spring Harbor Laboratory	Jun-12	Develop a human short hairpin RNA (shRNA) library to discover cancer therapeutics	Cancer
	Sanford-Burnham Medical Research Institute	Aug-13	Identify and validate new drug targets to prevent and treat insulin resistance in obesity and diabetes	Metabolic diseases
	University of Texas MD Anderson Cancer Center	Jan-14	Develop personalized immunotherapies for cancer immunotherapies and identify combination therapies and biomarkers for products in the pharma's pipeline	Cancer
	Massachusetts Institute of Technology (MIT)	Mar-14	Three-year collaboration to translate synthetic biology discoveries to advance drug development	Multiple
Sanofi	University of California	Jan-12	Identify siRNA targets for Type I and Type II diabetes	Metabolic diseases
	Joslin Diabetes Center	Jun-12	Identify compounds to treat late complications of diabetes in addition to targeted insulin analogs	Metabolic diseases
	Brigham and Women's Hospital	Jul-12	Research on an immunomodulatory approach to treat Type I diabetes	Metabolic diseases
	Massachusetts General Hospital	Oct-12	Research on hematological malignancies and solid tumors	Cancer
	Institut Pasteur Korea	Apr-13	Discover and develop first-in-class drugs for undisclosed indications	Multiple
	Institut Curie	Jun-13	Identify targets for ovarian cancer	Cancer
	Fraunhofer Institute for Molecular Biology and Applied Ecology	Jan-14	Identify and optimize naturally occurring compounds to treat infectious diseases	Infectious disease
Takeda Pharmaceutical Co. Ltd.	Osaka University	Jan-12	Research and develop hydrophobic nanoparticle-delivered adjuvant vaccines	Cancer; other
	British Columbia Cancer Agency	Sep-12	Discover and develop cancer targets based on gene analysis	Cancer
UCB Group	University of Oxford	Mar-12	Research projects in immunology and neurology	Immunology; neurology
	Harvard University	Jun-12	Develop small molecules for induction of autophagy to treat neurodegenerative diseases	Neurology
	Harvard University	Oct-12	Research on the human intestinal microbiome	Gastrointestinal diseases

Source: BioCentury (2014)

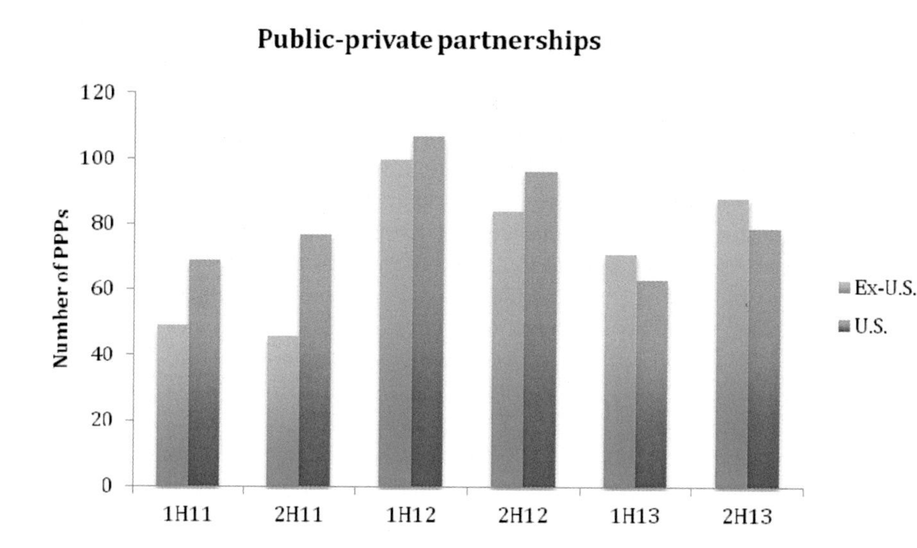

FIGURE 9.2 Public–private partnerships (PPPs) formed from 2011 to 2013. PPPs include licensing deals and ongoing collaborations between public institutions such as universities, hospitals, or nonprofit foundations and commercial organizations including pharma companies and biotech companies. *(Data sourced from BCIQ: BioCentury Online Intelligence database.)*

PRECOMPETITIVE CONSORTIA: THE ROAD AHEAD

While PPPs have moved the field forward by enabling greater integration of different sectors, they still represent business-focused strategic alliances that require detailed agreements to balance factors such as commercial interests, the need for intellectual freedom, and the desire to publish. Their effect has partly been to whet the appetite for greater collaboration, and there is an emerging new trend of precompetitive consortia that share information and knowledge—much of which was previously held proprietary—in the public domain.

Pharma companies have actively participated in these consortia, but agreements for these are no less complex than those for other alliances or PPPs, as companies need to find ways to share data and enable new discoveries to be made, and yet ultimately obtain a return on their investment. Such precompetitive consortia have sometimes been driven by government initiatives, such as with Europe's Innovative Medicines Initiative, which was founded in 2008 by the European Commission and the European Federation of Pharmaceutical Industries and Associations (Vaudano, 2013). In other cases consortia have been established between pharma companies—with no public organization involved—and use a third party to house the information shared. For example, AstraZeneca plc has partnered with Roche in a precompetitive alliance to create an *in silico* drug design algorithm that can better predict structure function than available programs. Medchemica Ltd. serves as an intermediary and collects fragment information from the compound libraries of both organizations. By creating a larger set of source data pooled from both companies' libraries, the partners hope to increase the statistical power and thus improve the fidelity of the prediction (Fishburn, 2013b).

Precompetitive consortia generally differ from other PPPs, as they focus on addressing specific problems too large for any single organization to tackle, or pool libraries or data from multiple organizations (Denee et al., 2012; Mittelman, Neil, and Cutcher-Gershenfeld (2013). The consortia cover a variety of goals, from solving specific development problems shared across the industry to addressing specific therapeutic areas, or sharing compound data to drive the generation of new compounds at the earliest stages of the drug development pathway.

For example, the Biomarkers Consortium (2014) was established by the Foundation for the National Institutes of Health (FNIH) to develop and validate biomarkers in multiple disease areas. Biomarkers represents a new area of vital importance to the entire industry, but there have been multiple parallel efforts to search for biomarkers for different diseases and no clear standards for validating them or obtaining regulatory approval. While there is not yet a specific FDA-approved biomarker that can trace its roots to the consortium, the alliance has launched over 15 projects in different therapeutic areas and has led to constructive engagement between industry and regulatory authorities on a path forward for qualifying biomarkers as drug development tools (Mittelman, Neil, and Cutcher-Gershenfeld, 2013). One tangible success of the consortium is the identification of adiponectin as a predictor of metabolic response to peroxisome proliferator-activated receptor agonists in Type II diabetes patients (Wagner et al., 2010). Other examples of consortia aimed at addressing industrywide problems include the Predictive Safety and Toxicology Consortium (PSTC, 2014) formed in 2006 to identify new and improved safety testing methods and obtain regulatory approval for them, and the Patient-Reported Outcomes Consortium (PROC, 2014) launched in 2009 to develop and gain approval for reliable methods of gathering patient-reported data for use in clinical trials. In these cases, where industry, academia, hospitals, and regulatory authorities are involved and have a stake in the outcome, an independent third party coordinates the consortium. The Biomarkers Consortium is run by the FNIH, and the PSTC and PROC are run by the Critical Path Institute.

Disease-focused consortia are becoming particularly relevant in fields such as neuroscience where drug development has become alarmingly stalled (Cain, 2012). The CommonMind Consortium was launched by Takeda in 2012 together with Sage Bionetworks, the Mount Sinai School of Medicine, the University of Pennsylvania, and the NIH's National Institute of Mental Health to generate and analyze genomic data from patients with neuropsychiatric diseases. In 2013 Roche and two more universities joined the consortium. The pharma companies, NIMH, and Sage contributed funding of over $3 million, and the universities and hospitals provide patient and control postmortem brain samples for genomic and SNP analysis. The data will be made publicly available through Sage's synapse web portal. Other disease-focused consortia include the NIH's Accelerating Medicines Partnership (AMP) set up in early 2014 to focus on Alzheimer's disease, Type II diabetes, rheumatoid arthritis, and systemic lupus erythematosus. The AMP includes 10 pharma and biotech companies and five disease-oriented nonprofit foundations, and has a $230 million 5-year budget to fund the identification of new targets and biomarkers in the noted disease areas. The results of the consortium will be made publicly available to advance drug development by anyone in the sector (Fishburn, 2014b).

While the therapeutic area-focused consortia focus on finding targets or pathways for specific diseases, other types of consortia aim to improve the efficiency of drug development by improving the design of candidate molecules. The Structural Genomics Consortium (SGC, 2014) generates three-dimensional protein structures and chemical tool compounds as probes to explore how subtle changes between proteins affect their functional properties. The consortium was established in 2003 and has nine pharma partners and four partners from nonprofit and Canadian government organizations. The SGC has thus far produced structures for over 1,500 proteins, which represent about 15% of the human proteome, and has made them publicly available.

Thus, as translational research has grown, pharma has responded by engaging in different forms of partnerships with public and private institutions to re-energize its drug discovery efforts.

SUMMARY

Pharma has gone through dramatic changes in the decades since the birth of biotech. The pharmaceutical companies no longer own a monopoly on drug development and have all but acknowledged that they are not the primary source of innovation. As biotech companies have emerged on the scene and academia has embraced translational research as a way to help advance their own discoveries, pharma companies have begun to foster relationships with those sectors as part of a strategy to externalize innovation. PPPs offer a way for pharma companies to tap into academic talent and stay abreast of cutting-edge research into new targets, technologies, and tools for advancing drug development. Precompetitive consortia represent perhaps the biggest turnaround from the early days of insular attitudes in which company knowledge was kept strictly under wraps, and companies are embracing the notion that, by sharing data and expertise, they can solve industrywide problems more efficiently than by acting independently.

Translational research, public–private partnerships, and precompetitive consortia are still an experiment. They have not yet had time to yield the dominant metric—more drugs per year reaching patients to solve unmet needs.

Several problems still need to be addressed. Much of the innovation landscape involves breakthroughs made in academia—but much of the research published in academia has proven not to be reproducible in pharma companies' hands (Begley and Ellis, 2012; Collins and Tabak, 2014; Prinz, Schlange, and Asadullah, 2010). The root of this is thought to lie in different standards of proof operating in industry and academia, in addition to inherent problems in the world of academia such as pressure to publish to obtain grants and tenure.

Nevertheless, a snapshot of the ecosystem in 2014 shows emerging new technologies and potentially transformative areas of research such as RNA-based therapeutics, CRISPR, epigenetics, and the microbiome that can lead to new drug classes and new therapeutic modalities. Discoveries in these areas have already transitioned to the commercial environment with biotech companies founded on their basis and pharma companies establishing internal and external programs to capitalize on the findings.

The ongoing challenge for pharma will be to identify the next "white space" area of innovation coming out of academia or biotech companies, and to determine the optimal structures to foster and capitalize on those developments.

REFERENCES

Begley, C.G., Ellis, L.M., 2012. Raise standards for preclinical cancer research. Nature 483, 531–533.

Biocentury. BioCentury and BCIQ online database. http://www.biocentury.com/bciq/ (accessed 02.09.14.).

Biomarkers Consortium. http://www.biomarkersconsortium.org/ (accessed 02.09.14.).

Booth, B., Zemmel, R., 2004. Prospects for productivity. Nat. Rev. Drug Discov. 3, 451–456.

Butler, D., 2008. Crossing the valley of death. Nature 453, 840–842.

Cain, C., 2010. Pfizer goes back to school. SciBX 3. http://dx.doi.org/10.1038/scibx.2010.1371.

Cain, C., 2012. A mind for precompetitive collaboration. SciBX 5. http://dx.doi.org/10.1038/scibx.2012.483.

Center for Drug Evaluation and Research (CDER)/Federal Drug Administration: Drug and Biologic Approval and IND Activity Reports. http://www.fda.gov/Drugs/DevelopmentApprovalProcess/HowDrugsareDevelopedandApproved/DrugandBiologicApprovalReports/default.htm (accessed 02.09.14).

Check, E., 2003. NIH 'roadmap' charts course to tackle big research issues. Nature 425, 438.

Collins, F.S., 1999. Medical and societal consequences of the Human Genome Project. N. Engl. J. Med. 341, 28–37.

Collins, F.S., Tabak, L.A., 2014. NIH plans to enhance reproducibility. Nature 505, 612–613.

Cook, D., Brown, D., Alexander, R., March, R., Morgan, P., Satterthwaite, G., et al., 2014. Lessons learned from the fate of AstraZeneca drug pipeline: a five-dimensional framework. Nat. Rev. Drug Discov. 13, 419–431.

Cuatrecasas, P., 2006. Drug discovery in jeopardy. J. Clin. Invest. 116, 2837–2842.

Denee, T.R., Sneekes, A., Stolk, P., Juliens, A., Raaijmakers, J.A.M., Goldman, M., et al., 2012. Measuring the value of public–private partnerships in the pharmaceutical sciences. Nat. Rev. Drug Discov. 11, 419.

Fishburn, C.S., 2013a. Translational research: the changing landscape of drug discovery. Drug Discov. Today 18, 487–494.

Fishburn, C.S., 2013b. Attenuating attrition. SciBX 6. http://dx.doi.org/10.1038/scibx.2013.647.

Fishburn, C.S., 2014a. Teaching translation. SciBX 7. http://dx.doi.org/10.1038/scibx.2014.333.

Fishburn, C.S., 2014b. AMPlifying targets. SciBX 7. http://dx.doi.org/10.1038/scibx.2014.215.

Fletcher, L., 2001. Eli Lilly enters venture capital arena. Nat. Biotechnol. 19, 997–998.

Haas, M.J., 2011. From Scotland to MaRS. SciBX 4. http://dx.doi.org/10.1038/scibx.2011.700.

Hodgson, J., 2006. Private biotech 2004—the numbers. Nat. Biotechnol. 24, 635–641.

Kesselbaum, A.S., 2011. An empirical view of major legislation affecting drug development: past experiences, effects, and unintended consequences. Milbank. Q 89, 450–502.

Kling, J., 2011. Biotechs follow big pharma lead back into academia. Nat. Biotechnol. 29, 555–556.

Kneller, R., 2010. The importance of new companies for drug discovery: origins of a decade of new drugs. Nat. Rev. Drug Discov. 9, 869–882.

Lähteenmäki, R., Lawrence, S., 2005. Public biotechnology 2004—the numbers. Nat. Biotechnol. 23, 663–671.

LaMattina, J.L., 2011. The impact of mergers on pharmaceutical R&D. Nat. Rev. Drug Discov. 10, 559–560.

Lou, K.-J., 2013. Evotec's growing ivy. SciBX 6. http://dx.doi.org/10.1038/scibx.2013.80.

Markel, H., 2013. Patents, profits, and the American people—the Bayh–Dole Act of 1980. N. Engl. J. Med. 369, 794–796.

May, M., 2011. Stanford program gives discoveries a shot in the arm. Nature 17, 1326–1327.

Menzel, G.E., Xanthopoulos, K.G., 2012. The art of the alliance. Nat. Biotechnol. 30, 313–315.

Meslin, E.M., 2013. Mapping the translational science policy 'valley of death. Clin. Trans. Med. 2, 14.

Mittelman, B., Neil, G., Cutcher-Gershenfeld, J., 2013. Precompetitive consortia in biomedicine—how are we doing? Nat. Biotechnol. 31, 979–985.

Munos, B., 2009. Lessons from 60 years of pharmaceutical innovation. Nat. Rev. Drug Discov. 8, 959–968.

Munos, B., 2010. Can open source drug R&D repower pharmaceutical innovation? Clin. Pharm. Ther. 87, 534–536.

Pammoli, F., Magazzini, L., Riccaboni, M., 2011. The productivity crisis in pharmaceutical R&D. Nat. Rev. Drug Discov. 10, 428–438.

Papadopoulos, S., 2000. Business models in biotech. Nat. Biotechnol. 18 (Suppl), IT3–IT4.

Paul, S.M., Mytelka, D.S., Dunwiddie, C.T., Persinger, C.C., Munos, B.H., Lindborg, S.R., et al., 2010. How to improve R&D productivity: the pharmaceutical industry's grand challenge. Nat. Rev. Drug Discov. 9, 203–214.

PhRMA, 2011. Pharmaceutical industry profile 2011. http://www.phrma-jp.org/archives/pdf/profile/PhRMA%20Profile%202011%20FINAL.pdf (accessed 02.09.14.).

Prinz, F., Schlange, T., Asadullah, K., 2010. Believe it or not: how much can we rely on published data on potential drug targets? Nat. Rev. Drug Discov. 10, 712.

PSTC (Predictive Safety and Toxicology Consortium). http://c-path.org/programs/pstc/ (accessed 02.09.14.).

PROC (Patient-Reported Outcomes Consortium). http://c-path.org/programs/pro/ (accessed 02.09.14.).

Ratner, M., 2011. Pfizer reaches out to academia—again. Nat. Biotechnol. 29, 3–4.

Senior, M., 2013. J&J courts biotech in clusters. Nat. Biotechnol. 31, 769–770.

SGC (Structural Genomics Consortium). http://www.thesgc.org/ (accessed 02.09.14.).

Tralau-Stewart, C.J., Wyatt, C.A., Kleyn, D.E., Ayad, A., 2009. Drug discovery: new models for industry-academic partnerships. Drug Discov. Today 14, 95–101.

Vaudano, E., 2013. The Innovative Medicines Initiative: a public–private partnership model to foster drug discovery. Comput. Struct. Biotechnol. J. 6, 1–7. http://dx.doi.org/10.5936/csbj.201303017.

Wagner, J.A., Prince, M., Wright, E.C., Ennis, M.M., Kochan, J., Nunez, D.J.R., et al., 2010. The Biomarkers Consortium: practice and pitfalls of open-source precompetitive collaboration. Clin. Pharm. Ther. 87, 539–542.

Williams, M., 2011. Productivity shortfalls in drug discovery: contributions from the preclinical sciences? J. Pharm. Exp. Ther. 336, 3–8.

Chapter 10

Translational Science in Medicine: Putting the Pieces Together

Biomarkers, Early Human Trials, Networking, and Translatability Assessment

Martin Wehling

The preceding chapters of this book detail the manifold aspects of translational processes in biomedical research. As stated repeatedly, these toolboxes, algorithms, and networking aspects need to be integrated into a meaningful, comprehensive, and thus structured way. In this respect, it is important to note that what is frequently simply called *translational medicine* should be developed into a solid science; the title of this book emphasizes this by inclusion of the word *science* in the title: *Translational SCIENCE in Medicine* (a science that should also exist in, for instance, physics).

This novel science needs to be developed as biostatistics had to be developed in the 1970s in reflection of the rapidly growing demands in biomedical research, in particular clinical studies driven by the claims to generate "evidence-based medicine." This evidence had to be derived from experimental (both preclinical and clinical) data by ever more sophisticated statistical analysis; a new science emerged although its roots could be tracked back to the seventeenth century. Today, almost all medical faculties, all major pharmaceutical companies, and biomedical research institutions provide the required expertise and practical work in independent departments or institutes.

A similar path is envisioned for translational science in medicine, which has just started to exist but still requires major developments of scientific content and institutional representation. Promising developments, mainly in the United States, are detailed in Chapter 1.

While conceptionally still in its infancy, this novel science requires major scientific input both at the conceptional and implementational level, with this aim being a major target of this book.

To more clearly structure the scope and key elements of translational science in medicine, four major tasks are proposed as cornerstone assets of this novel science:

Task 1: Need for rigorous science-based biomarker classification in regard to predictivity in translational processes
Task 2: Need for rigorous science-based design of early-stage, exploratory human trials
Task 3: Need for careful, foresightful planning and coordination of major players in translational processes: academia/industry interactions
Task 4: Need for rigorous science-based predictivity scoring in translational processes

Details of three of these essential tasks are given in Subchapters 3.1 through 3.5 (task 1), 4.1 through 4.6 (task 2), and 2.2 (task 4). Task 3 has not been explicitly elaborated yet but is as essential in that it creates the necessary planning and networking as an indispensable prerequisite of translational success. Because tasks 1, 2, and 4 may flourish independently in separate "silos" and, thus, may not be seen as integral components, their carefully planned integration may be considered as the main challenge of translational processes.

A key element of bringing together all the different disciplines that are necessarily involved in translation (e.g., molecular biologists, pharmacologists, animal researchers, PK/PD experts, human trialists, clinicians) is foresightful planning. A document needs to be created on day one of a relevant discovery, for example, of a novel receptor or a novel function of a known receptor. Translation will work only if clinical researchers but also preclinical scientists have an educated vision on how to translate early findings (mostly in vitro findings) into useful, human medicines. They should have an idea about unmet medical needs ("interest of patients otherwise not optimally treated") and about disease constituents, including the major

Principles of Translational Science in Medicine. http://dx.doi.org/10.1016/B978-0-12-800687-0.00033-5

Use the translational medicine plan (TMP) as facilitator and guidance for improved networking at the various interfaces, (discovery, translational medicine and clinical) to enable early input from clinical into discovery and vice versa

Create a concise, short document detailing the studies planned to identify and evaluate biomarkers and human exploratory studies

Content

1. Introduction

2. Plans to support

- **a. Target Identification and Validation e.g., by human exploratory studies**
- **b. Diagnostics: Identification of Targeted Population**
- **c. Biomarkers/Surrogates**
    - **i. Pharmacodynamic Activity**
    - **ii. Safety**

3. Project planning—timelines and resource needs (GANNT-chart, FTEs, $)

FIGURE 10.1 Template for an early translational plan.

biomarkers in the field. They should know about the obstacles and limitations of human experiments ("serial brain slicing" as only readout = no way into human development).

It is obvious that basic scientists need to develop and exercise

1. Interest, maybe driven by awards through funding restrictions if translation is not convincingly envisioned
2. Education in the major areas of translational science of medicine, for example, all four major tasks described previously
3. Advice from translational scientists, mainly from distant stages of development, typically clinicians if preclinical researchers are concerned

In an ideal world, basic researchers would be able to set up a basic translational plan from day one; as a living document, this plan requires updating at all major milestones. It has to describe decision trees (if this happens, then this will be done next; if not, that), milestone achievements, and all major constituents of translational success. A template for an early translational plan is shown in Figure 10.1.

This translational planning has to embrace and integrate the potential collaborators and should be based on established translational networks. There are various models of the organization of such a translational network. Its major feature is the coordinating unit or chair, which functions as network organizer and administrator. This capacity has to identify resources (potential network participants), identify gaps to be closed by scouting for suitable novel participants, connect distant players, and amalgamate interface resistances, in particular relating to the cooperation of industry and academia (public–private partnerships, or PPPs). It also has to provide support by legal, in particular, intellectual property specialists, and to organize administrative tasks coping with, for example, variable compensation schemes in private and public entities.

A potential networking model is proposed in Figure 10.2 (for comparison see fig. 9.1 which is similar but more comprehensive). This is only one of many sketches, all aiming at connecting and intertwining often-distant players in the translational plot that is indispensable if translation should happen.

An established networking structure is probably the single most important prerequisite of translational success; planning for translation will be done in a consortial and thus integrative approach. Those structured and educated approaches will aim at establishing high expertise and skills in addressing the other three major tasks.

It is important to note that during a typical translational development, public and private partners will have uneven shares of contribution. At certain stages, public partners are leading; at other stages, private partners. Typically, academia is strong at very early stages of target discovery/validation (the first highly visible paper in leading basic science journals such as *Nature* or *Science*), whereas lead optimization is better placed in the pharmaceutical industry with access to huge compound libraries. Lead identification may be done by either player. This scheme may be different if biologicals are involved, which may be designed and even produced at a small scale by academia.

As development progresses, early human trials ("phase 0, exploratory trials") are clearly a domain of public contributors, as they often require complex human instrumentation, especially imaging, which is available only in major hospitals. There is a large yet unexploited potential for academic hospitals to venture into this area in particular, as industry attempts at establishing human imaging centers (e.g., by Glaxo in London from 2005) largely failed; this is mainly due to the fact that only co-utilization of expensive human diagnostic machines by studies and regular patient care renders their investment profitable.

Academia may progress their own drug developments at maximum to classic phase II trials and achieve PoC as a major value increment. Certainly, beyond this, any drug development will have to be passed on to industry because phase III trials are very large and expensive, which almost excludes them from public financing.

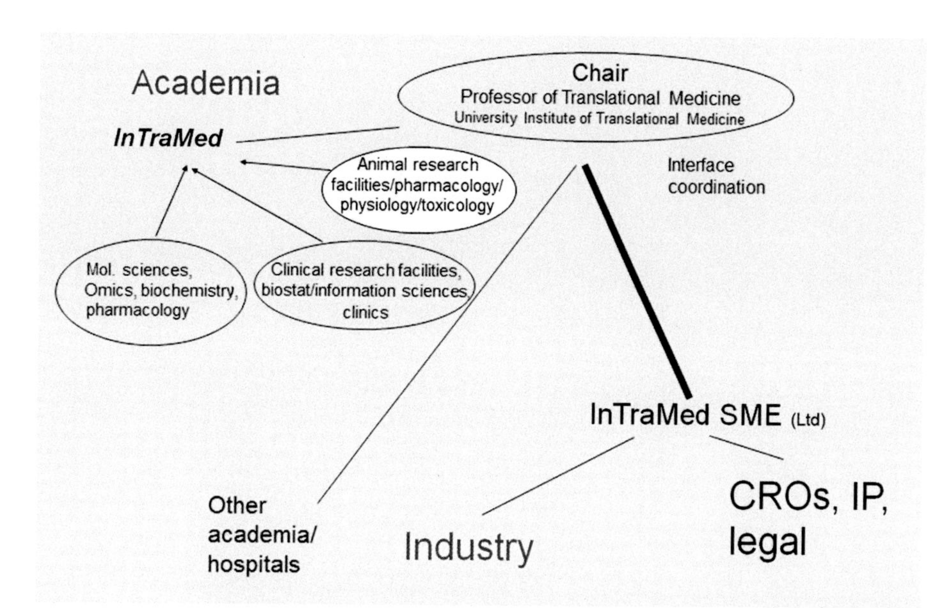

FIGURE 10.2 Sketch of a translational medicine network of public and private partners who have hands-on involvement in translational actions. It is based on the cooperative Institute of Translational Medicine (InTraMed), backed by a small–medium enterprise (SME) limited company, to bring academic resources, contract-research organizations (CROs), intellectual property (IP) management, and legal backup in contact with industry. *(From Wehling, 2011, by kind permission.)*

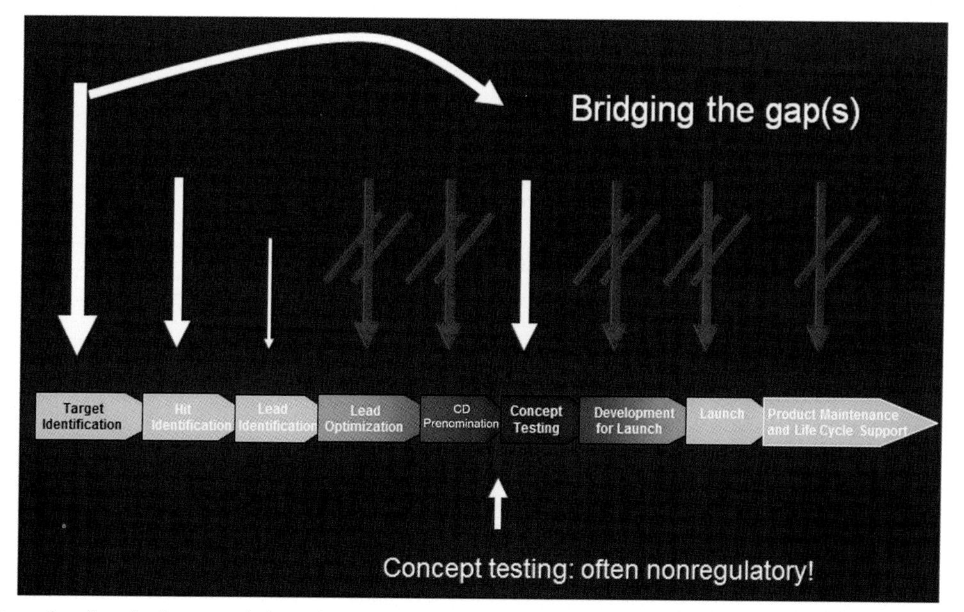

FIGURE 10.3 Strengths of academia versus industry in translational science. Academia is strong in target identification and/or validation, possibly lead identification, and early ("smart") nonregulatory clinical trials (red arrows). By contrast, lead optimization, regulatory clinical trial development, particularly phase III trials, and launch and product maintenance need to be done by industry (red arrows). *(From Wehling, 2011, by kind permission.)*

These preferences of private and public contributions are depicted in Figure 10.3.

Only if properly organized and planned, translational science in medicine will yield successfully translated projects. Task 4 (translational assessment) will guide researchers, investors, or funding agencies to the potentially promising projects and unravel weaknesses of others to be amended by efforts directed to those deficiencies. This assessment, thus, is a central obligation and best placed in the core structures of the translational network, in most cases to be covered by the translational department or chair. This central institution is, thus, pivotal not only in organizing translation, but also adjudicating funds and resources in a robust and transparent process of translatability assessment.

REFERENCE

Wehling, M., 2011. Drug development in the light of translational science: shine or shade? Drug Discov. Today 16, 1076–1083.

Chapter 11

Learning by Experience

Examples of Translational Processes in the Cardiovascular Field

Martin Wehling

EXAMPLE OF A SMART, SUCCESSFUL TRANSLATIONAL PROCESS

One has to dig deep and go far back in time to find an example of smart and successful translational processes in the area of cardiovascular therapy. This shows that true innovations have become rare in this field. One of the more recent successes and harbingers of progress is only indirectly tied to cardiovascular diseases: smoking cessation is a powerful tool to prevent cardiovascular disease (and—to a certain extent—treat it). As a central task in overall patient management attempts, however, smoking cessation is probably one of the biggest beneficial interventions possible. It has been shown to reduce incidence of cardiovascular disease and lung and other cancers, and it significantly reduces overall mortality. As the last direct innovation in cardiovascular treatment, statins could be considered, although one could claim they belong to lipidology. Thus, it does not seem as much of a stretch to consider aids to stop smoking as cardiovascular interventions. Therefore, the example of varenicline, the best currently available drug aid to reduce the craving for and satisfaction of smoking, is chosen here.

There have been various attempts to support people willing to stop smoking, and they have been accentuated as the harmful side effects of smoking become more and more evident. In 2008, the Centers for Disease Control and Prevention reported that cigarette smoking and exposure to secondhand smoke resulted in an estimated 443,000 deaths per year in the United States (Centers for Disease Control and Prevention, 2008). Quitting smoking is beneficial regardless of time point and age (U.S. Department of Health and Human Services, 1990). Thus, all methods for helping people quit have significant bearings on public health.

Until recently, the most commonly used pharmacological aids were nicotine itself and bupropion, a psychotropic drug. Nicotine is available in six different preparations (tablets, ointment, lozenges, etc.); it occupies and stimulates central nervous system (CNS) receptors ($\alpha 4\beta 2$ nicotinic receptors in the ventral tegmental area), whose stimulation results in dopamine release in the nucleus accumbens through the mesolimbic system. Because it is the natural stimulatory component of cigarette smoke, exogenous nicotine has the same effect as smoke-borne nicotine: it causes dopamine release at a site that is known to be the convergence site of almost all types of addiction. Dopamine release in the nucleus accumbens is induced by alcohol, cocaine, and amphetamines, and also food intake and nicotine. Addiction to those agents and habits thus is similar, and treatment strategies also are similar to a certain extent. Thus, pharmaceutical strategies for helping patients quit smoking were aimed at a controlled reduction of dopamine release, and this can be achieved principally through nicotine replacement therapy (NRT), which can replace the cigarette-based nicotine supply—which is tied to the poisonous tar components that cause most of the health problems related to smoking—with "clean" nicotine. Because the smoking habit itself will taper off if the act of smoking is not repeated frequently, the habit slowly disappears, and many components of craving a cigarette fade away. Though principally feasible, it is obvious that replacing a poison with a different version of the same poison may not be a very smart way to cure addiction, as the addictive agent is still being supplied.

The next incremental gain was the introduction of bupropion, which is clinically used as an antidepressant. It was noted that dopamine release can be attenuated by this drug and that this could be beneficial in addiction treatment. Thus, it gained market approval as a smoking cessation drug. However, its chemical similarity to amphetamines may be the reason for a multitude of side effects, including sleeplessness and seizures that are not trivial. Thus, there is a strong demand for more effective and safer drugs in this area.

Another approach starting from compounds similar to nicotine has been in practice since the 1960s: cytisine is a partial agonist at the nicotinic receptor and was tried as a smoking cessation aid in the 1960s and 1970s. To cut the long cytisine story short, it was not an international success because its deployment was locally restricted to Europe, and at that time, there were no clinical trials fulfilling current quality criteria. It is worth mentioning, however, that it is claimed even

Principles of Translational Science in Medicine. http://dx.doi.org/10.1016/B978-0-12-800687-0.00034-7

now—even after market approval of its expensive congener, varenicline—that performing such trials with this inexpensive drug may be warranted. This drug, if profiled to be similarly effective as varenicline, could save many lives in developing countries (Etter et al., 2008).

Which features of the foregoing developments made translation of varenicline successful? Which biomarkers could be used?

The main translational discovery dates back to the early cytisine experiences in humans: partial agonism at the nicotinic receptor (the particular type identified later as $\alpha4\beta2$) does produce the expected result. As receptors are occupied, the nicotine antagonistic properties are evident in that a new cigarette and its nicotine are ineffective, and the partial activation does release enough dopamine to prevent people from experiencing heavy craving. Quitting cold turkey is avoided, and satisfaction after a new cigarette is absent.

The translational evidence was based on isolated receptor subtypes cloned from animals and humans and on valid animal models of nicotine dependence and nicotine side effects (e.g., hypothermia), which are absent in the limited ceiling effects of partial agonists. A congener compound had produced beneficial clinical effects, and biomarkers at the animal level (behavioral measures) and comparably simple clinical studies with smoking habits and psychometric scales for craving and satisfaction were established and validated at the time of varenicline translation. Thus, not only have there been highly developed in vitro test systems to analyze binding and agonist or antagonist properties, but animal models were sensitive and predictive for both efficacy and safety aspects, and they seemed to reflect the benefits of partial agonism. They clearly showed that unwanted nicotine effects were absent with varenicline and thus demonstrated the low ceiling effect of partial agonism as mentioned previously. Thus, in the predictivity scoring system for biomarkers described in this book, the preclinical biomarkers (in animal models) would have achieved comparably high scores, had they been retrograde validated by the human experiences with cytosine and other similar compounds. The human biomarkers to measure craving and satisfaction had been validated by other model compounds (more by bupropion than cytisine, as the latter had probably been in "clinical development" when legal requirements were absent or low).

Thus, this combination of human data for drugs aiming at both the same clinical entity (bupropion) and the same principle mechanism (cytosine and, for a second similar compound, lobeline); animal data from very predictive models; and in vitro receptor data made the tailored development of varenicline possible and is the base of its clinical success (for a review, see Keating and Siddiqui, 2006).

This example shows that experiences from suboptimal development processes (mostly dating back to preregulatory periods in which "clinical development" was random and—possibly—dangerous to the patients) are valuable tools for defining translational strategies and assessing the biomarkers to be trusted in present processes. Many of the approaches at translational promotion have been utilized in this way (using experiences obtained from model compounds, reverse pharmacology, and meticulously developed animal models).

One potentially painful question remains: If cytisine had been developed the same way as varenicline, how much better would varenicline be than cytisine? Would it be any better? This question cannot be answered unless those human trials claimed to be necessary by Etter and colleagues (2008) are performed with cytisine. Under the current conditions, this is very unlikely, as cytisine has no genuine patent protection, and no one will ever invest in such a compound. This irony—that the model compound has helped the expensive patent-protected compound to go through a successful translational program but has never proved its own strength—is among the seemingly inevitable strange or even bizarre features of drug development processes.

EXAMPLE OF A FAILED TRANSLATIONAL PROCESS

Let us consider as an example of a failed translation process the development of torcetrapib. The torcetrapib case had already cost Pfizer about $800 million when the ILLUSTRATE trial showed excess mortality of high-risk cardiovascular patients versus the placebo, leading to the cessation of drug development (Barter et al., 2007). The secondary costs of this failure could even be counted in billions of dollars and could well affect the future well-being of the entire company. It was hoped that this compound would become a blockbuster and—as a combination product—prolong the life cycle of the company's biggest seller, atorvastatin, whose patent was due to expire in 2010 or 2011. Not only did mortality significantly increase; atherosclerosis progression under the drug was even worse than under the placebo as measured by intima-media thickness (Nissen et al., 2007). This biomarker represents a very reliable, high-scoring parameter (see Subchapter 3.5). Why were expectations running high, why was disappointment tremendous, and what role did translational processes play?

Torcetrapib is a small molecule aiming at the inhibition of the cholesteryl ester transfer protein (CETP), which is known to be involved in cholesterol transport and thus seems to be a key enzyme in the metabolism of this number-one cardiovascular killer. As reviewed by Nissen and colleagues (2007), CETP inhibition is thought to result in increased concentrations of HDL-cholesterol ("good" cholesterol), which is normally degraded by transferring cholesterol esters into VLDL- or LDL-particles. Increasing HDL-cholesterol was thought to represent an appropriate way to further reduce cardiovascular risk, which can be reduced, for example, by statins by roughly 20%–30%. This shows that there is a lot of suffering and premature cardiovascular deaths, despite the success story of statins and other cardiovascular preventive measures (e.g., acetylsalicylic acid). As early as the 1950s and—20 years later—with the first publication of the famous Framingham study (Gordon et al., 1977; Nikkila, 1953), evidence of a strong negative correlation between cardiovascular morbidity and mortality and HDL-cholesterol concentration was emerging, and epidemiological (that is, noninterventional) trials always confirmed that relationship. There were interventions in which HDL-cholesterol increases were associated with decreased cardiovascular endpoints, namely by exercise and by drugs: nicotinic acid (Brown et al., 2001; Taylor et al., 2004) and gemfibrozil, a fibrate drug (Frick et al., 1987). This evidence was the basis of the generalizing assumption that any intervention leading to increased HDL-cholesterol values should be beneficial to patients.

In light of this background, CETP inhibitors appeared to be an interesting principle; in various animal models and early human trials, they did what they were expected to do: they increased HDL-levels (Brousseau et al., 2004) and could even prevent diet-induced atherosclerosis in New Zealand White (NZW) rabbits (Morehouse et al., 2007).

Up to this point, the reasoning behind the torcetrapib trials seemed sound. Yet the resultant big clinical trial led to an excess of 34 deaths in the torcetrapib group; thus, a critical reappraisal is not only warranted but desperately needed to do justice to those patients who had to die in the context of drug development.

There are two major tracks of reasoning to explain the failure of torcetrapib that should be clearly separated, especially in regard to translational issues:

1. The drug did something else than just CETP inhibition, and this is the cause of failure—the so-called off-target effects of the drug.
2. CETP inhibition, as such, is unexpectedly harmful to humans.

The main off-target effect of torcetrapib was an elevation of blood pressure that amounted to 5.4 mmHg systolic in the ILLUSTRATE trial. CETP inhibition is not thought to be causal in this regard—although the facts are not yet entirely clear—as patients with genetic CETP deficiency do not show elevated blood pressure (Tall, Yvan-Charvet, and Wang, 2007). Other companies (e.g., Merck, Roche) are still pursuing the development of CETP inhibitors and hope that such an effect is absent in their setting, as shown for JTT-705 (Kuivenhoven et al., 2005).

Where could translational processes have intervened in regard to blood pressure? This effect was not apparent in animal experiments, as far as data in the public domain are concerned. It is not rare that side effects cannot be detected in animals, and translational toxicology is a major area urgently requiring substantial development. However, there will be gaps in our grasp of side effects, and surprises will be inevitable. In the torcetrapib case, one wonders how such an essential effect on blood pressure could have been overlooked or misinterpreted in the early human trials. In 2006, two similar trials were published (Davidson et al., 2006; McKenney et al., 2006) in which blood pressure effects were inconsistent: in the first trial, blood pressure values (systolic or diastolic, mmHg, group size around 30 patients) were 123.0/78.4, 118.7/76.6, 119.0/78.0, 120.0/77.6, and 120.4/76.5 for the placebo and featured 10, 30, 60, and 90 mg/day of torcetrapib. Corresponding figures for the second trial were 118.5/77.7, 120.7/76.9, 117.1/75.1, 119.4/79.4, and 121.4/78.8. At first glance, one would not suspect a substantial effect on blood pressure, with the second trial showing a mean increase of 2.9 mmHg systolic at the highest dose. Apparently, the rise in blood pressure in the pivotal trial, the ILLUSTRATE trial, was much larger than in phase II trials (Tall et al., 2007). Obviously, at the moment, no plausible explanation for this discrepancy can be given. Small sample sizes with considerable scattering (as is visible in these two examples of blood pressure readings) may have contributed to the failure to detect an effect on blood pressure, but blood pressure may not have been a major readout, and its acquisition could have been less accurate than in the larger ILLUSTRATE trial. This, however, remains speculative, although the description of blood pressure measurements is explicit in the ILLUSTRATE trial (Barter et al., 2007), whereas it is not mentioned in the phase II trials. This may indicate, although it does not prove, that blood pressure reading was not very standardized in these trials. In addition, patient populations were different, and concomitant medications were as well. From the translational perspective, it may have been better to have looked at the very important, surrogate-type biomarker blood pressure in a more detailed way in the earlier trials. The dramatic blood pressure rise in ILLUSTRATE was unforeseeable based on the data obtained prior to this trial. The only translational shortcoming at this point may have been the undervaluation of this surrogate in the early trials, and future drug developments should take this into account as a translational learning experience. Thus, a rise in blood pressure that, at an average of 2 mmHg systolic, has been considered insignificant could still indicate a rise of 5 mmHg in the pivotal trial, and bad luck seems to have played a partial role in this context. On the other hand, there is a tendency to suppress unfavorable effects such as blood pressure increases if they are not dramatic enough to be unavoidably seen even in small trials.

As calculated by Nissen and colleagues (2007), it is highly unlikely that the negative impact of the rise in blood pressure of systolic 5.4 mmHg not only balances a putative beneficial effect of CETP inhibition, but even outweighs it and causes an increase in mortality. Therefore, there is good reason to assume that CETP inhibition per se may not keep up with the promises generated by in vitro and animal in vivo data, although it excessively increases HDL-cholesterol levels in humans: 60 mg/day of torcetrapib raised HDL-levels by 72% in the ILLUSTRATE trial (15,000 patients total, treatment for 12 months). Here is the true translational problem. The data summarized previously have clearly demonstrated that low HDL is associated with high cardiovascular risk and vice versa, but there has been comparably little evidence that interventional changes of this biomarker are associated with beneficial risk changes. According to the scientific systematic approach, HDL-cholesterol so far has not been considered a risk factor, only a risk marker (see previous discussion). Or in other words, it has been established as a biomarker that seems to be associated with disease but lacks the major components of being a surrogate such as LDL-cholesterol, in that interventional data have been sparse and as yet inconclusive. Thus, it is not close to becoming a surrogate marker and would score below 40 in the biomarker predictivity score described in Subchapter 3.4. In major recommendations on cardiovascular prevention (e.g., the ATP-III guidelines, see National Cholesterol Education Program Expert Panel, 1991), HDL-cholesterol is still seen as a risk marker, but not a treatment goal, although options such as nicotinic acid or fibrates increase its levels by 10%–20%.

This formalistic approach is supported by several facts that need to be considered: HDL-cholesterol is a major component of reverse cholesterol transport (RCT), which describes the removal of cholesterol from, for example, plaque macrophages through HDL down the fecal route. CETP inhibition blocks HDL-cholesterol-ester transport, and fractional HDL-cholesterol-clearance is diverted to the liver, but total clearance remains constant. Thus, cholesterol excretion in feces is not increased. Whether HDL increases by this mechanism are equivalent to those by exercise or niacin is questionable: CETP inhibition may alter the important partial functions of HDL, namely antioxidant and antithrombotic activities as known beneficial properties of "native" HDL. Obviously, HDL does not equal HDL, and HDL may become dysfunctional under CETP inhibition (Nissen et al., 2007). The situation would be different if the turnover of HDL is stimulated and serum levels are elevated, or if HDL degradation is slowed and turnover is even reduced thereby. Further important doubts about CETP inhibition as a beneficial intervention are derived from human genetic data: in humans genetically deficient of CETP, no longevity is described (Hirano et al., 1997).

Which translational paradigm was neglected in the translational process? The assessment of HDL-cholesterol as a biomarker was too positive. The truth that its interventional record was not nearly as good as that of LDL-cholesterol was ignored. Even before statins had been introduced, the translational predictive value of LDL had been established beyond mere epidemiological evidence; it further dramatically increased in the light of all the positive interventional trials with statins, leading to its status as a surrogate. It was highly speculative to assume that any intervention raising HDL would be beneficial, because this had been shown in only a few trials before that, however, were confounded by the fact that the agents tested (niacin and fibrates) affect not only HDL but also LDL, and a separation of the contribution of either effect is difficult. In this situation, it should have been accepted that this biomarker, as it was determined at that time (just concentration), did not warrant a pivotal mortality trial. At this stage, more reliable biomarkers should have been tested prior to the pivotal mortality trial in much smaller trials, putting fewer patients at risk. Imaging techniques—including intima-media thickness assessment or intravascular ultrasound, magnetic resonance imaging techniques in groups of 100–200 patients—should have been applied in phase II trials to get enough evidence for atherosclerosis improvement to support a large mortality trial, a view largely shared by Nissen and colleagues (2007). At present, it seems that there is enormous competitive and economical pressure in pharmaceutical companies, leading to "missed" phase II trials. These phase II trials—if carefully designed—would be essential for successful translation, and in the given example, powerful biomarkers for the proof-of-principle stage are available. For torcetrapib, they have either been missed or explored in parallel to the mortality trial. The sequential character of biomarker-driven intermediaries in novel areas without established surrogates needs to be reinstituted to protect both patients and companies. If pharmaceutical companies continue to lose big bets such as this one—especially those whose risk could have been significantly reduced by 1–2 years of preparatory clinical trials—economic disaster and public judgment of moral incompetence are inevitable.

Even worse, because those preparatory studies are still insufficient, as off-target effects cannot be excluded as major confounders, similar drugs will go through the same process and little will learned from the failed example. It is sincerely hoped that a more systematic, step-wise approach will be chosen for the still-ongoing CETP inhibitor projects.

These results of the failed torcetrapib development should have been reflected into the development compounds, such as dalcetrapib by Roche. If the translatability scoring instrument described in Subchapter 2.2 would have been applied to assess translational risk at a fictive date before commencing the pivotal phase III trials, torcetrapib would have been given a comparably low score (= high translational risk) of 1.95 compared to 4.14 for varenicline. The individual score items are shown in Table 11.1a. For comparison, the score is also shown for dalcetrapib, which is even much lower (1.06) than that for torcetrapib, indicating the highest possible level of translational risk (Table 11.1b).

TABLE 11.1 Translatability scores[1] derived from specified items for torcetrapib, varenicline (a),[2] and dalcetrapib (b)

(a)

	Torcetrapib		Varenicline	
	Score 1–5	Σ (score × weight/100)	Score 1–5	Σ (score × weight/100)
Starting evidence				
In vitro including animal genetics	5	0.1	5	0.1
In vivo including animal genetics (e.g., knock-out. Overexpression models)	5	0.15	5	0.15
Animal disease model	4	0.12	5	0.15
Multiple species	1	0.03	5	0.15
Human evidence				
Genetics	1	0.05	1	0.05
Model compounds			5	0.65
Clinical trials	2	0.26	4	0.52
Biomarkers for efficacy/safety prediction				
Biomarker grading	2	0.48	5	1.2
Biomarker development	2	0.26	4	0.52
Concept for proof-of-mechanism, proof-of-principle, and proof-of-concept testing				
Biomarker strategy	2	0.1	5	0.25
Surrogate/endpoint strategy	2	0.16	4	0.32
Personalized medicine aspects				
Disease subclassification. Responder concentration	3	0.09	1	0.03
Pharmacogenomics	3	0.15	1	0.05
SUM		**1.95**		**4.14**

(b)

	Dalcetrapib	
	Score 1–5	Σ (score × weight/100)
Starting evidence		
In vitro including animal genetics	5	0.1
In vivo including animal genetics (e.g., knock-out. Overexpression models)	5	0.15
Animal disease model	4	0.12
Multiple species	1	0.03
Human evidence		
Genetics	1	0.05
Model compounds		

Continued

TABLE 11.1 Translatability scores[1] derived from specified items for torcetrapib, varenicline (a),[2] and dalcetrapib (b)—cont'd

(b)		
Clinical trials	2	0
Biomarkers for efficacy/safety prediction		
Biomarker grading	2	0
Biomarker development	2	0.26
Concept for proof-of-mechanism, proof-of-principle, and proof-of-concept testing		
Biomarker strategy	2	0.1
Surrogate/endpoint strategy	2	0.16
Personalized medicine aspects		
Disease subclassification. Responder concentration	3	0.09
Pharmacogenomics	3	0.15
SUM		**1.06**

[1]*Wehling, 2009; Wendler and Wehling, 2012.*
[2]*Data in (a) from Wehling (2011) with kind permission.*

Coming back to the principles of translatability scoring, the score was already low, the risk high for torcetrapib. After the excess mortality in the ILLUMINATE trial and the insufficient explanation for beneficial effects as being overcompensated by off-target effects, a congener based on the same principle was doomed to fail at an even greater likelihood. The score of 1.06 for dalcetrapib should have been seen as a stopper for consequent development—in particular for the initiation of the expensive phase III trial. This almost lowest possible score mostly reflected the devalidation of HDL concentration as a risk factor and therapeutic target in the ILLUMINATE trial. The related items in the biomarker segment of the score panel were seen to be zero, reflecting this fact. Zero values should always be discussed as stoppers of a program in its present form. With this information, further investments should have been made into the development of prospectively useful biomarkers rather than a phase III trial. Coming as no surprise, Roche had to stop the development of dalcetrapib after the phase III trial (Schwartz et al., 2012) failed to show any beneficial effect of the drug on hard cardiovascular endpoints. The trial, however, proved that the excess mortality seen with torcetrapib was likely to reflect only off-target effects of that particular compound.

In summary, a very important, highly predictive biomarker with surrogate status (blood pressure) was undervalued, and a weak biomarker (HDL-cholesterol) was not properly developed and originally rated too high with no subsequent reflection of its further devalidation during the course of the development of CETP inhibitors.

This example shows the importance of appropriate evaluation of the predictive value of a given biomarker. It underpins the dramatic need for a systematic approach to assessing biomarkers objectively. HDL functionality is now a major biomarker development area, and some time from now, a particular fraction or property of HDL may become a fantastic biomarker, leading to the development of novel therapies that will greatly benefit patients. It is still doubtful whether CETP inhibitors will be among them.

REFERENCES

Barter, P.J., Caulfield, M., Eriksson, M., Grundy, S.M., Kastelein, J.J., Komajda, M., et al., 2007. ILLUMINATE investigators: effects of torcetrapib in patients at high risk for coronary events. N. Engl. J. Med. 357, 2109–2122.

Brousseau, M.E., Schaefer, E.J., Wolfe, M.L., Bloedon, L.T., Digenio, A.G., Clark, R.W., et al., 2004. Effects of an inhibitor of cholesteryl ester transfer protein on HDL cholesterol. N. Engl. J. Med. 350, 1505–1515.

Brown, B.G., Zhao, X.Q., Chait, A., Fisher, L.D., Cheung, M.C., Morse, J.S., et al., 2001. Simvastatin and niacin, antioxidant vitamins, or the combination for the prevention of coronary disease. N. Engl. J. Med. 345, 1583–1592.

Centers for Disease Control and Prevention, 2008. Smoking-attributable mortality, years of potential life lost, and productivity losses—United States, 2000–2004. Morb. Mort. Weekly Rep. 57 (45), 1226–1228.

Davidson, M.H., McKenney, J.M., Shear, C.L., Revkin, J.H., 2006. Efficacy and safety of torcetrapib, a novel cholesteryl ester transfer protein inhibitor, in individuals with below-average high-density lipoprotein cholesterol levels. J. Am. Coll. Cardiol. 48, 1774–1781.

Etter, J.F., Lukas, R.J., Benowitz, N.L., West, R., Dresler, C.M., 2008. Cytisine for smoking cessation: a research agenda. Drug Alcohol Depend 92, 3–8.

Frick, M.H., Elo, O., Haapa, K., Heinonen, O.P., Heinsalmi, P., Helo, P., et al., 1987. Helsinki Heart Study: primary-prevention trial with gemfibrozil in middle-aged men with dyslipidemia. Safety of treatment, changes in risk factors, and incidence of coronary heart disease. N. Engl. J. Med. 317, 1237–1245.

Gordon, T., Castelli, W.P., Hjortland, M.C., Kannel, W.B., Dawber, T.R., 1977. High-density lipoprotein as a protective factor against coronary heart disease: the Framingham Study. Am. J. Med. 62, 707–714.

Hirano, K., Yamashita, S., Nakajima, N., Arai, T., Maruyama, T., Yoshida, Y., et al., 1997. Genetic cholesteryl ester transfer protein deficiency is extremely frequent in the Omagari area of Japan: marked hyperalphalipoproteinemia caused by CETP gene mutation is not associated with longevity. Arterioscler. Thromb. Vasc. Biol. 17, 1053–1059.

Keating, G.M., Siddiqui, M.A.A., 2006. Varenicline—a review of its use as an aid to smoking cessation therapy. CNS Drugs 20, 945–960.

Kuivenhoven, J.A., de Grooth, G.J., Kawamura, H., Klerkx, A.H., Wilhelm, F., Trip, M.D., et al., 2005. Effectiveness of inhibition of cholesteryl ester transfer protein by JTT-705 in combination with pravastatin in type II dyslipidemia. Am. J. Cardiol. 95, 1085–1088.

McKenney, J.M., Davidson, M.H., Shear, C.L., Revkin, J.H., 2006. Efficacy and safety of torcetrapib, a novel cholesteryl ester transfer protein inhibitor, in individuals with below-average high-density lipoprotein cholesterol levels on a background of atorvastatin. J. Am. Coll. Cardiol. 48, 1782–1790.

Morehouse, L.A., Sugarman, E.D., Bourassa, P.-A., Sand, T.M., Zimetti, F., Gao, F., et al., 2007. Inhibition of CETP activity by torcetrapib reduces susceptibility to diet-induced atherosclerosis in New Zealand White rabbits. J. Lipid Res. 48, 1263–1272.

National Cholesterol Education Program Expert Panel on Detection, 1991. Evaluation, and Treatment of High Blood Cholesterol in Adults (Adult Treatment Panel III). ATP-III guidelines. http://www.nhlbi.nih.gov/guidelines/cholesterol/atp3xsum.pdf (accessed 22.01.08.).

Nikkila, E., 1953. Studies on the lipid-protein relationship in normal and pathological sera and the effect of heparin on serum lipoproteins. Scand. J. Clin. Lab. Invest. 5 (Suppl. 8), 9–100.

Nissen, S.E., Tardif, J.C., Nicholls, S.J., Revkin, J.H., Shear, C.L., Duggan, W.T., et al., 2007. Effect of torcetrapib on the progression of coronary atherosclerosis. N. Engl. J. Med. 356, 1304–1316.

Schwartz, G.G., Olsson, A.G., Abt, M., Ballantyne, C.M., Barter, P.J., Brumm, J., et al., 2012. Effects of dalcetrapib in patients with a recent acute coronary syndrome. N. Engl. J. Med. 367, 2089–2099.

Tall, A.R., Yvan-Charvet, L., Wang, N., 2007. The failure of torcetrapib: was it the molecule or the mechanism? Arterioscler. Thromb. Vasc. Biol. 27, 257–260.

Taylor, A.J., Sullenberger, L.E., Lee, H.J., Lee, J.K., Grace, K.A., 2004. Arterial Biology for the Investigation of the Treatment Effects of Reducing Cholesterol (ARBITER) 2: a double-blind, placebo-controlled study of extended-release niacin on atherosclerosis progression in secondary prevention patients treated with statins. Circulation 110, 3512–3517.

U.S. Department of Health and Human Services, 1990. The health benefits of smoking cessation: a report of the Surgeon General. Centers for Disease Control and Prevention (CDC), Office on Smoking and Health. http://profiles.nlm.nih.gov/NN/B/B/C/T (accessed 10.02.07.).

Wehling, M., 2009. Assessing the translatability of drug projects: what needs to be scored to predict success? Nature Rev. Drug Discov. 8, 541–546.

Wehling, M., 2011. Drug development in the light of translational science: shine or shade? Drug Discov. Today 16, 1076–1083.

Wendler, A., Wehling, M., 2012. Translatability scoring in drug development: eight case studies. J. Transl. Med. 10, 39.

Index

Note: Page numbers followed by "b", "f" and "t" indicate boxes, figures and tables respectively.

A

A2M. *See* Alfa-2-macroglobulin
ABPI. *See* Association of the British Pharmaceutical Industry
Absorption, distribution, metabolism, and elimination (ADME), 59–60, 261
ABSSSI. *See* Acute Bacterial Skin and Skin Structure Infections
ACCE implications. *See* Analytical validity, clinical validity, clinical usefulness, and ethical implications
Accelerating Medicines Partnership (AMP), 324
Accelerator mass spectrometry (AMS), 230
  microdosing, 230–231
Accuracy indicators, 19
Acetylsalicylic acid, 110
ACG. *See* Anterior cingulate gyrus
Acute Bacterial Skin and Skin Structure Infections (ABSSSI), 105
Acute lymphocytic leukemia (ALL), 41, 66
AD. *See* Alzheimer's disease
Adaptive trial design
  adaptive design clinical studies, 236
  AWARD-5, 237
  benefits and risks comparison for, 237t
  classification, 235
  clinical trial, 235
  design modifications in, 236t
  with individualized dosing, 238f
  stages of clinical development, 235
  13-cis-retinoic acid, 237
ADME. *See* Absorption, distribution, metabolism, and elimination
ADNI. *See* Alzheimer's Disease Neuroimaging Initiative
Adverse drug reaction (ADR), 59
AFP. *See* Alpha-fetoprotein
Aging, 299
  animal models, 301–302
  translational research in, 306–308
Alfa-2-macroglobulin (A2M), 164–165
ALL. *See* Acute lymphocytic leukemia
Alpha-fetoprotein (AFP), 36
ALS. *See* Amyotrophic lateral sclerosis
Alzheimer's disease (AD), 105, 120
Alzheimer's Disease Neuroimaging Initiative (ADNI), 105
AM. *See* Amygdala
AMP. *See* Accelerating Medicines Partnership
AMS. *See* Accelerator mass spectrometry
Amygdala (AM), 203
Amyotrophic lateral sclerosis (ALS), 85
Analysis of covariance (ANCOVA), 278
Analysis of variance (ANOVA), 278
Analytical validity, clinical validity, clinical usefulness, and ethical implications (ACCE implications), 20–21, 21f
Analytical variables, 160
ANCOVA. *See* Analysis of covariance
Animal models
  aging, 301–302
  atherosclerosis, 171
  in heart failure, 174–175
  human tumors, 86
  hypertension, 167–168, 168t
  limitations for, 305–306
  remedies for failed translation
    improving models, 87
    improving rigor of preclinical studies, 87–88
  for translational research, 83–84
    clinical trials, 84
    disease modeling, 84
    modeling care of patients, 84–85
    modeling comorbidities, 84
    preclinical research, 84, 85f
  translational value, 85
    ALS, 85
    drug discovery, 86
    extrapolation of preclinical findings, 86
    HD, 86
    lost in translation, 86
    neurodegenerative disorders, 86
    neuropathic pain research, 86
    reverse-translational studies, 86
    TGN1412, 85
  value of, 83
Animal toxicity testing, 258–259
ANOVA. *See* Analysis of variance
Anterior cingulate gyrus (ACG), 200
Antiatherosclerotic drug, 244f
Antihypertensive drugs, 167
Antipsychotic drug effects characterization, 205–206
APCI. *See* Atmospheric pressure chemical ionization
Apolipoprotein A1 (ApoA1), 164–165
APRI. *See* AST-to-platelet ratio index
Area under curve (AUC), 277–278
Area-focused consortia, 324
Arrays, 28
  cDNA microarrays, 30–31
  oligonucleotide microarrays, 30
  and target preparation, 30f
Association of the British Pharmaceutical Industry (ABPI), 260
AST-to-platelet ratio index (APRI), 164–165
AstraZeneca plc (AZ), 315
Athero-Express biobank study, 78–79
Atherosclerosis, 77–78, 170, 244
  animal models, 171
  blood-borne biomarkers, 171
    CRP, 171
    GPx, 172
    IL-18, 171
    IL-6, 171
    Lp-PLA2, 172
    MCP-1, 172
    MMPs, 172
    OxLDL, 172
  imaging, 172
    CEUS, 173
    CIMT, 172–173
    CT, 173
    IVUS, 173
    MRI, 173
    OCT, 173
  in mice, 173
Atmospheric pressure chemical ionization (APCI), 35
Atypical antipsychotics, 205
AUC. *See* Area under curve
Auditory and language processing, 200–201
AVANTRA® Biomarker Workstation, 96, 96f
AZ. *See* AstraZeneca plc

B

BACE. *See* Beta-Site APP Cleaving Enzyme
Backward translational activities, 112
Baclofen, 107
  $GABA_B$ receptor agonism, 107
  spinal cord–related disorders, 107–108
  target validation, 108
  translation in medicine, 109
Bacterial angina lacunaris, 129
Balanced innovation environment, 294
Baltimore Longitudinal Study on Aging (BLSA), 303

Basic fibroblast growth factor (bFGF), 186
Bayh–Dole Act of 1980, 315
BBB. *See* Blood-brain barrier
Bead-based assays, 95
Bedside devices, 95–96
Beta blockers, 61–62
  CYP2D6 polymorphism, 62–63
  racemate carvedilol, 62
Beta-Site APP Cleaving Enzyme (BACE), 105
bFGF. *See* Basic fibroblast growth factor
Big pharma, 313
  Big pharmaceutical companies, 5
  history, 313
    biotech industry, 313, 314t
    business-focused solutions, 315–316
    causes of change, 314–315
    pharmas, 313b
    productivity crunch, 316
  precompetitive consortia, 323–324
  public–private partnerships, 317
  from 2011 to 2013, 323f
    biotech companies and academia, 319
    partnerships between pharma companies, 320t–322t
    pharma and academia, 317–319
  translational solutions, 316
    integrated discovery nexuses, 317, 318f
    NIH, 317
Biological marker. *See* Biomarkers
Biological modeling, 269, 275
  one-compartment model, 276f
  pharmacodynamics, 276
  pharmacokinetics, 276–278
Biomarker-assisted development, 186–187
Biomarkers, 7, 127, 127t, 131, 187, 274–275, 288, 304, 305f. *See also* Blood-borne biomarkers; Cardiovascular biomarkers
  attributes, 133–143
  categorization, 143t
  change in tumor size, 151
  characteristics, 185
  classification, 131, 131t, 143
  comparison of international agencies, 134f
  conceptual model, 134f
  consortia, 103
    Critical Path Institute's PSTC, 103
    innovative medicines initiative in Europe, 104
    iSAEC, 105
    PhRMA biomarkers consortium, 104–105
  critical path initiative, 101
  for decision making, 132t
  genomic, 135t–142t
  grading scheme for, 155
  for heart failure, 176
  hypertension, 168
    blood-borne biomarkers, 168–169
    clinical markers of, 168
  ideal longevity, 305b
  impact and remits, 128f
  importance in translational plot, 130f
  inherent value, 129, 131
  new technologies, health authorities, and regulatory decision making, 101–102
    guidances for qualification and validation, 102
    VGDS, 102
    VXDS, 102
  in oncology
    biomarker-assisted development, 186–187
    clinical decision making, 185
    molecular genetics, 185
    tissue biopsies, 187–188
  percentage of drug recommendations, 133f
  pharmaceutical toxicology, 260
  pharmacogenomic biomarker development
    proposal for, 152
  predictive profiling of atherosclerosis, 156t
  predictivity classification, 151–152
    predictive profiling, 153t
    scoring system, 152–154
  private–public partnerships, 102–103
  quality assessment, 151
  risk factor, 144
  risk marker, 144
  role, 129–130
  subclassification, 133
  surrogates, 132
  for translation in psychiatry, 197–198
  uses, 128–129
Biomarkers development, 18, 145
  for chronic graft rejection biomarkers, 148, 148f
  components for detection and, 148
  discovery, 18
  five-step scheme, 145
  identification/characterization, 18–19
  pharmacogenomic biomarker development
    FDA recommendation, 147f
    proposal for, 146–147
  PSTC, 149
  ROC curves, 146f
  sensitivity and specificity, 145
  stages, 146f, 147–148
  standardization/harmonization, 19
    analytes, 19
    analytical and postanalytical issues, 19
    analytical performances, 19
    optimal materials and methods, 19
  validation, 19
Biomedical research, 302–303
Biorepositories, 75–76
Biotech companies and academia, 319
Biotech industry, 313, 313b
BioVacSafe, 104
Blood pressure, 168, 336
Blood urea nitrogen (BUN), 96–99, 101–102
Blood-borne biomarkers. *See also* Biomarkers; Cardiovascular biomarkers
  for atherosclerosis, 171–172
  for heart failure, 176–178
  hypertension
    coagulation, 169
    endothelial dysfunction, 169
    inflammation, 169
    MicroRNA, 169
    RAAS activity, 168–169
    vascular resistance, 169
Blood-brain barrier (BBB), 87
BLSA. *See* Baltimore Longitudinal Study on Aging
Body mass index (BMI), 307–308
Bone marrow stromal cells (BMSC), 41–42, 52, 53t
  least-angle regression algorithm, 53–54
  from marrow aspirates, 53
  senescence, 53
Bonferroni's method, 273
Bovine serum albumin (BSA), 94
BRAF, 187
Breast cancer
  multiple logistic regression, 162–163
  neural network, 163
Breast cancer, 246
BSA. *See* Bovine serum albumin
BUN. *See* Blood urea nitrogen
Business-focused solutions, 315–316

C

c-kit protein detection, 93
C-Path Institute. *See* Critical Path Institute
C-reactive protein (CRP), 127–128, 171, 177
CA 125 marker, 162
CABP. *See* Community-Acquired Bacterial Pneumonia
California Institute of Quantitative Biosciences (QB3), 317
Calorie restriction (CR), 306
  biomarker approach, 308
  chronic, 307–308
  middle-aged individuals, 308
  mimetics, 307
Cambridge Healthtech Institute (CHI), 244
CAMD. *See* Coalition Against Major Diseases
CAR. *See* Chimeric antigen receptor
Cardiac arrhythmia suppression trial (CAST), 132–133
Cardiac remodeling, markers for, 176–177
Cardiovascular biomarkers. *See also* Biomarkers; Blood-borne biomarkers
  atherosclerosis, 170–173
  heart failure, 173–178
  hypertension, 167–170
Cardiovascular disease, 61. *See also* Malignant diseases
  beta blockers, 61–62
    CYP2D6 polymorphism, 62–63
    racemate carvedilol, 62
  vitamin K antagonists, 63
Cardiovascular Health Study (CHS), 304, 309
Cardiovascular medicine, 192
Carotid intima-media thickness (CIMT), 172–173
CART. *See* Classification and regression trees
CAST. *See* Cardiac arrhythmia suppression trial
Catechol-O-methyltransferase (COMT), 203–204
Catheter-based intravascular ultrasound (Catheter-based IVUS), 173
CATIE trial. *See* Clinical Antipsychotic Treatment of Intervention Effectiveness trial
CDC. *See* Centers for Disease Control and Prevention

cDNA. *See* Complementary DNA
Celecoxib, 65
Cellular therapies, 41–42
  adoptive cellular therapy, 41
  complexity, 41
  GMP, 42
  HSC transplants, 41–42
  pluripotent embryonic and iPSCs, 42
  potency testing, 42–43
    BMSCs, 52–54
    complexities association with, 43, 43t
    cultured CD4+ cells, 52
    factors affecting, 44, 44t
    gene expression arrays, 45–46
    measurement, 44–45
    miRNAs, 54, 55f
  regenerative medicine, 42
Centers for Disease Control and Prevention (CDC), 133
Centers for Therapeutic Innovation (CTI), 318
Central nervous system (CNS), 4, 198–199
Centre for Medicines Research International (CMR)
Cerebrospinal fluid (CSF), 105, 199–200, 230
CETP. *See* Cholesteryl ester transfer protein
CEUS. *See* Contrast-enhanced ultrasound
CFAST. *See* Coalition For Accelerating Standards & Therapies
Chain termination methods. *See* DNA sequencing
Chemical ionization (CI), 35
CHI. *See* Cambridge Healthtech Institute
Chimeric antigen receptor (CAR), 41
Cholesteryl ester transfer protein (CETP), 171, 332–333
Chronic obstructive pulmonary disease (COPD), 117
CHS. *See* Cardiovascular Health Study
CI. *See* Chemical ionization
CIMT. *See* Carotid intima-media thickness
13-cis-retinoic acid, 237
Clarification of Optimal Anticoagulation (COAG), 63
Classification and regression trees (CART), 164
Clinical and translational science awards (CTSA), 9, 317
Clinical Antipsychotic Treatment of Intervention Effectiveness trial (CATIE trial), 195–196
Clinical Pharmacogenetics Implementation Consortium (CPIC), 61
Clinical trial application (CTA), 258
Clopidogrel, 110
CMR. *See* Centre for Medicines Research International
CMV. *See* Cytomegalovirus
CNS. *See* Central nervous system
COAG. *See* Clarification of Optimal Anticoagulation
Coagulation biomarkers, 169
Coalition Against Major Diseases (CAMD), 103
Coalition For Accelerating Standards & Therapies (CFAST), 103
Codeine, 65
Commercialization, 282–283
Common toxicity criteria (CTC), 220
CommonMind Consortium, 324
Community-Acquired Bacterial Pneumonia (CABP), 105
Comparability testing, 49
Competitive ELISAs, 94
Complementary DNA (cDNA), 288–289
  microarrays, 30–31
Compound-inherent biological profile, 249–250
Computed tomography (CT), 173, 190
COMT. *See* Catechol-O-methyltransferase
Consortium for Resourcing and Evaluating AMS Microdosing (CREAM), 231
Construct validity, 87
Continuous performance test (CPT), 202
Contract research organization (CRO), 317
Contrast-enhanced ultrasound (CEUS), 173
Conventional antipsychotics. *See* First-generation antipsychotics
Cooperative research and development agreement contract (CRADA contract), 102
COPD. *See* Chronic obstructive pulmonary disease
Cost Utility of the Latest Antipsychotic Drugs in Schizophrenia Study trial (CUTLASS trial), 195–196
Costimulation, 49
Coumarin derivatives, 63
CPIC. *See* Clinical Pharmacogenetics Implementation Consortium
CPT. *See* Continuous performance test
CPTR. *See* Critical Path to Drug TB Regimen initiative
CR. *See* Calorie restriction
CRADA contract. *See* Cooperative research and development agreement contract
CREAM. *See* Consortium for Resourcing and Evaluating AMS Microdosing
Critical path initiative, 9, 10f, 101, 228
Critical Path Institute (C-Path Institute), 256
  PSTC, 103
Critical Path to Drug TB Regimen initiative (CPTR), 103
CRO. *See* Contract research organization
CRP. *See* C-reactive protein
CSF. *See* Cerebrospinal fluid
CT. *See* Computed tomography
CTA. *See* Clinical trial application
CTC. *See* Common toxicity criteria
CTI. *See* Centers for Therapeutic Innovation
CTSA. *See* Clinical and translational science awards
Cultured CD4+ cells, 52
CUTLASS trial. *See* Cost Utility of the Latest Antipsychotic Drugs in Schizophrenia Study trial
CYP. *See* Cytochrome *P*450
CYP2D6 genotype, 61
Cytochrome *P*450 (CYP), 32, 61
Cytogenetics, 92
Cytomegalovirus (CMV), 49
Cytomics, 15
Cytotoxicity assays, 45

D

DA. *See* Dopamine
Dabigatran, 120
DARPP-32 protein, 204
DC. *See* Dendritic cells
DCM. *See* Dilated cardiomyopathy
ddNTP. *See* dideoxynucleotide triphosphate
Decision Biomarker assay principle, 97f
Dendritic cells (DC), 46, 48–49
  plasmacytoid, 49
  potency testing, 48–49, 51
    antigens, 49
    candidate markers, 50–51
    comparability testing, 49
    gene expression profiling, 49, 50f
    genes with least and greatest index of variability, 51t–52t
    measurement, 49
    PBMCs, 50
    T cells, 49
Deoxynucleotide triphosphate (dNTP), 25–26
Depression, 60, 196
  CYP2D6, 61
  inadequate drug exposure, 61
  mirtazapine, 61
  pharmacokinetic-based dose-adjustment approach, 61
  psychiatry drugs, 60–61
  tricyclic antidepressants, 61
DIA. *See* Drug Information Association
Diabetes mellitus (DM), 164
Diagnostic
  algorithms, 159
  performance, 20–21
Diagnostic immunoassay devices
  bedside devices, 95–96
  high-throughput immunoassay platforms, 94
    bead-based assays, 95
    dynamic arrays, 95
    electrochemiluminescence-based arrays, 95
Dideoxy sequencing. *See* DNA sequencing
dideoxynucleotide triphosphate (ddNTP), 25–26
Digit Symbol Substitution Test (DSST), 305
Digoxin, 110
Dihydropyrimidine dehydrogenase (DPD), 32–33, 66
Dihydrouracil dehydrogenase. *See* Dihydropyrimidine dehydrogenase (DPD)
Dilated cardiomyopathy (DCM), 174
Direct ELISAs, 94
Discovery Partnerships with Academia (DPAc), 319
Discovery-based research, 17, 159–160
"Disease free", 301, 303
Disease induction models, 83
Disease modeling, 84
Disease-focused consortia, 324
DLPFC. *See* Dorsolateral prefrontal cortex
DLT. *See* Dose-limiting toxicity
DM. *See* Diabetes mellitus
DME. *See* Drug metabolizing enzymes
DMPK. *See* Drug metabolism and pharmacokinetics

DNA sequencing, 25–26
by ligation, 27
by synthesis, 27–28
dNTP. *See* Deoxynucleotide triphosphate
DOCA-salt method, 168
Donor T-cell responses, 52
Dopamine (DA), 201, 331
Dopamine receptor D2 (DRD2), 203
Dorsolateral prefrontal cortex (DLPFC), 201–202
Dose escalation, 219
Dose-limiting toxicity (DLT), 219–220
Dose-response relationship, 252–253
DPAc. *See* Discovery Partnerships with Academia
DPD. *See* Dihydropyrimidine dehydrogenase
DRD2. *See* Dopamine receptor D2
Drop-the loser-design, 236
Drug development, 267, 313b
pathway, 257
traditional, 227, 228f
Drug discovery, 224, 261
Drug Information Association (DIA), 260
Drug metabolism and pharmacokinetics (DMPK), 119
Drug metabolizing enzymes (DME), 60–61
Drug therapies, 60
Druggability, 115
DSST. *See* Digit Symbol Substitution Test
Dutch Athero-Express study, 78–79
Dynamic arrays, 95

E

Early Detection Research Network (EDRN), 18
Early translational assessment dimensions, 115, 116t, 117
animal models for normal physiology, 116
biomarkers for efficacy or safety prediction, 117
FLAP, 117
generic and proprietary domains, 116
model compounds, 117
personalized medicine aspects, 117
Eastern Cooperative Oncology Group (ECOG), 218
EBV antigens. *See* Epstein–Barr virus antigens
ECG. *See* Electrocardiogram
ECOG. *See* Eastern Cooperative Oncology Group
ECVAM. *See* European Centre for the Validation of Alternative Methods
ED. *See* Emergency department
EDRN. *See* Early Detection Research Network
EFPIA. *See* European Federation of Pharmaceutical Industries and Associations
EGF. *See* Epidermal growth factor
EGFR. *See* Epidermal growth factor receptor
EI. *See* Electron ionization
eINDs. *See* Experimental investigational new drugs
Electrocardiogram (ECG), 119
Electrochemiluminescence-based arrays, 95
Electron ionization (EI), 35
Electronic Patient-Reported Outcomes Consortium (EPRO), 103
Electrospray ionization (ESI), 35
ELISA. *See* Enzyme-linked immunosorbent assay
EM. *See* Extensive metabolizers
EMA. *See* European Medicines Agency
Emax model, 276
EMEA. *See* European Medicines Agency (EMA)
Emergency department (ED), 19
Endothelial dysfunction, 169
Endoxifen. *See* 4-hydroxy-N-desmethyl tamoxifen
Enzyme-linked immunosorbent assay (ELISA), 36, 93
competitive, 94
direct, 94
indirect, 94
sandwich, 94
EORTC. *See* European Organization for Research and Treatment of Cancer
Epidermal growth factor (EGF), 186
Epidermal growth factor receptor (EGFR), 120, 186, 218
EPO. *See* European Patent Office
EPRO. *See* Electronic Patient-Reported Outcomes Consortium
EPS. *See* Extrapyramidal side effects
Epstein–Barr virus antigens (EBV antigens), 49
ESI. *See* Electrospray ionization
Estrogen receptor (ER), 65, 246
ETS. *See* European Teratology Society
EU Microdosing AMS Partnership Programme (EUMAPP), 231
EU-PACT study. *See* European Pharmacogenetics of AntiCoagulant Therapy study
EUFEPS. *See* European Federation for Pharmaceutical Sciences
EUFEST. *See* European First Episode Schizophrenia Trial
EUMAPP. *See* EU Microdosing AMS Partnership Programme
European Centre for the Validation of Alternative Methods (ECVAM), 260
European Federation for Pharmaceutical Sciences (EUFEPS), 243
European Federation of Pharmaceutical Industries and Associations (EFPIA), 104, 323
European First Episode Schizophrenia Trial (EUFEST), 195–196
European Medicines Agency (EMA), 101–102, 260
European Organization for Research and Treatment of Cancer (EORTC), 10
European Patent Office (EPO), 284
European Pharmacogenetics of AntiCoagulant Therapy study (EU-PACT study), 63
European Teratology Society (ETS), 260
European Union (EU), 10
Evotec AG company, 319
Exenatide, 110
Experimental investigational new drugs (eINDs), 7
rules, 108
Exploratory clinical studies, 229
exploratory IND, 229
microdosing, 230–232
practical applications, 232–233
types, 229–230, 230f
Exploratory IND approach, 259
Exploratory trials, 239
exploratory aspects from regulatory trials, 240
exploratory data generation, 240t
FDA document, 239
Kaplan–Meier plots, 242f
liver metastasis assessment, 241f
phase 0 studies, 239
understanding of disease, 240
Exposure, 251
comparative assessment, 254
duration, 251
human, 259
Extensive metabolizers (EM), 61
Extrapyramidal side effects (EPS), 205

F

Face validity, 87
Facilitated serendipity, 109, 110f
False discovery rate (FDR), 51, 274
Familywise error rate (FWER), 273
FDA. *See* Food and Drug Administration; U.S. Food and Drug Administration
FDR. *See* False discovery rate
FEV_1. *See* Forced expiratory volume in 1 second
First-generation antipsychotics, 205
FISH. *See* Fluorescence in situ hybridization
“FISH-ing” translocations, 92
Fixation, 92
FL. *See* Follicular lymphoma
FLAP. *See* 5-lipoxygenase-activating protein
Fluorescence in situ hybridization (FISH), 92, 187
5-fluorouracil (5-FU), 32–33, 66–67
fMRI. *See* Functional magnetic resonance imaging
FNIH. *See* Foundation for National Institutes of Health
Follicular lymphoma (FL), 92
Food and Drug Administration (FDA), 9
Forced expiratory volume in 1 second (FEV_1), 302–303
Foundation for National Institutes of Health (FNIH), 104, 323
Fourier transform ion cyclotron resonance (FT-ICR), 35
FT-ICR. *See* Fourier transform ion cyclotron resonance
5-FU. *See* 5-fluorouracil
Functional assessment, 309
Functional impairment, 302–303
Functional limitation, 302–303
Functional magnetic resonance imaging (fMRI), 200
FWER. *See* Familywise error rate

G

G-CSF. *See* Granulocyte colony-stimulating factor
Galectin-3, 177

Gamma camera, 191
Gas chromatography (GC), 35
Gas chromatography-mass spectrometry (GC-MS), 33
Gastroesophageal reflux disease (GERD), 64
Gastrointestinal stromal tumors (GIST), 93
GC. *See* Gas chromatography
GC-MS. *See* Gas chromatography-mass spectrometry
GCP. *See* Good clinical practice
GDF-15, 177
Gefitinib, 120
Gemcitabine, 217–218
Gene sequences, 288
Generalized linear model (GLM), 279
Genetic patents, 291
  gene sequences
    patentability, 291–292
    patenting and promotion of innovation, 292–293
Genetic susceptibility factors imaging, 203
  COMT, 204
  DARPP-32, 204
  depiction of complex path, 203f
  GWAS, 205
  one-to-one mapping, 204
Genetic tests, patents on, 289
Genomewide association study (GWAS), 197–198, 205
Genomics, 15, 25. *See also* Metabolomics
  applications, 31
    molecular diagnostics, 31–32
    pharmacogenomics, 32–33
  genome sizes of humans and model organisms, 26t
  genomic tools
    arrays, 28–31
    sequencing, 25–28
  specialists in, 25
  strategy for automating Sanger DNA sequencing system, 26f
"Genomics bubble", 314–315
GERD. *See* Gastroesophageal reflux disease
Geriatrics, gerontology *vs.*, 300–301
Gerontology, geriatrics *vs.*, 300–301
GIST. *See* Gastrointestinal stromal tumors
GlaxoSmith Kline (GSK), 315, 319
GLM. *See* Generalized linear model
GLP. *See* Good laboratory practice
GLP-1. *See* Glucagons-like peptide type 1
Glucagons-like peptide type 1 (GLP-1), 110
Glutathion peroxidase (GPx), 172
GM. *See* Gray matter
GMP. *See* Good manufactory product
Go-or-no-go decisions, 222
Good clinical practice (GCP), 239
Good laboratory practice (GLP), 258
Good manufactory product (GMP), 42
GPx. *See* Glutathion peroxidase
Graded dose-response relationship, 252, 253f
Granulocyte colony-stimulating factor (G-CSF), 47
Gray matter (GM), 199
GSK. *See* GlaxoSmith Kline
GWAS. *See* Genomewide association study

H

Haber's Law of toxicology, 251
Hatch–Waxman Act of 1984, 315
HbA_1C. *See* Hemoglobin A_1C
HCC. *See* Hepatocellular carcinoma
HCG. *See* Human chorionic gonadotropin
HD. *See* Huntington's disease
HDL-cholesterol, 123, 334
Health Aging and Body Composition Study, 304
Health technology assessment (HTA), 133
Healthspan
  age-related physiologic change effect, 303f
  determinants identification, 303–304
  lifespan *vs.*, 299–300
  testing treatments to extending, 304–305
Healthy survivor effect, 301
Heart failure, 173–174
  animal models in, 174, 175t
    DCM, 174
    hypertensive heart disease, 174
    RCM, 175
    valvular lesions, 174
  biomarkers for, 176f, 178t
    6MWD, 176
    echocardiography, 176
  blood-borne biomarkers, 176
    for cardiac remodeling, 176–177
    inflammation, 177
    for myocardial stretch, 177
    neurohormonal activation, 177
    renal injury, 177–178
HED. *See* Human equivalent dose
HEK 293 cells. *See* Human embryonic kidney 293 cells
Helicobacter pylori (*H. pylori*), 64
Hematologic malignancies, 92–93
Hematopoietic stem cells (HSC), 41
  potency testing, 47
    CD34 antigen expression, 47
    CD34+ cells, 47–48
    gene expression profiles, 48f
    LTC-IC, 47
    methylcellulose culture systems, 47
    plerixafor, 48
    total nucleated cells counts, 47
    UCB CD34+ cells, 48
Hemoglobin A_1C (HbA_1C), 276
Hepatocellular carcinoma (HCC), 36
Her2/neu diagnostic test, 93
Herceptest™, 93
Herceptin®, 92
hERG. *See* Human ether-a-go-go related gene
hERG channel, 263
hESC. *See* Human embryonic stem cells
HF. *See* Hippocampal formation
High-throughput immunoassay platforms, 94
  bead-based assays, 95
  dynamic arrays, 95
  electrochemiluminescence-based arrays, 95
High-throughput technologies, 159
Hippocampal formation (HF), 203
Horizon 2020 Program, 10
HSC. *See* Hematopoietic stem cells
HTA. *See* Health technology assessment
HUGO. *See* Organisation of the Human Genome
Human chorionic gonadotropin (HCG), 18–19
Human embryonic kidney 293 cells (HEK 293 cells), 46
Human embryonic stem cells (hESC), 46–47
Human equivalent dose (HED), 219
Human ether-a-go-go related gene (hERG), 119
Human models, limitations for, 305–306
Human Proteome Organization (HUPO), 159–160
Human studies as target information source, 107
  baclofen, 107
    $GABA_B$ receptor agonism, 107
    spinal cord–related disorders, 107–108
    target validation, 108
    translation in medicine, 109
  reverse pharmacology, 110, 112t
    backward translational activities, 112
    eINDs, 112
    facilitated serendipity, 111
    feedback route, 113
    GLP-1, 110
    historic examples, 110
    incretin-analogues, 111
    Kaplan–Meier curves, 111f
    plant preparations, 112
    plant-derived principles, 111
    serendipity findings, 112
    shamanism, 111–112
    TNFalpha, 111
  sildenafil, 109
Huntington's disease (HD), 86
HUPO. *See* Human Proteome Organization
Hybridization, 92
4-hydroxy tamoxifen (4-OH-TAM), 65
3-hydroxy-3-methylglutaryl coenzyme A reductase inhibitors. *See* Statins
4-hydroxy-N-desmethyl tamoxifen, 65
Hypertension, 167
  animal models, 167–168, 168t
  biomarkers of, 170t
    blood-borne biomarkers, 168–169
    clinical markers of hypertension, 168
    renin, 170
  single disorder, 167

I

ICAM. *See* Intercellular adhesion molecule
ICC. *See* Interclass correlation coefficient
ICCVAM. *See* Interagency Coordinating Committee on Validation of Alternative Methods
IDI. *See* Integrated discrimination improvement
IGF. *See* Insulin-like growth factor
IHC. *See* Immunohistochemistry
IL-2. *See* Interleukin-2
Imaging, 189
  genetics approach, 198
  modalities, 190
    characteristics, 191–192
    CT, 190
    gamma camera, 191
    MRI, 190

Imaging (*Continued*)
MRS, 191
PET, 191
ultrasound imaging, 190
X-ray imaging, 190
validation, 189–190
IMI. *See* Innovative Medicines Initiative
IMIDIA, 104
Immunoassays, 93
development, 93
diagnostic immunoassay devices
bedside devices, 95–96
high-throughput immunoassay platforms, 94–95
ELISA, 93
competitive, 94
direct, 94
indirect, 94
sandwich, 94
Immunohistochemistry (IHC), 93
Immunohistochemistry, 92
IMP. *See* Investigational medicinal product
In situ hybridization (ISH), 91–92
In vitro assays, 44
cell function assays, 45
In vitro diagnostic industry (IVD industry), 19
In vitro diagnostic multivariate index assay (IVDMIA), 162
Inbred strains, 83
Incretin-analogues, 111
IND. *See* Investigational New Drug
Indirect ELISAs, 94
induced pluripotent stem cells (iPSCs), 42
Inferential statistical modeling, 268
Inflammation, 177
biomarkers, 169
Innovation ecosystem, 313
Innovative Medicines Initiative (IMI), 10, 104, 243, 323
INR. *See* International normalized ratio
Insulin-like growth factor (IGF), 306
Integrated discovery nexuses, 317, 318f
Integrated discrimination improvement (IDI), 161
Interagency Coordinating Committee on Validation of Alternative Methods (ICCVAM), 260
Intercellular adhesion molecule (ICAM), 169
Interclass correlation coefficient (ICC), 50–51
Interleukin-2 (IL-2), 45
Interleukin-6 (IL-6), 171
International normalized ratio (INR), 63
international Serious Adverse Event Consortium (iSAEC), 105
International Union of Toxicology (IUTOX), 260
Intravascular ultrasound (IVUS), 155, 173, 192
Invention, 283
Investigational medicinal product (IMP), 219
Investigational New Drug (IND), 227
application, 228
Ion sources, 35
Ion traps (IT), 35
iPSCs. *See* induced pluripotent stem cells
Irinotecan, 67
iSAEC. *See* international Serious Adverse Event Consortium
Ischemic penumbra, 83
ISH. *See* In situ hybridization
IT. *See* Ion traps
IUTOX. *See* International Union of Toxicology
IVD industry. *See* In vitro diagnostic industry
IVDMIA. *See* In vitro diagnostic multivariate index assay
IVUS. *See* Intravascular ultrasound

J

Johnson & Johnson (J&J), 318–319

K

Kidney injury, biomarker screening for, 96–99
Kidney injury molecule-1 (Kim-1), 96–99, 98f
Known valid biomarker, 147–148
KRAS gene, 223–224

L

l'art-pour-l'art approach, 3
LAK. *See* Leukocyte activated killer cells
"Lame-duck" drugs, 222
Late-stage attrition, 4
LC. *See* Liquid chromatography
LC-MS. *See* Liquid chromatography-mass spectrometry
LC-MS/MS. *See* Liquid chromatography-tandem mass spectrometry
LDL-cholesterol, 123, 129–130
Lead identification (LI), 3
Least-angle regression algorithm, 53–54
Left ventricular hypertrophy (LVH), 176
Leukocyte activated killer cells (LAK), 41
LI. *See* Lead identification
Licensing, 315–316
Licofelone, 117
Lifespan
aging and, 299
healthspan *vs.*, 299–300
testing treatments to extending, 304–305
Limits of detection, 20
Lipopolysaccharide (LPS), 50–51, 302
Lipoprotein-associated phospholipase (Lp-PLA2), 172
5-lipoxygenase-activating protein (FLAP), 117
Liquid chromatography (LC), 34–35
Liquid chromatography-mass spectrometry (LC-MS), 33
Liquid chromatography-tandem mass spectrometry (LC-MS/MS), 230
Liver cirrhosis (LvC), 36
Liver fibrosis, 164–165
Localization technologies
genetics and genomics, 91–92
biopsies with Her2 and CEP 17 signal ratio, 92
challenges, 92
"FISH-ing" translocations, 92
hematologic malignancies, 92–93
using ISH for medical diagnoses, 92
protein localization, 93
Long-term culture initiating cells (LTC-IC), 47
Longevity determinants identification, 303–304
Low-density arrays, 31
Lp-PLA2. *See* Lipoprotein-associated phospholipase
LPS. *See* Lipopolysaccharide
LTC-IC. *See* Long-term culture initiating cells
Luminex® xMAP® technology, 95
LvC. *See* Liver cirrhosis
LVH. *See* Left ventricular hypertrophy

M

M&A. *See* Mergers and acquisitions
MAD. *See* Multiple ascending dose
Magnetic resonance (MR), 199–200
Magnetic resonance imaging (MRI), 78, 173, 190, 198
Magnetic resonance spectroscopy (MRS), 191
Magnetic resonance tomography (MRT), 244
Major tranquilizers. *See* First-generation antipsychotics
MALDI. *See* Matrix-assisted laser desorption/ionization
Malignant diseases, 65
5-fluorouracil, 66–67
irinotecan, 67
tamoxifen, 65
thiopurines, 66
Mantle cell lymphoma (MCL), 92
MAPK. *See* Mitogen-activated protein kinase
MARCAR, 104
Mass analyzers, 35
Mass spectrometry, 34–35
ion sources, 35
mass analyzers, 35
"NMR-invisible" moieties, 35
UPLC, 35
Matrix metalloproteinase (MMP), 172
Matrix-assisted laser desorption/ionization (MALDI), 35
MAX BIOCHIP®, 97f
Maximum administered dose (MAD). *See* Maximum tolerated dose (MTD)
Maximum recommended starting dose (MRSD), 219
Maximum tolerated dose (MTD), 221, 229
Maximum-likelihood (ML), 277
MB. *See* Midbrain
MCI. *See* Mild Cognitive Impairment
MCL. *See* Mantle cell lymphoma
MCP-1. *See* Monocyte chemoattractant protein-1
MDSCs. *See* Myeloid-derived suppressor cells
Mergers and acquisitions (M&A), 315–316
MESA. *See* Multi-Ethnic Study on Atherosclerosis
Mesenchymal stem cell (MSC). *See* Bone marrow stromal cells (BMSC)

Mesenchymal stromal cells. *See* Bone marrow stromal cells (BMSC)
Meso Scale Diagnostics (MSD), 95
Metabolic footprinting, 34
Metabolic medicine, 192
Metabolome, 33
Metabolomics, 15, 33–34. *See also* Proteomics
analytical techniques
mass spectrometry, 34–35
NMR, 34
in clinical use, 36–37
metabolic footprinting, 34
metabolite
fingerprinting, 33–34
profiling, 33
"omics" and biomarkers, 35–36
Metabonomics, 15, 33–34
Methodological studies
early-phase drug development paradigm, 223f
endpoints measurement, 220
dose-limiting toxicities, 221t
pharmacodynamic endpoints, 221
toxicity, 220–221
go-or-no-go decisions, 222
mechanism-oriented trial design, 221
PoC, 222
PoM, 221–222
PoP, 222
open access clinical trials, 224
personalized medicine, 223–224
phase I trial methodology, 217–220
phase II trials, 223
Methylcellulose culture systems, 47
Methylenetetrahydrofolate reductase (MTHFR), 67
Metoprolol, 61–62
MI. *See* Myocardial infarction
Michaelis–Menten equation, 276
Microarray technology, 32
Microdosing, 230
AMS, 230–231
exploratory clinical studies, 232
PET, 231–232
repeated dosing, 232
semi-log plot comparing elimination of drug, 231f
MicroRNAs (miRNAs), 54, 55f
biomarkers, 169
Microscopy techniques, 92
Mid-regional pro-atrial natriuretic peptide (MR-proANP), 177
Midbrain (MB), 203
Midregional pro-adrenomedullin (MR-proADM), 177
Mild Cognitive Impairment (MCI), 105
Minimal residual disease (MRD), 66
Minoxidil, 6–7
MiR-155, 54
MiR-223, 54
miRNAs. *See* MicroRNAs
Mirtazapine, 61
Mismatch repair deficiency (MMR-D), 185
Mitogen-activated protein kinase (MAPK), 186
ML. *See* Maximum-likelihood
MMP. *See* Matrix metalloproteinase
MMR-D. *See* Mismatch repair deficiency
Model compounds, 117
Modeling care of patients, 84–85
Modeling comorbidities, 84
Molecular diagnostics, 31–32
Monocyte chemoattractant protein-1 (MCP-1), 172
Morphine, 110
Motor functioning, 201
MR. *See* Magnetic resonance
MR-proADM. *See* Midregional pro-adrenomedullin
MR-proANP. *See* Mid-regional pro-atrial natriuretic peptide
MRD. *See* Minimal residual disease
MRI. *See* Magnetic resonance imaging
MRS. *See* Magnetic resonance spectroscopy
MRSD. *See* Maximum recommended starting dose
MRT. *See* Magnetic resonance tomography
MSD. *See* Meso Scale Diagnostics
MSOAC. *See* Multiple Sclerosis Outcomes Assessments Consortium
MTD. *See* Maximum tolerated dose
MTHFR. *See* Methylenetetrahydrofolate reductase
Multi-Ethnic Study on Atherosclerosis (MESA), 309
Multimarker panel, 162, 164–165
Multiparameter approach, 161
breast cancer, 162–163
liver fibrosis, 164–165
ovarian cancer, 161–162
pancreatic cancer, 164
prostate cancer, 163–164
Multiple ascending dose (MAD), 119, 239
Multiple logistic regression, 162–164
Multiple Sclerosis Outcomes Assessments Consortium (MSOAC), 103
Multiplicity, 273–274
Multivariate statistical methods, 33–34
Muscle weakness, 303
6MWD. *See* Six-minute walking distance
Myeloid-derived DCs, 49
Myeloid-derived suppressor cells (MDSCs), 41
Myocardial infarction (MI), 177
Myocardial stretch, biomarkers, 177
Myriad Case, 291–292

N

N-acetylfelinine, 37
n-back tasks, 201–202
N-terminal pro B-type natriuretic peptide (NT-proBNP), 177
Naïve approach, 272
National Cancer Institute (NCI), 220
National Center for Advancing Translational Science (NCATS), 9, 317
National Institute on Aging (NIA), 304
National Institutes of Health (NIH), 104, 317
NCATS. *See* National Center for Advancing Translational Science
NCI. *See* National Cancer Institute
Net Reclassification Improvement (NRI), 161
Neural network (NNW), 163
Neurohormonal activation, 177
Neuroimaging studies, 206
Neuroscience, 193
Neutrophil gelatinase-associated lipocalin (NGAL), 177–178
New medical/biological entities (NMEs/NBEs), 6
New molecular entities (NMEs), 227, 315
New Zealand White (NZW), 333
Next-generation sequencing (NGS), 27, 29t
NGAL. *See* Neutrophil gelatinase-associated lipocalin
NGS. *See* Next-generation sequencing
NHP. *See* Nonhuman primate
NIA. *See* National Institute on Aging
Nicotine, 331
Nicotine replacement therapy (NRT), 331
NIH. *See* National Institutes of Health
Nitric oxide (NO), 169
NMEs. *See* New molecular entities
NMEs/NBEs. *See* New medical/biological entities
NMR. *See* Nuclear magnetic resonance
"NMR-invisible" moieties, 35
NNT. *See* Number needed to treat
NNW. *See* Neural network
NO. *See* Nitric oxide
No observed adverse effect level (NOAEL), 219, 253
NOAEL. *See* No observed adverse effect level
NOD. *See* Nonobese diabetic
Non-small cell lung cancer (NSCLC), 186
Nonhuman primate (NHP), 309
NONMEM software, 277
Nonobese diabetic (NOD), 47
Nonregulatory trial, 239
Nonsteroidal anti-inflammatory drugs (NSAID), 65
Novel cancer therapies, 32–33
Novel translatability scoring instrument, 117–118, 120
assessment of translatability for drugs, 121t–122t
DMPK, 119
drug project approaches phase II, 119
me-too drugs, 118
"natural" course of translational risk, 120f
PK/PD modeling, 118–119
proof-of-concept validation, 119
proposal for, 118, 118t
to real-life experiences, 120–123
target validation process, 119f
NRI. *See* Net Reclassification Improvement
NRT. *See* Nicotine replacement therapy
NSAID. *See* Nonsteroidal anti-inflammatory drugs
NSCLC. *See* Non-small cell lung cancer
NT-proBNP. *See* N-terminal pro B-type natriuretic peptide
Nuclear magnetic resonance (NMR), 33–34
Null hypothesis, 272

Number needed to treat (NNT), 86
NZW. *See* New Zealand White

O

OATP1B1 transporter, 63
OCT. *See* Optical coherence tomography
OECD. *See* Organisation for Economic Co-Operation and Development
OFC. *See* Orbitofrontal cortex
Off-target effects, 333
4-OH-TAM. *See* 4-hydroxy tamoxifen
Older adults, 299
  age-related physiologic changes, 303f
  aging, 299
    animal models, 301–302
    translational research in, 306–308
  gerontology *vs.* geriatrics, 300–301
  human approaches to translational aging research, 302–304
  lifespan *vs.* healthspan, 299–300
  limitations for animal and human models, 305–306
  testing treatments to extend health-and lifespan, 304–305
  translational aging resources, 309
Oligonucleotide microarrays, 30
"Omics" revolution, 15
"Omics" technologies, 25
  genomics, 25–33
  metabolomics, 33–37
"Omics" translation, 15. *See also* Translational medicine (TM)
  ACCE framework for laboratory tests evaluation, 21f
  biomarkers development, 18
  clinical association and benefit, 20–21
  into clinical practice, 21
  continuum of translation research and, 21–22
  proteomics, 15–16
    discovery-based research, 17
    lack of standardization, 16
    non-disease-associated factors, 16
    obstacles in translation, 17t
    reliability, 17
Onco Track, 104
Oncology
  biomarkers
    biomarker-assisted development, 186–187
    clinical decision making, 185
    molecular genetics, 185
    tissue biopsies, 187–188
  imaging, 192–193
Open access clinical trials, 224
Open innovation, 287
Open science, 285
  benefits of, 286
  open innovation, 287
  principles of, 285–286
  renewed interest in, 286–287
Optical coherence tomography (OCT), 173
Orbitofrontal cortex (OFC), 203
Organisation for Economic Co-Operation and Development (OECD), 260, 286
Organisation of the Human Genome (HUGO), 286
ORION. *See* *O*utcome of *r*osuvastatin treatment on carotid artery atheroma: a magnetic resonance *i*maging *o*bservatio*n*
*O*utcome of *r*osuvastatin treatment on carotid artery atheroma: a magnetic resonance *i*maging *o*bservatio*n* (ORION), 155
OVA1 panel, 162
Ovarian cancer
  multimarker panel, 162
  score, 161–162
Overfitting, 269
Oxidized LDL (OxLDL), 172

P

Pain treatment, 65
Pancreatic cancer, 164
Patent(s), 284
  on genetic tests, 289
  pool, 282–283
  on risk prediction models, 290
  system importance, 285
  thicket, 293
Patient consent, 219
Patient Outcomes Research Team (PORT), 197
Patient Reported Outcomes initiative (PRO), 103
Patient Reported Outcomes Measurement Information System (PROMIS®), 309
PBL. *See* Peripheral blood leukocyte
PBSC. *See* Peripheral blood stem cell
PCR. *See* Polymerase chain reaction
PD. *See* Pharmacodynamics
PDGFR. *See* Platelet-derived growth factor receptor
Peripheral blood leukocyte (PBL), 54
Peripheral blood stem cell (PBSC), 47
Permeabilization, 92
Peroxisome proliferator-activated receptor (PPAR), 323
Personalized medicine, 7, 223–224, 288. *See also* Translational medicine (TM)
Personalized therapies, 290
PET. *See* Positron emission tomography
PFC. *See* Prefrontal cortex
Pharma companies, 317–319. *See also* Big pharma
Pharmaceutical industry, 316
Pharmaceutical R&D productivity crisis, 227
  IND application, 228
  opportunities for earlier decision making, 228
  traditional drug development, 227, 228f
Pharmaceutical Research and Manufacturers of America (PhRMA), 104, 260
  biomarkers consortium, 104–105
Pharmaceutical toxicology
  biomarkers, 260
  elements of toxicity, 249
    differences among species, 253–254
    dose, 251
    dose-response relationship, 252–253
    duration of exposure, 251–252
    repair and defense, 252
    test substance, 249–250
  links, 260–261
  practice of discovery safety assessment, 261–264
  preclinical safety, 264–265
  regulatory toxicology, 255–259
  risk assessment, 255
  sixteenth-century depiction of medical research, 250f
  viewpoints of toxicity, 254–255
Pharmaceuticals, 260
Pharmacodynamics (PD), 217, 229, 229b, 239, 276
  endpoints, 220–221
  test, 147–148
Pharmacogenetics
  cohort studies, 69, 69f
  drug therapies, 60
    cardiovascular disease, 61–63
    depression, 60–61
    malignant diseases, 65–67
    pain treatment, 65
    proton pump inhibitors, 63–64
    statins, 63–64
  for individual drug therapy improvement, 59–60
  principle of panel study design, 68f
  randomized controlled clinical trial, 69
  study designs, 68
  use of pharmacogenetic markers, 67
Pharmacogenomics, 32
  biomarker development
    FDA recommendation, 147f
    proposal for, 146–147, 152
  disorders, 32
  key proteins, 32
  novel cancer therapies, 32–33
  positive impact, 32
Pharmacokinetic or pharmacodynamic modeling (PK/PD modeling), 118–119
Pharmacokinetic-based dose-adjustment approach, 61
Pharmacokinetics (PK), 217, 227, 229, 229b, 239, 276–278
Phase 0 trials. *See* Exploratory clinical studies
Phase I trial methodology, 217
  aims, 217
  design, 217–218
  dose escalation, 219
  patient consent, 219
  patient entry criteria, 218
    cancer type, 218
    laboratory investigations, 218
    novel agents and targets, 218t
    performance status, 218
  patients requirement for dose administration, 220
  special drug administration, 219
  starting dose calculation, 219
  stopping rules, 220
Phase II trials, 223
PhRMA. *See* Pharmaceutical Research and Manufacturers of America
Physiological study, 239

Pipeline, 18
Pivotal trial, 239
PK. *See* Pharmacokinetics
PK/PD modeling. *See* Pharmacokinetic or pharmacodynamic modeling
PKD. *See* Polycystic Kidney Disease Outcome Consortium
Placebo group, 271
Plant-derived principles, 111
Plasmacytoid DCs, 49
Platelet-derived growth factor receptor (PDGFR), 187
Plerixafor, 48
PM. *See* Poor metabolizers
PoC. *See* Proof-of-concept
Polycystic Kidney Disease Outcome Consortium (PKD), 103
Polymerase chain reaction (PCR), 25
  methods, 92
PoM. *See* Proof-of-mechanism
Poor metabolizers (PM), 61
PoP. *See* Proof-of-principle
PORT. *See* Patient Outcomes Research Team
Positive predictive value, 274
Positron emission tomography (PET), 78, 155, 190–191, 230, 244
  microdosing, 231–232
Post-analytical variables, 160
Post-translational modification (PTM), 18–19
Potency testing, 42–43
  applications of gene expression profiling
    HEK 293 cells, 46
    hESC, cell differentiation status analysis of, 46–47
  assays for, 45t
  BMSCs, 52–54
  complexities association with, 43, 43t
  cultured CD4+ cells, 52
  DCs, 48–49, 51
    antigens, 49
    candidate markers, 50–51
    comparability testing, 49
    gene expression profiling, 49, 50f
    genes with least and greatest index of variability, 51t–52t
    measurement, 49
    PBMCs, 50
    plasmacytoid DCs, 49
    T cells, 49
  factors affecting, 44, 44t
  gene expression arrays, 45
    critical functions, 46
    limitations, 46
    microarrays, 46
    quantitative real-time PCR, 45
  HSCs, 47
    CD34 antigen expression, 47
    CD34+ cells, 47–48
    gene expression profiles of, 48f
    LTC-IC, 47
    methylcellulose culture systems, 47
    Plerixafor, 48
    total nucleated cells counts, 47
    UCB CD34+ cells, 48
  measurement, 44
    cell counts and viability measurements, 45
    mechanism of action, 44–45
    in vitro assays, 44
    in vitro cell function assays, 45
  miRNAs, 54, 55f
PPAR. *See* Peroxisome proliferator-activated receptor
PPI. *See* Proton pump inhibitors
PPP. *See* Public–private partnership
PPP1R1B protein. *See* DARPP-32 protein
PR. *See* Progesterone receptor
pRCTs. *See* preclinical randomized controlled trials
Pre-analytical conditions, 160
PRECISESADS, 104
preclinical randomized controlled trials (pRCTs), 84
Precompetitive consortia, 323–324
Predictive Safety Testing Consortium (PSTC), 101, 149, 260
  collaboration model, 103
  Critical Path Institute's, 103
Predictive validity, 87
Prefrontal cortex (PFC), 201
Primary translation, secondary translation *vs.*, 2–3
Private–public partnerships, 102–103
PRO. *See* Patient Reported Outcomes initiative
Probable valid biomarker, 147–148
Productivity crunch, 316
Progesterone receptor (PR), 246
PROMIS®. *See* Patient Reported Outcomes Measurement Information System
Proof-of-concept (PoC), 119–120, 222, 239, 268. *See also* Smart early clinical trials
  study, 243
  testing, 117
Proof-of-mechanism (PoM), 221–222, 229b, 243
  testing, 117
Proof-of-principle (PoP), 159–160, 222, 243
  testing, 117
Prostacyclin, 169
Prostate cancer, 163–164
Prostate-specific antigen (PSA), 18–19, 36, 145
Protein localization, 93
Proteomics, 15–16. *See also* Biomarkers
  bioinformatics methods, 162f
  CA125, HE4, and ROMA-derived composite score comparison, 163t
  discovery-based research, 17
  lack of standardization, 16
  multiparameter approach, 161
    breast cancer, 162–163
    liver fibrosis, 164–165
    ovarian cancer, 161–162
    pancreatic cancer, 164
    prostate cancer, 163–164
  non-disease-associated factors, 16
  NRI, 161
  obstacles in translation, 17t
  overfitting issues, 161
  reliability, 17
  ROC analysis, 161
  source of errors, 159–160
  supervised computational methods, 160–161
  unsupervised computational methods, 160–161
Proton pump inhibitors (PPI), 63–64
PSA. *See* Prostate-specific antigen
PSTC. *See* Predictive Safety Testing Consortium
Psychiatric disorders biological treatment, 195–196
Psychiatry. *See also* Translational medicine (TM)
  antipsychotic drug effects characterization, 205–206
  challenges of translation in psychiatry
    second translation, 197
    stigma, 197
    unknown pathophysiology, 196–197
  genetic susceptibility factors imaging, 203–205
  imaging biomarkers in schizophrenia, 198–202
  new biomarkers for translation in, 197–198
  psychiatric disorders biological treatment, 195–196
PTM. *See* Post-translational modification
Public–private partnership (PPP), 287, 317, 328
  from 2011 to 2013, 323f
  benefits of, 287
  biotech companies and academia, 319
  partnerships between pharma companies, 320t–322t
  pharma and academia, 317–319

Q

22q11 deletion syndrome (22q11DS), 204
QB3. *See* California Institute of Quantitative Biosciences
Quantitative reverse transcription PCR (qPCR), 31
Quantitative structure activity relationship (QSAR), 115

R

R&D. *See* Research & development
RAAS. *See* Renin angiotensin aldosterone system
Racemate carvedilol, 62
Radiotelemetry techniques, 168
Random effects, 277
RBC. *See* Red blood cells
RCM. *See* Restrictive cardiomyopathy
RCT. *See* Reverse cholesterol transport
RD. *See* Recommended phase II dose
Receiver operating characteristics (ROC), 145
  analysis, 161
  curve, 19, 274
Receptor tyrosine kinase (RTK), 186
Recommended phase II dose (RD), 221
Red blood cells (RBC), 66
Reduce, refine, replace principles (3R principles), 258
Regions of interest (ROI), 199
Regulatory authorities, 239

Regulatory toxicology, 255
animal models, 258–259
Exploratory IND approach, 259
GLP, 258
ICH guidance SAFETY documents, 257t
phases of toxicity testing, 258f
studies, 256–257
Regulatory trials, 239. *See also* Exploratory trials
Reliability, 17
Renal injury, 177–178
Renin, 168–169
Renin angiotensin aldosterone system (RAAS), 168–169
Repeatability, 20
Repositioned drugs, 290–291
Reproducibility, 20
Resampling approach, 273
Research & development (R&D), 3, 102
Restrictive cardiomyopathy (RCM), 175
Reverse cholesterol transport (RCT), 334
Reverse pharmacology, 107, 110, 112t
backward translational activities, 112
eINDs, 112
facilitated serendipity, 111
feedback route, 113
GLP-1, 110
historic examples, 110
incretin-analogues, 111
Kaplan–Meier curves, 111f
plant preparations, 112
plant-derived principles, 111
serendipity findings, 112
shamanism, 111–112
TNFalpha, 111
Reverse pharmacology approach, 6–7
Risk factors, 144, 301
Risk marker, 144
Risk of Ovarian Malignancy Algorithm (ROMA), 162
Risk prediction models, 290
Risk-benefit-based approach, 228
ROC. *See* Receiver operating characteristics
ROI. *See* Regions of interest
ROMA. *See* Risk of Ovarian Malignancy Algorithm
RTK. *See* Receptor tyrosine kinase

S

SAD. *See* Single ascending dose
SAEs. *See* Serious adverse events
Safer and faster evidence-based translation (SAFE-T), 104
Safety assessment, 261
ADME optimization, 261
safety-directed drug design, 262–264
target-based, 261–262
Safety biomarkers, 260
Safety-directed drug design, 262–264
Sandwich ELISAs, 94
SAS PROC NLMIXED software, 277
SASP. *See* Senescence-associated secretory phenotype
scFv. *See* single chain fragment variable
Schizophrenia, 197
antipsychotic medication, 205
highly heritable mental disorder, 203
imaging biomarkers in, 198
auditory and language processing, 200–201
functional imaging markers in, 200
motor functioning, 201
selective attention, 202
structural brain biomarkers, 198–200
WM, 201–202
positive psychotic symptoms, 195–196
symptoms, 196
SCID. *See* Severe combined immunodeficiency
"Scientifically novel" drugs, 316
Scoring system, 152–155
grading, 155
intra-plaque lipid core, 155
Second-generation antipsychotic drugs, 195–196
Secondary hypertension, 168
Secondary translation, 197
primary translation *vs.*, 2–3
Secrecy, 291
SELDI. *See* Surface-enhanced laser desorption ionization
Selective attention, 202
Selective serotonin reuptake inhibitors (SSRI), 61
Senescence-associated secretory phenotype (SASP), 301–302
Sensitivity, 274
Sequencing, 25–26
DNA sequencing methodologies, 27
Illumina sequencing process, 29f
by ligation, 27, 28f
NGS, 27, 29t
by synthesis, 27–28
Serendipity findings, 112
Serious adverse events (SAEs), 105
Severe combined immunodeficiency (SCID), 47
SGC. *See* Structural Genomics Consortium
Shamanism, 111–112
SHR. *See* Spontaneous hypertensive rat
Sildenafil, 109
Single ascending dose (SAD), 119, 239
single chain fragment variable (scFv), 41
Single nucleotide polymorphism (SNP), 63
Single photon emission computed tomography (SPECT), 155, 190–191
sIVPA. *See* Spectroscopic intravascular photoacoustics
Six-minute walking distance (6MWD), 176
SLCO1B1 polymorphisms, 64
SLE. *See* Systemic lupus erythematosus
Small- and medium-sized enterprise (SME), 104, 315
Smart early clinical trials, 243
antiatherosclerotic drug, 244f
breast cancer, 246
disease classification, 246
human trials, 243
mutational status of melanomas, 245t
on new antiatherosclerotic drug, 244
phase II traditional scenario, 244f
PoC, 244
PoM markers, 244
targeted approach, 245
SME. *See* Small-and medium-sized enterprise
Smoking cessation, 331
SNP. *See* Single nucleotide polymorphism
SOLiD™. *See* Support oligonucleotide ligation detection
SOP. *See* Standard operating procedure
SOT. *See* U.S. Society of Toxicology
Specificity, 274
SPECT. *See* Single photon emission computed tomography
Spectroscopic intravascular photoacoustics (sIVPA), 78
Spontaneous hypertensive rat (SHR), 167, 174
SSRI. *See* Selective serotonin reuptake inhibitors
Standard operating procedure (SOP), 159–160
Standards for Reporting of Diagnostic Accuracy (STARD), 17
Statins, 63–64
Statistical models, 278–279
STG. *See* Superior temporal gyrus
Stigma, 197
Stopping rules, 220
Structural brain biomarkers, 198–200
Structural Genomics Consortium (SGC), 224
Study designs, 68
Superior temporal gyrus (STG), 200
Support oligonucleotide ligation detection (SOLiD™), 27
Support vector machines, 163–164
Surface-enhanced laser desorption ionization (SELDI), 16
Surrogates, 132, 155
Systemic lupus erythematosus (SLE), 324

T

T-cell receptors (TCRs), 41
T-Rapa cells, 52
t-test, 278
TAC. *See* Transverse aortic constriction
Target of rapamycin (TOR), 306
Target validation, 108
TBI. *See* Traumatic brain injury
TCRs. *See* T-cell receptors
Temozolomide, 86
Test research, 19
Test substance, 249–250
6-TGN. *See* 6-thioguanine nucleotide
TGN1412, 85
6-thioguanine nucleotide (6-TGN), 66
Thiopurine S-methyltransferase (TPMT), 66
Thiopurines, 66
3R principles. *See* Reduce, refine, replace principles
Thrombospondin 1 (TSP1), 186
Thymidylate synthase (TYMS), 67
TILs. *See* Tumor infiltrating leukocytes
Time-of-flight (TOF), 35
Tiotropium, 117
Tissue biobanks, 75
developments in vascular biobanking research, 77–79
medical specialties, 75

principles and types, 75–77
biological samples, 76
determinants, 76–77
tissue biopsies, 76
translational tissue biobanks
vascular tissue biobanking research, 75
Tissue biopsies, 187–188
TLESR. *See* Transient lower esophageal sphincter relaxations
TM. *See* Translational medicine
TNFalpha, 111
Tobacco, 303
TOF. *See* Time-of-flight
TOR. *See* Target of rapamycin
Torcetrapib, 123, 332–333
off-target effect, 333
Toxicity, 220–221
Toxicology, 249
Toxnet, 260
ToxPortal, 260
TPMT. *See* Thiopurine S-methyltransferase
Trade secrecy, 293
Transcriptomics, 15
Transgenic models, 83
Transient lower esophageal sphincter relaxations (TLESR), 107
Translation
discovery, 332
evidence, 332
failed example, 332
blood pressure, 336
CETP inhibitors, 333
HDL-cholesterol, 334
torcetrapib off-target effect, 333
translatability scores, 335t–336t
process, 281
smart and successful example, 331–332
Translational aging research. *See also* Aging
human approaches, 302–303
age-related physiologic change effect, 303f
longevity and healthspan determinants identification, 303–304
translational aging resources
animals and animal tissues, 309
cohorts/populations, 309
tools and toolboxes, 309
Translational imaging research
back-translation, 189–190
and conventional imaging, 189
in disease areas, 192
cardiovascular medicine, 192
metabolic medicine, 192
neuroscience, 193
oncology, 192–193
imaging, 189
modalities, 190–192
validation, 189–190
Translational medicine (TM), 1, 189, 243, 281–282, 327
aspects, 2f
balanced innovation environment, 294
do's and don'ts, 6–8
explanatory sentences, 1
history, 3
chance findings, 3
drug discovery and development process, 3
l'art-pour-l'art approach, 3
late-stage attrition, 4
NME and NBE, 6
persistence of low-output syndrome, 5
R&D, 3, 4f
reasons for termination of drug development, 5f
shrinkage, 4
success rates from first-in-man to registration, 5f
intellectual property and, 282
basic functioning of patents, 282–285
patent system importance, 285
open science, 285–287
perspective on genetic patents future, 291
gene sequences patentability, 291–292
gene sequences patenting, 292–293
promotion of innovation, 292–293
PPP, 287–288
present status, 8–11
new pathways to discovery, 8
re-engineering the clinical research enterprise, 9–11
research teams of future, 8–9
primary *vs.* secondary translation, 2–3
"pseudo"-linear model, 8f
remits, 6t
trade secrecy, 293
translational science in, 11–12
trends in translational intellectual property, 288
genetic tests, 289
patents, 288–289
patents on new and repositioned drugs, 290–291
personalized therapies, 290
research tools, 288–289
risk prediction models, 290
secrecy, 291
Translational research, 301
in aging, 306–308
animal models role, 301–302
Translational science biostatistics
biological modeling, 275–278
biomarkers, 274–275
experiment design and interpretation, 270
Bayesian concepts, 273
cross-over designs, 271
dose-response relationship estimation, 271
drug development project, 272
hypothesis tests, 272
pharmaceutical research, 271
proof-of-concept study, 270
multiplicity, 273–274
statistical models, 278–279
statistical
inference, 268–270
models, 268–270
problems, 267–268
Translational SCIENCE in Medicine, 327
Translational science in medicine, 327
template for early translational plan, 328f
translational
assessment, 329
medicine network, 329f
planning, 328
Translational tissue biobanks, 76
Translational toxicology, 333
Transverse aortic constriction (TAC), 174
Traumatic brain injury (TBI), 84
Triage Cardiac Panel, 95–96
Triage MeterPlus, 95–96
Triage® Parasite Panel, 96
Tricyclic antidepressants, 61
Triple negative breast cancer, 246, 246f
Trueness, 19
TSP1. *See* Thrombospondin 1
Tumor biopsy, 187
Tumor infiltrating leukocytes (TILs), 41
TYMS. *See* Thymidylate synthase

U

U-BIOPRED. *See* Unbiased biomarkers for the prediction of respiratory disease
U.S. Food and Drug Administration (FDA), 16, 18, 101, 260
U.S. Society of Toxicology (SOT), 260
UCB CD34+ cells, 48
UGT. *See* Uridine diphosphate glucuronosyltransferase
Ultra performance liquid chromatography (UPLC), 35
Ultrarapid metabolizers (UM), 61
Ultrasound (US), 155
imaging, 190
UM. *See* Ultrarapid metabolizers
Unbiased biomarkers for the prediction of respiratory disease (U-BIOPRED), 104
UPLC. *See* Ultra performance liquid chromatography
Uracil reductase. *See* Dihydropyrimidine dehydrogenase (DPD)
Uridine diphosphate glucuronosyltransferase (UGT), 67
US. *See* Ultrasound

V

Valid biomarker, 147–148
Validation, 91
Valley of death, 316
Varenicline, 331–332
Vascular cell adhesion molecule (VCAM), 169
Vascular endothelial growth factor (VEGF), 186, 222
Vascular endothelial growth factor receptor (VEGFR), 186–187
Vascular resistance biomarkers, 169
Vascular tissue biobanking research, 75, 77
Athero-Express biobank study, 78–79
atherosclerosis, 77–78
clinical value of vascular biobanks and diagnostic imaging, 78
VBM. *See* Voxel-based morphometry
VCAM. *See* Vascular cell adhesion molecule
VEGF. *See* Vascular endothelial growth factor
VEGFR. *See* Vascular endothelial growth factor receptor
Ventricular premature beats (VPBs), 132–133
Ventrolateral prefrontal cortex (VLPFC), 201–202
VeraCode®, 95

VGDS. *See* Voluntary genomics data submission
Virtual-histology IVUS (VH-IVUS), 173
Vitamin K antagonists, 63
Vitamin K epoxide reductase (VKOR), 63
VLPFC. *See* Ventrolateral prefrontal cortex
Voluntary exploratory data submission (VXDS), 102
Voluntary genomics data submission (VGDS), 102
Voxel-based morphometry (VBM), 199
  meta-analysis, 199–200
VPBs. *See* Ventricular premature beats
Vulnerable plaque, 78
VXDS. *See* Voluntary exploratory data submission

W

Within-class biological variability, 160
Women's Health Initiative (WHI), 309
Working memory (WM), 201–202
World Health Organization (WHO), 260

X

Xanthin oxidase (XO), 66
Xenobiotics, 249–250
Xenograft animal models, 83

Z

Zileuton, 117

Printed in the United States
By Bookmasters